BASIC DATA ANALYSIS RELATIONS

PROPAGATION OF UNCERTAINTY

General

$$R = f(x_1, x_2, \ldots x_n); \quad w_R = \left[\left(\frac{\partial R}{\partial x_1} \right)^2 w_{x_1}^2 + \cdots \left(\frac{\partial R}{\partial x_n} \right)^2 w_{x_n}^2 \right]^{1/2}$$

Product Function

$$R = x_1^{a_1}, x_2^{a_2} \ldots x_n^{a_n} = \prod_{i=1}^{n} x_i^{a_i}; \quad \frac{w_R}{R} = \left[\sum_{i=1}^{n} \left(a_i \frac{w_{x_i}}{x_i} \right)^2 \right]^{1/2}$$

Arithmetic Function

$$R = a_1 x_1 + a_2 x_2 + \cdots a_n x_n = \sum_{i=1}^{n} a_i x_i; \quad w_R = \left[\sum_{i=1}^{n} \left(a_i w_{x_i} \right)^2 \right]^{1/2}$$

POPULATION STANDARD DEVIATION

$$\sigma = \left[\frac{\sum_{i=1}^{n} (x_i - x_m)^2}{n} \right]^{1/2}, \quad \text{where } x_m = \frac{\sum x_i}{n}$$

SAMPLE STANDARD DEVIATION

$$\sigma = \left[\frac{\sum_{i=1}^{n} (x_i - x_m)^2}{n - 1} \right]^{1/2}$$

STANDARD DEVIATION OF MEAN

$$\sigma_m = \frac{\sigma}{n^{1/2}}$$

COEFFICIENT OF DETERMINATION r^2 (square of correlation coefficient)

$$r^2 = 1 - \sigma_{y,x}^2 / \sigma_y^2; \quad \sigma_y = \left[\frac{\sum (y_i - y_m)^2}{n - 1} \right]^{1/2}; \quad \sigma_{y,x} = \left[\frac{\sum (y_i - y_{ic})^2}{n - 2} \right]^{1/2}$$

Computer Equation for r^2

$$r^2 = \frac{[n \Sigma x_i y_i - (\Sigma x_i)\,(\Sigma y_i)]^2}{\left[n \Sigma x_i - (\Sigma x_i)^2 \right] \left[n \Sigma y_i - (\Sigma y_i)^2 \right]}$$

NORMAL DISTRIBUTION

x	Fraction of Data within Range of $\pm x$
$0.6745\,\sigma$	0.5
$1.0\,\sigma$	0.6827
$2.0\,\sigma$	0.9545
$3.0\,\sigma$	0.9973
$4.0\,\sigma$	0.999937

POWER FUNCTION CORRELATION

$y = ax^b$; $\ln y = \ln a + b \ln x$; plots as straight line on loglog chart

EXPONENTIAL CORRELATION

$y = ae^{bx}$; $\ln y = \ln a + bx$; plots as straight line on semilog chart

EXPERIMENTAL METHODS FOR ENGINEERS

EIGHT EDITION

McGraw-Hill Series in Mechanical Engineering

Alciatore	*Introduction to Mechatronics and Measurement Systems*
Anderson	*Fundamentals of Aerodynamics*
Anderson	*Introduction to Flight*
Anderson	*Modern Compressible Flow*
Beer/Johnston	*Mechanics of Materials*
Beer/Johnston	*Vector Mechanics for Engineers*
Budynas	*Advanced Strength and Applied Stress Analysis*
Budynas/Nisbett	*Shigley's Mechanical Engineering Design*
Byers/Dorf/Nelson	*Technology Ventures: From Idea to Enterprise*
Cengel	*Introduction to Thermodynamics and Heat Transfer*
Cengel/Boles	*Thermodynamics: An Engineering Approach*
Cengel/Cimbala	*Fluid Mechanics: Fundamentals and Applications*
Cengel/Ghajar	*Heat and Mass Transfer: Fundamentals and Applications*
Cengel/Turner/Cimbala	*Fundamentals of Thermal-Fluid Sciences*
Dieter/Schmidt	*Engineering Design*
Finnemore/Franzini	*Fluid Mechanics with Engineering Applications*
Heywood	*Internal Combustion Engine Fundamentals*
Holman	*Experimental Methods for Engineers*
Holman	*Heat Transfer*
Kays/Crawford/Weigand	*Convective Heat and Mass Transfer*
Norton	*Design of Machinery*
Palm	*System Dynamics*
Plesha/Grey/Costanzo	*Engineering Mechanics: Statics and Dynamics*
Reddy	*An Introduction to Finite Element Method*
Schey	*Introduction to Manufacturing Processes*
Smith/Hashemi	*Foundations of Materials Science and Engineering*
Turns	*An Introduction to Combustion: Concepts and Applications*
Ullman	*The Mechanical Design Process*
White	*Fluid Mechanics*
White	*Viscous Fluid Flow*
Zeid	*Mastering CAD/CAM*

EXPERIMENTAL METHODS FOR ENGINEERS

EIGHTH EDITION

J. P. Holman
Department of Mechanical Engineering
Southern Methodist University

The McGraw·Hill Companies

EXPERIMENTAL METHODS FOR ENGINEERS, EIGHTH EDITION

ISBN 978-0-07-352930-1
MHID 0-07-352930-3

Vice President & Editor-in-Chief: *Marty Lange*
Vice President & Director of Specialized Publishing: *Janice M. Roerig-Blong*
Publisher: *Raghothaman Srinivasan*
Executive Editor: *Bill Stenquist*
Senior Marketing Manager: *Curt Reynolds*
Developmental Editor: *Lorraine Buczek*
Project Manager: *Erin Melloy*
Design Coordinator: *Brenda A. Rolwes*
Cover Designer: *Studio Montage, St. Louis, Missouri*
Buyer: *Sandy Ludovissy*
Media Project Manager: *Balaji Sundararaman*
Compositor: *S4Carlisle Publishing Services*
Typeface: *10/12 Times New Roman*
Printer: *R. R. Donnelley*

Library of Congress Cataloging-in-Publication Data

Holman, J. P. (Jack Philip)
   Experimental methods for engineers / J.P. Holman.—8th ed.
      p. cm.—(McGraw-Hill series in mechanical engineering)
   Includes bibliographical references and index.
   ISBN-13: 978-0-07-352930-1
   ISBN-10: 0-07-352930-3
  1. Engineering—Laboratory manuals. 2. Engineering instruments. I. Title.
   TA152.H6 2011
   620.0078—dc23
                                                          2011015004

www.mhhe.com

ABOUT THE AUTHOR

J. P. Holman received the Ph.D. in mechanical engineering from Oklahoma State University. After two years as a research scientist at the Wright Aerospace Research Laboratory, he joined the faculty of Southern Methodist University, where he is presently Professor Emeritus of Mechanical Engineering. He has also held administrative positions as Director of the Thermal and Fluid Sciences Center, Head of the Civil and Mechanical Engineering Department, and Assistant Provost for Instructional Media. During his tenure at SMU he has been voted the outstanding faculty member by the student body 13 times.

Dr. Holman has published over 30 papers in several areas of heat transfer and his three widely used textbooks, *Heat Transfer* (10th edition, 2010), *Experimental Methods for Engineers* (7th edition, 2001), and *Thermodynamics* (4th edition, 1988), all published by McGraw-Hill, have been translated into Spanish, Portuguese, Japanese, Chinese, Korean, and Indonesian, and are distributed worldwide. He is also the author of the utilitarian monograph *What Every Engineer Should Know About EXCEL* (2006), published by CRC Press. Dr. Holman also consults for industry in the fields of heat transfer and energy systems.

A member of ASEE, he is past Chairman of the National Mechanical Engineering Division and past chairman of the ASME Region X Mechanical Engineering Department Heads. Dr. Holman is a Fellow of ASME and recipient of several national awards: the George Westinghouse Award from ASEE for distinguished contributions to engineering education (1972), the James Harry Potter Gold Medal from ASME for contributions to thermodynamics (1986), the Worcester Reed Warner Gold Medal from ASME for outstanding contributions to the permanent literature of engineering (1987), and the Ralph Coats Roe Award from ASEE as the outstanding mechanical engineering educator of the year (1995). In 1993 he was the recipient of the Lohmann Medal from Oklahoma State University, awarded annually to a distinguished engineering alumnus of that institution.

CONTENTS

LIST OF WORKED EXAMPLES

PREFACE

Experimental measurements can be vexatious, and a textbook about experimental methods cannot alleviate all the problems that are perplexing to the experimental engineer. Engineering education has placed an increased emphasis on the ability of an individual to perform a theoretical analysis of a problem. Experimental methods are not unimportant, but analytical studies have, at times, seemed to deserve more emphasis, particularly with the enormous computing power that is available. Laboratory work has also become more sophisticated in the modern engineering curricula. Conventional laboratory courses have consistently been changed to include experiments with rather elaborate electronic instrumentation and microprocessor- or computer-based data acquisition systems. Surprisingly enough, however, many engineering graduates do not seem capable of performing simple engineering measurements with acceptable precision. Furthermore, they are amazingly inept when asked the question: How *good* is the measurement? They automatically assume the results are accurate to the number of digits displayed in the computer printout.

This book represents a first survey of experimental methods for undergraduate students. As such, it covers a broad range of topics and may be lacking in depth on certain topics. In these instances, the reader is referred to more detailed treatments in specialized monographs.

It is important that engineers be able to perform successful experiments, and it is equally important that they know or be able to estimate the accuracy of their measurements. This book discusses a rather broad range of instruments and experimental measurement techniques. Strong emphasis is placed on problem solving, and the importance of accuracy, error, and uncertainty in experimental measurements is stressed throughout all the discussions. The book is generally suitable as an accompaniment to laboratory sessions oriented around the specific experiments available at a particular institution. Portions of the text material may be covered in a lecture session. The lectures would be concerned with the principles of instrumentation, whereas the laboratory periods would afford the student an opportunity to use some of the devices discussed in this text and laboratory manuals that may be available to faculty planning the course. The particular experiments, or the instruments used in the laboratory periods, will depend on the facilities available and the objectives set by each curriculum. A mathematical background through ordinary differential equations is assumed for the text developments, and it is expected that basic courses in thermodynamics, engineering mechanics, and electric-circuit theory will precede a course based on this text.

Whatever the course arrangement for which this text is applied, it is strongly recommended that the problems at the end of each chapter receive careful attention. These problems force the student to examine several instruments to determine their accuracy and the uncertainties that might result from faulty measurement techniques. In many instances the problems are very similar to numerical examples in the text. Other problems require the student to extend the text material through derivations, design of experiments, and so on. The selection of problems for a typical course will depend, naturally, on the types of experiments and laboratory facilities available for use with the course.

A few remarks concerning the arrangement of the text material are in order. A brief presentation of all topics was desired so that a rather broad range of experimental methods could be discussed within the framework of a book of modest length. Chapters 1 and 2 provide initial motivation remarks and some brief definitions of important terms common to all measurement systems. This discussion includes basic concepts of dynamic response in zeroth-, first-, and second-order systems.

Next, a simple presentation of some of the principles of statistical data analysis is given in Chapter 3. Some of the concepts in Chapter 3 are used in almost every subsequent chapter in the book, particularly the concept of experimental uncertainty.

Chapter 4 presents several simple electrical-measurement and amplifier circuits and the principles of operation of typical electric transducers. Many of these transducers are applicable to measurement problems described in later chapters. Chapters 5 and 6, concerning dimensional and pressure measurements, offer fairly conventional presentations of their subject matter, except that numerical examples and problems are included to emphasize the importance of experimental uncertainty in the various devices. Flow measurement is discussed in Chapter 7 in a rather conventional manner. A notable feature of this chapter is the section on flow-visualization techniques. Again, the examples and problems illustrate some of the advantages and shortcomings of the various experimental techniques. Chapter 8 is quite specific in its discussion of temperature-measurement devices. Strong emphasis is placed on the errors that may arise from conduction, convection, and heat transfer between the temperature-sensing device and its thermal environment. Methods are presented for correcting these errors. Chapter 9 is brief but gives the reader an insight into the problems associated with transport-property measurements. The material in this chapter is dependent on the measurement techniques discussed in Chapters 6, 7, and 8. The material in Chapter 9 could be dispersed through the three previous chapters and still achieve a balanced presentation; however, it was believed best to bring transport properties and thermal measurements into sharper focus by grouping them in one chapter.

Static force, torque, and strain measurements are discussed in Chapter 10. The strain measurements are related to some elementary principles of experimental stress analysis, and the operation of the electrical strain gage is emphasized.

Some of the elementary principles of motion- and vibration-measurement devices are discussed in Chapter 11. Included in this presentation is a discussion of sound waves, sound-pressure level, and acoustic measurements. The inclusion of the acoustics material in Chapter 11 is somewhat arbitrary since this material would be equally pertinent in Chapter 6.

Chapter 12 discusses thermal- and nuclear-radiation measurements. The presentation is brief, but some of the more important detection techniques are examined, and examples are given to illustrate the important principles. A short presentation of the statistics of counting illustrates the importance of background activity in nuclear-radiation detection. The thermal-radiation measurements are properly related to the temperature measurement material in Chapter 8.

Chapter 13 presents some of the measurement techniques that are used in air-pollution-control applications. Such measurements make use of the basic pressure, flow, and temperature measurement techniques discussed in Chapters 6, 7, and 8. The importance of electronic data processing and its relation to the basic electrical measuring devices of Chapter 4 are discussed in a general way in Chapter 14. Because the fields of electronic microprocessors and data acquisition systems change so rapidly, the discussion follows a fairly general pattern. A glossary of terms is given that is applicable to a number of acquisition systems.

Since all experimental work must be reported in some form, Chapter 15, on reports and presentations, has been given in a general format that will apply to several applications. This material includes information on graphical and oral presentations as well.

Some remarks concerning units are in order. There is no question regarding the desirability to move toward the adoption of SI (metric) units wherever possible. Accordingly, the educational system must anticipate this movement and teach students to operate in SI. Courses in analysis (fluid mechanics, heat transfer, mechanics, and so on) can adopt SI as the primary system with appropriate conversions to the old English system. The experimental engineer may have a more cumbersome task. One does not buy new instruments, gages, or meters just to change to SI units. Unhappily, a common practice is to operate in a mixed system of units that combines SI and non-SI metric with English units, which are coupled to empirical design relationships at hand. This means that the experimental engineer will be operating in a "bilingual" mode for many years. In addition, many field engineers do not *want* to change to SI and will not change until they are forced to do so. Of course, companies heavily involved with export markets, like the auto industry, are already heavily into SI units. It is in recognition of these facts that I have chosen to mix SI and English units throughout the book, even though my personal desire is to change over to SI completely as soon as possible. There is another problem that the engineer must deal with. Some work is performed in metric units that are not in the SI system. Examples are the calorie unit for energy and the kilogram *force* per square centimeter for pressure. Alas, these are hazards to be encountered, and one must adapt as best as one can.

In this edition, pruning of comments and topics has been applied throughout. Recognition is made of the fact that electronic instrumentation and computer-based data acquisition systems change so rapidly that the experimental engineer is best served by timely consultation with well-disposed manufacturers of the necessary instrumentation for their particular application. There are some things that do not change so rapidly though. Uncertainty analysis and data correlations are still important. Increased computer power just makes the tasks easier to perform. The discussions in Chapter 3 concerning data correlations, regression analysis, and graphical presentations have been expanded considerably, with emphasis on computer generation of trendlines and least-squares analysis for a number of functional relationships. Particular attention is devoted to the type of graphical displays that should be used for different data presentations. This emphasis on adaptation of computer-generated graphs is carried through to the report writing discussions of Chapter 15 in which specific examples of graph constructions are displayed that will be applicable to reports and oral presentations.

Chapter 16 on design of experiments remains intact and the author urges the reader to consult the protocols contained in this chapter in conjunction with studies of uncertainty propagation in Chapter 3 and the specific measurement techniques discussed in later chapters. The information in Chapter 16 is quite general as illustrated by five diverse examples (case studies) which are offered as applications of the protocol.

NEW TO THIS EDITION:

- A new section on Fourier analysis has been added to Chapter 2 to reflect the importance of spectrum analysis of vibrating systems, and to emphasize the importance of extended frequency response beyond the fundamental.

- A new section on causation, correlations, and curve fits has been added to Chapter 3 to emphasize the distinction between the three terms, and the respective utility of each.

- A new section on radiation effects in humans has been added to Chapter 12 both to indicate the specific physical impacts of different types of nuclear radiation and the unit systems employed for measurement of these effects.

- A brief appendix (Appendix B) has been added describing the basics of digital imaging (photographic) systems. This appendix is then referenced at several relevant places in the measurement chapters.

- New Examples and 15 percent more problems have been added throughout.

ACKNOWLEDGMENTS

With a book at this stage of development, the list of persons who have been generous with their comments and suggestions has grown very long indeed. The author hopes that a blanket note of thanks for all these people's contributions will suffice. As in the past, all comments from users will be appreciated and acknowledged. The McGraw-Hill editorial staff appreciates the comments and suggestions of the following people who reviewed the new edition:

David Baldwin, University of Oklahoma
Ian Kennedy, University of California–David
Bernard Koplik, New Jersey Institute of Technology
Terry Ng, University of Toledo
Marcio de Queiroz, Louisiana State University
Chiang Shih, Florida State University
Charles Smith, Lehigh University

J. P. Holman

1

INTRODUCTION

There is no such thing as an easy experiment, nor is there any substitute for careful experimentation in many areas of basic research and applied product development. Because experimentation is so important in all phases of engineering, there is a very definite need for the engineer to be familiar with methods of measurement as well as analysis techniques for interpreting experimental data.

Experimental techniques have changed quite rapidly with the development of electronic devices for sensing primary physical parameters and for controlling process variables. In many instances more precision is now possible in the measurement of basic physical quantities through the use of these new devices. Further development in instrumentation techniques is certain because of the increasing demand for measurement and control of physical variables in a wide variety of applications.

Obviously, a sound knowledge of many engineering principles is necessary to perform successful experiments; it is for this reason that experimentation is so difficult. To design the experiment, the engineer must be able to specify the physical variables to be investigated and the role they will play in later analytical work. Then, to design or procure the instrumentation for the experiment, the engineer must have a knowledge of the governing principles of a broad range of instruments. Finally, to analyze the data, the experimental engineer must have a combination of keen insight into the physical principles of the processes being investigated and a knowledge of the limitations of the data.

Research involves a combination of analytical and experimental work. The theoretician strives to explain or predict the results of experiments on the basis of analytical models which are in accordance with fundamental physical principles that have been well established over the years. When experimental data are encountered which do not fit into the scheme of existing physical theories, a skeptical eye is cast first at the experimental data and then at appropriate theories. In some cases the theories are modified or revised to take into account the results of the new experimental data, after being sure that the validity of the data has been ascertained. In any event, all physical theories must eventually rely upon experiment for verification.

Whether the research is of a basic or developmental character, the dominant role of experimentation is still present. The nuclear physicist must always test theories in the laboratory to be sure of their validity, just as the engineer who conducts research on a new electronic circuit or a new type of hydraulic flow system certainly must perform a significant number of experiments in order to establish the usefulness of the device. Physical experimentation is the ultimate test of most theories.

In many engineering applications certain basic physical phenomena are well known and experience with devices using these phenomena is already available. Examples are radiation detectors, solid-state electronics, flowmeters, and certain mechanisms. There will always arise, however, new uses for such devices in combination with other devices, for example, a new type of amplifier or a new flow-control system. In these cases the engineer must utilize all available experience with the previous devices to design the apparatus for a new application. No matter how reliable the information on which the design is based, the engineer must insist on thorough experimental tests of the new device before the design is finalized and production is initiated.

There is a whole spectrum of tests and experiments which an engineer may be called upon to perform. These vary from a crude test to determine the weight of a device to precise electronic measurements of nuclear radioactivity. Since the range of experiments is so broad, the experimental background of the engineer must be correspondingly diverse in order to function effectively in many experimental situations. Obviously, it is quite hopeless to expect any one person to operate at a high level of effectiveness in all areas of experimental work. The primary capability of a particular individual will necessarily be developed in an area of experimentation closely connected with that person's analytical and theoretical capability and related interests. The broader the interests of an individual, the more likely it is that that individual will develop broad experimental interests and capabilities.

In the past there have been some engineers who were primarily experimentalists— those individuals who designed devices by trial and error with very little analytical work as a preliminary to the experimentation. There are some older areas of engineering where this technique may still prevail, primarily where years of experience have built up a background knowledge to rely upon. But, in new fields more emphasis must be placed on a combination of theory and experimentation. To create a rather absurd example, we cite the development of a rocket engine. It would be possible to build different sizes of rockets and test them until a lucky combination of design parameters was found; however, the cost would be prohibitive. The proper approach is one of test and theoretical study where experimental data are constantly evaluated and compared with theoretical estimates. New theories are formulated on the basis of the experimental measurements, and these theories help to guide further tests and the final design.

The engineer should know what to look for before beginning the experiments. The objective of the experiments will dictate the accuracy required, expense justified, and level of human effort necessary. A simple calibration check of a mercury-in-glass thermometer would be a relatively simple matter requiring a limited amount of equipment and time: however, the accurate measurement of the temperature of a high-speed gas stream at 1600°C (2912°F) would involve more thought and care. A test of an amplifier for a home music system might be less exacting than a test of an amplifier to be used as part of the electronic equipment in a satellite, and so on.

The engineer is not only interested in the measurement of physical variables but also concerned with their control. The two functions are closely related, however, because one must be able to measure a variable such as temperature or flow in order to control it. The accuracy of control is necessarily dependent on the accuracy of measurement. Hence, we see that a good knowledge of measurement techniques is necessary for the design of control systems. A detailed consideration of control systems is beyond the scope of our discussion, but the applicability of specific instruments and sensing devices to control systems will be indicated from time to time.

It is not enough for the engineer to be able to measure skillfully certain physical variables. For the data to have maximum significance the engineer must be able to specify the degree of accuracy with which a certain variable has been measured. To specify this accuracy the limitations of the apparatus must be understood and full account must be taken of certain random and/or regular errors which may occur in the experimental data. Statistical techniques are available for analyzing data to determine expected errors and deviations from the true measurements. The engineer must be familiar with these techniques in order to analyze the data effectively.

All too frequently the engineer embarks on an experimental program in a stumbling blind-faith manner. Data are collected at random, many of which are not needed for later analysis. Certain ranges of operation are not investigated thoroughly enough, resulting in the collection of data which may have limited correlative value. The engineer must be sure to take enough data but should not waste time and money by taking too many. The obvious point is that experiments should be carefully planned. Most experimentalists do indeed plan tests with respect to the range of certain variables that they will want to investigate. But they often neglect the fact that more data points may be necessary in certain ranges of operation than in others in order to ensure the same degree of accuracy in the final data evaluation. In other words, the anticipated methods of data analysis, statistical or otherwise, should be taken into account in planning the experiment, just as one would take into account certain variables in designing the physical size of the experimental apparatus. The engineer should always ask the question: How many data do I need to ensure that my data are not just the result of luck? We will have more to say about experiment planning throughout the book, and the reader should consider these opening remarks as only the initial motivation.

A few remarks concerning experimental research are in order at this point. It is very difficult to describe the atmosphere and technique of performing research. For, unlike standard performance testing where experiments are conducted according to some well-established procedure, in research there is seldom a clear-cut way of proceeding. Each problem is a different one, and if the research is worthwhile, it has not been attacked extensively before. This means that the engineer engaged in research must be prepared to face numerous experimental difficulties of varying complexities. Some desirable objectives of the research may have to be relaxed because of the unavailability of instrumentation to measure the variables involved. Many seemingly trivial details become significant problems before a new experimental apparatus is functioning properly. One of the most basic problems is that the engineer seldom gets to measure in a direct manner the variable he or she really wants. There are always corrections to apply to the measurements, and seldom do they fall into the category of "standard" corrections. One trivial detail piles on another until the whole experimental problem takes on a complexity which is usually not anticipated at the start of the research. Again, we state the truism: There is no such thing as an *easy* experiment.

Neophyte experimentalists frequently assume that a certain experiment will be easy to perform. All they need to do is hook up the apparatus and flip the switch, and out will come reams of significant data which will startle colleagues (or supervisors). They do not realize that one simple instrument may not work and thus spoil the experiment. Once this instrument is functioning properly, another may go bad, and so on. When the apparatus is functioning, neophytes are then tempted to take data at random without giving much consideration to the results that they will want to derive from the data. They try to solve all problems at once and vary many parameters at the same time, so that little control is exerted on the data and it eventually becomes necessary to go back and do some of the work over. The important point, once again, is that careful planning is called for. In experimental research great care and patience will usually produce the best results in the *quickest* possible way.

The above remarks may appear discouraging to beginners for whom this book is written. On the contrary, they are intended to advise beginners so that they can avoid some of the more obvious pitfalls. And, even more important, the intent is to let the beginner know that some troubles combined with intelligent planning will almost always lead to the desired results—accurate and meaningful data.

Certainly, not all experimental work is of a "research" nature. The majority of measurements are performed in routine practical industrial applications. Such measurements call for skill and communication ability as much as research because they may be very directly related to the profit and loss of a company.

The objective of the presentation in this book is to impart a broad knowledge of experimental methods and measurement techniques. To accomplish this object a rather large number of instruments will be discussed from the standpoint of both theory of operation and specific functional characteristics. Emphasis will be placed on analytical calculations to familiarize the reader with important points in the theoretical development as well as in the descriptive information pertaining to operating characteristics. As a further means of emphasizing the discussions, the uncertainties which may arise in the various instruments are given particular attention.

The study of experimental methods is a necessary extension of all analytical subjects. A knowledge of the methods of verifying analytical work injects new life and vitality into the theories, and a clear understanding of the difficulties of experimental measurements creates a careful attitude in the theoretician which cannot be generated in any other way.

chapter

2

BASIC CONCEPTS

2.1   INTRODUCTION

In this chapter we seek to explain some of the terminology used in experimental methods and to show the generalized arrangement of an experimental system. We shall also discuss briefly the standards which are available and the importance of calibration in any experimental measurement. A major portion of the discussion on experimental errors is deferred until Chap. 3, and only the definition of certain terms is given here.

2.2   DEFINITION OF TERMS

We are frequently concerned with the *readability* of an instrument. This term indicates the closeness with which the scale of the instrument may be read; an instrument with a 12-in scale would have a higher readability than an instrument with a 6-in scale and the same range. The *least count* is the smallest difference between two indications that can be detected on the instrument scale. Both readability and least count are dependent on scale length, spacing of graduations, size of pointer (or pen if a recorder is used), and parallax effects.

For an instrument with a digital readout the terms "readability" and "least count" have little meaning. Instead, one is concerned with the display of the particular instrument.

The *sensitivity* of an instrument is the ratio of the linear movement of the pointer on an analog instrument to the change in the measured variable causing this motion. For example, a 1-mV recorder might have a 25-cm scale length. Its sensitivity would be 25 cm/mV, assuming that the measurement was linear all across the scale. For a digital instrument readout the term "sensitivity" does not have the same meaning because different scale factors can be applied with the push of a button. However, the manufacturer will usually specify the sensitivity for a certain scale setting, for example, 100 nA on a 200-μA scale range for current measurement.

5

An instrument is said to exhibit *hysteresis* when there is a difference in readings depending on whether the value of the measured quantity is approached from above or below. Hysteresis may be the result of mechanical friction, magnetic effects, elastic deformation, or thermal effects.

The *accuracy* of an instrument indicates the deviation of the reading from a known input. Accuracy is frequently expressed as a percentage of full-scale reading, so that a 100-kPa pressure gage having an accuracy of 1 percent would be accurate within ± 1 kPa over the entire range of the gage.

In other cases accuracy may be expressed as an absolute value, over all ranges of the instrument.

The *precision* of an instrument indicates its ability to reproduce a certain reading with a given accuracy. As an example of the distinction between precision and accuracy, consider the measurement of a known voltage of 100 volts (V) with a certain meter. Four readings are taken, and the indicated values are 104, 103, 105, and 105 V. From these values it is seen that the instrument could not be depended on for an accuracy of better than 5 percent (5 V), while a precision of ± 1 percent is indicated since the maximum deviation from the mean reading of 104 V is only 1 V. It may be noted that the instrument could be *calibrated* so that it could be used dependably to measure voltages within ± 1 V. This simple example illustrates an important point. Accuracy can be improved up to but not beyond the precision of the instrument by calibration. The precision of an instrument is usually subject to many complicated factors and requires special techniques of analysis, which will be discussed in Chap. 3.

We should alert the reader at this time to some data analysis terms which will appear in Chap. 3. Accuracy has already been mentioned as relating the deviation of an instrument reading from a *known* value. The deviation is called the *error*. In many experimental situations we may not have a known value with which to compare instrument readings, and yet we may feel fairly confident that the instrument is within a certain plus or minus range of the true value. In such cases we say that the plus or minus range expresses the *uncertainty* of the instrument readings. Many experimentalists are not very careful in using the words "error" and "uncertainty." As we shall see in Chap. 3, uncertainty is the term that should be most often applied to instruments.

2.3 CALIBRATION

The calibration of all instruments is important, for it affords the opportunity to check the instrument against a known standard and subsequently to reduce errors in accuracy. Calibration procedures involve a comparison of the particular instrument with either (1) a primary standard, (2) a secondary standard with a higher accuracy than the instrument to be calibrated, or (3) a known input source. For example, a flowmeter might be calibrated by (1) comparing it with a standard flow-measurement facility of the National Institute for Standards and Technology (NIST), (2) comparing it with another flowmeter of known accuracy, or (3) directly calibrating with a primary measurement such as weighing a certain amount of water in a tank and recording the time elapsed for this quantity to flow through the meter. In item 2 the keywords are "known accuracy." The meaning here is that the accuracy of the meter must be specified by a reputable source.

The importance of calibration cannot be overemphasized because it is calibration that firmly establishes the accuracy of the instruments. Rather than accept the reading of an instrument, it is usually best to make at least a simple calibration check to be sure of the validity of the measurements. Not even manufacturers' specifications or calibrations can always be taken at face value. Most instrument manufacturers are reliable; some, alas, are not. We shall be able to give more information on calibration methods throughout the book as various instruments and their accuracies are discussed.

2.4 STANDARDS

In order that investigators in different parts of the country and different parts of the world may compare the results of their experiments on a consistent basis, it is necessary to establish certain standard units of length, weight, time, temperature, and electrical quantities. NIST has the primary responsibility for maintaining these standards in the United States.

The meter and the kilogram are considered fundamental units upon which, through appropriate conversion factors, the English system of length and mass is based. At one time, the standard meter was defined as the length of a platinum-iridium bar maintained at very accurate conditions at the International Bureau of Weights and Measures in Sèvres, France. Similarly, the kilogram was defined in terms of a platinum-iridium mass maintained at this same bureau. The conversion factors for the English and metric systems in the United States are fixed by law as

$$1 \text{ meter} = 39.37 \text{ inches}$$

$$1 \text{ pound-mass} = 453.59237 \text{ grams}$$

Standards of length and mass are maintained at NIST for calibration purposes. In 1960 the General Conference on Weights and Measures defined the standard meter in terms of the wavelength of the orange-red light of a krypton-86 lamp. The standard meter is thus

$$1 \text{ meter} = 1,650,763.73 \text{ wavelengths}$$

In 1983 the definition of the meter was changed to the distance light travels in 1/299,792,458ths of a second. For the measurement, light from a helium-neon laser illuminates iodine which fluoresces at a highly stable frequency.

The inch is exactly defined as

$$1 \text{ inch} = 2.54 \text{ centimeters}$$

Standard units of time are established in terms of known frequencies of oscill
of certain devices. One of the simplest devices is a pendulum. A torsional vibr
system may also be used as a standard of frequency. Prior to the introd
quartz oscillator–based mechanisms, torsional systems were widely use
and watches. Ordinary 60-hertz (Hz) line voltage may be used as a freque
under certain circumstances. An electric clock uses this frequency

because it operates from a synchronous electric motor whose speed depends on line frequency. A tuning fork is a suitable frequency source, as are piezoelectric crystals. Electronic oscillators may also be designed to serve as precise frequency sources.

The fundamental unit of time, the second(s), has been defined in the past as $\frac{1}{86400}$ of a mean solar day. The solar day is measured as the time interval between two successive transits of the sun across a meridian of the earth. The time interval varies with location of the earth and time of year; however, the *mean solar day* for one year is constant. The *solar year* is the time required for the earth to make one revolution around the sun. The mean solar year is 365 days 5 h 48 min 48 s.

The above definition of the second is quite exact but is dependent on astronomical observations in order to establish the standard. In October 1967 the Thirteenth General Conference on Weights and Measures adopted a definition of the second as the duration of 9,192,631,770 periods of the radiation corresponding to the transition between the two hyperfine levels of the fundamental state of the atom of cesium-133, Ref. [7]. This standard can be readily duplicated in standards laboratories throughout the world. The estimated accuracy of this standard is 2 parts in 10^9.

Standard units of electrical quantities are derivable from the mechanical units of force, mass, length, and time. These units represent the absolute electrical units and differ slightly from the international system of electrical units established in 1948. A detailed description of the previous international system is given in Ref. [1]. The main advantage of this sytem is that it affords the establishment of a standard cell, the output of which may be directly related to the absolute electrical units. The conversion from the international system was established by the following relations:

1 international ohm = 1.00049 absolute ohms

1 international volt = 1.000330 absolute volts

1 international ampere = 0.99835 absolute ampere

The standard for the volt was changed in 1990 to relate to a phenomenon called the Josephson effect which occurs at liquid helium temperatures. At the same time resistance standards were based on a quantum Hall effect. These standards, and a historical perspective as well as the use of standard cells, are discussed in Refs. [15], [16], and [17].

Laboratory calibration is usually made with the aid of secondary standards such as standard cells for voltage sources and standard resistors as standards of comparison for measurement of electrical resistance.

An absolute temperature scale was proposed by Lord Kelvin in 1854 and forms the basis for thermodynamic calculations. This absolute scale is so defined that particular meaning is given to the second law of thermodynamics when this temperature scale is used. The International Practical Temperature Scale of 1968 (IPTS-68) [2] furnishes an experimental basis for a temperature scale which approximates as closely as possible the absolute thermodynamic temperature scale. In the international scale primary points are established as shown in Table 2.1. Secondary fixed points are established as given in Table 2.2. In addition to the fixed points, precise points

Table 2.1 Primary points for the International Practical Temperature Scale of 1968

Point	Temperature	
Normal Pressure = 14.6959 psia = 1.0132×10^5 Pa	°C	°F
Triple point of equilibrium hydrogen	−259.34	−434.81
Boiling point of equilibrium hydrogen at 25/76 normal pressure	−256.108	−428.99
Normal boiling point (1 atm) of equilibrium hydrogen	−252.87	−423.17
Normal boiling point of neon	−246.048	−410.89
Triple point of oxygen	−218.789	−361.82
Normal boiling point of oxygen	−182.962	−297.33
Triple point of water	0.01	32.018
Normal boiling point of water	100	212.00
Normal freezing point of zinc	419.58	787.24
Normal freezing point of silver	961.93	1763.47
Normal freezing point of gold	1064.43	1947.97

Table 2.2 Secondary fixed points for the International Practical Temperature Scale of 1968

Point	Temperature, °C
Triple point, normal H_2	−259.194
Boiling point, normal H_2	−252.753
Triple point, Ne	−248.595
Triple point, N_2	−210.002
Boiling point, N_2	−195.802
Sublimation point, CO_2 (normal)	−78.476
Freezing point, Hg	−38.862
Ice point	0
Triple point, phenoxibenzene	26.87
Triple point, benzoic acid	122.37
Freezing point, In	156.634
Freezing point, Bi	271.442
Freezing point, Cd	321.108
Freezing point, Pb	327.502
Freezing point, Hg	356.66
Freezing point, S	444.674
Freezing point, Cu-Al eutectic	548.23
Freezing point, Sb	630.74
Freezing point, Al	660.74
Freezing point, Cu	1084.5
Freezing point, Ni	1455
Freezing point, Co	1494
Freezing point, Pd	1554
Freezing point, Pt	1772
Freezing point, Rh	1963
Freezing point, Ir	2447
Freezing point, W	3387

Table 2.3   Interpolation procedures for International Practical Temperature Scale of 1968

Range, °C	Procedure
−259.34–0	Platinum resistance thermometer with cubic polynomial coefficients determined from calibration at fixed points, using four ranges
0–630.74	Platinum resistance thermometer with second-degree polynomial coefficients determined from calibration at three fixed points in the range
630.74–1064.43	Standard platinum–platinum rhodium (10%) thermocouple with second-degree polynomial coefficients determined from calibration at antimony, silver, and gold points
Above 1064.43	Temperature defined by: $$\frac{J_t}{J_{\text{Au}}} = \frac{e^{C_2/\lambda(T_{\text{Au}}+T_0)} - 1}{e^{C_2/\lambda(T+T_0)} - 1}$$ J_t, J_{Au} = radiant energy emitted per unit time, per unit area, and per unit wavelength at wavelength λ, at temperature T, and gold-point temperature T_{Au}, respectively $C_2 = 1.438 \text{ cm} - \text{K}$ $T_0 = 273.16 \text{ K}$ $\lambda = \text{wavelength}$

are also established for interpolating between these points. These interpolation procedures are given in Table 2.3.

More recently, the International Temperature Scale of 1990 (ITS-90) has been adopted as described in Ref. [13].

The fixed points for ITS-90 that are shown in Table 2.4 differ only slightly from IPTS-68. For ITS-90 a platinum resistance thermometer is used for interpolation between the triple point of hydrogen and the solid equilibrium for silver, while above the silver point blackbody radiation is used for interpolation. Reference [14] gives procedures for converting between ITPS-68 calibrations and those under ITS-90. In many practical situations the errors or uncertainties in the primary sensing elements will overshadow any differences in the calibration. According to Ref. [13] the differences between IPTS-68 and ITS-90 are less than 0.12°C between −200°C and 900°C.

Both the Fahrenheit (°F) and Celsius (°C) temperature scales are in wide use, and the experimentalist must be able to work in either. The absolute Fahrenheit scale is called the Rankine (°R) scale, while absolute Celsius has been designated the Kelvin (K) scale. The relationship between these scales is as follows:

$$K = °C + 273.15$$

$$°R = °F + 459.67$$

$$°F = \tfrac{9}{5}°C + 32.0$$

The absolute thermodynamic temperature scale is discussed in Ref. [4].

Table 2.4   Fixed points for International Temperature Scale of 1990

| | Temperature | |
Defining State	°C	K
Triple point of hydrogen	−259.3467	13.8033
Liquid/vapor equilibrium for hydrogen at $\frac{25}{76}$ atm	−256.15	≈17
Liquid/vapor equilibrium for hydrogen at 1 atm	≈ −252.87	≈20.3
Triple point of neon	−248.5939	24.5561
Triple point of oxygen	−218.7916	54.3584
Triple point of argon	−189.3442	83.8058
Triple point of water	0.01	273.16
Solid/liquid equilibrium for gallium at 1 atm	29.7646	302.9146
Solid/liquid equilibrium for tin at 1 atm	231.928	505.078
Solid/liquid equilibrium for zinc at 1 atm	419.527	692.677
Solid/liquid equilibrium for silver at 1 atm	961.78	1234.93
Solid/liquid equilibrium for gold at 1 atm	1064.18	1337.33
Solid/liquid equilibrium for copper at 1 atm	1084.62	1357.77

2.5   DIMENSIONS AND UNITS

Despite strong emphasis in the professional engineering community on standardizing units with an international system, a variety of instruments will be in use for many years, and an experimentalist must be conversant with the units which appear on the gages and readout equipment. The main difficulties arise in mechanical and thermal units because electrical units have been standardized for some time. It is hoped that the SI (Système International d'Unités) set of units will eventually prevail, and we shall express examples and problems in this system as well as in the English system employed in the United States for many years.

Although the SI system is preferred, one must recognize that the English system is still very popular.

One must be careful not to confuse the meaning of the term "units" and "dimensions." A *dimension* is a physical variable used to specify the behavior or nature of a particular system. For example, the length of a rod is a dimension of the rod. In like manner, the temperature of a gas may be considered one of the thermodynamic dimensions of the gas. When we say the rod is so many meters long, or the gas has a temperature of so many degrees Celsius, we have given the units with which we choose to measure the dimension. In our development we shall use the dimensions

L = length
M = mass
F = force
τ = time
T = temperature

All the physical quantities used may be expressed in terms of these fundamental dimensions. The units to be used for certain dimensions are selected by somewhat arbitrary definitions which usually relate to a physical phenomenon or law. For example, Newton's second law of motion may be written

$$\text{Force} \sim \text{time rate of change of momentum}$$

$$F = k\frac{d(mv)}{d\tau}$$

where k is the proportionality constant. If the mass is constant,

$$F = kma \qquad\qquad \textbf{[2.1]}$$

where the acceleration is $a = dv/d\tau$. Equation (2.1) may also be written

$$F = \frac{1}{g_c}ma \qquad\qquad \textbf{[2.2]}$$

with $1/g_c = k$. Equation (2.2) is used to define our systems of units for mass, force, length, and time. Some typical systems of units are:

1. 1 pound-force will accelerate 1 pound-mass 32.174 feet per second squared.
2. 1 pound-force will accelerate 1 slug-mass 1 foot per second squared.
3. 1 dyne-force will accelerate 1 gram-mass 1 centimeter per second squared.
4. 1 newton (N) force will accelerate 1 kilogram-mass 1 meter per second squared.
5. 1 kilogram-force will accelerate 1 kilogram-mass 9.80665 meter per second squared.

The kilogram-force is sometimes given the designation *kilopond* (kp).

Since Eq. (2.2) must be dimensionally homogeneous, we shall have a different value of the constant g_c for each of the unit systems in items 1 to 5 above. These values are:

1. $g_c = 32.174 \text{ lbm} \cdot \text{ft/lbf} \cdot \text{s}^2$
2. $g_c = 1 \text{ slug} \cdot \text{ft/lbf} \cdot \text{s}^2$
3. $g_c = 1 \text{ g} \cdot \text{cm/dyn} \cdot \text{s}^2$
4. $g_c = 1 \text{ kg} \cdot \text{m/N} \cdot \text{s}^2$
5. $g_c = 9.80665 \text{ kgm} \cdot \text{m/kgf} \cdot \text{s}^2$

It does not matter which system of units is used so long as it is consistent with the above definitions.

Work has the dimensions of a product of force times a distance. Energy has the same dimensions. Thus the units for work and energy may be chosen from any of the systems used above as:

1. lbf · ft
2. lbf · ft
3. dyn · cm $= 1$ erg

4. N · m = 1 joule (J)

5. kgf · m = 9.80665 J

In addition, we may use the units of energy which are based on thermal phenomena:

1 British thermal unit (Btu) will raise 1 pound-mass of water 1 degree Fahrenheit at 68°F.

1 calorie (cal) will raise 1 gram of water 1 degree Celsius at 20°C.

1 kilocalorie will raise 1 kilogram of water 1 degree Celsius at 20°C.

The conversion factors for the various units of work and energy are

1 Btu = 778.16 lbf · ft
1 Btu = 1055 J
1 kcal = 4182 J
1 lbf · ft = 1.356 J
1 Btu = 252 cal

Additional conversion factors are given in Appendix A.

The weight of a body is defined as the force exerted on the body as a result of the acceleration of gravity. Thus

$$W = \frac{g}{g_c} m \qquad\qquad \text{[2.3]}$$

where W is the weight and g is the acceleration of gravity. Note that the weight of a body has the dimensions of a force. We now see why systems 1 and 5 above were devised; 1 lbm will weigh 1 lbf at sea level, and 1 kgm will weigh 1 kgf.

Unfortunately, *all* the above unit systems are used in various places throughout the world. While the foot-pound force, pound-mass, second, degree Fahrenheit, Btu system is still widely used in the United States, there should be increasing impetus to institute the SI units as a worldwide standard. In this system the fundamental units are meter, newton, kilogram-mass, second, and degree Celsius; a "thermal" energy unit is not used; that is, the joule (N · m) becomes the energy unit used throughout. The watt (J/s) is the unit of power in this system. In SI the concept of g_c is not normally used, and the newton is defined as

$$1 \text{ newton} \equiv 1 \text{ kilogram-meter per second squared} \qquad\qquad \text{[2.4]}$$

Even so, one should keep in mind the physical relation between force and mass as expressed by Newton's second law of motion.

Despite the present writer's personal enthusiasm for complete changeover to the SI system, the fact is that a large number of engineering practitioners still use English units and will continue to do so for some time to come. Therefore, we shall use both English and SI units in this text so that the reader can operate effectively in the expected industrial environment.

Table 2.5 lists the basic and supplementary SI units, while Table 2.6 gives a list of derived SI units for various physical quantities. The SI system also specifies standard multiplier prefixes, as shown in Table 2.7. For example, 1 atm pressure is

Table 2.5   Basic and supplemental SI units

Quantity	Unit	Symbol
Basic units		
Length	meter	m
Mass	kilogram	kg
Time	second	s
Electric current	ampere	A
Temperature	kelvin	K
Luminous intensity	candela	cd
Supplemental units		
Plane angle	radian	rad
Solid angle	steradian	sr

1.0132×10^5 N/m^2 (Pa), which could be written 1 atm = 0.10132 MN/m^2 (MPa). Conversion factors are given in Appendix A as well as in sections of the text that discuss specific measurement techniques for pressure, energy flux, and so forth.

2.6   THE GENERALIZED MEASUREMENT SYSTEM

Most measurement systems may be divided into three parts:

1. *A detector-transducer stage,* which detects the physical variable and performs either a mechanical or an electrical transformation to convert the signal into a more usable form. In the general sense, a transducer is a device that transforms one physical effect into another. In most cases, however, the physical variable is transformed into an electric signal because this is the form of signal that is most easily measured. The signal may be in digital or analog form. Digital signals offer the advantage of easy storage in memory devices, or manipulations with computers.

2. *Some intermediate stage,* which modifies the direct signal by amplification, filtering, or other means so that a desirable output is available.

3. *A final or terminating stage,* which acts to indicate, record, or control the variable being measured. The output may also be digital or analog.

As an example of a measurement system, consider the measurement of a low-voltage signal at a low frequency. The detector in this case may be just two wires and possibly a resistance arrangement, which are attached to appropriate terminals. Since we want to indicate or record the voltage, it may be necessary to perform some amplification. The amplification stage is then stage 2, designated above. The final stage of the measurement system may be either a voltmeter or a recorder that operates in the range of the output voltage of the amplifier. In actuality, an electronic

Table 2.6   Derived SI units

Quantity	Name(s) of Unit	Unit Symbol or Abbreviation, Where Differing from Basic Form	Unit Expressed in Terms of Basic or Supplementary Units
Area	square meter		m^2
Volume	cubic meter		m^3
Frequency	hertz, cycle per second	Hz	s^{-1}
Density, concentration	kilogram per cubic meter		kg/m^3
Velocity	meter per second		m/s
Angular velocity	radian per second		rad/s
Acceleration	meter per second squared		m/s^2
Angular acceleration	radian per second squared		rad/s^2
Volumetric flow rate	cubic meter per second		m^3/s
Force	newton	N	$kg \cdot m/s^2$
Surface tension	newton per meter, joule per square meter	N/m, J/m^2	kg/s^2
Pressure	newton per square meter, pascal	N/m^2, Pa	$kg/m \cdot s^2$
Viscosity, dynamic	newton-second per square meter poiseuille	$N \cdot s/m^2$, Pl	$kg/m \cdot s$
Viscosity, kinematic; diffusivity; mass conductivity	meter square per second		m^2/s
Work, torque, energy, quantity of heat	joule, newton-meter, watt-second	J, $N \cdot m$, $W \cdot s$	$kg \cdot m^2/s^2$
Power, heat flux	watt, joule per second	W, J/s	$kg \cdot m^2/s^3$
Heat flux density	watt per square meter	W/m^2	kg/s^3
Volumetric heat release rate	watt per cubic meter	W/m^3	$kg/m \cdot s^3$
Heat-transfer coefficient	watt per square meter degree	$W/m^2 \cdot$ deg	$kg/s^3 \cdot$ deg
Latent heat, enthalpy (specific)	joule per kilogram	J/kg	m^2/s^2
Heat capacity (specific)	joule per kilogram degree	$J/kg \cdot$ deg	$m^2/s^2 \cdot$ deg
Capacity rate	watt per degree	W/deg	$kg \cdot m^2/s^3 \cdot$ deg
Thermal conductivity	watt per meter degree	$W/m \cdot$ deg, $j \cdot m/s^2 \cdot$ deg	$kg \cdot m/s^3 \cdot$ deg
Mass flux, mass flow rate	kilogram per second		kg/s
Mass flux density, mass flow rate per unit area	kilogram per square meter-second		$kg/m^2 \cdot s$
Mass-transfer coefficient	meter per second		m/s
Quantity of electricity	coulomb	C	$A \cdot s$
Electromotive force	volt	V, W/A	$kg \cdot m^2/A \cdot s^3$
Electric resistance	ohm	Ω, V/A	$kg \cdot m^2/A \cdot s^3$
Electric conductivity	ampere per volt meter	$A/V \cdot m$	$A^2 \cdot s^3/kg \cdot m^3$
Electric capacitance	farad	F, $A \cdot s/V$	$A^3 \cdot s^4/kg \cdot m^2$
Magnetic flux	weber	Wb, $V \cdot s$	$kg \cdot m^2/A \cdot s^2$
Inductance	henry	H, $V \cdot s/A$	$kg \cdot m^2/A^2 \cdot s^2$
Magnetic permeability	henry per meter	H/m	$kg \cdot m/A^2 \cdot s^2$
Magnetic flux density	tesla, weber per square meter	T, Wb/m^2	$kg/A \cdot s^2$
Magnetic field strength	ampere per meter		A/m
Magnetomotive force	ampere		A
Luminous flux	lumen	lm	$cd \cdot sr$
Luminance	candela per square meter		cd/m^2
Illumination	lux, lumen per square meter	lx, lm/m^2	$cd \cdot sr/m^2$

Table 2.7   Standard prefixes and multiples in SI units

Multiples and Submultiples	Prefixes	Symbols	Pronunciations
10^{12}	tera	T	tĕr'á
10^9	giga	G	jĭ'gá
10^6	mega	M	mĕg'á
10^3	kilo	k	kĭl'ô
10^2	hecto	h	hek'tô
10	deka	da	dĕk'á
10^{-1}	deci	d	dĕs'ĭ
10^{-2}	centi	c	sĕn'tĭ
10^{-3}	milli	m	mĭl ĭ
10^{-6}	micro	μ	mī' krô
10^{-9}	nano	n	năn'ô
10^{-12}	pico	p	pē'cô
10^{-15}	femto	f	fĕm'tô
10^{-18}	atto	a	ăt'tô

voltmeter is a measurement system like the one described here. The amplifier and the readout voltmeter are contained in one package, and various switches enable the user to change the range of the instrument by varying the input conditions to the amplifier.

Consider the simple bourdon-tube pressure gage shown in Fig. 2.1. This gage offers a mechanical example of the generalized measurement system. In this case the bourdon tube is the detector-transducer stage because it converts the pressure signal into a mechanical displacement of the tube. The intermediate stage consists of the gearing arrangement, which amplifies the displacement of the end of the tube so that a relatively small displacement at that point produces as much as three-quarters of a revolution of the center gear. The final indicator stage consists of the pointer and the dial arrangement, which, when calibrated with known pressure inputs, gives an indication of the pressure signal impressed on the bourdon tube. A schematic diagram of the generalized measurement system is shown in Fig. 2.2.

When a control device is used for the final measurement stage, it is necessary to apply some feedback signal to the input signal to accomplish the control objectives. The control stage compares the signal representing the measured variable with some other signal in the same form representing the assigned value the measured variable *should have*. The assigned value is given by a predetermined setting of the controller. If the measured signal agrees with the predetermined setting, then the controller does nothing. If the signals do not agree, the controller issues a signal to a device which acts to alter the value of the measured variable. This device can be many things, depending on the variable which is to be controlled. If the measured variable is the flow rate of a fluid, the control device might be a motorized valve placed in the flow system. If the measured flow rate is too high, then the controller would cause the motorized valve to close, thereby reducing the flow rate. If the flow rate were too low, the valve would be opened. Eventually the operation would cease when the desired flow rate was achieved. The control feedback function is indicated in Fig. 2.2.

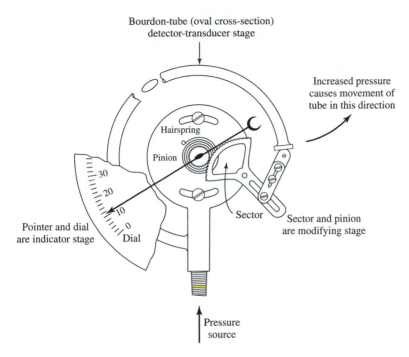

Figure 2.1 Bourdon-tube pressure gage as the generalized measurement system.

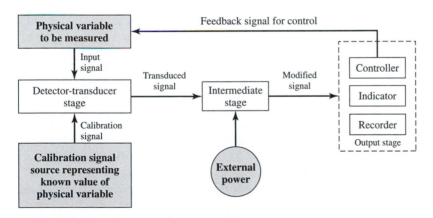

Figure 2.2 Schematic of the generalized measurement system.

It is very important to realize that the accuracy of control cannot be any better than the accuracy of the measurement of the control variable. Therefore, one must be able to measure a physical variable accurately before one can hope to control the variable. In the flow system mentioned above the most elaborate controller could not control the flow rate any more closely than the accuracy with which the primary sensing element measures the flow. We shall have more to say about some simple control systems in a later chapter. For the present we want to emphasize the importance of the measurement system in any control setup.

The overall schematic of the generalized measurement system is quite simple, and, as one might suspect, the difficult problems are encountered when suitable devices are sought to fill the requirements for each of the "boxes" on the schematic diagram. Most of the remaining chapters of the book are concerned with the types of detectors, transducers, modifying stages, and so forth, that may be used to fill these boxes.

2.7 BASIC CONCEPTS IN DYNAMIC MEASUREMENTS

A *static* measurement of a physical quantity is performed when the quantity is not changing with time. The deflection of a beam under a constant load would be a static deflection. However, if the beam were set in vibration, the deflection would vary with time, and the measurement process might be more difficult. Measurements of flow processes are much easier to perform when the fluid is in a nice steady state and become progressively more difficult to perform when rapid changes with time are encountered.

Many experimental measurements are taken under such circumstances that ample time is available for the measurement system to reach steady state, and hence one need not be concerned with the behavior under non-steady-state conditions. In many other situations, however, it may be desirable to determine the behavior of a physical variable over a period of time. Sometimes the time interval is short, and sometimes it may be rather extended. In any event, the measurement problem usually becomes more complicated when the transient characteristics of a system need to be considered. In this section we wish to discuss some of the more important characteristics and parameters applicable to a measurement system under dynamic conditions.

ZEROTH-, FIRST-, AND SECOND-ORDER SYSTEMS

A system may be described in terms of a general variable $x(t)$ written in differential equation form as

$$a_n \frac{d^n x}{dt^n} + a_{n-1} \frac{d^{n-1}}{dt^{n-1}} + \cdots + a_1 \frac{dx}{dt} + a_0 x = F(t) \qquad \textbf{[2.5]}$$

where $F(t)$ is some forcing function imposed on the system. The *order* of the system is designated by the order of the differential equation.

A *zeroth-order* system would be governed by

$$a_0 x = F(t) \qquad \textbf{[2.6]}$$

a *first-order* system by

$$a_1 \frac{dx}{dt} + a_0 x = F(t) \qquad \textbf{[2.7]}$$

and a *second-order* system by

$$a_2 \frac{d^2 x}{dt^2} + a_1 \frac{dx}{dt} + a_0 x = F(t) \qquad \text{[2.8]}$$

We shall examine the behavior of these three types of systems to study some basic concepts of dynamic response. We shall also give some concrete examples of physical systems which exhibit the different orders of behavior.

The zeroth-order system described by Eq. (2.6) indicates that the system variable $x(t)$ will track the input forcing function instantly by some constant value: that is,

$$x = \frac{1}{a_0} F(t)$$

The constant $1/a_0$ is called the *static sensitivity* of the sytem. If a constant force were applied to the beam mentioned above, the static deflection of the beam would be F/a_0.

The first-order system described by Eq. (2.7) may be expressed as

$$\frac{a_1}{a_0} \frac{dx}{dt} + x = \frac{F(t)}{a_0} \qquad \text{[2.9]}$$

The ratio a_1/a_0 has the dimension of time and is usually called the *time constant* of the system. If Eq. (2.9) is solved for the case of a sudden constant (step) input $F(t) = A$ at time zero, we express the condition as

$$F(t) = 0 \qquad \text{at } t = 0$$
$$F(t) = A \qquad \text{for } t > 0$$

along with the initial condition

$$x = x_0 \qquad \text{at } t = 0$$

The solution to Eq. (2.9) is then

$$x(t) = \frac{A}{a_0} + \left(x_0 - \frac{A}{a_0} \right) e^{-t/\tau} \qquad \text{[2.10]}$$

where, now, we set $\tau = a_1/a_0$. The *steady-state response* is the first term on the right, or the value of x, which will be obtained for large values of time. The second term, involving the exponential decay term, represents the *transient response* of the system. Designating the steady-state value as x_∞, Eq. (2.10) may be written in dimensionless form as

$$\frac{x(t) - x_\infty}{x_0 - x_\infty} = e^{-t/\tau} \qquad \text{[2.11]}$$

When $t = \tau$, the value of $x(t)$ will have responded to 63.2 percent of the step input, so the time constant is frequently called the time to achieve this value. The *rise time* is the time required to achieve a response of 90 percent of the step input. This requires

$$e^{-t/\tau} = 0.1$$

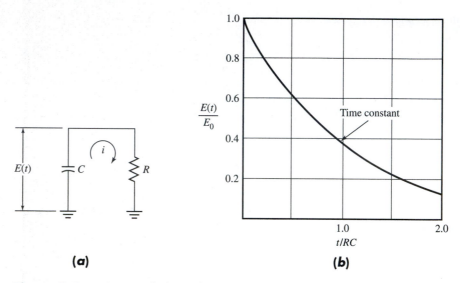

(a)   **(b)**

Figure 2.3   Capacity discharge through a resistance. (a) Schematic; (b) plot of voltage.

or $t = 2.303\tau$. A response is usually assumed to be complete after 5τ since $1 - e^{-5} = 0.993$.

Systems that exhibit first-order behavior usually involve storage and dissipation capabilities such as an electric capacitor discharging through a resistor, as shown in Fig. 2.3.

The voltage varies with time according to

$$\frac{E(t)}{E_0} = e^{-(1/RC)t} \qquad\qquad \textbf{[2.12]}$$

In this system

$$\tau = RC$$

where R is the value of the external resistance and C is the capacitance. The voltage across the capacitor as a function of time is $E(t)$, and the initial voltage is E_0. Some types of thermal systems also display this same kind of response. The temperature of a hot block of metal allowed to cool in a room varies with approximately the same kind of relation as shown in Eq. (2.12).

For a thermal system we shall have analogous concepts of thermal capacity and resistance. In the capacitance system we speak of voltage, in a thermal system we speak of temperature, and in some mechanical systems we might speak of velocity or displacement as the physical variables that change with time. The time constant would then be related to the initial and steady-state values of these variables. A plot of the voltage decay is shown in Fig. 2.3b, illustrating the position of the time constant.

We could say that the term on the right-hand side of Eq. (2.11) represents the *error* in achieving the steady-state value $x_\infty = A/a_0$.

First-order systems may also be subjected to harmonic inputs. We may have

Initial condition:

$$x = x_0 \qquad \text{at } t = 0$$

$$F(t) = A \sin \omega t \qquad \text{for } t > 0$$

The solution to Eq. (2.9) is, then,

$$x(t) = Ce^{-t/\tau} + \frac{A/a_0}{[1 + (\omega\tau)^2]^{1/2}} \sin(\omega t - \tan^{-1} \omega\tau) \qquad \textbf{[2.13]}$$

where $\tau = a_1/a_0$ is the time constant as before. At this point we define the *phase-shift angle* ϕ as

$$\phi(\omega) = -\tan^{-1} \omega\tau \qquad \textbf{[2.14]}$$

where ϕ is in radians. We see that the steady-state response [the last term of Eq. (2.13)] lags by a *time delay* of

$$\Delta t = \frac{\phi(\omega)}{\omega} \qquad \textbf{[2.15]}$$

where ω is the frequency of the input signal in rad/s. We may also see from the last term of Eq. (2.13) that the steady-state amplitude response decreases with an increase in input frequency through the term

$$\frac{1}{[1 + (\omega\tau)^2]^{1/2}}$$

The net result is that a first-order system will respond to a harmonic input in a harmonic fashion with the same frequency, but with a phase shift (time delay) and reduced amplitude. The larger the time constant, the greater the phase lag and amplitude decrease. In physical systems the constant a_1 is usually associated with storage (electric or thermal capacitance) and the constant a_0 is associated with dissipation (electric or thermal resistance), so the most rapid response is obtained in systems of low capacitance and high dissipation (low resistance). As an example, a small copper bead (high conductivity, low resistance to heat flow) will cool faster than a large ceramic body (high capacity, low conductivity).

In Chap. 8 we shall examine the response of thermal systems in terms of the significant heat-transfer parameters and relate the response to temperature-measurement applications.

STEP RESPONSE OF FIRST-ORDER SYSTEM. A certain thermometer has a time con- | **Example 2.1**
stant of 15 s and an initial temperature of 20°C. It is suddenly exposed to a temperature of 100°C. Determine the rise time, that is, the time to attain 90 percent of the steady-state value, and the temperature at this time.

Solution

The thermometer is a first-order system which will follow the behavior in Eq. (2.11). The variable in this case is the temperature, and we have

$$T_0 = 20°C = \text{temperature at } t = 0$$

$$T_\infty = 100°C = \text{temperature at steady state}$$

$$\tau = 15 \text{ s} = \text{time constant}$$

For the 90 percent rise time

$$e^{-t/\tau} = 0.1$$

and

$$\ln(0.1) = \frac{-t}{15}$$

so that

$$t = 34.54 \text{ s}$$

Then, at this time Eq. (2.11) becomes

$$\frac{T(t) - 100}{20 - 100} = 0.1$$

and

$$T(t) = 92°C$$

Example 2.2 | **PHASE LAG IN FIRST-ORDER SYSTEM.**   Suppose the thermometer in Example (2.1) was subjected to a very slow harmonic disturbance having a frequency of 0.01 Hz. The time constant is still 15 s. What is the time delay in the response of the thermometer and how much does the steady-state amplitude response decrease?

Solution

We have

$$\omega = 0.01 \text{ Hz} = 0.06283 \text{ rad/s}$$

$$\tau = 15 \text{ s}$$

so that

$$\omega\tau = (0.06283)(15) = 0.9425$$

From Eq. (2.14) the phase angle is

$$\phi(\omega) = -\tan^{-1}(0.9425)$$

$$= -43.3° = -0.756 \text{ rad}$$

so that the time delay is

$$\Delta t = \frac{\phi(\omega)}{\omega} = \frac{-0.756}{0.06283} = -12.03 \text{ s}$$

The amplitude response decreases according to

$$\frac{1}{[1 + (\omega\tau)^2]^{1/2}} = \frac{1}{[1 + (0.9425)^2]^{1/2}} = 0.7277$$

HARMONIC RESPONSE OF FIRST-ORDER SYSTEM.   A first-order system experi- | **Example 2.3**
ences a phase shift of $-45°$ at a certain frequency ω_1. By what fraction has the amplitude
decreased from a frequency of one-half this value?

Solution

We have

$$\phi(\omega) = -45° = -\tan^{-1}(\omega_1\tau)$$

which requires that $\omega_1\tau = 1.0$. The amplitude factor is thus

$$\frac{A/a_0}{[1 + (1)^2]^{1/2}} = 0.707\frac{A}{a_0} \qquad \qquad \textbf{[a]}$$

The time constant τ does not depend on frequency, so halving the frequency produces a value
of $\omega\tau = 0.5$, which gives an amplitude factor of

$$\frac{A/a_0}{[1 + (0.5)^2]^{1/2}} = 0.894\frac{A}{a_0} \qquad \qquad \textbf{[b]}$$

The value in Eq. (a) is 7.91 percent below this value.

Second-order systems described by Eq. (2.8) are those that have mass inertia or
electric inductance. There is no thermal analogy to inertia because of the second law of
thermodynamics. We shall illustrate second-order system behavior with a mechanical
example.

To initiate the discussion, let us consider a simple spring-mass damper system,
as shown in Fig. 2.4. We might consider this as a simple mechanical-measurement
system where $x_1(t)$ is the input displacement variable which acts through the spring-
mass damper arrangement to produce an output displacement $x_2(t)$. Both x_1 and x_2
vary with time. Suppose we wish to find $x_2(t)$, knowing $x_1(t)$, m, k, and the damping
constant c. We assume that the damping force is proportional to velocity so that the
differential equation governing the system is obtained from Newton's second law of
motion as

$$k(x_1 - x_2) + c\left(\frac{dx_1}{dt} - \frac{dx_2}{dt}\right) = m\frac{d^2x_2}{dt^2} \qquad \qquad \textbf{[2.16]}$$

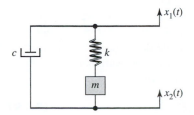

Figure 2.4   Simple spring-mass damper system.

Written in another form,

$$m\frac{d^2x_2}{dt^2} + c\frac{dx_2}{dt} + kx_2 = c\frac{dx_1}{dt} + kx_1 \qquad \textbf{[2.17]}$$

Now, suppose that $x_1(t)$ is the harmonic function

$$x_1(t) = x_0 \cos \omega_1 t \qquad \textbf{[2.18]}$$

where x_0 is the amplitude of the displacement and ω_1 is the frequency.

We might imagine this simple vibrational system as being similar to a simple spring scale. The mass of the scale is m, the spring inside the scale is represented by the spring constant k, and whatever mechanical friction may be present is represented by c. We are subjecting the scale to an oscillating-displacement function and wish to know how the body of the scale will respond: that is, we want to know $x_2(t)$. We might imagine that the spring scale is shaken by hand. When the oscillation $x_1(t)$ is very slow, we would note that the scale body very nearly follows the applied oscillation. When the frequency of the oscillation is increased, the scale body will react more violently until, at a certain frequency called the *natural frequency,* the amplitude of the displacement of the spring will take on its maximum value and could be *greater* than the amplitude of the impressed oscillation $x_1(t)$. The smaller the value of the damping constant, the larger the maximum amplitude of the natural frequency. If the impressed frequency is increased further, the amplitude of the displacement of the spring body will decrease rather rapidly. The reader may conduct an experiment to verify this behavior using a simple spring-mass system shaken by hand.

Clearly, the displacement function $x_2(t)$ depends on the frequency of the impressed function $x_1(t)$. We say that the system *responds* differently depending on the input frequency, and the overall behavior is designated as the *frequency response* of the system.

A simple experiment with the spring-mass system will show that the displacement of the mass is not in *phase* with the impressed displacement; that is, the maximum displacement of the mass does not occur at the same time as the maximum displacement of the impressed function. This phenomenon is described as *phase shift*. We could solve Eq. (2.16) and determine the detailed characteristics of $x_2(t)$, including the frequency response and phase-shift behavior. However, we shall defer the solution for this system until Chap. 11, where vibration measurements are discussed. At this point in our discussion we have used this example because it is easy to visualize in a physical sense. Now, we shall consider a practical application which involves a transformation of a *force-input function* into a displacement function. We shall present the solution to this problem because it shows very clearly the nature of frequency response and phase shift in a second-order system, while emphasizing a practical system that might be used for a transient force or pressure measurement.

This system is shown in Fig. 2.5. The forcing function

$$F(t) = F_0 \cos \omega_1 t \qquad \textbf{[2.19]}$$

is impressed on the spring-mass system, and we wish to determine the displacement

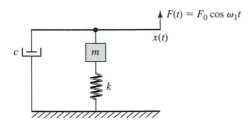

Figure 2.5   Spring-mass damper system subjected to a force input.

of the mass $x(t)$ as a function of time. The differential equation for the system is

$$m\frac{d^2x}{dt^2} + c\frac{dx}{dt} + kx = F_0 \cos \omega_1 t \qquad [2.20]$$

Equation (2.20) has the solution

$$x = \frac{(F_0/k) \cos (\omega_1 t - \phi)}{\{[1 - (\omega_1/\omega_n)^2]^2 + [2(c/c_c)(\omega_1/\omega_n)]^2\}^{1/2}} \qquad [2.21]$$

where

$$\phi = \tan^{-1} \frac{2(c/c_c)(\omega_1/\omega_n)}{1 - (\omega_1/\omega_n)^2} \qquad [2.22]$$

$$\omega_n = \sqrt{\frac{k}{m}} \qquad [2.23]$$

$$c_c = 2\sqrt{mk} \qquad [2.24]$$

ϕ is called the *phase angle,* ω_n is the natural frequency, and c_c is called the *critical damping coefficient.*

The ratio of output to input amplitude $x_0/(F_0/k)$, where x_0 is the amplitude of the motion given by

$$x_0 = \frac{F_0/k}{\{[1 - (\omega_1/\omega_n)^2]^2 + [2(c/c_c)(\omega_1/\omega_n)]^2\}^{1/2}} \qquad [2.25]$$

is plotted in Fig. 2.6 to show the frequency response of the system, and the phase angle ϕ is plotted in Fig. 2.7 to illustrate the phase-shift characteristics. From these graphs we make the following observations:

1. For low values of c/c_c the amplitude is very nearly constant up to a frequency ratio of about 0.3.

2. For large values of c/c_c (overdamped systems) the amplitude is reduced substantially.

3. The phase-shift characteristics are a strong function of the damping ratio for all frequencies.

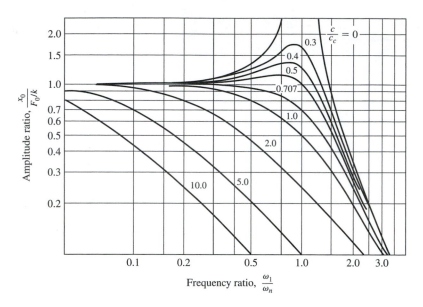

Figure 2.6   Frequency response of the system in Fig. 2.5.

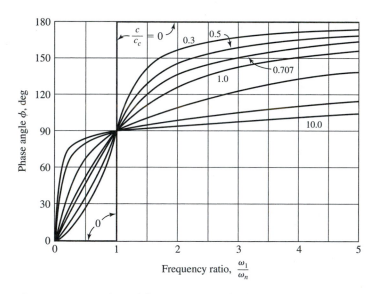

Figure 2.7   Phase-shift characteristics of the system in Fig. 2.5.

We might say that the system has good *linearity* for low damping ratios and up to a frequency ratio of 0.3 since the amplitude is essentially constant in this range.

The *rise time* for a second-order system is still defined as the time to attain a value of 90 percent of a step input. It may be reduced by reducing the damping ratio only for values of c/c_c below about 0.7. In these cases a *ringing* phenomenon is

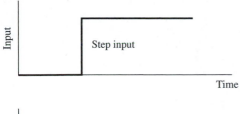

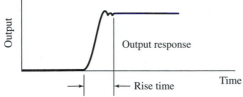

Figure 2.8   Effect of rise time and ringing on output response to a step input.

experienced having a frequency of

$$\omega_r = \omega_n[1 - (c/c_c)^2]^{1/2} \qquad [2.26]$$

The rise time and ringing are illustrated in Fig. 2.8. The *response time* is usually stated as the time for the system to settle to within ± 10 percent of the steady-state value. The damping characteristics of a second-order system may be studied by examining the solutions for Eq. (2.20) for the case of a step input instead of the harmonic forcing function. With the initial conditions

$$x = 0 \qquad \text{at } t = 0$$

$$\frac{dx}{dt} = 0 \qquad \text{at } t = 0$$

four solution forms may be obtained. Using the nomenclature

$$\zeta = c/c_c$$

and

$$x = x_s \qquad \text{as } t \to \infty$$

we obtain:

For $\zeta = 0$,

$$\frac{x(t)}{x_s} = 1 - \cos(\omega_n t) \qquad [2.27]$$

For $0 < \zeta < 1$,

$$\frac{x(t)}{x_s} = 1 - \exp(-\zeta\omega_n t) \times \left\{ \left[\frac{\zeta}{(1 - \zeta^2)^{1/2}} \right] \sin[\omega_n t(1 - \zeta^2)^{1/2}] \right.$$

$$\left. + \cos[\omega_n t(1 - \zeta^2)^{1/2}] \right\} \qquad [2.28]$$

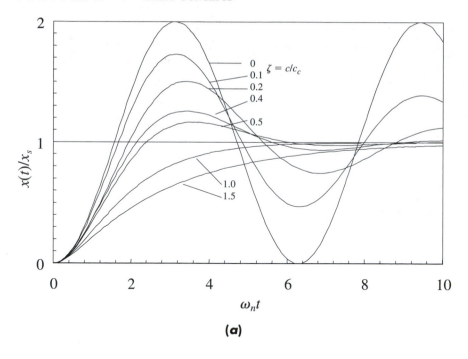

(a)

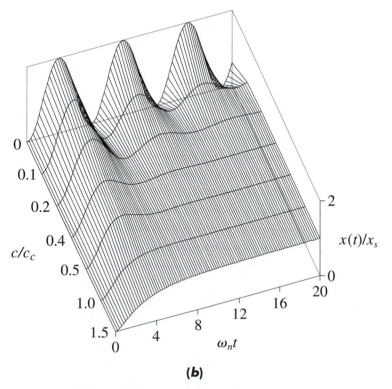

(b)

Figure 2.9   (a) Response of second-order system to step input. (b) Three-dimensional representation of second-order system response to step input.

For $\zeta = 1$,

$$\frac{x(t)}{x_s} = 1 - (1 + \omega_n t) \exp(-\omega_n t) \qquad \textbf{[2.29]}$$

For $\zeta > 1$,

$$\frac{x(t)}{x_s} = 1 - \frac{\zeta + K}{2K} \times \exp(-\zeta + K\omega_n t) - \frac{\zeta - K}{2K} \times \exp(-\zeta - K\omega_n t) \quad \textbf{[2.30]}$$

where $K = (\zeta^2 - 1)^{1/2}$.

We note that x_s is the steady-state displacement obtained after a long period of time. A plot of these equations is shown in Fig. 2.9a for several values of the damping ratio. We observe that:

(a) For an *undamped* system ($c = 0, \zeta = 0$) the initial disturbance produces a harmonic response which continues indefinitely.

(b) For an *underdamped* system ($\zeta < 1$) the displacement response *overshoots* the steady-state value initially, and then eventually decays to the value of x_s. The smaller the value of ζ, the larger the overshoot.

(c) For *critical damping* ($\zeta = 1, c = c_c$) an exponential rise occurs to approach the steady-state value without overshoot.

(d) For *overdamping* ($\zeta > 1$) the system also approaches the steady-state value without overshoot, but at a slower rate.

The damping action of Fig. 2.9a is shown over double the number of cycles in the three-dimensional format of Fig. 2.9b. The latter format illustrates perhaps more graphically the contrast between harmonic behavior for underdamped systems and exponential approach to steady state for the overdamped situation.

While this brief discussion has been concerned with a simple mechanical system, we may remark that similar frequency and phase-shift characteristics are exhibited by electrical and thermal systems as well, and whenever time-varying measurements are made, due consideration must be given to these characteristics. Ideally, we should like to have a system with a linear frequency response over all ranges and with zero phase shift, but this is never completely attainable, although a certain instrument may be linear over a range of operation in which we are interested so that the behavior is good enough for the purposes intended. There are methods of providing compensation for the adverse frequency-response characteristics of an instrument, but these methods represent an extensive subject in themselves and cannot be discussed here. We shall have something to say about the dynamic characteristics of specific instruments in subsequent chapters. For electrical systems and recording of dynamic signals digital methods can eliminate most adverse frequency-response problems.

Example 2.4	**SELECTION OF SECOND-ORDER SYSTEM.** A second-order system is to be subjected to inputs below 75 Hz and is to operate with an amplitude response of ± 10 percent. Select appropriate design parameters to accomplish this goal.

Solution

The problem statement implies that the amplitude ratio $x_0 k/F$ must remain between 0.9 and 1.10. There are many combinations of parameters which can be used. Examining Fig. 2.6, we see that the curve for $c/c_c = 0.707$ has a flat behavior and drops off at higher frequencies. If we take the ordinate value of 1.0 as the mean value, then the maximum frequency limit will be obtained from Eq. (2.25), where

$$0.9 = \frac{1}{\{[1 - (\omega_1/\omega_n)^2]^2 + [2(c/c_c)(\omega_1/\omega_n)]^2\}^{1/2}}$$

This requires that

$$\frac{\omega_1}{\omega_n} = 0.696$$

We want to use the system up to 75 Hz = 471 rad/s so that the minimum value of ω_n is

$$\omega_n(\text{min}) = \frac{471}{0.696} = 677\,\text{rad/s}$$

Example 2.5	**RESPONSE OF PRESSURE TRANSDUCER.** A certain pressure transducer has a natural frequency of 5000 Hz and a damping ratio c/c_c of 0.4. Estimate the resonance frequency and amplitude response and phase shift at a frequency of 2000 Hz.

Solution

We have

$$\omega_n = 5000\,\text{Hz}$$
$$c/c_c = 0.4$$

From Fig. 2.6 we estimate that the maximum amplitude point for these conditions occurs at

$$\frac{\omega_1}{\omega_n} \sim 0.8$$

So, $\omega_1 \sim (0.8)(5000) = 4000$ Hz for resonance. At 2000 Hz we obtain

$$\frac{\omega_1}{\omega_n} = \frac{2000}{5000} = 0.4$$

which may be inserted into Eq. (2.23) to give the phase shift as

$$\phi = -\tan^{-1}\left[\frac{(2)(0.4)(0.4)}{1 - (0.4)^2}\right]$$
$$= -20.9° = -0.364\,\text{rad}$$

The amplitude ratio is obtained from Eq. (2.15) as

$$\frac{x_0}{F_0/k} = \frac{1}{\{[1 - (0.4)^2]^2 + [(2)(0.4)(0.4)]^2\}^{1/2}}$$
$$= 1.112$$

The dynamic error in this case would be

$$1.112 - 1 = 0.112 = \pm 11.2\%$$

RISE TIME FOR DIFFERENT NATURAL FREQUENCIES.   Determine the rise time for | **Example 2.6**
a critically damped second-order system subjected to a step input when the natural frequency
of the system is (a) 10 Hz, (b) 100 kHz, (c) 50 MHz.

Solution
We have the natural frequencies of

$$\omega_n = 10 \text{ Hz} = 62.832 \text{ rad/s}$$
$$\omega_n = 100 \text{ kHz} = 6.2832 \times 10^5 \text{ rad/s}$$
$$\omega_n = 50 \text{ MHz} = 3.1416 \times 10^8 \text{ rad/s}$$

For a critically damped system subjected to a step input, $\zeta = c/c_c = 1.0$ and Eq. (2.29) applies

$$x(t)/x_s = 1 - (1 + \omega_n t) \exp(-\omega_n t) \qquad\qquad \textbf{[a]}$$

The rise time is obtained when $x(t)/x_s = 0.9$ so that Eq. (a) becomes

$$0.1 = (1 + \omega_n t) \exp(-\omega_n t) \qquad\qquad \textbf{[b]}$$

which has the solution

$$\omega_n t = 3.8901$$

Solving for the rise time at each of the given natural frequencies

$$t_{\text{rise}} = 3.8901/\omega_n$$
$$t_{10 \text{ Hz}} = 0.06191 \text{ s}$$
$$t_{100 \text{ kHz}} = 6.191 \text{ } \mu\text{s}$$
$$t_{50 \text{ MHz}} = 0.01238 \text{ } \mu\text{s}$$

2.8   SYSTEM RESPONSE

We have already discussed the meaning of frequency response and observed that in
order for a system to have good response, it must treat all frequencies the same within
the range of application so that the ratio of output-to-input amplitude remains the
same over the frequency range desired. We say that the system has *linear frequency
response* if it follows this behavior.

Amplitude response pertains to the ability of the system to react in a linear way to various input amplitudes. In order for the system to have *linear amplitude response,* the ratio of output-to-input amplitude should remain constant over some specified range of input amplitudes. When this linear range is exceeded, the system is said to be *overdriven,* as in the case of a voltage amplifier where too high an input voltage is used. Overdriving may occur with both analog and digital systems.

We have already noted the significance of phase-shift response and its relation to frequency response. Phase shift is particularly important where complex waveforms are concerned because severe distortion may result if the system has poor phase-shift response.

2.9   DISTORTION

Suppose a harmonic function of a complicated nature, that is, composed of many frequencies, is transmitted through the mechanical system of Figs. 2.4 and 2.5. If the frequency spectrum of the incoming waveform were sufficiently broad, there would be different amplitude and phase-shift characteristics for each of the input-frequency components, and the output waveform might bear little resemblance to the input. Thus, as a result of the frequency-response characteristics of the system, distortion in the waveform would be experienced. *Distortion* is a very general term that may be used to describe the variation of a signal from its true form. Depending on the system, the distortion may result from either poor frequency response or poor phase-shift response. In electronic devices various circuits are employed to reduce distortion to very small values. For pure electrical measurements distortion is easily controlled by analog or digital means. For mechanical systems the dynamic response characteristics are not as easily controlled and remain a subject for further development. For example, the process of sound recording may involve very sophisticated methods to eliminate distortion in the electronic signal processing; however at the origin of the recording process, complex room acoustics and microphone placement can alter the reproduction process beyond the capabilities of electronic correction. Finally, at the terminal stage, the loudspeaker and its interaction with the room acoustics can introduce distortions and unwanted effects. The effects of poor frequency and phase-shift response on a complex waveform are illustrated in Fig. 2.10.

2.10   IMPEDANCE MATCHING

In many experimental setups it is necessary to connect various items of electrical equipment in order to perform the overall measurement objective. When connections are made between electrical devices, proper care must be taken to avoid impedance

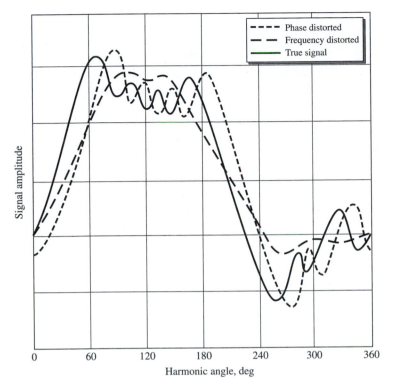

Figure 2.10   Effects of frequency response and phase-shift response on complex waveform.

mismatching. The input impedance of a two-terminal device may be illustrated as in Fig. 2.11. The device behaves as if the internal resistance R_i were connected in series with the internal voltage source E. The connecting terminals for the instrument are designated as A and B, and the open-circuit voltage presented at these terminals is the internal voltage E. Now, if an external load R is connected to the device and the internal voltage E remains constant, the voltage presented at the output terminals A and B will be dependent on the value of R. The potential presented at the output

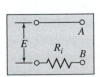
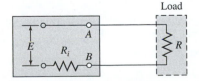

Figure 2.11   Two-terminal device with internal impedance R_i.

terminals is

$$E_{AB} = E \frac{R}{R + R_i} \qquad \textbf{[2.31]}$$

The larger the value of R, the more closely the terminal voltage approaches the internal voltage E. Thus, if the device is used as a voltage source with some internal impedance, the external impedance (or load) should be large enough that the voltage is essentially preserved at the terminals. Or, if we wish to measure the internal voltage E, the impedance of the measuring device connected to the terminals should be large compared with the internal impedance.

Now, suppose that we wish to deliver power from the device to the external load R. The power is given by

$$P = \frac{E_{AB}^2}{R} \qquad \textbf{[2.32]}$$

We ask for the value of the external load that will give the maximum power for a constant internal voltage E and internal impedance R_i. Equation (2.32) is rewritten

$$P = \frac{E^2}{R} \left(\frac{R}{R + R_i} \right)^2 \qquad \textbf{[2.33]}$$

and the maximizing condition

$$\frac{dP}{dR} = 0 \qquad \textbf{[2.34]}$$

is applied. There results

$$R = R_i \qquad \textbf{[2.35]}$$

That is, the maximum amount of power may be drawn from the device when the impedance of the external load just matches the internal impedance. This is the essential principle of impedance matching in electric circuits.

Example 2.7 | **POWER SUPPLY.**   A power supply has an internal impedance of 10 Ω and an internal voltage of 50 V. Calculate the power which will be delivered to external loads of 5 and 20 Ω.

Solution

For this problem we apply Eq. (2.33) with $E = 50$ V, $R_i = 10$ Ω, and $R = 5$ or 20 Ω.

$$R = 5\,\Omega: \qquad P = \frac{(50)^2}{5} \left(\frac{5}{15} \right)^2 = 55.55\,\text{W}$$

$$R = 15\,\Omega: \qquad P = \frac{(50)^2}{15} \left(\frac{15}{25} \right)^2 = 60.0\,\text{W}$$

For maximum power delivery $R = R_i$ and

$$P_{\text{max}} = 62.5\,\text{W}$$

Clearly, the internal impedance and external load of a complicated electronic device may contain inductive and capacitive components that will be important in alternating current transmission and dissipation. Nevertheless, the basic idea is the same. The general principles of matching, then, are that the external impedance should match the internal impedance for maximum energy transmission (minimum attenuation), and the external impedance should be large compared with the internal impedance when a measurement of internal voltage of the device is desired. It is this latter principle that makes an electronic voltmeter essential for measurement of voltages in electronic circuits. The electronic voltmeter has a very high internal impedance so that little current is drawn and the voltage presented to the terminals of the instrument is not altered appreciably by the measurement process. Many such voltmeters today operate on digital principles but all have the capacity of very high-input impedance.

Impedance-matching problems are usually encountered in electrical systems but can be important in mechanical systems as well. We might imagine the simple spring-mass system of the previous section as a mechanical transmission system. From the curves describing the system behavior it is seen that frequencies below a certain value are transmitted through the system; that is, the force is converted to displacement with little attenuation. Near the natural frequency undesirable amplification of the signal is performed, and above this frequency severe attenuation is present. We might say that this system exhibits a behavior characteristic of a variable impedance that is frequency-dependent. When it is desired to transmit mechanical motion through a system, the natural-frequency and dampling characteristics must be taken into account so that good "matching" is present. The problem is an impedance-matching situation, although it is usually treated as a subject in mechanical vibrations.

2.11 FOURIER ANALYSIS

In the early 19th century Joseph Fourier introduced the notion that a piecewise continuous function that is periodic may be represented by a series of sine and cosine functions [19]. Thus

$$
\begin{aligned}
y(x) = {}& a_0 + a_1\cos x + a_2\cos 2x + \cdots + a_n\cos nx \\
& + b_1\sin x + b_2\sin 2x + \cdots + b_n\sin nx
\end{aligned}
$$

or, more compactly,

$$
y(x) = a_0 + \sum_{n=1}^{\infty}(a_n\cos nx + b_n\sin nx) \qquad \textbf{[2.36]}
$$

The values of the constants a_0, a_n, and b_n may be determined by integrating Eq. (2.36) over the period from $-\pi < x < \pi$.

First,

$$\int_{-\pi}^{\pi} y(x)dx = 2\pi a_0 + 0 + 0 = 2\pi a_0$$

since the integrals of the sine and cosine functions over the interval $-\pi$ to π are both zero. We thus have,

$$a_0 = (1/2\pi) \int_{-\pi}^{\pi} y(x)dx \qquad\qquad \textbf{[2.37]}$$

To determine b_n we multiply both sides of Eq. (2.36) by $\sin mx$ where m is an integer that may or may not be equal to the summation index n. We have

$$\int_{-\pi}^{\pi} \sin mx dx = 0$$

$$\int_{-\pi}^{\pi} \cos nx \sin mx dx = 0 \text{ for } m = n$$

$$\int_{-\pi}^{\pi} b_n \sin^2 nx dx = \pi b_n \text{ for } m = n$$

so that,

$$b_n = (1/\pi) \int_{-\pi}^{\pi} y(x)\sin nx dx \qquad\qquad \textbf{[2.38]}$$

In a similar manner, we can determine the coefficients a_n by multiplying Eq. (2.36) by $\cos mx$ and integrating over the interval. This results in

$$a_n = (1/\pi) \int_{-\pi}^{\pi} y(x)\cos nx \, dx \qquad\qquad \textbf{[2.39]}$$

Note that when $y(x)$ is an even function, that is, $f(x) = f(-x)$, the function is represented by cosines alone, and if the function is odd, that is, $f(x) = -f(-x)$, it is represented by sines alone. If the function is neither even nor odd, all terms must be employed.

The interval for expansion of the function may be expressed in a more general sense by making a variable substitution. Let

$$u = (\pi/L)x \qquad\qquad \textbf{[2.40]}$$

Then the coefficients a_n and b_n become

$$a_n = (1/L) \int_{-L}^{L} y(x) \cos (n\pi x/L) \, dx \qquad\qquad \textbf{[2.41]}$$

$$b_n = (1/L) \int_{-L}^{L} y(x) \sin (n\pi x/L) \, dx \qquad\qquad \textbf{[2.42]}$$

$$a_0 = (1/2L) \int_{-L}^{L} y(x) \, dx \qquad\qquad \textbf{[2.43]}$$

If the function repeats over a period T, then L becomes the half period. The Fourier representation may be further generalized by replacing the integral limits in Eqs. (2.41), (2.42), (2.43) by p and $p + T$ where p is an arbitrary value for the start of the repeating function. Of course, one may just consider that a single sample is taken and obtain the function over the range T.

Fourier series representations of functions are useful in experimental measurements in those applications where oscillatory phenomena are observed, as for the vibrating spring mass of Fig. 2.4, in a resonant electric circuit, or in applications concerned with propagation of sound waves and their absorption. The measurement that is usually made is one of the amplitude of vibration as a function of time. The independent variable in the Fourier series would then be time. We may rearrange the Fourier series in terms of a circular frequency ω related to the period T through

$$\omega = 2\pi/T = 2\pi f \qquad \textbf{[2.44]}$$

where ω is expressed in radian/sec and f is expressed in cycles/sec or Hertz.

The summation index n is then said to determine the harmonics of the wave: $n = 1$ represents the fundamental value or the lowest frequency, $n = 2$ the second harmonic, $n = 3$ the third harmonic, and so on.

We may also introduce the concept of the phase angle presented in Eq. (2.14) to combine the sine and cosine terms into the following form

$$y(t) = a_0 + \sum_{n=1}^{\infty}(a_n \cos n\omega t + b_n \sin n\omega t)$$

$$= a_0 + \sum_{n=1}^{\infty} C_n \cos (n\omega t - \phi_n) \qquad \textbf{[2.45]}$$

where the new constant C_n is determined from

$$C_n = (a_n^2 + b_n^2)^{1/2} \qquad \textbf{[2.46]}$$

and the phase angle is determined from

$$\tan \phi_n = b_n/a_n \qquad \textbf{[2.47]}$$

If one were to apply Eqs. (2.41), (2.42), and (2.43) for the constants a_0, a_n, and b_n to the trigonometric function

$$y(x) = 3.5 + 2 \sin x + 6.3 \cos 5x$$

the simple result

$$a_0 = 3.5$$
$$a_n = 2 \text{ for } n = 1$$
$$= 0 \text{ for } n \neq 1$$
$$b_n = 6.3 \text{ for } n = 5$$
$$= 0 \text{ for } n \neq 5$$

would be obtained.

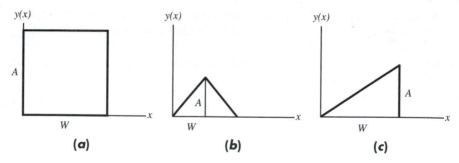

Figure 2.12   (a) Square wave function. (b) Sawtooth function. (c) Ramp function.

Now let us consider three waveforms as examples of Fourier series. In Fig. 2.12 we have (a) a square wave of amplitude A and width W, (b) a sawtooth wave of amplitude A and width W, and (c) a ramp function having amplitude A and duration W. In an experiment the amplitude might be displacement in a mechanical system, voltage in an electric circuit, or sound pressure level in an acoustic system. The duration of the wave would be some unit of time.

The square wave is described by

$$y(x) = A \text{ for } 0 < x < W \qquad\qquad \textbf{[2.48]}$$
$$= 0 \text{ for } x < 0$$
$$= 0 \text{ for } x > W$$

Later we will substitute ωt for the displacement function x. This is an odd function so we expect only sine terms in the series. Inserting the constant amplitude in Eq. (2.42) we obtain

$$b_n = (2A/n\pi)[1 - \cos(n\pi x/W)] \qquad\qquad \textbf{[2.49]}$$

which reduces to

$$b_n = 4A/n\pi \quad \text{for } n = \text{odd} \qquad\qquad \textbf{[2.50]}$$
$$= 0 \qquad\qquad \text{for } n = \text{even}$$

The final series representation for the square wave is thus

$$y(x) = (2A/\pi) \sum_{n=1}^{\infty} \{[(-1)^{n+1} + 1]/n\}\sin(n\pi x/W) \qquad\qquad \textbf{[2.51]}$$

This series may be summed as indicated or an alternate index, which automatically leaves out the zero terms, may be used to obtain

$$y(x) = (4A/\pi) \sum_{N=1}^{\infty} [\sin(2N - 1)\pi x/W]/(2N - 1) \qquad\qquad \textbf{[2.52]}$$

Note that the index n, and not N, represents the harmonics: $n = 1$ $(N = 1)$ represents the fundamental, $n = 3$ $(N = 2)$ represents the third harmonic, $n = 5$ $(N = 3)$ represents the fifth harmonic, and so on.

Figure 2.13 shows the summation represented by Eq. (2.52) for $N = 1$ to 50 ($n = 1$ to 99). In Fig. 2.13a the lines become quite cluttered, but are shown more clearly in Fig. 2.13b. A three-dimensional diagram is shown in Fig. 2.13c that illustrates how the bumps in the square wave are smoothed out as the number of terms in the series is increased.

Now, let us examine the other two cases in Fig. 2.12.

The sawtooth wave is described by

$$y(x) = (2A/W)x \quad \text{for } 0 < x < W/2 \qquad \textbf{[2.53a]}$$

$$= -(2A/W)x + 2A \quad \text{for } W/2 < x < W \qquad \textbf{[2.53b]}$$

Applying Eq. (2.43) for the Fourier coefficient a_0 as before gives

$$a_0 = (4/W) \int_0^W (2A/W)x\,dx + \int_0^W [(-2A/W)x + 2A]\,dx = A/2 \quad \textbf{[2.54]}$$

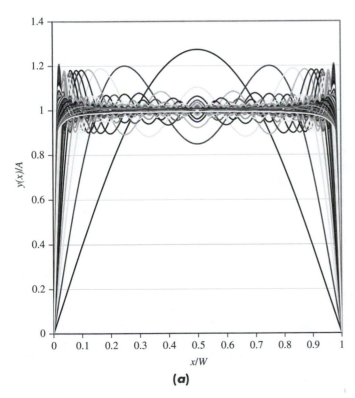

(a)

Figure 2.13   (a) Square wave for $n = 1$ to 99; (b) Square wave for five values of index n; (c) Three-dimensional view of square wave Fourier series.

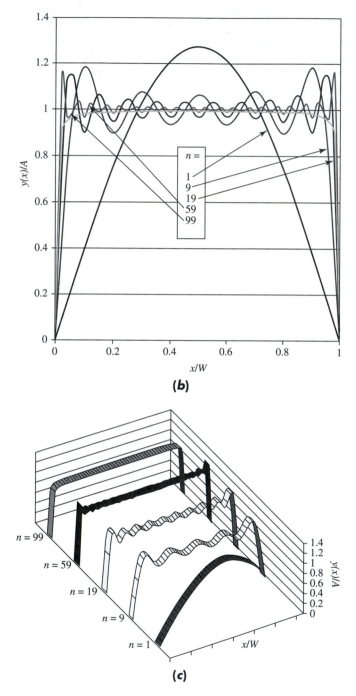

(b)

(c)

Figure 2.13   (*Continued*)

Using Eq. (2.41) for a_n and Eq. (2.42) for b_n results in

$$a_n = (2A/n^2\pi^2)(\cos n\pi - 1) \qquad \text{[2.55]}$$
$$= 0 \text{ for } n \text{ even}$$
$$= -(4A/n^2\pi^2) \text{ for } n \text{ odd} \qquad \text{[2.56]}$$

and

$$b_n = 0$$

Making the same index substitution as before to automatically omit the zero terms gives for the final series

$$y(x) = A/2 - (4A/\pi^2)\sum_{N=1}^{\omega}\cos[(2N-1)2\pi x/W]/(2N-1)^2 \qquad \text{[2.57]}$$

For the ramp function shown in Fig. 2.13c the Fourier series becomes

$$y(x) = (2A/\pi)\sum_{n=1}^{\omega}(-1)^{n+1}[\sin(n\pi x/W)]/n \qquad \text{[2.58]}$$

The index n in this series does represent the harmonics. Equations (2.57) and (2.58) are plotted in Figs. 2.14 and 2.15.

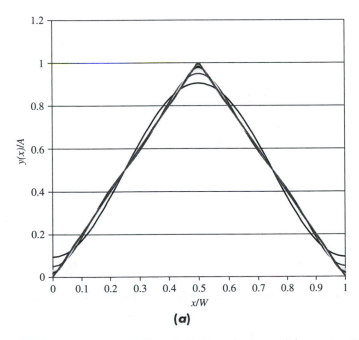

(a)

Figure 2.14   (a) Fourier representation for Sawtooth for $n = 1$ to 30;
(b) Three-dimensional representation for sawtooth $n = 1$ to 30.

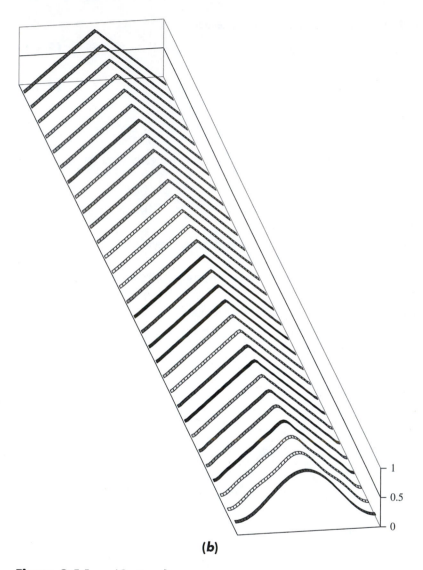

(b)

Figure 2.14   (Continued)

THE DECIBEL

As we shall see in later sections, one is frequently interested in the value of a certain parameter as related to a reference value of that parameter. The electric power dissipated in a resistor is

$$P = E^2/R$$

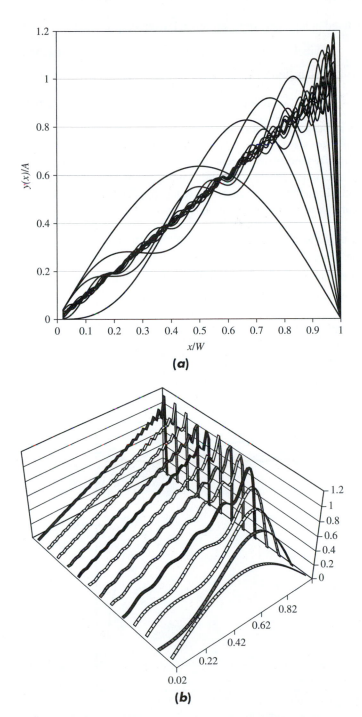

Figure 2.15 (a) Ramp function; (b) Three-dimensional representation of ramp function.

The decibel is defined in terms of some reference Power P_{ref} as

$$\text{decibel} = \text{dB} = 10 \log(P/P_{ref})$$ **[2.59]**

In terms of the voltage the decibel level would be

$$\text{dB} = 20 \log(E/E_{ref})$$ **[2.60]**

since the power varies as the square of the voltage.

If we identify the square wave of Fig. 2.13 with a voltage pulse and the Fourier series as a representation of that pulse over a certain frequency range, we could express the accuracy of the representation in terms of decibel units by choosing the reference value as $[y(x)]_{ref} = 1.0$. Then the decibel response would appear as in Fig. 2.16. We could say that the square wave is faithfully reproduced within

± 0.4 dB for a frequency response up to the 99th harmonic

± 0.6 dB for a frequency response up to the 59th harmonic

± 1.4 dB for a frequency response up to the 19th harmonic

± 4.1 dB for a frequency response up to the 9th harmonic.

All of these values are within the range $0.02 < x/W < 0.98$.

This means that an electronic amplifier to reproduce step or square wave pulse functions must have frequency-response capabilities that far exceed the fundamental frequency of the basic wave.

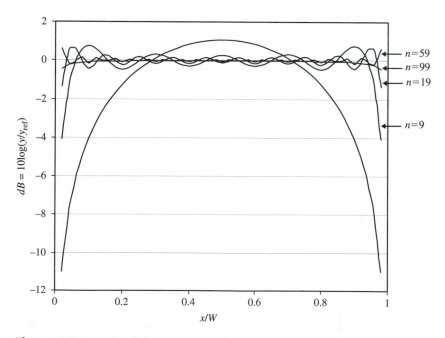

Figure 2.16   Decibel representation of square wave.

THE FOURIER TRANSFORM

We have already noted that a Fourier series representation of a function breaks down the function into harmonic components. For the case of the square wave the Fourier coefficients are shown in Fig. 2.17 as a function of the harmonic index n. Thus, an oscillatory displacement behavior of a spring-mass damper system could be represented as a summation of frequency components. Or, we might say that the physical representation in the displacement-time domain could be transformed into a frequency-time domain through the use of Fourier series.

In practice, a signal may be sampled in discrete time increments Δt and stored in a digital computer. If N increments in Δt are taken over the signal, the period T becomes

$$T = N\Delta t$$

To evaluate the integrals in Eqs. (2.41), (2.42), and (2.43) the integration must be performed numerically.

The signal is $y(t_k)$ where k is a new index indicating the time increment so that

$$t_k = k\Delta t \text{ for } k = 1, 2, \cdots, N \qquad \textbf{[2.61]}$$

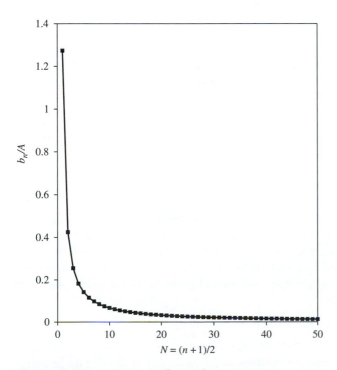

Figure 2.17   Fourier coefficients for square wave.

The Fourier coefficients thus are expressed in terms of summations instead of integrals to give

$$a_n = (2/N) \sum_{k=1}^{N} y(k\Delta t)\cos(2\pi kn/N) \quad \text{for } n = 0 \text{ to } N/2 \qquad \textbf{[2.62]}$$

$$b_n = (2/N) \sum_{k=1}^{N} y(k\Delta t)\sin(2\pi kn/N) \quad \text{for } n = 1 \text{ to } N/2 - 1 \qquad \textbf{[2.63]}$$

where N, the number of time increments, is selected as an even number.

The coefficient a_o is obtained by performing a numerical integration of the discrete function by the most appropriate method. (See Sect. 5.8.)

The operations performed in the foregoing summations are called a *Discrete Fourier transform* (DFT) and may be performed either with separate software packages or with built-in computation algorithms in the particular device employed for the data collection. Hardware dedicated to specific applications may have integrated circuits that output information in the exact format needed by the user.

As the number of samples increases, the number of computations required for the discrete Fourier transform increases by N^2. The *fast Fourier transform* (FFT) employs a special algorithm to speed the calculation for large data samples so that the number of computations varies as $N \log_2 N$ instead of N^2. For samples with N about 1000 the FFT is about 100 times faster than the DFT scheme. The reader may consult Refs. [19], [20], [21] for further information on the development of the DFT and FFT, and Ref. [22] for specific calculation techniques and examples. In Sect. 14.5 we shall see that it is necessary to sample at a frequency at least twice the anticipated frequency content of the signal to avoid the problem of aliasing. The highest frequency that may be resolved is called the Nyquist frequency and is given by

$$f_{\text{Nyq}} = f_s/2 = 1/2\Delta t \qquad \textbf{[2.64]}$$

where f_s is the sampling frequency and Δt, as before, is the sampling time increment.

2.12   EXPERIMENT PLANNING

The key to success in experimental work is to ask continually: What am I looking for? Why am I measuring this—does the measurement really answer any of my questions? What does the measurement tell me? These questions may seem rather elementary, but they should be asked frequently throughout the progress of any experimental program. Some particular questions that should be asked in the initial phases of experiment planning are:

1. What primary variables shall be investigated?

2. What *control* must be exerted on the experiment?

3. What ranges of the primary variables will be necessary to describe the phenomena under study?

4. How many data points should be taken in the various ranges of operation to ensure good sampling of data considering instrument accuracy and other factors? (See Chap. 3.)

5. What instrument accuracy is required for each measurement?

6. If a dynamic measurement is involved, what frequency response must the instruments have?

7. Are the instruments available commercially, or must they be constructed especially for the particular experiment?

8. What safety precautions are necessary if some kind of hazardous operation is involved in the experiment?

9. What financial resources are available to perform the experiment, and how do the various instrument requirements fit into the proposed budget?

10. What provisions have been made for recording the data?

11. What provisions have been made for either on-line or subsequent computer reduction of data?

12. If the data reduction is not of a "research" nature where manipulation and calculations depend somewhat on the results of measurements, what provisions are made to have direct output of a data acquisition system available for the final report? In many cases appropriate graphical results may be obtained with digital data acquisition systems as the experiment progresses or shortly thereafter.

The importance of control in any experiment should always be recognized. The physical principle, apparatus, or device under investigation will dictate the variables which must be controlled carefully. For example, a heat-transfer test of a particular apparatus might involve some heat loss to the surrounding air in the laboratory where the test equipment is located. Consequently, it would be wise to maintain (control) the surrounding temperature at a reasonably constant value. If one run is made with the room temperature at 90°C and another at 50°F, large unwanted effects may occur in the measurements. Or, suppose a test is to be made of the effect of cigarette smoke on the eating habits of mice. Clearly, we would want to control the concentration of smoke inhaled by the mice and also observe another group of mice which were not exposed to cigarette smoke at all. All other environmental variables should be the same if we are to establish the effect of the cigarette smoke on eating habits.

In the case of the heat-transfer test we make a series of measurements of the characteristics of a device under certain specified operating conditions—no comparison with other devices is made. For the smoke test with mice it is necessary to measure the performance of the mice under specified conditions and *also* to compare this performance with the performance of another group under different controlled conditions. For the heat-transfer test we establish an absolute measurement of performance, but for the mice a *relative performance* is all that can be ascertained. We have chosen two diverse examples of absolute and relative experiments, but the lesson is clear. Whenever a comparison test is performed to establish relative performance, control must be exerted over more than one experimental setup in order for the comparison to be significant.

It would seem obvious that very careful provisions should be made to record the data and all ideas and observations concerned with the experiment. Yet, many experimenters record data and important sketches on pieces of scratch paper or in such a disorganized manner that they may be lost or thrown away. In some experiments the readout instrument is a recording type so that a record is automatically obtained and there is little chance for loss. For many experiments, however, visual observations must be made and values recorded on an appropriate data sheet. This data sheet should be very carefully planned so that it may subsequently be used, if desired, for data reduction. Frequently, much time may be saved in the reduction process by eliminating unnecessary transferal of data from one sheet to another. If a computer is to be used for data reduction, then the primary data sheet should be so designed that the data may be easily transferred to the input device of the computer. Even with digital readout systems the printout must be carefully labeled, either in the machine programming or by hand.

A bound notebook should be maintained to record sketches and significant observations of an unusual character which may occur during both the planning and the execution stages of the experiment. The notebook is also used for recording thoughts and observations of a theoretical nature as the experiment progresses. Upon the completion of the experimental program the well-kept notebook forms a clear and sequential record of the experiment planning, observations during the experiment, and, where applicable, correspondence of important observations with theoretical predictions. Every experimenter should get into the habit of keeping a good notebook. In some cases the output of various instrument transducers will be fed directly to a data acquisition system and computer which processes the data and furnishes a printed output of the desired results. Then, the notebook becomes the repository of important sketches and listings of computer programs, as well as documentation which may be needed later to analyze the data or repeat the experiment. Sample printouts or computer-generated graphs may be taped in the notebook along with program documentation.

For those engineers conducting tests or experiments that are an essential part of product development, which may lead to patent application, the bound notebook is an essential legal element in the documentation that may be required in securing and defending the patent. The well-documented bound notebook is an obvious advantage to the person(s) who may be called upon to continue the work. One may suggest that computer records and documentation substitute for the notebook and offer the ability for rapid transfer of information to other interested parties. For confidential information the need for stringent security measures in any electronic transmission is obvious.

As a summary of our remarks on experimental planning we present the generalized experimental procedure given in Table 2.8. This procedure is, of course, a flexible one, and the reader should consider the importance of each item in relation to the entire experimental program. Notice particularly item 1*a*. The engineer should give careful thought to the *need* for the experiment. Perhaps after some sober thinking the engineer will decide that a previously planned experiment is really not necessary at all and that the desired information could be found from an analytical study or from

Table 2.8   Generalized experimental procedure

---

1. *a.* Establish the need for the experiment.
   b. Establish the optimum budgetary, manpower, and time requirements, including time sequencing of the project. Modify scope of the experiment to *actual* budget, manpower, and time schedule which are allowable.
2. Begin detail planning for the experiment; clearly establish objectives of experiment (verify performance of production model, verify theoretical analysis of particular physical phenomenon, etc.). If experiments are similar to those of previous investigators, be sure to make use of experience of the previous workers. *Never* overlook the possibility that the work may have been done before and reported in the literature.
3. Continue planning by performing the following steps:
   a. Establish the primary variables which must be measured (force, strain, flow, pressure, temperature, etc.).
   b. Determine as nearly as possible the accuracy which may be required in the primary measurements and the number of such measurements which will be required for proper data analysis.
   c. Set up date reduction calculations *before* conducting the experiments to be sure that adequate data are being collected to meet the objectives of the experiment.
   d. Analyze the possible errors in the anticipated results *before* the experiments are conducted so that modifications in accuracy requirements on the various measurements may be changed if necessary.
4. Select instrumentation for the various measurements to match the anticipated accuracy requirements. Modify the instrumentation to match budgetary limitations if necessary.
5. Collect a few data points and conduct a preliminary analysis of these data to be sure that the experiment is going as planned.
6. Modify the experimental apparatus and/or procedure in accordance with the findings in item 5.
7. Collect the bulk of experimental data and analyze the results.
8. Organize, discuss, and publish the findings and results of the experiments, being sure to include information pertaining to all items 1 to 7, above.

---

the results of experiments already conducted. ***Do not take this item lightly.*** A great amount of money is wasted by individuals who rush into a program only to discover later that the experiments were unnecessary for their own particular purposes.

THE ROLE OF UNCERTAINTY ANALYSIS IN EXPERIMENT PLANNING

Items 3*b* and *d* in Table 2.8 note the need to perform preliminary analyses of experimental uncertainties in order to effect a proper selection of instruments and to design the apparatus to meet the overall goals of the experiment. These items are worthy of further amplification.

Recall our previous comments about the terms *accuracy*, *error*, and *uncertainty*. We noted that many persons use the term "error" when "uncertainty" is the proper nomenclature. As promised before, we shall clarify this matter in Chap. 3.

In Chap. 3 we shall see how one goes about estimating uncertainty in an experimental measurement. For now, let us consider how these estimates can aid our experiment planning. It is clear that certain variables we wish to measure are set by the particular experimental objectives, but there may be several choices open in the method we use to measure these variables. An electric-power measurement could be performed by measuring current and voltage and taking the product of these variables. The power might also be calculated by measuring the voltage drop across a known resistor, or possibly through some calorimetric determination of the heat dissipated from a resistor. The choice of the method used can be made on the basis of an uncertainty analysis, which indicates the relative accuracy of each method. A flow measurement might be performed by sensing the pressure drop across an obstruction meter, or possibly by counting the number of revolutions of a turbine placed in the flow (see Chap. 7). In the first case the overall uncertainty depends on the accuracy of a measurement of pressure differential and other variables, such as flow area, while in the second case the overall uncertainty depends on the accuracy of counting and a time determination.

The point is that a careful uncertainty analysis during the experiment planning period may enable the investigator to make a better selection of instruments for the program. Briefly, then, an uncertainty analysis enters into the planning phase with the following approximate steps:

1. Several alternative measurement techniques are selected once the variables to be measured have been established.

2. An uncertainty analysis is performed on each measurement technique, taking into account the estimated accuracies of the instruments that will actually be used.

3. The different measurement techniques are then compared on the basis of cost, availability of instrumentation, ease of data collection, and calculated uncertainty. The technique with the least uncertainty is clearly the most desirable from an experimental-accuracy standpoint, but it may be too expensive. Frequently, however, the investigator will find the cost is not a strong factor and that the technique with the smallest uncertainty (within reason) is as easy to perform as some other less accurate method.

Figure 2.18 divides the procedure of Table 2.8 into a graphical pattern of preliminary, intermediate, and final stages of an experimental program. The feedback blocks in these diagrams are very important because they illustrate the need to retrace continuously one's steps and modify the program in accordance with the most current information that is available.

In Chap. 16 we will return to the notions of experiment planning by examining a reasonable design protocol for experiments. This procedure will build upon the information pertaining to uncertainty analysis contained in Chap. 3, which we have already mentioned several times, as well as specific measurement techniques which will be explored in other chapters.

2.13   REVIEW QUESTIONS

2.1. What is meant by sensitivity; accuracy; precision?

2.2. Why is instrument calibration necessary?

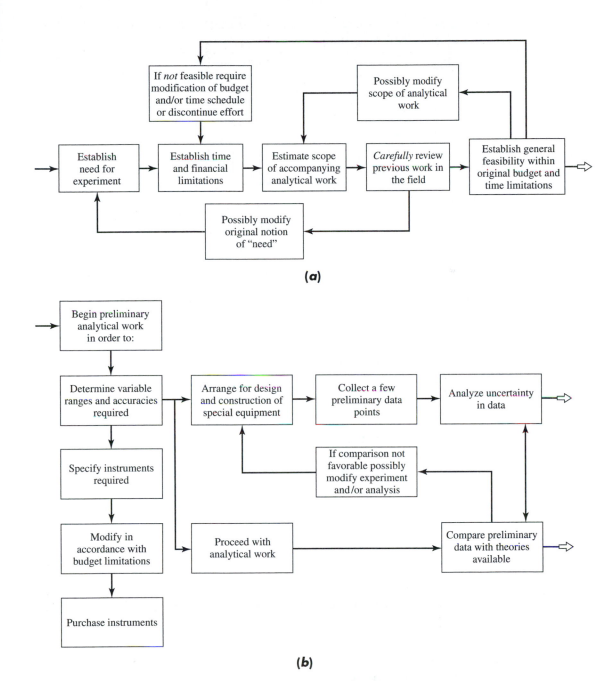

Figure 2.18   (a) Preliminary stages of experiment planning; (b) intermediate stages of experiment planning; (c) final stages of experimental program.

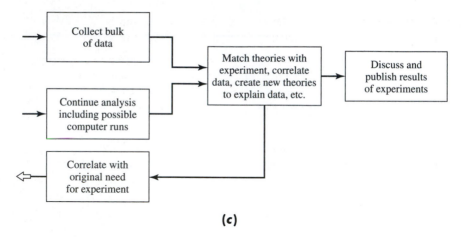

(c)

Figure 2.18   *(Continued)*

2.3. Why are standards necessary?

2.4. What is meant by frequency response?

2.5. Describe the meaning of phase shift.

2.6. Define time constant.

2.7. What kind of impedance matching is desired for (*a*) maximum power transmission and (*b*) minimum influence on the output of the system?

2.8. Why is a literature survey important in the preliminary stages of experiment planning?

2.9. Why is an uncertainty analysis important in the preliminary stages of experiment planning?

2.10. How can an uncertainty analysis help to reduce overall experimental uncertainty?

2.11. What is rise time?

2.12. What is meant by zeroth-, first-, and second-order systems?

2.13. What is meant by steady-state response?

2.14. What factors influence the time constant in first-order systems?

2.14   PROBLEMS

2.1. Consider an ordinary mercury-in-glass thermometer as a measurement system and indicate which parts of the thermometer correspond to the boxes in the diagram of Fig. 2.2.

2.2. A thermometer is used for the range of 200 to 400°F, and it is stated that its accuracy is one-quarter of 1 percent. What does this mean in terms of temperature?

2.3. A sinusoidal forcing function is impressed on the system in Fig. 2.5. The natural frequency is 100 Hz, and the damping ratio c/c_c is 0.7. Calculate the amplitude ratio and time lag of the system for an input frequency of 40 Hz. (The time lag is the time interval between the maximum force input and maximum displacement output.)

2.4. For a natural frequency of 100 Hz and a damping ratio of 0.7, compute the input-frequency range for which the system in Fig. 2.5 will have an amplitude ratio of 1.00 ± 0.01.

2.5. A thermometer is initially at a temperature of 70°F and is suddenly placed in a liquid which is maintained at 300°F. The thermometer indicates 200 and 270°F after time intervals of 3 and 5 s, respectively. Estimate the time constant for the thermometer.

2.6. Plot the power output of the circuit in Fig. 2.11 as a function of R/R_i. Assume R_i and E as constants and show the plot in dimensionless form; that is, use PR_i/E^2 as the ordinate for the curve.

2.7. A steel scale is graduated in increments of $\frac{1}{32}$ in. What is the readability and least count of such a scale?

2.8. A 10-μF capacitor is charged to a potential of 100 V. At time zero it is discharged through a 1-MΩ resistor. What is the time constant for this sytem?

2.9. The two-terminal device shown in Fig. 2.11 has an internal resistance of 5000 Ω. A meter with an impedance of 20,000 Ω is connected to the output to perform a voltage measurement. What is the percent error in determination of the internal voltage?

2.10. The device in Prob. 2.9 has an internal voltage of 100 V. Calculate the power output for the loading conditions indicated. What would be the maximum power output? What power output would result for a load resistance of 1000 Ω?

2.11. A 2-g mass is suspended from a simple spring. The deflection caused by this mass is 0.5 cm. What is the natural frequency of the system?

2.12. A 10-lbm turntable is placed on rubber supports such that a deflection of 0.25 in results from the weight. Calculate the natural frequency of such a system.

2.13. A resistance-capacitance system is to be designed with a time constant of 1 ms. Specify several combinations of R and C to accomplish this, and based on commercially available items, obtain the cost of each combination.

2.14. What is the minimum amplitude reduction to be expected for the system in Prob. 2.12 for frequencies greater than 20, 40, and 60 Hz?

2.15. A small tank contains 10 liters of water at 20°C which is allowed to discharge from an opening in the side at the initial rate of 6 liters/h. The discharge rate is directly proportional to the volume of water remaining in the tank. Calculate an equivalent time constant for this system.

2.16. Pressure is measured in units of lbf/in^2 in the English system of units. Derive factors to convert to units of N/m^2 (Pa) and $kilopond/cm^2$.

2.17. One gallon equals $231\ in^3$. Derive a conversion factor to convert automobile fuel economy from mi/gal to km/liter.

2.18. A unit for viscosity in the English system is $lbf \cdot s/ft^2$. Determine a factor to convert this to $kg/m \cdot s$.

2.19. A unit for specific heat in the SI system is $kJ/kg \cdot °C$. Derive a factor to convert this to $Btu/lbm \cdot °F$, and to $kcal/g \cdot °C$.

2.20. Derive a factor to convert density from g/m^3 to $slugs/ft^3$.

2.21. An English unit for thermal conductivity is $Btu/h \cdot ft \cdot °F$. Derive a factor to convert to $ergs/cm \cdot °C$.

2.22. A unit of kinematic viscosity in the metric system is the Stoke (St), defined as $1.0\ cm^2/s$. Derive a factor to convert to ft^2/s.

2.23. The SI unit for heat generation is W/m^3. Derive a factor to convert to $Btu/h \cdot ft^3$.

2.24. A unit for dynamic viscosity in the metric system is the poise (P) $= 1.0\ dyn \cdot s/cm^2$. Derive a factor to convert to $lbm/h \cdot ft$.

2.25. Heat flux may be expressed in units of W/cm^2. Derive a factor to convert to $Btu/h \cdot ft^2$.

2.26. The universal gas constant has a value of $1545\ ft \cdot lbf/lbm\ mol \cdot °R$. By applying appropriate conversion factors, obtain its value in SI units.

2.27. Derive a factor to convert volume flow rate from cm^3/s to gal/min.

2.28. How do you convert degrees Kelvin to degrees Rankine?

2.29. A thermometer has a time constant of 10 s and behaves as a first-order system. It is initially at a temperature 30°C and then suddenly subjected to a surrounding temperature of 120°C. Calculate the 90 percent rise time and the time to attain 99 percent of the steady-state temperature.

2.30. The thermometer in Prob. 2.29 is subjected to a harmonic temperature variation having an amplitude of 20°C and a frequency of 0.01 Hz. Determine the phase lag of the thermometer and the amplitude attenuation. The time constant is still taken as 10 s.

2.31. A pressure transducer operates as a second-order system having a natural frequency of 10,000 Hz. For damping ratios c/c_c of 0.3 and 0.4, determine the resonance frequencies.

2.32. For the transducer in Prob. 2.31, determine the amplitude response and dynamic error for frequencies of 2000 and 4000 Hz. Also, determine the phase lag for these frequencies.

2.33. A second-order system is to be designed for damping the amplitude response by 40 percent for input frequencies from 10 to 50 Hz. Select suitable natural frequencies and damping ratios c/c_c to accomplish this objective.

2.34. A first-order system has a phase shift of $-50°$ at a certain frequency. What will be the phase lag at a frequency of twice this value? What will be the relative amplitude responses at the two frequencies?

2.35. A first-order system is subjected to a harmonic input of 3 Hz. The system has a time constant of 0.5 s. Calculate the error of the amplitude response and the phase lag.

2.36. A thermometer acting as a first-order system is initially at a temperature of 35°C and is then suddenly subjected to a temperature of 110°C. After 8 s the thermometer indicates a temperature of 75°C. Calculate the time constant and the 90 percent rise time for the thermometer.

2.37. A first-order system has a time constant of 0.05 s. Over what frequency range will the amplitude response be within 10 percent?

2.38. A force transducer is connected such that the output registers a static sensitivity of 1.0 V/kgf input. What is the output for a force input of 10 kgf?

2.39. The rise time for a certain RC circuit is to be 0.003 ms. Determine suitable values of R and C to accomplish this.

2.40. A dynamic measurement device operating as a second-order system is to be designed to measure an input frequency of 60 Hz with an amplitude error of no greater than 5 percent. Determine appropriate design parameters which would accomplish this objective. Many answers are possible, so discuss what factors influenced your selection.

2.41. A pressure transducer has a damping ratio of 0.3 and a natural frequency of 12,000 Hz. Determine the frequency range for which the amplitude dynamic error will be less than 10 percent.

2.42. A small temperature sensor operates as a first-order system and is stated to have a time constant of 0.1 s. If it is initially at a temperature of 100°C and suddenly exposed to an environment temperature of 15°C, how long will it take to indicate a temperature of 17°C?

2.43. If the temperature sensor of Prob. 2.42 is exposed to a harmonic temperature source, for what frequency range will its amplitude response be within 10 percent? What will be the time delay under these circumstances?

2.44. A pressure transducer operating as a second-order system is to be used to measure a signal at 500 Hz. To select the transducer, we shall choose one with a natural frequency of 1500 Hz. What damping ratio c/c_c must be selected so that the dynamic error of the amplitude response is less than 2 percent?

2.45. An RC electric circuit is to have a rise time of one microsecond (1 μs). Select appropriate values of R and C to accomplish this objective.

2.46. A large building mass behaves approximately like a first-order system when responding to a harmonic thermal input. If the harmonic input follows the pattern of daily heating and cooling, that is, one cycle occurs over a 24-h period, and the time delay is 2 h, estimate the time constant for the system. Also, estimate how much the amplitude response decreases at this frequency.

2.47. A platform weighing 1.3 kg (force) is attached to a spring damper system with a spring constant of 100 N/m. The damper is adjustable. If the damper is adjusted to the critical value, how long will it take for the system to recover 90 percent of the displaced value of a step input?

2.48. The system of Prob. 2.47 is subjected to a step input and the damper is adjusted so that a damping ratio of 0.1 is obtained. Using the results of Fig. 2.9 or the appropriate equation, estimate the time required for the system to reach the first maximum point in overshoot.

2.49. Estimate the rise time for the system in Prob. 2.47 for an overdamped condition with $c/c_c = 1.5$.

2.50. Calculate the relative displacement from steady-state value for the system in Prob. 2.47 at $t = 1$ s and $c = 5.7$ kg/s.

2.51. A special oscilloscope used in high-power laser systems is stated to have a rise time of 1 picosecond for a step input. What does this imply in terms of frequency response?

2.52. Assume that the weight of a 3000-lb automobile is distributed evenly over the four wheels and associated spring shock absorber system. Also assume that the springs will be deflected 1.5 in when the auto is loaded with 1000 lb. What is the natural frequency of such a system? What is the rise time for critical damping and a step input to the system?

2.53. A temperature-sensing element is stated to behave as a first-order system with a rise time of 0.1 s. If the element is initially at a temperature of 20°C and suddenly subjected to a temperature of 125°C, what temperature will the element indicate after a time of 0.05 s?

2.54. A metric unit for dynamic viscosity is kg/m-s. Determine a conversion factor to convert to lbf-s/ft^2.

2.55. A dimensionless group used in fluid mechanics is the Reynolds number defined as

$$Re = \rho u x / \mu$$

where ρ is fluid density, u is velocity, x is a dimension, and μ is the dynamic viscosity. Determine sets of units for these four parameters in SI and English systems that will make Re dimensionless.

2.56. A dimensionless group used in free-convection heat-transfer problems is the Grashof number defined by

$$Gr = \frac{g\beta\rho^2 \Delta T x^3}{\mu^2}$$

where g is the acceleration of gravity, β is rate of change of fluid volume per unit change in temperature per unit volume, ΔT is a temperature difference, x is a distance parameter, and μ is the fluid dynamic viscosity having units of N-s/m^2 in the SI system. Determine suitable units for all parameters in the SI system which will cause Gr to be dimensionless. Repeat for the English system.

2.57. At a certain electronics company the units employed for thermal conductivity are W-cm/in^2-°F. Determine a factor to convert to the standard SI unit of W/m-°C. In the thermal conductivity parameter °C or °F represents a temperature *difference*.

2.58. A temperature measurement device is believed to behave as a first-order system with a rise time of 0.2. At time zero the device is at a temperature of 45°C. What temperature will be indicated by the device 0.1 s after it is suddenly exposed to a temperature of 100°C? Assume that the rise time remains constant.

2.59. An empty pickup truck weighs 4500 lbf with the weight distributed about equally to the four wheels. When the truck is loaded with 1500 lb distributed uniformly to all four wheels, the suspension springs deflect by 1 in. Calculate the natural frequency for the system. Assuming a step input, what would be the rise time for the system?

2.60. A pressure transducer is to be selected to measure a varying pressure at a frequency of 400 Hz. The device selected has a natural frequency of 1200 Hz. It is desired that the dynamic error of the amplitude response of the transducer not exceed 2 percent. What damping ratio is necessary to achieve this response?

2.61. Determine the Fourier sine series for the function $f(x) = x^3$ for $0 < x < \pi$
Ans. $x^3 = 2\Sigma(-1)^{n+1}\{[(n\pi)^2 - 6]/n^3\}\sin nx$

2.62. The thermal response of buildings behaves approximately as a first-order system when subjected to a harmonic temperature variation. Without imposition of a sudden "cold front" a single cycle is executed over a 24-h period. If the time delay for maximum and minimum points in the temperature wave lags by 1.5 h, calculate the time constant for the system.

2.63. A thermocouple behaves approximately as a first-order system. The element is initially at a temperature of 45°C and suddenly subjected to a temperature of 100°C. After a period of 6 s the thermocouple indicates a temperature of 70°C. Calculate the 90 percent rise time for the thermocouple.

2.64. Determine the Fourier series for the function $f(x) = x$ for $-\pi < x < \pi$
Ans. $x = 2\Sigma[(-1)^{n+1}/n]\sin nx$

2.65. A thermocouple has a time constant of 8 s and may be approximated as a first-order system. If it is initially at a temperature of 40°C and suddenly exposed to a surrounding temperature of 100°C, calculate the 90 percent rise time and the time to attain 99 percent of the steady-state temperature.

2.66. A certain first-order system is subjected to a harmonic input of 5 Hz, producing a time constant of 0.6 s. Calculate the error for the amplitude response and the phase lag.

2.67. The thermocouple of Prob. 2.65 is subjected to a harmonic temperature variation having an amplitude of 15°C and frequency of 0.01 Hz. Determine the phase lag of the thermocouple and the amplitude attenuation. Assume the time constant does not change.

2.68. Determine the Fourier series for the function $f(x) = x + x^3/3$ for $0 < x < \pi$. Hint: Combine solutions to Probs. 2.61 and 2.64.

2.15 REFERENCES

1. Silsbee, F. B.: "Extension and Dissemination of the Electric and Magnetic Units by the National Bureau of Standards," *Natl. Bur. Std. (U.S.), Circ.* 531, 1952.

2. "International Practical Temperature Scale of 1968," *Metrologia,* vol. 5, no. 2, April 1969.

3. Eshbach, O. N.: *Handbook of Engineering Fundamentals,* 2d ed., John Wiley & Sons, New York, 1952.

4. Holman, J. P.: *Thermodynamics,* 4th ed., McGraw-Hill, New York, 1988.

5. Constant, F. W.: *Theoretical Physics,* vol. 2, Addison-Wesley, Reading, MA, 1958.

6. Stimson, H. F.: "The International Temperature Scale of 1948," *J. Res. Natl. Bur. Std.* (paper 1962), vol. 4, p. 211, March 1949.

7. *NBS Tech. News Bull.,* vol. 52, no. 1, p. 10, January 1968.

8. Beckwith, T. G., and N. L. Buck: *Mechanical Measurements,* 2d ed., Addison-Wesley, Reading, MA, 1969.

9. Schenck, H., Jr.: *An Introduction to the Engineering Research Project,* McGraw-Hill, New York, 1969.

10. Doebelin, E. O.: *Measurement Systems: Application and Design,* 4th ed., McGraw-Hill, New York, 1990.

11. "Time Standards," *Instr. Cont. Syst.,* p. 87, October 1965.

12. Wilson, E. B.: *An Introduction to Scientific Research,* McGraw-Hill, New York, 1952.

13. Committee report, "The International Temperature Scale of 1990," *Metrologia,* vol. 27, no. 3, 1990.

14. Quinn, T. J.: "News from BIPM," *Metrologia,* vol. 26, pp. 69–74, 1989.

15. Taylor, B. N.: "New International Representations of the Volt and Ohm Effective January 1, 1990," *IEEE Trans. Instruments and Meas.,* vol. 39, p. 2, 1990.

16. Helfrick, A. D., and W. D. Cooper: *Modern Electronic Instrumentation and Control,* Prentice-Hall, Englewood Cliffs, NJ, 1990.

17. Northrop, R. B.: *Introduction to Instrumentation and Measurements,* CRC Press, Boca Raton, FL, 1997.

18. Vu, H. V., and R. Esfandiari: *Dynamic Systems Modeling and Analysis,* McGraw-Hill, New York, 1997.

19. Brown, J. A., and R.V. Churchill: *Fourier Series and Boundary Value Problems,* 7th ed., McGraw-Hill, 2008.

20. Bracewell, R. N.: *The Fourier Transform and Its Applications,* 3d ed., McGraw-Hill, 1999.

21. Cooley, J. W., P. A. Lewis, and P. D. Welch, "Historical Notes on the Fast Fourier Transform," *IEEE Trams. Audio Electronics,* vol. 15, no. 2, pp. 76–79, 1977.

22. Chapra, S. C., and R. P. Canale, *Numerical Methods for Engineers,* 6th ed., McGraw-Hill, 2010.

3

ANALYSIS OF EXPERIMENTAL DATA

3.1 INTRODUCTION

Some form of analysis must be performed on all experimental data. The analysis may be a simple verbal appraisal of the test results, or it may take the form of a complex theoretical analysis of the errors involved in the experiment and matching of the data with fundamental physical principles. Even new principles may be developed in order to explain some unusual phenomenon. Our discussion in this chapter will consider the analysis of data to determine errors, precision, and general validity of experimental measurements. The correspondence of the measurements with physical principles is another matter, quite beyond the scope of our discussion. Some methods of graphical data presentation will also be discussed. The interested reader should consult the monograph by Wilson [4] for many interesting observations concerning correspondence of physical theory and experiment.

The experimentalist should always know the validity of data. The automobile test engineer must know the accuracy of the speedometer and gas gage in order to express the fuel-economy performance with confidence. A nuclear engineer must know the accuracy and precision of many instruments just to make some simple radioactivity measurements with confidence. In order to specify the performance of an amplifier, an electrical engineer must know the accuracy with which the appropriate measurements of voltage, distortion, and so forth, have been conducted. Many considerations enter into a final determination of the validity of the results of experimental data, and we wish to present some of these considerations in this chapter.

Errors will creep into all experiments regardless of the care exerted. Some of these errors are of a random nature, and some will be due to gross blunders on the part of the experimenter. Bad data due to obvious blunders may be discarded immediately. But what of the data points that just "look" bad? We cannot throw out data because they do not conform with our hopes and expectations unless we see something obviously wrong. If such "bad" points fall outside the range of normally expected random deviations, they may be discarded on the basis of some consistent statistical data analysis. The keyword here is "consistent." The elimination of data points must be consistent and should not be dependent on human whims and bias based on what "ought to be." In many instances it is very difficult for the individual to be consistent and unbiased. The pressure of a deadline, disgust with previous experimental failures, and normal impatience all can influence rational thinking processes. However, the competent experimentalist will strive to maintain consistency in the primary data analysis. Our objective in this chapter is to show how one may go about maintaining this consistency.

3.2 CAUSES AND TYPES OF EXPERIMENTAL ERRORS

In this section we present a discussion of some of the types of errors that may be present in experimental data and begin to indicate the way these data may be handled. First, let us distinguish between single-sample and multisample data.

Single-sample data are those in which some uncertainties may not be discovered by repetition. Multisample data are obtained in those instances where enough experiments are performed so that the reliability of the results can be assured by statistics. Frequently, cost will prohibit the collection of multisample data, and the experimenter must be content with single-sample data and prepared to extract as much information as possible from such experiments. The reader should consult Refs. [1] and [4] for further discussions on this subject, but we state a simple example at this time. If one measures pressure with a pressure gage and a single instrument is the only one used for the entire set of observations, then some of the error that is present in the measurement will be sampled only once no matter how many times the reading is repeated. Consequently, such an experiment is a single-sample experiment. On the other hand, if more than one pressure gage is used for the same total set of observations, then we might say that a multisample experiment has been performed. The *number* of observations will then determine the success of this multisample experiment in accordance with accepted statistical principles.

An experimental error is an experimental error. If the experimenter knew what the error was, he or she would correct it and it would no longer be an error. In other words, the real errors in experimental data are those factors that are always vague to some extent and carry some amount of *uncertainty*. Our task is to determine just how uncertain a particular observation may be and to devise a consistent way of specifying the uncertainty in analytical form. A reasonable definition of experimental uncertainty may be taken as the *possible value* the error may have. This uncertainty may vary a great deal depending on the circumstances of the experiment. Perhaps it is better to speak of experimental uncertainty instead of experimental error because the magnitude of an error is always uncertain. Both terms are used in practice, however, so the reader should be familiar with the meaning attached to the terms and the ways that they relate to each other.

It is very common for people to speak of experimental errors when the correct terminology should be "uncertainty." Because of this common usage, we ask that the reader accept the faulty semantics when they occur and view each term in its proper context.

TYPES OF ERRORS

At this point we mention some types of errors that may cause uncertainty in an experimental measurement. First, there can always be those gross blunders in apparatus or instrument construction which may invalidate the data. Hopefully, the careful experimenter will be able to eliminate most of these errors. Second, there may be certain *fixed errors* which will cause repeated readings to be in error by roughly the same amount but for some unknown reason. These fixed errors are sometimes called *systematic errors,* or *bias errors*. Third, there are the *random errors,* which may be caused by personal fluctuations, random electronic fluctuations in the apparatus or instruments, various influences of friction, and so forth. These random errors usually follow a certain statistical distribution, *but not always*. In many instances it is very difficult to distinguish between fixed errors and random errors.

The experimentalist may sometimes use theoretical methods to estimate the magnitude of a fixed error. For example, consider the measurement of the temperature of a hot gas stream flowing in a duct with a mercury-in-glass thermometer. It is well known that heat may be conducted from the stem of the thermometer, out of the body, and into the surroundings. In other words, the fact that part of the thermometer is exposed to the surroundings at a temperature different from the gas temperature to be measured may influence the temperature of the stem of the thermometer. There is a heat flow from the gas to the stem of the thermometer, and, consequently, the temperature of the stem must be lower than that of the hot gas. Therefore, the temperature we read on the thermometer is not the true temperature of the gas, and it will not make any difference how many readings are taken—we shall always have an error resulting from the heat-transfer condition of the stem of the thermometer. This is a *fixed error,* and its magnitude may be estimated with theoretical calculations based on known thermal properties of the gas and the glass thermometer.

3.3   ERROR ANALYSIS ON A COMMONSENSE BASIS

We have already noted that it is somewhat more explicit to speak of experimental uncertainty than experimental error. Suppose that we have satisfied ourselves with the uncertainty in some basic experimental measurements, taking into consideration such factors as instrument accuracy, competence of the people using the instruments, and so forth. Eventually, the primary measurements must be combined to calculate a particular result that is desired. We shall be interested in knowing the uncertainty in the final result due to the uncertainties in the primary measurements. This may be done by a commonsense analysis of the data which may take many forms. One rule of thumb that could be used is that the error in the result is equal to the maximum error in any parameter used to calculate the result. Another commonsense analysis would combine all the errors in the most detrimental way in order to determine the maximum error in the final result. Consider the calculation of electric power from

$$P = EI$$

where E and I are measured as

$$E = 100 \text{ V} \pm 2 \text{ V}$$

$$I = 10 \text{ A} \pm 0.2 \text{ A}$$

The nominal value of the power is $100 \times 10 = 1000 \text{ W}$. By taking the worst possible variations in voltage and current, we could calculate

$$P_{max} = (100 + 2)(10 + 0.2) = 1040.4 \text{ W}$$

$$P_{min} = (100 - 2)(10 - 0.2) = 960.4 \text{ W}$$

Thus, using this method of calculation, the uncertainty in the power is $+4.04$ percent, -3.96 percent. It is quite unlikely that the power would be in error by these amounts because the voltmeter variations would probably not correspond with the ammeter variations. When the voltmeter reads an extreme "high," there is no reason that the ammeter must also read an extreme "high" at that particular instant; indeed, this combination is most unlikely.

The simple calculation applied to the electric-power equation above is a useful way of inspecting experimental data to determine what errors *could* result in a final calculation; however, the test is too severe and should be used only for rough inspections of data. It is significant to note, however, that if the results of the experiments appear to be in error by *more* than the amounts indicated by the above calculation, then the experimenter had better examine the data more closely. In particular, the experimenter should look for certain fixed errors in the instrumentation, which may be eliminated by applying either theoretical or empirical corrections.

As another example we might conduct an experiment where heat is *added* to a container of water. If our temperature instrumentation should indicate a *drop* in temperature of the water, our good sense would tell us that something is wrong and the data point(s) should be thrown out. No sophisticated analysis procedures are necessary to discover this kind of error.

The term "common sense" has many connotations and means different things to different people. In the brief example given above it is intended as a quick and expedient vehicle, which may be used to examine experimental data and results for gross errors and variations. In subsequent sections we shall present methods for determining experimental uncertainties in a more precise manner.

3.4 UNCERTAINTY ANALYSIS AND PROPAGATION OF UNCERTAINTY

A more precise method of estimating uncertainty in experimental results has been presented by Kline and McClintock [1]. The method is based on a careful specification of the uncertainties in the various primary experimental measurements. For example, a certain pressure reading might be expressed as

$$p = 100 \text{ kPa} \pm 1 \text{ kPa}$$

When the plus or minus notation is used to designate the uncertainty, the person making this designation is stating the degree of accuracy with which he or she *believes* the measurement has been made. We may note that this specification is in itself uncertain because the experimenter is naturally uncertain about the accuracy of these measurements.

If a very careful calibration of an instrument has been performed recently with standards of very high precision, then the experimentalist will be justified in assigning a much lower uncertainty to measurements than if they were performed with a gage or instrument of unknown calibration history.

To add a further specification of the uncertainty of a particular measurement, Kline and McClintock propose that the experimenter specify certain odds for the uncertainty. The above equation for pressure might thus be written

$$p = 100 \text{ kPa} \pm 1 \text{ kPa (20 to 1)}$$

In other words, the experimenter is willing to bet with 20 to 1 odds that the pressure measurement is within ± 1 kPa. It is important to note that the specification of such odds can *only* be made by the experimenter based on the total laboratory experience.

Suppose a set of measurements is made and the uncertainty in each measurement may be expressed with the same odds. These measurements are then used to calculate some desired result of the experiments. We wish to estimate the uncertainty in the calculated result on the basis of the uncertainties in the primary measurements. The result R is a given function of the independent variables $x_1, x_2, x_3, \ldots, x_n$. Thus,

$$R = R(x_1, x_2, x_3, \ldots, x_n) \qquad \textbf{[3.1]}$$

Let w_R be the uncertainty in the result and $w_1, w_2, \ldots, w_n$ be the uncertainties in the independent variables. If the uncertainties in the independent variables are all given with the same odds, then the uncertainty in the result having these odds is given in Ref. [1] as

$$w_R = \left[\left(\frac{\partial R}{\partial x_1} w_1 \right)^2 + \left(\frac{\partial R}{\partial x_2} w_2 \right)^2 + \cdots + \left(\frac{\partial R}{\partial x_n} w_n \right)^2 \right]^{1/2} \qquad \textbf{[3.2]}$$

If this relation is applied to the electric-power relation of the previous section, the expected uncertainty is 2.83 percent instead of 4.04 percent.

We should call the reader's attention to the requirement that all the uncertainties in Eq. (3.2) should be expressed with the same odds. As a practical matter, the relation is most often used without regard to a specification of the odds of the uncertainties w_n. The experimentalist conducting the experiments is the person best qualified to estimate such odds, so it not unreasonable to assign responsibility for relaxation of the equal-odds to him or her. Further information is given in Ref. [1].

UNCERTAINTIES FOR PRODUCT FUNCTIONS

In many cases the result function of Eq. (3.2) takes the form of a product of the respective primary variables raised to exponents and expressed as

$$R = x_1^{a_1} x_2^{a_2} \cdots x_n^{a_n} \qquad \text{[3.1a]}$$

When the partial differentiations are performed, we obtain

$$\frac{\partial R}{\partial x_i} = x_1^{a_1} x_2^{a_2} \left(a_i x_i^{a_i - 1} \right) \cdots x_n^{a_n}$$

Dividing by R from Eq. (3.1a)

$$\frac{1}{R} \frac{\partial R}{\partial x_i} = \frac{a_i}{x_i}$$

Inserting this relation in Eq. (3.2) gives

$$\frac{w_R}{R} = \left[\sum \left(\frac{a_i w_{x_i}}{x_i} \right)^2 \right]^{1/2} \qquad \text{[3.2a]}$$

The reader should note that this relation for the fractional uncertainty in the result may only be employed when the result function takes the product form indicated in Eq. (3.1a).

UNCERTAINTIES FOR ADDITIVE FUNCTIONS

When the result function has an additive form, R will be expressed as

$$R = a_1 x_1 + a_2 x_2 + \cdots + a_n x_n = \sum a_i x_i \qquad \text{[3.1b]}$$

and the partial derivatives for use in Eq. (3.2) are then

$$\frac{\partial R}{\partial x_i} = a_i$$

The uncertainty in the result may then be expressed as

$$w_R = \left\{ \sum \left[\left(\frac{\partial R}{\partial x_i} \right)^2 w_{x_i}^2 \right] \right\}^{1/2}$$

$$= \left[\sum \left(a_i w_{x_i} \right)^2 \right]^{1/2} \qquad \text{[3.2b]}$$

Equations (3.2a) and (3.2b) may be used in combination when the result function involves both product and additive terms.

UNCERTAINTY OF RESISTANCE OF A COPPER WIRE.   The resistance of a certain | **Example 3.1**

size of copper wire is given as

$$R = R_0[1 + \alpha(T - 20)]$$

where $R_0 = 6\ \Omega \pm 0.3$ percent is the resistance at 20°C, $\alpha = 0.004°C^{-1} \pm 1$ percent is the temperature coefficient of resistance, and the temperature of the wire is $T = 30 \pm 1°C$. Calculate the resistance of the wire and its uncertainty.

Solution

The nominal resistance is

$$R = (6)[1 + (0.004)(30 - 20)] = 6.24\ \Omega$$

The uncertainty in this value is calculated by applying Eq. (3.2). The various terms are

$$\frac{\partial R}{\partial R_0} = 1 + \alpha(T - 20) = 1 + (0.004)(30 - 20) = 1.04$$

$$\frac{\partial R}{\partial \alpha} = R_0(T - 20) = (6)(30 - 20) = 60$$

$$\frac{\partial R}{\partial T} = R_0\alpha = (6)(0.004) = 0.024$$

$$w_{R_0} = (6)(0.003) = 0.018\ \Omega$$

$$w_\alpha = (0.004)(0.01) = 4 \times 10^{-5}°C^{-1}$$

$$w_T = 1°C$$

Thus, the uncertainty in the resistance is

$$W_R = [(1.04)^2(0.018)^2 + (60)^2(4 \times 10^{-5})^2 + (0.024)^2(1)^2]^{1/2}$$
$$= 0.0305\ \Omega \quad \text{or} \quad 0.49\%$$

Example 3.2 | **UNCERTAINTY IN POWER MEASUREMENT.**   The two resistors R and R_s are connected in series as shown in the accompanying figure. The voltage drops across each resistor are measured as

$$E = 10\,V \pm 0.1\,V\ (1\%)$$
$$E_s = 1.2\,V \pm 0.005\,V\ (0.467\%)$$

along with a value of

$$R_s = 0.0066\ \Omega \pm 1/4\%$$

From these measurements determine the power dissipated in resistor R and its uncertainty.

Figure Example 3.2

Solution

The power dissipated in resistor R is

$$P = EI$$

The current through both resistors is $I = E_s/R_s$ so that

$$P = \frac{EE_s}{R_s} \qquad\qquad \text{[a]}$$

The nominal value of the power is therefore

$$P = (10)(1.2)/(0.0066) = 1818.2 \text{ W}$$

The relationship for the power given in Eq. (a) is a product function, so the fractional uncertainty in the power may be determined from Eq. (3.2a). We have

$$a_E = 1 \qquad a_{E_s} = 1 \qquad \text{and} \qquad a_{R_s} = -1$$

so that

$$\frac{w_P}{P} = \left[\left(\frac{a_E w_E}{E}\right)^2 + \left(\frac{a_{E_s} w_{E_s}}{E_s}\right)^2 + \left(\frac{a_{R_s} w_{R_s}}{R_s}\right)^2\right]^{1/2}$$

$$= \left[(1)^2\left(\frac{0.1}{10}\right)^2 + (1)^2\left(\frac{0.005}{1.2}\right)^2 + (-1)^2(0.0025)^2\right]^{1/2} = 0.0111$$

Then

$$w_P = (0.0111)(1818.2) = 20.18 \text{ W}$$

Particular notice should be given to the fact that the uncertainty propagation in the result w_R predicted by Eq. (3.2) depends on the squares of the uncertainties in the independent variables w_n. This means that if the uncertainty in one variable is significantly larger than the uncertainties in the other variables, say, by a factor of 5 or 10, then it is the largest uncertainty that predominates and the others may probably be neglected.

To illustrate, suppose there are three variables with a product of sensitivity and uncertainty $[(\partial R/\partial x)w_x]$ of magnitude 1, and one variable with a magnitude of 5. The uncertainty in the result would be

$$(5^2 + 1^2 + 1^2 + 1^2)^{1/2} = \sqrt{28} = 5.29$$

The importance of this brief remark concerning the relative magnitude of uncertainties is evident when one considers the design of an experiment, procurement of instrumentation, and so forth. Very little is gained by trying to reduce the "small" uncertainties. Because of the square propagation it is the "large" ones that predominate, and any improvement in the overall experimental result must be achieved by improving the instrumentation or technique connected with these relatively large uncertainties. In the examples and problems that follow, both in this chapter and throughout the book, the reader should always note the relative effect of uncertainties in primary measurements on the final result.

In Sec. 2.12 (Table 2.8) the reader was cautioned to examine possible experimental errors *before* the experiment is conducted. Equation (3.2) may be used very effectively for such analysis, as we shall see in the sections and chapters that follow. A further word of caution may be added here. It is just as unfortunate to overestimate uncertainty as to underestimate it. An underestimate gives false security, while an overestimate may make one discard important results, miss a real effect, or buy much too expensive instruments. The purpose of this chapter is to indicate some of the methods for obtaining reasonable estimates of experimental uncertainty.

In the previous discussion of experimental planning we noted that an uncertainty analysis may aid the investigator in selecting alternative methods to measure a particular experimental variable. It may also indicate how one may improve the overall accuracy of a measurement by attacking certain critical variables in the measurement process. The next three examples illustrate these points.

Example 3.3

SELECTION OF MEASUREMENT METHOD.   A resistor has a nominal stated value of $10\ \Omega \pm 1$ percent. A voltage is impressed on the resistor, and the power dissipation is to be calculated in two different ways: (1) from $P = E^2/R$ and (2) from $P = EI$. In (1) only a voltage measurement will be made, while both current and voltage will be measured in (2). Calculate the uncertainty in the power determination in each case when the measured values of E and I are

$$E = 100\ \text{V} \pm 1\% \qquad \text{(for both cases)}$$
$$I = 10\ \text{A} \pm 1\%$$

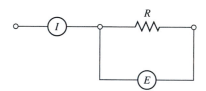

Figure Example 3.3   Power measurement across a resistor.

Solution

The schematic is shown in the accompanying figure. For the first case we have

$$\frac{\partial P}{\partial E} = \frac{2E}{R} \qquad \frac{\partial P}{\partial R} = -\frac{E^2}{R^2}$$

and we apply Eq. (3.2) to give

$$w_P = \left[\left(\frac{2E}{R} \right)^2 w_E^2 + \left(-\frac{E^2}{R^2} \right)^2 w_R^2 \right]^{1/2} \qquad \textbf{[a]}$$

Dividing by $P = E^2/R$ gives

$$\frac{w_P}{P} = \left[4\left(\frac{w_E}{E} \right)^2 + \left(\frac{w_R}{R} \right)^2 \right]^{1/2} \qquad \textbf{[b]}$$

Inserting the numerical values for uncertainty gives

$$\frac{w_P}{P} = [4(0.01)^2 + (0.01)^2]^{1/2} = 2.236\%$$

For the second case we have

$$\frac{\partial P}{\partial E} = I \qquad \frac{\partial P}{\partial I} = E$$

and after similar algebraic manipulation we obtain

$$\frac{w_P}{P} = \left[\left(\frac{w_E}{E} \right)^2 + \left(\frac{w_I}{I} \right)^2 \right]^{1/2} \qquad \text{[c]}$$

Inserting the numerical values of uncertainty yields

$$\frac{w_P}{P} = [(0.01)^2 + (0.01)^2]^{1/2} = 1.414\%$$

Comment

The second method of power determination provides considerably less uncertainty than the first method, even though the primary uncertainties in each quantity are the same. In this example the utility of the uncertainty analysis is that it affords the individual a basis for *selection of a measurement method* to produce a result with less uncertainty.

INSTRUMENT SELECTION.   The power measurement in Example 3.2 is to be conducted by measuring voltage and current across the resistor with the circuit shown in the accompanying figure. The voltmeter has an internal resistance R_m, and the value of R is known only approximately. Calculate the nominal value of the power dissipated in R and the uncertainty for the following conditions: **Example 3.4**

$$R = 100\,\Omega \qquad \text{(not known exactly)}$$
$$R_m = 1000\,\Omega \pm 5\%$$
$$I = 5\,\text{A} \pm 1\%$$
$$E = 500\,\text{V} \pm 1\%$$

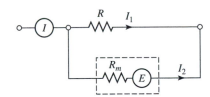

Figure Example 3.4   Effect of meter impedance on measurement.

Solution

A current balance on the circuit yields

$$I_1 + I_2 = I$$
$$\frac{E}{R} + \frac{E}{R_m} = I$$

and

$$I_1 = I - \frac{E}{R_m} \qquad \text{[a]}$$

The power dissipated in the resistor is

$$P = EI_1 = EI - \frac{E^2}{R_m} \qquad \qquad [b]$$

The nominal value of the power is thus calculated as

$$P = (500)(5) - \frac{500^2}{1000} = 2250 \, \text{W}$$

In terms of known quantities the power has the functional form $P = f(E, I, R_m)$, and so we form the derivatives

$$\frac{\partial P}{\partial E} = I - \frac{2E}{R_m} \qquad \frac{\partial P}{\partial I} = E$$

$$\frac{\partial P}{\partial R_m} = \frac{E^2}{R_m^2}$$

The uncertainty for the power is now written as

$$w_P = \left[\left(I - \frac{2E}{R_m} \right)^2 w_E^2 + E^2 w_I^2 + \left(\frac{E^2}{R_m^2} \right)^2 w_{R_m}^2 \right]^{1/2} \qquad \qquad [c]$$

Inserting the appropriate numerical values gives

$$w_P = \left[\left(5 - \frac{1000}{1000} \right)^2 5^2 + (25 \times 10^4)(25 \times 10^{-4}) + \left(25 \times \frac{10^4}{10^6} \right)^2 (2500) \right]^{1/2}$$

$$= [16 + 25 + 6.25]^{1/2} (5)$$

$$= 34.4 \, \text{W}$$

or

$$\frac{w_P}{P} = \frac{34.4}{2250} = 1.53\%$$

In order of influence on the final uncertainty in the power we have

1. Uncertainty of current determination
2. Uncertainty of voltage measurement
3. Uncertainty of knowledge of internal resistance of voltmeter

Comment

There are other conclusions we can draw from this example. The relative influence of the experimental quantities on the overall power determination is noted above. But this listing may be a bit misleading in that it implies that the uncertainty of the meter impedance does not have a large effect on the final uncertainty in the power determination. This results from the fact that $R_m \gg R$ ($R_m = 10R$). If the meter impedance were lower, say, 200 Ω, we would find that it was a dominant factor in the overall uncertainty. For a very *high* meter impedance there would be little influence, even with a very inaccurate knowledge of the exact value of R_m. Thus, we are led to the simple conclusion that we need not worry too much about the precise value of the internal impedance of the meter as long as it is very large compared with the resistance we are measuring the voltage across. This fact should influence *instrument selection* for a particular application.

Example 3.5

WAYS TO REDUCE UNCERTAINTIES. A certain obstruction-type flowmeter (orifice, venturi, nozzle), shown in the accompanying figure, is used to measure the flow of air at low velocities. The relation describing the flow rate is

$$\dot{m} = CA \left[\frac{2g_c p_1}{RT_1} (p_1 - p_2) \right]^{1/2} \qquad \text{[a]}$$

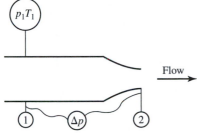

Figure Example 3.5 Uncertainty in a flowmeter.

where
- C = empirical-discharge coefficient
- A = flow area
- p_1 and p_2 = upstream and downstream pressures, respectively
- T_1 = upstream temperature
- R = gas constant for air

Calculate the percent uncertainty in the mass flow rate for the following conditions:

$C = 0.92 \pm 0.005$ (from calibration data)
$p_1 = 25$ psia ± 0.5 psia
$T_1 = 70°F \pm 2°F$ $T_1 = 530°R$
$\Delta p = p_1 - p_2 = 1.4$ psia ± 0.005 psia   (measured directly)
$A = 1.0$ in^2 ± 0.001 in^2

Solution

In this example the flow rate is a function of several variables, each subject to an uncertainty.

$$\dot{m} = f(C, A, p_1, \Delta p, T_1) \qquad \text{[b]}$$

Thus, we form the derivatives

$$\frac{\partial \dot{m}}{\partial C} = A \left(\frac{2g_c p_1}{RT_1} \Delta p \right)^{1/2}$$

$$\frac{\partial \dot{m}}{\partial A} = C \left(\frac{2g_c p_1}{RT_1} \Delta p \right)^{1/2}$$

$$\frac{\partial \dot{m}}{\partial p_1} = 0.5CA \left(\frac{2g_c}{RT_1} \Delta p \right)^{1/2} p_1^{-1/2} \qquad \text{[c]}$$

$$\frac{\partial \dot{m}}{\partial \Delta p} = 0.5CA \left(\frac{2g_c p_1}{RT_1} \right)^{1/2} \Delta p^{-1/2}$$

$$\frac{\partial \dot{m}}{\partial T_1} = -0.5CA \left(\frac{2g_c p_1}{R} \Delta p \right)^{1/2} T_1^{-3/2}$$

The uncertainty in the mass flow rate may now be calculated by assembling these derivatives in accordance with Eq. (3.2). Designating this assembly as Eq. (c) and then dividing by Eq. (a) gives

$$\frac{w_{\dot{m}}}{\dot{m}} = \left[\left(\frac{w_C}{C} \right)^2 + \left(\frac{w_A}{A} \right)^2 + \frac{1}{4} \left(\frac{w_{p_1}}{p_1} \right)^2 + \frac{1}{4} \left(\frac{w_{\Delta p}}{\Delta p} \right)^2 + \frac{1}{4} \left(\frac{w_{T_1}}{T_1} \right)^2 \right]^{1/2} \qquad \textbf{[d]}$$

We may now insert the numerical values for the quantities to obtain the percent uncertainty in the mass flow rate.

$$\frac{w_{\dot{m}}}{\dot{m}} = \left[\left(\frac{0.005}{0.92} \right)^2 + \left(\frac{0.001}{1.0} \right)^2 + \frac{1}{4} \left(\frac{0.5}{25} \right)^2 + \frac{1}{4} \left(\frac{0.005}{1.4} \right)^2 + \frac{1}{4} \left(\frac{2}{530} \right)^2 \right]^{1/2}$$

$$= [29.5 \times 10^{-6} + 1.0 \times 10^{-6} + 1.0 \times 10^{-4} + 3.19 \times 10^{-6} + 3.57 \times 10^{-6}]^{1/2}$$

$$= [1.373 \times 10^{-4}]^{1/2} = 1.172\% \qquad \textbf{[e]}$$

Comment

The main contribution to uncertainty is the p_1 measurement with its basic uncertainty of 2 percent. Thus, to improve the overall situation the accuracy of this measurement should be attacked first. In order of influence on the flow-rate uncertainty we have:

1. Uncertainty in p_1 measurement (± 2 percent)
2. Uncertainty in value of C
3. Uncertainty in determination of T_1
4. Uncertainty in determination of Δp
5. Uncertainty in determination of A

By inspecting Eq. (e) we see that the first and third items make practically the whole contribution to uncertainty. The value of the uncertainty analysis in this example is that it shows the investigator how to improve the overall measurement accuracy of this technique. First, obtain a more precise measurement of p_1. Then, try to obtain a better calibration of the device, that is, a better value of C. In Chap. 7 we shall see how values of the discharge coefficient C are obtained.

3.5   EVALUATION OF UNCERTAINTIES FOR COMPLICATED DATA REDUCTION

We have seen in the preceding discussion and examples how uncertainty analysis can be a useful tool to examine experimental data. In many cases data reduction is a rather complicated affair and is often performed with a computer routine written specifically for the task. An adaptation of the routine can provide for direct calculation of uncertainties without resorting to an analytical determination of the partial derivatives in Eq. (3.2). We still assume that this equation applies, although it could involve several computational steps. We also assume that we are able to obtain estimates by some means of the uncertainties in the primary measurements, that is, w_1, w_2, etc.

Suppose a set of data is collected in the variables $x_1, x_2, \ldots, x_n$ and a result is calculated. At the same time one may perturb the variables by $\Delta x_1, \Delta x_2$, etc., and calculate new results. We would have

$$R(x_1) = R(x_1, x_2, \ldots, x_n)$$
$$R(x_1 + \Delta x_1) = R(x_1 + \Delta x_1, x_2, \ldots, x_n)$$
$$R(x_2) = R(x_1, x_2, \ldots, x_n)$$
$$R(x_2 + \Delta x_2) = R(x_1, x_2 + \Delta x_2, \ldots, x_n)$$

For small enough values of Δx the partial derivatives can be approximated by

$$\frac{\partial R}{\partial x_1} \simeq \frac{R(x_1 + \Delta x_1) - R(x_1)}{\Delta x_1}$$

$$\frac{\partial R}{\partial x_2} \simeq \frac{R(x_2 + \Delta x_2) - R(x_2)}{\Delta x_2}$$

and these values could be inserted in Eq. (3.2) to calculate the uncertainty in the result.

At this point we must again alert the reader to the ways uncertainties or errors of instruments are normally specified. Suppose a pressure gage is available and the manufacturer states that it is accurate within ± 1.0 percent. This statement normally refers to *percent of full scale*. So a gage with a range of 0 to 100 kPa would have an uncertainty of ± 10 percent when reading a pressure of only 10 kPa. Of course, this means that the uncertainty in the calculated result, as either an absolute value or a percentage, can vary widely depending on the range of operation of instruments used to make the primary measurements. The above procedure can be used to advantage in complicated data-reduction schemes.

Evaluation of the partial derivatives in Eq. (3.2) in terms of finite differences is not as cumbersome as it might appear. A computer will normally be employed for data reduction and calculation of intermediate and final results from the primary experimental measurements. In other words, one will usually have a procedure in place for calculating $R(x_1, x_2, \ldots, x_n)$. The procedure may be executed with a spreadsheet or other software packages. It is a simple matter to modify the procedure to calculate perturbed values $R(x_1 + \Delta x_1, \ldots)$, etc., and then obtain the finite difference approximations to $\partial R/\partial x_1$. Furthermore, the procedure may be extended to evaluate the influence coefficients

$$\left(\frac{\partial R}{\partial x_n} w_{x_n} \right)^2$$

at various points in the data reduction process. A study of the relative values of these coefficients can then indicate the data points in the experiment which contribute most to the overall uncertainty in the results. An example of a spreadsheet procedure using Microsoft Excel is given in Ref. [28].

A very full description of this technique and many other considerations of uncertainty analysis are given by Moffat [7, 14]. An example of an industry standard on uncertainty analysis is given in Ref. [8].

| **Example 3.6** | **UNCERTAINTY CALCULATION BY RESULT PERTURBATION.**   Calculate the uncertainty of the wire resistance in Example 3.1 using the result-perturbation technique. |

Solution

In Example 3.1 we have already calculated the nominal resistance as 6.24 Ω. We now perturb the three variables R_0, α, and T by small amounts to evaluate the partial derivatives. We shall take

$$\Delta R_0 = 0.01 \qquad \Delta \alpha = 1 \times 10^{-5} \qquad \Delta T = 0.1$$

Then

$$R(R_0 + \Delta R_0) = (6.01)[1 + (0.004)(30 - 20)] = 6.2504$$

and the derivative is approximated as

$$\frac{\partial R}{\partial R_0} \approx \frac{R(R_0 + \Delta R_0) - R}{\Delta R_0} = \frac{6.2504 - 6.24}{0.01} = 1.04$$

or the same result as in Example 3.1. Similarly,

$$R(\alpha + \Delta \alpha) = (6.0)[1 + (0.00401)(30 - 20)] = 6.2406$$

$$\frac{\partial R}{\partial \alpha} \approx \frac{R(\alpha + \Delta \alpha) - R}{\Delta \alpha} = \frac{6.2406 - 6.24}{1 \times 10^{-5}} = 60$$

$$R(T + \Delta T) = (6)[1 + (0.004)(30.1 - 20)] = 6.2424$$

$$\frac{\partial R}{\partial T} \approx \frac{R(T + \Delta T) - R}{\Delta T} = \frac{6.2424 - 6.24}{0.1} = 0.24$$

All the derivatives are the same as in Example 3.1, so the uncertainty in R would be the same, or 0.0305 Ω.

3.6   STATISTICAL ANALYSIS OF EXPERIMENTAL DATA

We shall not be able to give an extensive presentation of the methods of statistical analysis of experimental data; we may only indicate some of the more important methods currently employed. First, it is important to define some pertinent terms.

When a set of readings of an instrument is taken, the individual readings will vary somewhat from each other, and the experimenter may be concerned with the *mean* of all the readings. If each reading is denoted by x_i and there are n readings, the *arithmetic mean* is given by

$$x_m = \frac{1}{n} \sum_{i=1}^{n} x_i \qquad\qquad \textbf{[3.3]}$$

The *deviation* d_i for each reading is defined by

$$d_i = x_i - x_m \qquad\qquad \textbf{[3.4]}$$

We may note that the average of the deviations of all the readings is zero since

$$\bar{d}_i = \frac{1}{n}\sum_{i=1}^{n} d_i = \frac{1}{n}\sum_{i=1}^{n}(x_i - x_m)$$

$$= x_m - \frac{1}{n}(nx_m) = 0 \qquad \textbf{[3.5]}$$

The average of the absolute values of the deviations is given by

$$|\bar{d}_i| = \frac{1}{n}\sum_{i=1}^{n} |d_i| = \frac{1}{n}\sum_{i=1}^{n}|x_i - x_m| \qquad \textbf{[3.6]}$$

Note that this quantity is not necessarily zero.

The *standard deviation* or *root-mean-square deviation* is defined by

$$\sigma = \left[\frac{1}{n}\sum_{i=1}^{n}(x_i - x_m)^2\right]^{1/2} \qquad \textbf{[3.7]}$$

and the square of the standard deviation σ^2 is called the *variance*. This is sometimes called the *population* or *biased* standard deviation because it strictly applies only when a large number of samples is taken to describe the population.

In many circumstances the engineer will not be able to collect as many data points as necessary to describe the underlying population. Generally speaking, it is desired to have at least 20 measurements in order to obtain reliable estimates of standard deviation and general validity of the data. For small sets of data an *unbiased* or *sample standard deviation* is defined by

$$\sigma = \left[\frac{\sum_{i=1}^{n}(x_i - x_m)^2}{n - 1}\right]^{1/2} \qquad \textbf{[3.8]}$$

Note that the factor $n - 1$ is used instead of n as in Eq. (3.7). The sample or unbiased standard deviation should be used when the underlying population is not known. However, when comparisons are made against a known population or standard, Eq. (3.7) is the proper one to use for standard deviation. An example would be the calibration of a voltmeter against a known voltage source.

There are other kinds of mean values of interest from time to time in statistical analysis. The *median* is the value that divides the data points in half. For example, if measurements made on five production resistors give 10, 12, 13, 14, and 15 kΩ, the median value would be 13 kΩ. The *arithmetic* mean, however, would be

$$R_m = \frac{10 + 12 + 13 + 14 + 15}{5} = 12.8 \text{ k}\Omega$$

In some instances it may be appropriate to divide data into quartiles and deciles also. So, when we say that a student is in the upper quartile of the class, we mean that that student's grade is among the top 25 percent of all students in the class.

Sometimes it is appropriate to use a *geometric mean* when studying phenomena which grow in proportion to their size. This would apply to certain biological processes and to growth rates in financial resources. The geometric mean is defined by

$$x_g = [x_1 \cdot x_2 \cdot x_3 \cdots x_n]^{1/n}$$ [3.9]

As an example of the use of this concept, consider the 5-year record of a mutual fund investment:

Year	Asset Value	Rate of Increase over Previous Year
1	1000	
2	890	0.89
3	990	1.1124
4	1100	1.1111
5	1250	1.1364

The average growth rate is therefore

$$\text{Average growth} = [(0.89)(1.1124)(1.1111)(1.1364)]^{1/4}$$
$$= 1.0574$$

To see that this is indeed a valid average growth rate, we can observe that

$$(1000)(1.0574)^4 = 1250$$

Example 3.7   **CALCULATION OF POPULATION VARIABLES.**   The following readings are taken of a certain physical length. Compute the mean reading, standard deviation, variance, and average of the absolute value of the deviation, using the "biased" basis:

Reading	x, cm
1	5.30
2	5.73
3	6.77
4	5.26
5	4.33
6	5.45
7	6.09
8	5.64
9	5.81
10	5.75

Solution

The mean value is given by

$$x_m = \frac{1}{n} \sum_{i=1}^{n} x_i = \frac{1}{10}(56.13) = 5.613 \text{ cm}$$

The other quantities are computed with the aid of the following table:

Reading	$d_i = x_i - x_m$	$(x_i - x_m)^2 \times 10^2$
1	−0.313	9.797
2	0.117	1.369
3	1.157	133.865
4	−0.353	12.461
5	−1.283	164.609
6	−0.163	2.657
7	0.477	22.753
8	0.027	0.0729
9	0.197	3.881
10	0.137	1.877

$$\sigma = \left[\frac{1}{n}\sum_{i=1}^{n}(x_i - x_m)^2\right]^{1/2} = \left[\frac{1}{10}(3.533)\right]^{1/2} = 0.5944 \text{ cm}$$

$$\sigma^2 = 0.3533 \text{ cm}^2$$

$$|\bar{d}_i| = \frac{1}{n}\sum_{i=1}^{n}|d_i| = \frac{1}{n}\sum_{i=1}^{n}|x_i - x_m|$$

$$= \tfrac{1}{10}(4.224) = 0.4224 \text{ cm}$$

SAMPLE STANDARD DEVIATION.   Calculate the best estimate of standard deviation | **Example 3.8**
for the data of Example 3.7 based on the "sample" or unbiased basis.

Solution

The calculation gives

$$\sigma = \left[\frac{1}{10-1}(3.536)\right]^{1/2} = (0.3929)^{1/2} = 0.627 \text{ cm}$$

SIMPLE PROBABILITY CONCEPTS

Suppose an "honest" coin is flipped a large number of times. It will be noted that after a large number of tosses heads will be observed about the same number of times as tails. If one were to bet consistently on either heads or tails the best one could hope for would be a break-even proposition over a long period of time. In other words, the *frequency of occurrence* is the same for both heads or tails for a very large number of tosses. It is common knowledge that a few tosses of a coin, say, 5 or 10, may not be a break-even proposition, as a large number of tosses would be. This observation illustrates the fact that frequency of occurrence of an event may be dependent on the total number of events which are observed.

The *probability* that one will get a head when flipping an unweighted coin is $\frac{1}{2}$, regardless of the number of times the coin is tossed. The probability that a tail will occur is also $\frac{1}{2}$. The probability that either a head or a tail will occur is $\frac{1}{2} + \frac{1}{2}$ or unity. (We ignore the possibility that the coin will stand on edge.) *Probability* is a mathematical quantity that is linked to the frequency with which a certain phenomenon occurs after a large number of tries. In the case of the coin, it is the number of times heads would be expected to result in a large number of tosses divided by the total number of tosses. Similarly, the toss of an unloaded die results in the occurrence of each side one-sixth of the time. Probabilities are expressed in values of less than one, and a probability of unity corresponds to certainty. In other words, if the probabilities for all possible events are added, the result must be unity. For separate events the probability that one of the events will occur is the sum of the individual probabilities for the events. For a die the probability that any one side will occur is $\frac{1}{6}$. The probability for one of three given sides is $\frac{1}{6} + \frac{1}{6} + \frac{1}{6}$, or $\frac{1}{2}$, etc.

Suppose two dice are thrown and we wish to know the probability that both will display a 6. The probability for a 6 on a single die is $\frac{1}{6}$. By a short listing of the possible arrangements that the dice may have, it can be seen that there can be 36 possibilities and that the desired result of two 6s represents only one of these possibilities. Thus, the probability is $\frac{1}{36}$. For a throw of 7 or 11 there are 6 possible ways of getting a 7; thus, the probability of getting a 7 is $\frac{6}{36}$ or $\frac{1}{6}$. There are only 2 ways of getting an 11; thus, the probability is $\frac{2}{36}$ or $\frac{1}{18}$. The probability of getting *either* a 7 or an 11 is $\frac{6}{36} + \frac{2}{36}$.

If several *independent* events occur at the same time such that each event has a probability p_i, the probability that all events will occur is given as the product of the probabilities of the individual events. Thus, $p = \Pi p_i$, where the Π designates a product. This rule could be applied to the problem of determining the probability of a double 6 in the throw of two dice. The probability of getting a 6 on each die is $\frac{1}{6}$, and the total probability is therefore $(\frac{1}{6})(\frac{1}{6})$, or $\frac{1}{36}$. This reasoning could not be applied to the problem of obtaining a 7 on the two dice because the number on each die is not *independent* of the number on the other die, since a 7 can be obtained in more than one way.

As a final example we ask what the chances are of getting a royal flush in the first five cards drawn off the top of the deck. There are 20 suitable possibilities for the first draw (4 suits, 5 possible cards per suit) out of a total of 52 cards. On the second draw we have fixed the suit so that there are only 4 suitable cards out of the 51 remaining. There are three suitable cards on the third draw, two on the fourth, and only one on the fifth draw. The total probability of drawing the royal flush is thus the product of the probabilities of each draw, or

$$\frac{20}{52} \times \frac{4}{51} \times \frac{3}{50} \times \frac{2}{49} \times \frac{1}{48} = \frac{1}{649,740}$$

In the above discussion we have seen that the probability is related to the number of ways a certain event may occur. In this case we are assuming that all events are equally likely, and hence the probability that an event will occur is the number of ways the event may occur divided by the number of possible events. Our primary concern is the application of probability and statistics to the analysis of experimental data. For this purpose we need to discuss next the meaning and use of *probability distributions*. We shall be concerned with a few particular distributions that are directly applicable to experimental data analysis.

3.7 PROBABILITY DISTRIBUTIONS

Suppose we toss a horseshoe some distance x. Even though we make an effort to toss the horseshoe the same distance each time, we would not always meet with success. On the first toss the horseshoe might travel a distance x_1, on the second toss a distance of x_2, and so forth. If one is a good player of the game, there would be more tosses which have an x distance equal to that of the objective. Also, we would expect fewer and fewer tosses for those x distances which are farther and farther away from the target. For a large number of tosses the probability that it will travel a distance is obtained by dividing the number traveling this distance by the total number of tosses. Since each x distance will vary somewhat from other x distances, we might find it advantageous to calculate the probability of a toss landing in a certain increment of x between x and $x + \Delta x$. When this calculation is made, we might get something like the situation shown in Fig. 3.1. For a good player the maximum probability is expected to surround the distance x_m designating the position of the target.

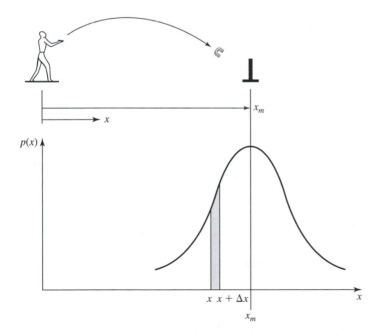

Figure 3.1   Distribution of throws for a "good" horseshoes player.

The curve shown in Fig. 3.1 is called a *probability distribution*. It shows how the probability of success in a certain event is distributed over the distance x. Each value of the ordinate $p(x)$ gives the probability that the horseshoe will land between x and $x + \Delta x$, where Δx is allowed to approach zero. We might consider the deviation from x_m as the error in the throw. If the horseshoe player has good aim, large errors are less likely than small errors. The area under the curve is unity since it is certain that the horseshoe will land somewhere.

We should note that more than one variable may be present in a probability distribution. In the case of the horseshoes player a person might throw the object an exact distance of x_m and yet to one side of the target. The sideways distance is another variable, and a large number of throws would have some distribution in this variable as well.

A particular probability distribution is the *binomial distribution*. This distribution gives the number of successes n out of N possible independent events when each event has a probability of success p. The probability that n events will succeed is given in Ref. [2] as

$$p(n) = \frac{N!}{(N-n)!n!} p^n (1-p)^{N-n} \qquad \textbf{[3.10]}$$

It will be noted that the quantity $(1 - p)$ is the probability of failure of each independent event. Now, suppose that the number of possible independent events N is very large and the probability of occurrence of each p is very small. The calculation of the probability of n successes out of the N possible events using Eq. (3.10) would be most cumbersome because of the size of the numbers. The limit of the binomial distribution as $N \to \infty$ and $p \to 0$ such that

$$Np = a = \text{const}$$

is called the *Poisson distribution* and is given by

$$p_a(n) = \frac{a^n e^{-a}}{n!} \qquad \textbf{[3.11]}$$

The Poisson distribution is applicable to the calculation of the decay of radioactive nuclei, as we shall see in a subsequent chapter. It may be shown that the standard deviation of the Poisson distribution is

$$\sigma = \sqrt{a} \qquad \textbf{[3.12]}$$

Example 3.9 | **TOSSING A COIN—BINOMIAL DISTRIBUTION.**   An unweighted coin is flipped three times. Calculate the probability of getting zero, one, two, or three heads in these tosses.

Solution

The binomial distribution applies in this case since the probability of each flip of the coin is independent of previous or successive flips. The probability of getting a head on each throw is $p = \frac{1}{2}$ and $N = 3$, while n takes on the values 0, 1, 2, and 3. The probabilities are calculated as

$$p(0) = \frac{3!}{(3!)(0!)} \left(\frac{1}{2}\right)^0 \left(\frac{1}{2}\right)^3 = \frac{1}{8}$$

$$p(1) = \frac{3!}{(2!)(1!)} \left(\frac{1}{2}\right)^1 \left(\frac{1}{2}\right)^2 = \frac{3}{8}$$

$$p(2) = \frac{3!}{(1!)(2!)} \left(\frac{1}{2}\right)^2 \left(\frac{1}{2}\right)^1 = \frac{3}{8}$$

$$p(3) = \frac{3!}{(0!)(3!)} \left(\frac{1}{2}\right)^3 \left(\frac{1}{2}\right)^0 = \frac{1}{8}$$

Comment

Note that the sum of the four probabilities, that is, $\frac{1}{8} + \frac{3}{8} + \frac{3}{8} + \frac{1}{8}$ is unity or *certainty* because there are no other possibilities. Heads must come up zero, one, two, or three times in three flips. Of course, one would obtain the same result for probabilities of obtaining zero, one, two, or three tails in three flips.

HISTOGRAMS

We have noted that a probability distribution like Fig. 3.1 is obtained when we observe frequency of occurrence over a large number of observations. When a limited number of observations is made and the raw data are plotted, we call the plot a *histogram*. For example, the following distribution of throws might be observed for a horseshoes player:

Distance from Target, cm	Number of Throws
0–10	5
10–20	15
20–30	13
30–40	11
40–50	9
50–60	8
60–70	10
70–80	6
80–90	7
90–100	5
100–110	5
110–120	3
Over 120	2
Total	99

These data are plotted in Fig. 3.2 using increments of 10 cm in Δx. The same data are plotted in Fig. 3.3 using a Δx of 20 cm. The *relative frequency,* or fraction of

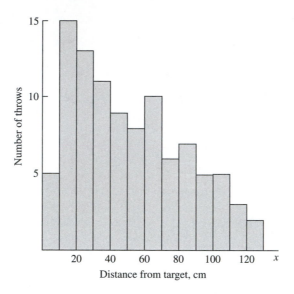

Figure 3.2   Histogram with △x = 10 cm.

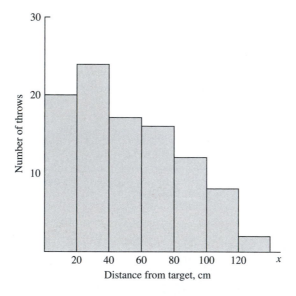

Figure 3.3   Histogram with △x = 20 cm.

throws in each △x increment, could also be used to convey the same information. A *cumulative frequency* diagram could be employed for these data, as shown in Fig. 3.4. If this figure had been constructed on the basis of a very large number of throws, then we could appropriately refer to the ordinate as the probability that the horseshoe will land within a distance x of the target.

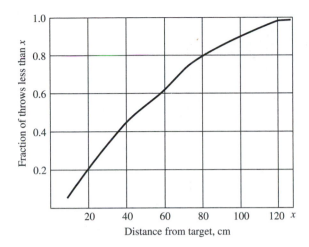

Figure 3.4   Cumulative frequency diagram.

3.8   THE GAUSSIAN OR NORMAL ERROR DISTRIBUTION

Suppose an experimental observation is made and some particular result is recorded. We know (or would strongly suspect) that the observation has been subjected to many random errors. These random errors may make the final reading either too large or too small, depending on many circumstances which are unknown to us. Assuming that there are many small errors that contribute to the final error and that each small error is of equal magnitude and equally likely to be positive or negative, the *gaussian* or *normal error distribution* may be derived. If the measurement is designated by x, the gaussian distribution gives the probability that the measurement will lie between x and x + dx and is written

$$P(x) = \frac{1}{\sigma\sqrt{2\pi}}e^{-(x-x_m)^2/2\sigma^2}$$ **[3.13]**

In this expression x_m is the mean reading and σ is the standard deviation. Some may prefer to call $P(x)$ the *probability density*. The units of $P(x)$ are those of $1/x$ since these are the units of $1/\sigma$. A plot of Eq. (3.13) is given in Fig. 3.5. Note that the most probable reading is x_m. The standard deviation is a measure of the width of the distribution curve; the larger the value of σ, the flatter the curve and hence the larger the expected error of all the measurements. Equation (3.13) is normalized so that the total area under the curve is unity. Thus,

$$\int_{-\infty}^{+\infty} P(x)\,dx = 1.0$$ **[3.14]**

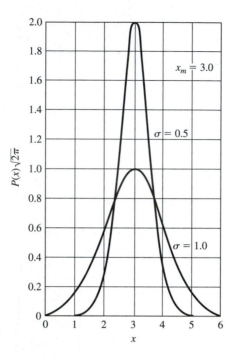

Figure 3.5   The gaussian or normal error distribution for two values
of the standard deviation.

At this point we may note the similarity between the shape of the normal error
curve and the expected experimental distribution for tossing horseshoes, as shown in
Fig. 3.1. This is what we would expect because the good horseshoes player's throws
will be bunched around the target. The better the player is at the game, the more
closely the throws will be grouped around the mean and the more probable will be
the mean distance x_m. Thus, in the case of the horseshoes player a smaller standard
deviation would mean a larger percentage of "ringers."

We may quickly anticipate the next step in the analysis as one of trying to determine the precision of a set of experimental measurements through an application of the
normal error distribution. One may ask: But how do you know that the assumptions
pertaining to the derivation of the normal error distribution apply to experimental
data? The answer is that for sets of data where a large number of measurements is
taken, experiments indicate that the measurements do indeed follow a distribution
like that shown in Fig. 3.5 when the experiment is under control. If an important
parameter is not controlled, one gets just scatter, that is, no sensible distribution at all.
Thus, as a matter of experimental verification, the gaussian distribution is believed
to represent the *random* errors in an adequate manner for a properly controlled
experiment.

By inspection of the gaussian distribution function of Eq. (3.13) we see that the maximum probability occurs at $x = x_m$, and the value of this probability is

$$P(x_m) = \frac{1}{\sigma\sqrt{2\pi}} \qquad \text{[3.15]}$$

It is seen from Eq. (3.15) that smaller values of the standard deviation produce larger values of the maximum probability, as would be expected in an intuitive sense. $P(x_m)$ is sometimes called a *measure of precision* of the data because it has a larger value for smaller values of the standard deviation.

We next wish to examine the gaussian distribution to determine the likelihood that certain data points will fall within a specified deviation from the mean of all the data points. The probability that a measurement will fall within a certain range x_1 of the mean reading is

$$P = \int_{x_m-x_1}^{x_m+x_1} \frac{1}{\sigma\sqrt{2\pi}} e^{-(x-x_m)^2/2\sigma^2} \, dx \qquad \text{[3.16]}$$

Making the variable substitution

$$\eta = \frac{x - x_m}{\sigma}$$

Eq. (3.16) becomes

$$P = \frac{1}{\sqrt{2\pi}} \int_{-\eta_1}^{+\eta_1} e^{-\eta^2/2} \, d\eta \qquad \text{[3.17]}$$

where

$$\eta_1 = \frac{x_1}{\sigma} \qquad \text{[3.18]}$$

Values of the gaussian normal error function

$$\frac{1}{\sqrt{2\pi}} e^{-\eta^2/2}$$

and integrals of the gaussian function corresponding to Eq. (3.17) are given in Tables 3.1 and 3.2.

If we have a sufficiently large number of data points, the error for each point should follow the gaussian distribution and we can determine the probability that certain data fall within a specified deviation from the mean value. Example 3.10 illustrates the method of computing the chances of finding data points within one or two standard deviations from the mean. Table 3.3 gives the chances for certain deviations from the mean value of the normal distribution curve.

Table 3.1   Values of the gaussian normal error distribution

Values of the function $(1/\sqrt{2\pi})e^{-\eta^2/2}$ for different values of the argument η. Each figure in the body of the table is preceded by a decimal point.

η	0.00	0.01	0.02	0.03	0.04	0.05	0.06	0.07	0.08	0.09
0.0	39894	39892	39886	39876	39862	39844	39822	39797	39767	39733
0.1	39695	39654	39608	39559	39505	39448	39387	39322	39253	39181
0.2	39104	39024	38940	38853	38762	38667	38568	38466	38361	38251
0.3	38139	38023	37903	37780	37654	37524	37391	37255	37115	36973
0.4	36827	36678	36526	36371	36213	36053	35889	35723	35553	35381
0.5	35207	35029	34849	34667	34482	34294	34105	33912	33718	33521
0.6	33322	33121	32918	32713	32506	32297	32086	31875	31659	31443
0.7	31225	31006	30785	30563	30339	30114	29887	29658	29430	29200
0.8	28969	28737	28504	28269	28034	27798	27562	27324	27086	26848
0.9	26609	26369	36129	25888	25647	25406	25164	24923	24681	24439
1.0	24197	23955	23713	23471	23230	22988	22747	22506	22265	22025
1.1	21785	21546	21307	21069	20831	20594	20357	20121	19886	19652
1.2	19419	19186	18954	18724	18494	18265	18037	17810	17585	17360
1.3	17137	16915	16694	16474	16256	16038	15822	15608	15395	15183
1.4	14973	14764	14556	14350	14146	13943	13742	13542	13344	13147
1.5	12952	12758	12566	12376	12188	12001	11816	11632	11450	11270
1.6	11092	10915	10741	10567	10396	10226	10059	09893	09728	09566
1.7	09405	09246	09089	08933	08780	08628	08478	08329	08183	08038
1.8	07895	07754	07614	07477	07341	07206	07074	06943	06814	06687
1.9	06562	06438	06316	06195	06077	05959	05844	05730	05618	05508
2.0	05399	05292	05186	05082	04980	04879	04780	04682	04586	04491
2.1	04398	04307	04217	04128	04041	03955	03871	03788	03706	03626
2.2	03547	03470	03394	03319	03246	03174	03103	03034	02965	02898
2.3	02833	02768	02705	02643	02582	02522	02463	02406	02349	02294
2.4	02239	02186	02134	02083	02033	01984	01936	01888	01842	01797
2.5	01753	01709	01667	01625	01585	01545	01506	01468	01431	01394
2.6	01358	01323	01289	01256	01223	01191	01160	01130	01100	01071
2.7	01042	01014	00987	00961	00935	00909	00885	00861	00837	00814
2.8	00792	00770	00748	00727	00707	00687	00668	00649	00631	00613
2.9	00595	00578	00562	00545	00530	00514	00499	00485	00470	00457
3.0	00443									
3.5	008727									
4.0	0001338									
4.5	0000160									
5.0	000001487									

Table 3.2 Integrals of the gaussian normal error function

Values of the integral $(1/\sqrt{2\pi})\int_0^{\eta_1} e^{-\eta^2/2}d\eta$ are given for different values of the argument η_1. It may be observed that

$$\frac{1}{\sqrt{2\pi}}\int_{-\eta_1}^{+\eta_1} e^{-\eta^2/2}d\eta = 2\frac{1}{\sqrt{2\pi}}\int_0^{\eta_1} e^{-\eta^2/2}d\eta$$

The values are related to the error function since

$$\text{erf } \eta_1 = \frac{1}{\sqrt{\pi}}\int_{-\eta_1}^{+\eta_1} e^{-\eta^2}d\eta$$

so that the tabular values are equal to $\frac{1}{2}\text{erf}(\eta_1/\sqrt{2})$. Each figure in the body of the table is preceded by a decimal point.

η_1	0.00	0.01	0.02	0.03	0.04	0.05	0.06	0.07	0.08	0.09
0.0	0000	00399	00798	01197	01595	01994	02392	02790	03188	03586
0.1	03983	04380	04776	05172	05567	05962	06356	06749	07142	07355
0.2	07926	08317	08706	09095	09483	09871	10257	10642	11026	11409
0.3	11791	12172	12552	12930	13307	13683	14058	14431	14803	15173
0.4	15554	15910	16276	16640	17003	17364	17724	18082	18439	18793
0.5	19146	19497	19847	20194	20450	20884	21226	21566	21904	22240
0.6	22575	22907	23237	23565	23891	24215	24537	24857	25175	25490
0.7	25084	26115	26424	26730	27035	27337	27637	27935	28230	28524
0.8	28814	29103	29389	29673	29955	30234	30511	30785	31057	31327
0.9	31594	31859	32121	32381	32639	32894	33147	33398	33646	33891
1.0	34134	34375	34614	34850	35083	35313	35543	35769	35993	36214
1.1	36433	36650	36864	37076	37286	37493	37698	37900	38100	38298
1.2	38493	38686	38877	39065	39251	39435	39617	39796	39973	40147
1.3	40320	40490	40658	40824	40988	41198	41308	41466	41621	41774
1.4	41924	42073	42220	42364	42507	42647	42786	42922	43056	43189
1.5	43319	43448	43574	43699	43822	43943	44062	44179	44295	44408
1.6	44520	44630	44738	44845	44950	45053	45154	45254	45352	45449
1.7	45543	45637	45728	45818	45907	45994	46080	46164	46246	46327
1.8	46407	46485	46562	46638	46712	46784	46856	46926	46995	47062
1.9	47128	47193	47257	47320	47381	47441	47500	47558	47615	47670
2.0	47725	47778	47831	47882	47932	47962	48030	48077	48124	48169
2.1	48214	48257	48300	48341	48382	48422	48461	48500	48537	48574
2.2	48610	48645	48679	48713	48745	48778	48809	48840	48870	48899
2.3	48928	48956	48983	49010	49036	49061	49086	49111	49134	49158
2.4	49180	49202	49224	49245	49266	49286	49305	49324	49343	49361
2.5	49379	49296	49413	49430	49446	49461	49477	49492	49506	49520
2.6	49534	49547	49560	49573	49585	49598	49609	49621	49632	49643
2.7	49653	49664	49674	49683	49693	49702	49711	49720	49728	49736
2.8	49744	49752	49760	49767	49774	49781	49788	49795	49801	49807
2.9	49813	49819	49825	49831	49836	49841	49846	49851	49856	49861
3.0	49865									
3.5	4997674									
4.0	4999683									
4.5	4999966									
5.0	4999997133									

Table 3.3   Chances for deviations from mean value of normal distribution curve

Deviation	Chances of Results Falling within Specified Deviation
$\pm 0.6745\sigma$	1–1
σ	2.15–1
2σ	21–1
3σ	369–1

Example 3.10 | **PROBABILITY FOR DEVIATION FROM MEAN VALUE.**   Calculate the probabilities that a measurement will fall within one, two, and three standard deviations of the mean value and compare them with the values in Table 3.3.

Solution

We perform the calculation using Eq. (3.17) with $\eta_1 = 1, 2$, and 3. The values of the integral may be obtained from Table 3.2. We observe that

$$\int_{-\eta_1}^{+\eta_1} e^{-\eta^2/2}\, d\eta = 2\int_0^{\eta_1} e^{-\eta^2/2}\, d\eta$$

so that

$$P(1) = (2)(0.34134) = 0.6827$$

$$P(2) = (2)(0.47725) = 0.9545$$

$$P(3) = (2)(0.49865) = 0.9973$$

Using the odds given in Table 3.3, we would calculate the probabilities as

$$P(1) = \frac{2.15}{2.15 + 1} = 0.6827$$

$$P(2) = \frac{21}{21 + 1} = 0.9545$$

$$P(3) = \frac{369}{369 + 1} = 0.9973$$

Comment

This example shows how the concept of probability in the gaussian distribution is related to the "odds" concept mentioned in the previous discussion of uncertainty specifications.

CONFIDENCE INTERVAL AND LEVEL OF SIGNIFICANCE

The *confidence interval* expresses the probability that the mean value will lie within a certain number of σ values and is given by the symbol z. Thus,

$$\bar{x} = \bar{x} \pm z\sigma \qquad (\% \text{ confidence level})$$

Table 3.4

Confidence Interval	Confidence Level, %	Level of Significance, %
3.30	99.9	0.1
3.0	99.7	0.3
2.57	99.0	1.0
2.0	95.4	4.6
1.96	95.0	5.0
1.65	90.0	10.0
1.0	68.3	31.7

and using the procedure of Example 3.10, the confidence level in percent could be expressed as in Table 3.4. For small data samples z should be replaced by

$$\Delta = \frac{z\sigma}{\sqrt{n}} \qquad \qquad [3.19]$$

We thus expect that the mean value will lie within $\pm 2.57\sigma$ with less than 1 percent error (confidence level of 99 percent). The *level of significance* is 1 minus the confidence level. Thus, for $z = 2.57$ the level of significance is 1 percent.

DETERMINATION OF NUMBER OF MEASUREMENTS TO ASSURE A SIGNIFI- | **Example 3.11**
CANCE LEVEL.   A certain steel bar is measured with a device which has a known precision of ± 0.5 mm when a large number of measurements is taken. How many measurements are necessary to establish the mean length $\bar{x}$ with a 5 percent level of significance such that

$$\bar{x} = \bar{x} \pm 0.2 \text{ mm}$$

Solution
For a large number of measurements the 5 percent level of significance is obtained at $z = 1.96$ and for the population here

$$\Delta = \frac{z\sigma}{\sqrt{n}} = 0.2 \text{ mm} = \frac{(1.96)(0.5 \text{ mm})}{\sqrt{n}}$$

which yields

$$n = 24.01$$

So, for 25 measurements or more we could state with a confidence level of 95 percent that the population mean value will be within ± 0.2 mm of the sample mean value.

POWER SUPPLY.   A certain power supply is stated to provide a constant voltage output | **Example 3.12**
of 10.0 V within ± 0.1 V. The output is assumed to have a normal distribution. Calculate the probability that a single measurement of voltage will lie between 10.1 and 10.2 V.

Solution

For this problem $\sigma = \pm 0.1$ V. The probability that the voltage will lie between 10.0 and 10.1 V ($+1\sigma$) is, from Table 3.2,

$$P(+0.1) = 0.34134$$

while the probability it will lie between 10.0 and 10.2 V ($+2\sigma$) is

$$P(+0.2) = 0.47725$$

The probability that it will lie between 10.1 and 10.2 V is therefore

$$P(10.1 \text{ to } 10.2) = 0.47725 - 0.34134 = 0.13591$$

CHAUVENET'S CRITERION

It is a rare circumstance indeed when an experimenter does not find that some of the data points look bad and out of place in comparison with the bulk of the data. The experimenter is therefore faced with the task of deciding if these points are the result of some gross experimental blunder and hence may be neglected or if they represent some new type of physical phenomenon that is peculiar to a certain operating condition. The engineer cannot just throw out those points that do not fit with expectations—there must be some consistent basis for elimination.

Suppose n measurements of a quantity are taken and n is large enough that we may expect the results to follow the gaussian error distribution. This distribution may be used to compute the probability that a given reading will deviate a certain amount from the mean. We would not expect a probability much smaller than $1/n$ because this would be unlikely to occur in the set of n measurements. Thus, if the probability for the observed deviation of a certain point is less than $1/n$, a suspicious eye would be cast at that point with an idea toward eliminating it from the data. Actually, a more restrictive test is usually applied to eliminate data points. It is known as *Chauvenet's criterion*[1] and specifies that a reading may be rejected if the probability of obtaining the particular deviation from the mean is less than $1/2n$. Table 3.5 lists values of the ratio of deviation to standard deviation for various values of n according to this criterion with Fig. 3.6 furnishing a graphical representation.

In applying Chauvenet's criterion to eliminate dubious data points, one first calculates the mean value and standard deviation using all data points. The deviations of the individual points are then compared with the standard deviation in accordance with the information in Table 3.5 (or by a direct application of the criterion), and the dubious points are eliminated. For the final data presentation a new mean value and standard deviation are computed with the dubious points eliminated from the

---

[1] History is not clear on the matter, but it appears [29] that the criterion stated in the foregoing paragraph and Table 3.5 is really due to Pierce [30], as confirmed by Chauvenet [32] and early tables were presented by Gould [31] that are not exactly in agreement with Table 3.5. However, the term *Chauvenet's criterion* as described herein has become so ubiquitous in various applications that it would not seem appropriate to change the name at this time.

Table 3.5   Chauvenet's criterion for rejecting a reading

Number of Readings, n	Ratio of Maximum Acceptable Deviation to Standard Deviation, d_{max}/σ
3	1.38
4	1.54
5	1.65
6	1.73
7	1.80
10	1.96
15	2.13
25	2.33
50	2.57
100	2.81
300	3.14
500	3.29
1000	3.48

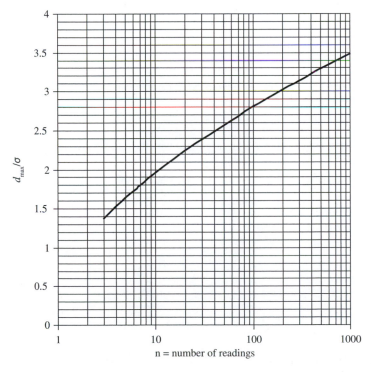

Figure 3.6   Chauvenet's criterion.

calculation. Note that Chauvenet's criterion might be applied a second or third time to eliminate additional points; but this practice is unacceptable, and only the first application may be used.

Example 3.13 | **APPLICATION OF CHAUVENET'S CRITERION.**   Using Chauvenet's criterion, test the data points of Example 3.7 for possible inconsistency. Eliminate the questionable points and calculate a new standard deviation for the adjusted data.

Solution

The best estimate of the standard deviation is given in Example 3.8 as 0.627 cm. We first calculate the ratio d_i/σ and eliminate data points in accordance with Table 3.5.

Reading	d_i/σ
1	0.499
2	0.187
3	1.845
4	0.563
5	2.046
6	0.260
7	0.761
8	0.043
9	0.314
10	0.219

In accordance with Table 3.5, we may eliminate only point number 5. When this point is eliminated, the new mean value is

$$x_m = \tfrac{1}{9}(51.80) = 5.756 \text{ cm}$$

The new value of the standard deviation is now calculated with the following table:

Reading	$d_i = x_i - x_m$	$(x_i - x_m)^2 \times 10^2$
1	−0.456	20.7936
2	−0.026	0.0676
3	1.014	102.8196
4	−0.496	24.602
6	−0.306	9.364
7	0.334	11.156
8	−0.116	1.346
9	0.054	0.292
10	−0.006	0.0036

$$\sigma = \left[\frac{1}{n-1} \sum_{i=1}^{n} (x_i - x_m)^2 \right]^{1/2} = \left[\tfrac{1}{8}(1.7044) \right]^{1/2} = (0.213)^{1/2} = 0.4615 \text{ cm}$$

Thus, by the elimination of the one point the standard deviation has been reduced from 0.627 to 0.462 cm. This is a 26.5 percent reduction.

Comment

Please note that for the revised calculation of standard deviation a new mean value must be computed leaving out the excluded data point.

The Chauvenet's criterion we have applied in this example is

$$d_{max}/\sigma = 1.96 \text{ for } n = 10$$

This value may be calculated directly from the gaussian distribution shown in Table 3.2 in the following way. The criterion is that the probability of a point lying outside the normal distribution should not exceed $1/2n$ or $1/20$. The probability of the point lying *inside* the normal distribution would then be

$$P(n) = 1 - 1/20 = 0.95$$

The entry point for Table 3.2 is half this value or 0.475. We obtain

$$\eta = 1.96$$

which agrees, of course, with Table 3.5 and Fig. 3.6.

3.9   COMPARISON OF DATA WITH NORMAL DISTRIBUTION

We have seen that the normal error distribution offers a means for examining experimental data for statistical consistency. In particular, it enables us to eliminate questionable readings with the Chauvenet criterion and thus obtain a better estimate of the standard deviation and mean reading. If the distribution of random errors is not *normal,* then this elimination technique will not apply. It is to our advantage, therefore, to determine if the data are following a normal distribution before making too many conclusions about the mean value, variances, and so forth. Specially constructed probability graph paper is available for this purpose and may be purchased from a technical drawing shop. The paper uses the coordinate system shown in Fig. 3.7. The ordinate has the percent of readings at or below the value of the abscissa, and the abscissa is the value of a particular reading. The ordinate spacings are arranged so that the gaussian-distribution curve will plot as a straight line on the graph. In addition, this straight line will intersect the 50 percent ordinate at an abscissa equal to the arithmetic mean of the data.

Thus, to determine if a set of data points is distributed normally, we plot the data on probability paper and see how well they match with the theoretical straight line. It is to be noted that the largest reading cannot be plotted on the graph because the ordinate does not extend to 100 percent. In assessing the validity of the data, we should not place as much reliance on the points near the upper and lower ends of the curve since they are closer to the "tails" of the probability distribution and are thus less likely to be valid.

An alternative approach is to plot the cumulative frequencies for the normal distribution using the tabular values from Table 3.2 or computer software. This plot

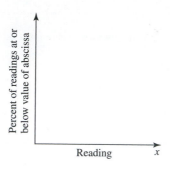

Figure 3.7   Probability graph paper.

is then displayed and compared with the actual frequency distribution to make an evaluation of the "normality" of the data.

Example 3.14 | **USE OF PROBABILITY GRAPH PAPER AND COMPUTER COMPARISON.**   The following data are collected for a certain measurement. Plot the data on probability paper and comment on the normality of the distribution: Make the same comparison with a computer-generated display.

Reading	x_i, cm
1	4.62
2	4.69
3	4.86
4	4.53
5	4.60
6	4.65
7	4.59
8	4.70
9	4.58
10	4.63
	$\sum x_i = 46.45$

From these data the mean value is calculated as

$$x_m = \frac{1}{10}\sum x_i = \frac{1}{10}(46.45) = 4.645 \text{ cm}$$

The population standard deviation is

$$\sigma = \left[\frac{1}{n}\sum (x_i - x_m)^2\right]^{1/2} = 0.0864$$

The data are plotted in the example figure (*a*) indicating a reasonably normal distribution. It should be noted that the straight line crosses the 50 percent ordinate at a value of approximately $x = 4.62$, which is not in agreement with the calculated value of x_m. Note that point 3, $x = 4.86$, does not appear on the plot since it would represent the 100 percent ordinate.

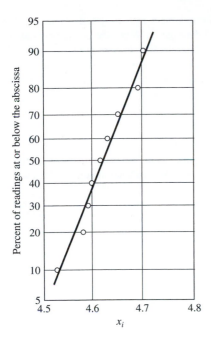

Figure Example 3.14(a)

A computer-generated comparison may be made by listing the values of x in ascending order as shown in the table below.

x	Actual Frequency	Normal Distribution Frequency
4.3	0.00	0.00003
4.35	0.00	0.00032
4.4	0.00	0.00229
4.45	0.00	0.01201
4.5	0.00	0.04665
4.53	0.10	0.09159
4.58	0.20	0.22593
4.59	0.30	0.26220
4.6	0.40	0.30124
4.62	0.50	0.38616
4.63	0.60	0.43109
4.65	0.70	0.52307
4.69	0.80	0.69876
4.7	0.90	0.73780
4.75	0.90	0.88787
4.8	0.90	0.96359
4.85	0.90	0.99117
4.86	1.00	0.99358
4.9	1.00	0.99842
4.95	1.00	0.99979
5	1.00	0.99998

The values of x are extended below and above the minimum and maximum value of 4.53 and 4.86 in order to pick up the tails of the normal distribution. Next, the actual cumulative frequencies are listed in the second column of the table. Note that there are no data points below $x = 4.53$ so these entries for the actual frequencies are zero. Likewise, all points have been observed at $x = 4.86$ or greater, so the actual frequencies are 1.0 at this point and above. The values for the normal distribution frequencies are computed with

$$\eta = \frac{x - x_m}{\sigma}$$

and Table 3.2, or a computer function. In this case the values were obtained with the probability functions in Microsoft Excel. Note the behavior of the normal distribution at large deviations from the mean value of x_m.

A graphical display of the actual and normal distribution frequencies is shown in example figure (*b*). The 50 percent value for the normal distribution occurs at $x = x_m = 4.645$. The actual frequency curve deviates substantially from the normal distribution in some regions of the chart.

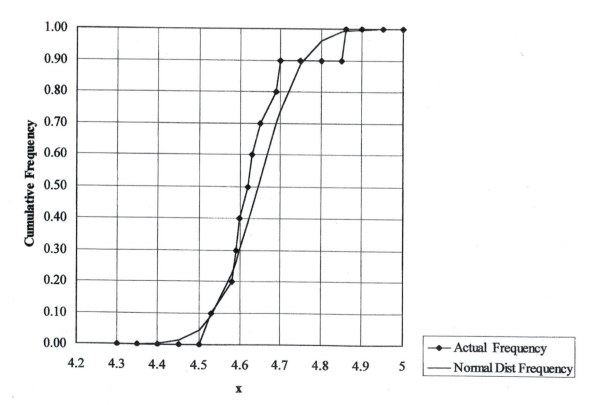

Figure Example 3.14(*b*)

3.10   THE CHI-SQUARE TEST OF GOODNESS OF FIT

In the previous discussion we have noted that random experimental errors would be expected to follow the gaussian distribution, and the examples illustrated the method of calculating the probability of occurrence of a particular experimental determination. We might ask how it is known that the random errors or deviations do approximate a gaussian distribution. In general, we may ask how we can determine if experimental observations match some particular expected distribution for the data. As a simple example, consider the tossing of a coin. We would like to know if a certain coin is "honest," that is, unweighted toward either heads or tails. If the coin is unweighted, then heads should occur half the time and tails should occur half the time. But suppose we do not want to take the time to make thousands of tosses to get a frequency distribution of heads and tails for a large number of tosses. Instead, we toss the coin a few times and wish to infer from these few tosses whether the coin is unweighted or weighted. Common sense tells us not to expect exactly six heads and six tails out of, say, 12 tosses. But how much deviation from this arrangement could we tolerate and still expect the coin to be unweighted? The chi-square test of goodness of fit is a suitable way of answering this question. It is based on a calculation of the quantity chi squared, defined by

$$\chi^2 = \sum_{i=1}^{n} \frac{[(\text{observed value})_i - (\text{expected value})_i]^2}{(\text{expected value})_i} \qquad \textbf{[3.20]}$$

where n is the number of cells or groups of observations. The expected value is the value which would be obtained if the measurements matched the expected distribution perfectly.

The chi-square test may be applied to check the validity of various distributions. Calculations have been made [2] of the probability that the actual measurements match the expected distribution, and these probabilities are given in Table 3.6. In this table F represents the number of degrees of freedom in the measurements and is given by

$$F = n - k \qquad \textbf{[3.21]}$$

where n is the number of cells and k is the number of imposed conditions on the expected distribution. A plot of the chi-square function is given in Fig. 3.8.

While we initiated the discussion on the chi-square test in terms of random errors following the gaussian distribution, the test is an important tool for testing any expected experimental distribution. In other words, we might use the test to analyze random errors or to check the adherence of certain data to an expected distribution. We interpret the test by calculating the number of degrees of freedom and χ^2 from the experimental data. Then, consulting Table 3.6, we obtain the probability P that this value of χ^2 or higher value could occur by chance. If $\chi^2 = 0$, then the assumed or expected distribution and measured distribution match exactly. The larger the value of χ^2, the larger is the disagreement between the assumed distribution and the observed values, or the smaller the probability that the observed distribution matches the

Table 3.6 Chi-squared. P is the probability that the value in the table will be exceeded for a given number of degrees of freedom F†

$F =$ \ P	0.995	0.990	0.975	0.950	0.900	0.750	0.500	0.250	0.100	0.050	0.025	0.010	0.005
1	0.0^4393	0.0^3157	0.0^3982	0.0^2393	0.0158	0.102	0.455	1.32	2.71	3.84	5.02	6.63	7.88
2	0.0100	0.0201	0.0506	0.103	0.211	0.575	1.39	2.77	4.61	5.99	7.38	9.21	10.6
3	0.0717	0.115	0.216	0.352	0.584	1.21	2.37	4.11	6.25	7.81	9.35	11.3	12.8
4	0.207	0.297	0.484	0.711	1.06	1.92	3.36	5.39	7.78	9.49	11.1	13.3	14.9
5	0.412	0.554	0.831	1.15	1.61	2.67	4.35	6.63	9.24	11.1	12.8	15.1	16.7
6	0.676	0.872	1.24	1.64	2.20	3.45	5.35	7.84	10.6	12.6	14.4	16.8	18.5
7	0.989	1.24	1.69	2.17	2.83	4.25	6.35	9.04	12.0	14.1	16.0	18.5	20.3
8	1.35	1.65	2.18	2.73	3.49	5.07	7.34	10.2	13.4	15.5	17.5	20.1	22.0
9	1.73	2.09	2.70	3.33	4.17	5.90	8.34	11.4	14.7	16.9	19.0	21.7	23.6
10	2.16	2.56	3.25	3.94	4.87	6.74	9.34	12.5	16.0	18.3	20.5	23.2	25.2
11	2.60	3.05	3.82	4.57	5.58	7.58	10.3	13.7	17.3	19.7	21.9	24.7	26.8
12	3.07	3.57	4.40	5.23	6.30	8.44	11.3	14.8	18.5	21.0	23.3	26.2	28.3
13	3.57	4.11	5.01	5.89	7.04	9.30	12.3	16.0	19.8	22.4	24.7	27.7	29.8
14	4.07	4.66	5.63	6.57	7.79	10.2	13.3	17.1	21.1	23.7	26.1	29.1	31.3
15	4.60	5.23	6.26	7.26	8.55	11.0	14.3	18.2	22.3	25.0	27.5	30.6	32.8
16	5.14	5.81	6.91	7.96	9.31	11.9	15.3	19.4	23.5	26.3	28.8	32.0	34.3
17	5.70	6.41	7.56	8.67	10.1	12.8	16.3	20.5	24.8	27.6	30.2	33.4	35.7
18	6.26	7.01	8.23	9.39	10.9	13.7	17.3	21.6	26.0	28.9	31.5	34.8	37.2
19	6.84	7.63	8.91	10.1	11.7	14.6	18.3	22.7	27.2	30.1	32.9	36.2	38.6
20	7.43	8.26	9.59	10.9	12.4	15.5	19.3	23.8	28.4	31.4	34.2	37.6	40.0
21	8.03	8.90	10.3	11.6	13.2	16.3	20.3	24.9	29.6	32.7	35.5	38.9	41.4
22	8.64	9.54	11.0	12.3	14.0	17.2	21.3	26.0	30.8	33.9	36.8	40.3	42.8
23	9.26	10.2	11.7	13.1	14.8	18.1	22.3	27.1	32.0	35.2	38.1	41.6	44.2
24	9.89	10.9	12.4	13.8	15.7	19.0	23.3	28.3	33.2	36.4	39.4	43.0	45.6
25	10.5	11.5	13.1	14.6	16.5	19.9	24.3	29.3	34.4	37.7	40.6	44.3	46.9
26	11.2	12.2	13.8	15.4	17.3	20.8	25.3	30.4	35.6	38.9	41.9	45.6	48.3
27	11.8	12.9	14.6	16.2	18.1	21.7	26.3	31.5	36.7	40.1	43.2	47.0	49.6
28	12.5	13.6	15.3	16.9	18.9	22.7	27.3	32.6	37.9	41.3	44.5	48.3	51.0
29	13.1	14.3	16.0	17.7	19.8	23.6	28.3	33.7	39.1	42.6	45.7	49.6	52.3
30	13.8	15.0	16.8	18.5	20.6	24.5	29.3	34.8	40.3	43.8	47.0	50.9	53.7

† From C. M. Thompson: *Biometrika*, vol. 32, 1941, as abridged by A. M. Mood and F. A. Graybill, *Introduction to the Theory of Statistics*, 2d ed., McGraw-Hill, New York, 1963.

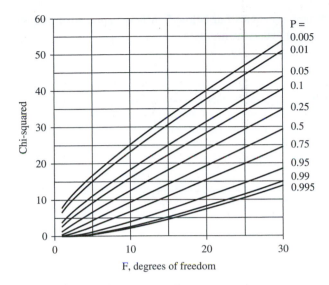

Figure 3.8   The chi-square function.

expected distribution. The reader should consult Refs. [2] and [4] for more specific information on the chi-square test and the derivation of the probabilities associated with it.

One may note that the heading of this section includes the term "goodness of fit." We see that the chi-square test may be used to determine how well a set of experimental observations fits an assumed distribution. In connection with this test we may remark that data may sometimes be "too good" or "too consistent." For example, we would be quite surprised if in the conduct of an experimental test, the results were found to check with theory *exactly* or to follow some well-defined relationship exactly. We might find, for instance, that a temperature controller maintained a set point temperature *exactly,* with no measurable deviation whatsoever. Experienced laboratory people know that controllers usually do not operate this way and would immediately suspect that the temperature recorder might be stuck or otherwise defective. The point of this brief remark is that one must be suspicious of high values of P as well as of low values. A good rule of thumb is that if P lies between 0.1 and 0.9, the observed distribution may be considered to follow the assumed distribution. If P is either less than 0.02 or greater than 0.98, the assumed distribution may be considered unlikely.

Let us return for a moment to the tossing of a coin. Suppose a coin is tossed twice, resulting in one head and one tail. This observation certainly matches exactly with what would be expected for an unweighted coin; however, our common sense tells us not to believe the coin is unweighted on the basis of only two tosses. In other words, we must have a certain minimum number of samples for statistics to apply. For the chi-square test the generally accepted minimum number of expected values for each ith cell is 5. If some frequencies fall below 5, it is recommended that the cells or groups be redefined to alleviate the problem.

Example 3.15

DEFECTS IN PLASTIC CUPS. A plastics company produces two types of styrofoam cups (call them A and B) which can experience eight kinds of defects. One hundred defective samples of each cup are collected and the number of each type of defect is determined. The following table results:

Type Defect	Cup A	Cup B
1	1	5
2	2	3
3	3	3
4	25	23
5	10	12
6	15	16
7	38	30
8	6	8
Total	100	100

We would like to know if the two cups have the same pattern of defects. To do this, we could compute chi-squared for cup B assuming cup A has the expected distribution. But we encounter a problem. Defects 1, 2, and 3 do not meet our criterion of a minimum of five expected values in each cell. So, we must reconstruct the cells by combining 1, 2, and 3 to obtain:

Type Defect	Cup A	Cup B
1, 2, 3	6	11
4	25	23
5	10	12
6	15	16
7	38	30
8	6	8
Total	100	100

For the former case we had eight cells or groups and one imposed condition (total observations $= 100$), so $F = 8 - 1 = 7$. After grouping defects 1, 2, and 3, we have $F = 6 - 1 = 5$. Using this new tabulation the value of chi-squared is calculated as 7.145. Consulting Table 3.6, we obtain the value of P as 0.43. Thus, we might expect that the two cups have approximately the same pattern of defects.

ROLLING THE DICE.   Two dice are rolled 300 times and the following results are | **Example 3.16**
noted:

Number	Number of Occurrences
2	6
3	9
4	27
5	36
6	39
7	57
8	45
9	39
10	24
11	12
12	6

Calculate the probability that the dice are unloaded.

Solution

Eleven cells have been observed with only one restriction: the number of rolls of the dice is fixed. Thus, $F = 11 - 1 = 10$. If the dice are unloaded, a short listing of the combinations of the dice will give the probability of occurrence for each number. The expected value of each number is then the probability multiplied by 300, the total number of throws. The values of interest are tabulated as follows:

Number	Observed	Probability	Expected
2	6	1/36	8.333
3	9	1/18	16.667
4	27	1/12	25.0
5	36	1/9	33.333
6	39	5/36	41.667
7	57	1/6	50.0
8	45	5/36	41.667
9	39	1/9	33.333
10	24	1/12	25.0
11	12	1/18	16.667
12	6	1/36	8.333

From these data the value of chi-squared is calculated as 8.034. If Table 3.6 is consulted, the probability is given as $P = 0.626$.

Comment

The value of $P = 0.626$ lies between our acceptable limits of 0.1 and 0.9, so we might conclude, based on these observations, that the dice are honest or unweighted.

Example 3.17 | **TOSS OF COIN: INFLUENCE OF ADDITIONAL DATA POINTS.**   A coin is tossed 20 times, resulting in 6 heads and 14 tails. Using the chi-square test, estimate the probability that the coin is unweighted. Suppose another set of tosses of the same coin is made and 8 heads and 12 tails are obtained. What is the probability of having an unweighted coin based on the information from both sets of data?

Solution

For each set of data we may make only two observations: the number of heads and the number of tails. Thus, $n = 2$. Furthermore, we impose one restriction on the data: the number of tosses is fixed. Thus, $k = 1$ and the number of degrees of freedom is

$$F = n - k = 2 - 1 = 1$$

The values of interest are:

	Observed	Expected
Heads	6	10
Tails	14	10

For these values χ^2 is calculated as

$$\chi^2 = \frac{(6 - 10)^2}{10} + \frac{(14 - 10)^2}{10} = 3.20$$

Consulting Table 3.6, we find $P = 0.078$; that is, there is an 8 percent chance that this distribution is just the result of random fluctuations and that the coin may be unweighted.

Now, consider the additional information we gain about the coin from the second set of observations. We then have four observations: the number of heads and tails in each set. There are only two restrictions on the data: the total number of tosses is fixed in each set. Thus, the number of degrees of freedom is

$$F = n - k = 4 - 2 = 2$$

For the second set of data the values of interest are:

	Observed	Expected
Heads	8	10
Tails	12	10

Chi-squared is now calculated on the basis of all four observations:

$$\chi^2 = \frac{(6 - 10)^2}{10} + \frac{(14 - 10)^2}{10} + \frac{(8 - 10)^2}{10} + \frac{(12 - 10)^2}{10} = 4.0$$

Consulting Table 3.6 again, we find $P = 0.15$. So, with the additional information we find a stronger likelihood that the tosses are following a random variation and that the coin is unweighted.

Comment

This example illustrates how the collection of additional data may strengthen a conclusion which may have been previously viewed as marginal.

EFFECT OF CIGARETTE SMOKE ON MICE.   A test is conducted to determine the effect of cigarette smoke on the eating habits and weight of mice. One group is fed a certain diet while being exposed to a controlled atmosphere containing cigarette smoke. A control group is fed the same diet but in the presence of clean air. The observations are given below. Does the presence of smoke cause a loss in weight?

Example 3.18

	Gained Weight	Lost Weight	Total
Exposed to smoke	61	89	150
Exposed to clean air	65	77	142
Total	126	166	292

Solution

Clearly, there are four observations in this experiment, but we are faced with the problem of deciding on the expected values. We cannot just take the "clean-air" data as the expected values because some of the behavior might be a result of the special diet that is fed to both groups of mice. Consequently, about the best estimate we can make is one based on the total sample of mice. Thus, the expected frequencies would be

$$\text{Expected fraction to gain weight} = \frac{126}{292}$$

$$\text{Expected fraction to lose weight} = \frac{166}{292}$$

The expected values for the groups would thus be:

	Gained Weight	Lost Weight
Exposed to smoke	$\frac{126}{292}\,150 = 64.7$	$\frac{166}{292}\,150 = 85.3$
Exposed to clean air	$\frac{126}{292}\,142 = 61.3$	$\frac{166}{292}\,142 = 80.7$

We observe that there are three restrictions on the data: (1) the number exposed to smoke, (2) the number exposed to clean air, and (3) the additional restriction involved in the calculation of the expected fractions which gain and lose weight. The number of degrees of freedom is thus

$$F = 4 - 3 = 1$$

The value of chi-squared is calculated from

$$\chi^2 = \frac{(61 - 64.7)^2}{64.7} + \frac{(89 - 85.3)^2}{85.3} + \frac{(65 - 61.3)^2}{61.3} + \frac{(77 - 80.7)^2}{80.7} = 0.767$$

From Table 3.6 we find $P = 0.41$, or there is a 41 percent chance that the difference in the observations for the two groups is just the result of random fluctuations. One may not conclude from this information that the presence of cigarette smoke causes a loss in weight for the mice.

3.11 METHOD OF LEAST SQUARES

Suppose we have a set of observations $x_1, x_2, \ldots, x_n$. The sum of the squares of their deviations from some mean value is

$$S = \sum_{i=1}^{n} (x_i - x_m)^2 \qquad \textbf{[3.22]}$$

Now, suppose we wish to minimize S with respect to the mean value x_m. We set

$$\frac{\partial S}{\partial x_m} = 0 = \sum_{i=1}^{n} -2(x_i - x_m) = -2 \left(\sum_{i=1}^{n} x_i - n x_m \right) \qquad \textbf{[3.23]}$$

where n is the number of observations. We find that

$$x_m = \frac{1}{n} \sum_{i=1}^{n} x_i \qquad \textbf{[3.24]}$$

or the mean value which minimizes the sum of the squares of the deviations is the arithmetic mean. This example might be called the simplest application of the method of least squares.

Suppose that the two variables x and y are measured over a range of values. Suppose further that we wish to obtain a simple analytical expression for y as a function of x. The simplest type of function is a linear one; hence, we might try to establish y as a linear function of x. (Both x and y may be complicated functions of other parameters so arranged that x and y vary approximately in a linear manner. This matter will be discussed later.) The problem is one of finding the *best* linear function, for the data may scatter a considerable amount. We could solve the problem rather quickly by plotting the data points on graph paper and drawing a straight line through them by eye. Indeed this is common practice, but the method of least squares gives a more reliable way to obtain a better functional relationship than the guesswork of plotting. We seek an equation of the form

$$y = ax + b \qquad \textbf{[3.25]}$$

We therefore wish to minimize the quantity

$$S = \sum_{i=1}^{n} [y_i - (ax_i + b)]^2 \qquad \textbf{[3.26]}$$

This is accomplished by setting the derivatives with respect to a and b equal to zero.

Performing these operations, there results

$$nb + a \sum x_i = \sum y_i \qquad \text{[3.27]}$$

$$b \sum x_i + a \sum x_i^2 = \sum x_i y_i \qquad \text{[3.28]}$$

Solving Eqs. (3.27) and (3.28) simultaneously gives

$$a = \frac{n \sum x_i y_i - \left(\sum x_i\right)\left(\sum y_i\right)}{n \sum x_i^2 - \left(\sum x_i\right)^2} \qquad \text{[3.29]}$$

$$b = \frac{\left(\sum y_i\right)\left(\sum x_i^2\right) - \left(\sum x_i y_i\right)\left(\sum x_i\right)}{n \sum x_i^2 - \left(\sum x_i\right)^2} \qquad \text{[3.30]}$$

Designating the computed value of y as $\hat{y}$, we have

$$\hat{y} = ax + b$$

and the standard error of estimate of y for the data is

$$\text{Standard error} = \left[\frac{\sum (y_i - \hat{y}_i)^2}{n - 2}\right]^{1/2} \qquad \text{[3.31]}$$

$$= \left[\frac{\sum (y_i - ax_i - b)^2}{n - 2}\right]^{1/2} \qquad \text{[3.32]}$$

The method of least squares may also be used for determining higher-order polynomials for fitting data. One only needs to perform additional differentiations to determine additional constants. For example, if it were desired to obtain a least-squares fit according to the quadratic function

$$y = ax^2 + bx + c$$

the quantity

$$S = \sum_{i=1}^{n} \left[y_i - \left(ax_i^2 + bx_i + c\right)\right]^2$$

would be minimized by setting the following derivatives equal to zero:

$$\frac{\partial S}{\partial a} = 0 = \sum 2\left[y_i - \left(ax_i^2 + bx_i + c\right)\right]\left(-x_i^2\right)$$

$$\frac{\partial S}{\partial b} = 0 = \sum 2\left[y_i - \left(ax_i^2 + bx_i + c\right)\right]\left(-x_i\right)$$

$$\frac{\partial S}{\partial c} = 0 = \sum 2\left[y_i - \left(ax_i^2 + bx_i + c\right)\right]\left(-1\right)$$

Expanding and collecting terms, we have

$$a \sum x_i^4 + b \sum x_i^3 + c \sum x_i^2 = \sum x_i^2 y_i \qquad \text{[3.33]}$$

$$a \sum x_i^3 + b \sum x_i^2 + c \sum x_i = \sum x_i y_i \qquad \text{[3.34]}$$

$$a \sum x_i^2 + b \sum x_i + cn = \sum y_i \qquad \textbf{[3.35]}$$

These equations may then be solved for the constants a, b, and c.

REGRESSION ANALYSIS

In the above discussion of the method of least squares no mention has been made of the influence of experimental uncertainty on the calculation. We are considering the method primarily for its utility in fitting an algebraic relationship to a set of data points. Clearly, the various x_i and y_i could have different experimental uncertainties. To take all these into account requires a rather tedious calculation procedure, which we shall not present here; however, we may state the following rules:

1. If the values of x_i and y_i are taken as the data value in y and the value of x *on the fitted curve for the same value of y,* then there is a presumption that the uncertainty in x is large compared with that in y.

2. If the values of x_i and y_i are taken as the data value in y and the value *on the fitted curve for the same value of x,* the presumption is that the uncertainty in y dominates.

3. If the uncertainties in x_i and y_i are believed to be of approximately equal magnitude, a special averaging technique must be used.

In rule 1 we say we are taking a *regression* of x on y, and in rule 2 there is a regression of y on x. In the second case we are minimizing the sum of the squares of the deviations of the actual points from the assumed curve and also assuming *that x does not vary appreciably at each point.* If we obtained

$$y = a + bx$$

and then solved to get

$$x = \frac{1}{b} y - \frac{a}{b}$$

this second relation would *not* necessarily give a good calculation for x since the minimization was carried out in the y direction and not in the x direction. In Example 3.19 rule 2 is assumed to apply.

Example 3.19 | **LEAST-SQUARES REGRESSION.**   From the following data obtain y as a linear function of x using the method of least squares:

y_i	x_i
1.2	1.0
2.0	1.6
2.4	3.4
3.5	4.0
3.5	5.2
$\sum y_i = 12.6$	$\sum x_i = 15.2$

Solution

We seek an equation of the form

$$y = ax + b$$

We first calculate the quantities indicated in the following table:

$x_i y_i$	x_i^2
1.2	1.0
3.2	2.56
8.16	11.56
14.0	16.0
18.2	27.04
$\sum x_i y_i = 44.76$	$\sum x_i^2 = 58.16$

We calculate the value of a and b using Eqs. (3.29) and (3.30) with $n = 5$:

$$a = \frac{(5)(44.76) - (15.2)(12.6)}{(5)(58.16) - (15.2)^2} = 0.540$$

$$b = \frac{(12.6)(58.16) - (44.76)(15.2)}{(5)(58.16) - (15.2)^2} = 0.879$$

Thus, the desired relation is

$$y = 0.540x + 0.879$$

A plot of this relation and the data points from which it was derived is shown in the accompanying figure.

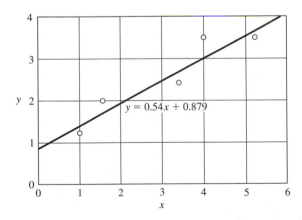

Figure Example 3.19

Comment

The least-squares minimization process has been carried out assuming that the uncertainty in y predominates, that is, x is known more exactly than y. If we write the correlating equation in the form

$$x = (1/b)y - (a/b)$$

and then performed a least-squares process we would obtain

$$x = 1.6353y - 1.0809$$

or

$$y = 0.6115x + 0.66098$$

In this case it is presumed that the uncertainty in x predominates or y is known more accurately than x. The difference in the two relations can be significant at the limits of the variables. When the analysis is applied to actual experimental data, the experimentalist should select the most appropriate one based on the total information available concerning the experiment.

3.12   THE CORRELATION COEFFICIENT

Let us assume that a suitable correlation between y and x has been obtained, by either least-squares analysis or graphical curve fitting. We want to know how good this fit is and the parameter which conveys this information is the *correlation coefficient r* defined by

$$r = \left[1 - \frac{\sigma_{y,x}^2}{\sigma_y^2} \right]^{1/2}$$ **[3.36]**

where σ_y is the standard deviation of y given as

$$\sigma_y = \left[\frac{\sum_{i=1}^{n}(y_i - y_m)^2}{n - 1} \right]^{1/2}$$ **[3.37]**

and

$$\sigma_{y,x} = \left[\frac{\sum_{i=1}^{n}(y_i - y_{ic})^2}{n - 2} \right]^{1/2}$$ **[3.38]**

The y_i are the actual values of y, and the y_{ic} are the values computed from the correlation equation for the same value of x.

The division by $n - 2$ results from the fact that we have used the two derived variables a and b in determining the value of y_{ic}. We might say that this removes 2 degrees of freedom from the system of data. The correlation coefficient r may also be written as

$$r^2 = \frac{\sigma_y^2 - \sigma_{y,x}^2}{\sigma_y^2}$$ **[3.39]**

where, now, r^2 is called the *coefficient of determination*. We note that for a perfect fit $\sigma_{y,x} = 0$ because there are no deviations between the data and the correlation. In this case $r = 1.0$. If $\sigma_y = \sigma_{y,x}$, we obtain $r = 0$, indicating a poor fit or substantial scatter around the fitted line. The reader must be cautioned about ascribing too much virtue to values of r close to 1.0. These values may occur when the data do *not* fit the line.

To be on the safe side, one should *never* accept a least-squares analysis based *only* on calculations. One should *always* plot the data to obtain a visual observation of the behavior. If the data points do indeed hug the least-squares line, then a high value of r will be indicative of a very good correlation. If the data scatter but still appear to follow the fitted relationship, then a small value of r will also be meaningful as a measure of poorer correlation.

At this point we must stress the need for graphical displays of data for other purposes. In our discussions of uncertainties and errors we noted that the experimentalist may be the best person to assess the uncertainties in the primary measurements. Sometimes during the course of the experiment the experimenter may note that a particular data point has an erratic behavior and so record the observation in the lab notebook. When the data are plotted, such a point may be excluded if it appears out of line with other data or retained if it appears satisfactory. If a least-squares analysis was performed which included all the data, the correlation might not be so good as could be obtained with exclusion of the questionable data point(s). For these reasons seasoned experimentalists like to get an "eyeball" plot of the correlating straight line before actually performing a least-squares analysis. In Sec. 3.15 we shall see how one may go about obtaining straight-line plots for different functional relationships.

It may be noted that most scientific calculators have built-in routines which calculate the correlation coefficient as well as other statistical functions. In addition, there are many computer software packages which accomplish these calculations, for example, those of Refs. [15], [16], and [28].

A relationship for the correlation coefficient which may be preferable to Eq. (3.36) for computer calculations is

$$r = \frac{n \sum x_i y_i - \left(\sum x_i \right) \left(\sum y_i \right)}{\left[n \sum x_i^2 - \left(\sum x_i \right)^2 \right]^{1/2} \left[n \sum y_i^2 - \left(\sum y_i \right)^2 \right]^{1/2}} \qquad \textbf{[3.40]}$$

In Eqs. (3.33) through (3.35) we noted the technique that one might apply for a least-squares fit to a quadratic function. In this case the correlation coefficient is still given by Eq. (3.36), but now

$$\sigma_{y,x} = \left[\frac{(y_i - y_{ic})^2}{n - 3} \right]^{1/2} \qquad \textbf{[3.41]}$$

In general, for fit with a polynomial of order m one would obtain

$$\sigma_{y,x} = \left[\frac{(y_i - y_{ic})^2}{n - (m + 1)} \right]^{1/2} \qquad \textbf{[3.42]}$$

Aside from the fact that one may anticipate the form of the functional data relationship from theory, we may sometimes try to fit the data with a polynomial through a least-squares analysis. For those who are computer inclined the tendency is to assume that the higher the order of the polynomial, the better the correlation will be. As a result, a certain overkill may be experienced. In some cases a higher-order polynomial may actually provide a poorer correlation than the simple quadratic.

Again, it is a good idea to plot the data first to get a visual idea of the behavior before performing analyses.

3.13   MULTIVARIABLE REGRESSION

The least-squares method may be extended to perform a regression analysis for more than one variable. In the linear case we would have the form

$$y = b + m_1 x_1 + m_2 x_2 + \cdots + m_n x_n \qquad \textbf{[3.43]}$$

where the x_n are the independent variables. For only two variables we form the sum of the squares

$$S = \sum (y_i - b - m_1 x_{1,i} - m_2 x_{2,i})^2 \qquad \textbf{[3.44]}$$

and minimize this sum with the differentiations:

$$\frac{\partial S}{\partial b} = -2 \sum (y_i - b - m_1 x_{1,i} - m_2 x_{2,i}) = 0$$

$$\frac{\partial S}{\partial m_1} = -2 \sum x_{1,i}(y_i - b - m_1 x_{1,i} - m_2 x_{2,i}) = 0$$

$$\frac{\partial S}{\partial m_2} = -2 \sum x_{2,i}(y_i - b - m_1 x_{1,i} - m_2 x_{2,i}) = 0$$

This set of linear equations may then be solved for the coefficients m_1, m_2, and b.

A further extension of the multivariable regression method may be made to an exponential form where

$$y = b m_1^{x1} m_2^{x2} \cdots m_n^{xn} \qquad \textbf{[3.45]}$$

The equation may be modified by taking the logarithm of each side to give

$$\log y = \log b + x_1 \log m_1 + x_2 \log m_2 + \cdots \qquad \textbf{[3.46]}$$

A least-squares analysis is then performed to determine values of the constants b, m_1, m_2, etc. The calculation of correlation coefficients for multivariable regressions is described in Ref. [12].

Computer software packages are available to perform the multivariable calculations indicated above. Microsoft Excel is one such package, and examples of its use in multivariable linear and exponential regression analysis, including computation of correlation coefficients, are given in Ref. [28]. The standard deviation for multivariable regression is computed with Eq. (3.42) with the term $(m + 1)$ replaced by the number of constants to be determined. For the two-variable linear relation in Eq. (3.44) there are three constants to be determined so the denominator of Eq. (3.42) would be $n - 3$.

Multivariable regression calculations can become rather involved and are best performed with a computer. Some words of caution are in order though. It is very easy to accept the computer results as gospel truth without checking against the original data points. As we have noted before, no correlation or regression analysis should be accepted without a direct comparison with the original data, preferably a graphical comparison which gives a visual feel for the data correlation independent of calculated values of correlation coefficients. Constructing such plots may be difficult for multivariable problems but is well worth the effort. The present writer has seen computer-generated multivariable regression results presented with favorable correlation coefficients. In some cases, these results, when graphed with the original data, passed distinctly outside the range of the data. Obviously, when such disagreement occurs, the results of the regression analysis must be discarded.

CORRELATION COEFFICIENT. Calculate the correlation coefficient for the least-square correlation of Example 3.19.

Example 3.20

Solution

From Example 3.19

$$y_m = \frac{\sum y_i}{n} = \frac{12.6}{5} = 2.52$$

and from the correlating equation $y_{ic} = 0.5490x + 0.879$:

i	y_i	y_{ic}	$(y_i - y_{ic})^2$
1	1.2	1.419	0.048
2	2.0	1.743	0.066
3	2.4	2.715	0.0992
4	3.5	3.039	0.212
5	3.5	3.687	0.035
			$\sum = 0.4607$

so that

$$\sigma_{y,x} = \left(\frac{0.4607}{3}\right)^{1/2} = 0.3919$$

In addition, we have

$$y_m = (\Sigma y_i)/5 = 2.52$$
$$\sigma_y = [\Sigma(y_i - y_m)^2/(5-1)]^{1/2} = 0.987$$

so that the correlation coefficient is

$$r = \left[1 - \left(\frac{0.3919}{0.987}\right)^2\right]^{1/2} = 0.9178$$

REGRESSION, RESIDUAL, AND TOTAL SUM OF SQUARES

Let us define the following terms:

$$S_{total} = \Sigma(y_i - y_m)^2$$
$$S_{regression} = \Sigma(y_{ic} - y_m)^2$$
$$S_{residual} = \Sigma(y_i - y_{ic})^2$$

Because

$$(y_i - y_m) = (y_i - y_{ic}) + (y_{ic} - y_m)$$

it follows that

$$S_{total} = S_{residual} + S_{regression}$$

A perfect "goodness of fit" will be obtained when $S_{residual} = 0$, that is, the regression curve passes through the data points. The measure of goodness of fit may then be taken as

$$r^2 = S_{regression}/S_{total} = (S_{total} - S_{residual})/S_{total}$$

3.14 STANDARD DEVIATION OF THE MEAN

We have taken the arithmetic mean value as the best estimate of the true value of a set of experimental measurements. Considerable discussion has been devoted to data subjected to random uncertainties and to an examination of the various types of errors and deviations that may occur in an experimental measurement. But one very important question has not yet been answered: How *good* (or precise) is this arithmetic mean value which is taken as the best estimate of the true value of a set of readings? To obtain an experimental answer to this question it would be necessary to repeat the set of measurements and to find a new arithmetic mean. In general, we would find that this new arithmetic mean would differ from the previous value, and thus we would not be able to resolve the problem until a large number of *sets of data* was collected. We would then know how well the mean of a single set approximated the mean which would be obtained with a large number of sets. The mean value of a large number of sets is presumably the true value. Consequently, we wish to know the standard deviation of the mean of a single set of data from this true value.

It turns out that the problem may be resolved with a statistical analysis which we shall not present here. The result is

$$\sigma_m = \frac{\sigma}{\sqrt{n}} \qquad \text{[3.47]}$$

where σ_m = standard deviation of the mean value

σ = standard deviation of the set of measurements

n = number of measurements in the set

We should note that the calculation of statistical parameters like standard deviation and least-square fits to data is easily performed with standard computer programs which are available on even small hand calculators.

UNCERTAINTY IN MEAN VALUE. For the data of Example 3.7, estimate the uncertainty in the calculated mean value of the readings. **Example 3.21**

Solution

We shall make this estimate for the original data and for the reduced data of Example 3.13. For the original data the standard deviation of the mean is

$$\sigma_m = \frac{\sigma}{\sqrt{n}} = \frac{0.627}{\sqrt{10}} = 0.198 \text{ cm}$$

The arithmetic mean value calculated in Example 3.7 was $x_m = 5.613$ cm. We could now specify the uncertainty of this value by using the odds of Table 3.3:

$$x_m = 5.613 \pm 0.198 \text{ cm} \qquad (2.15 \text{ to } 1)$$
$$= 5.756 \pm 0.396 \text{ cm} \qquad (21 \text{ to } 1)$$
$$= 5.613 \pm 0.594 \text{ cm} \qquad (369 \text{ to } 1)$$

Using the data of Example 3.13, where one point has been eliminated by Chauvenet's criterion, we may make a better estimate of the mean value with less uncertainty. The standard deviation of the mean is calculated as

$$\sigma_m = \frac{\sigma}{\sqrt{n}} = \frac{0.465}{\sqrt{9}} = 0.155 \text{ cm}$$

for the mean value of 5.756 cm. Thus, we would estimate the uncertainty as

$$x_m = 5.756 \pm 0.155 \text{ cm} \qquad (2.15 \text{ to } 1)$$
$$= 5.756 \pm 0.310 \text{ cm} \qquad (21 \text{ to } 1)$$
$$= 5.756 \pm 0.465 \text{ cm} \qquad (369 \text{ to } 1)$$

3.15 STUDENT'S t-DISTRIBUTION

In Sec. 3.14 we have used the relation

$$\sigma_m = \frac{\sigma}{\sqrt{n}}$$

to determine the standard deviation of the mean in terms of the standard deviation of the population. For small samples ($n < 10$) this relation has been shown to be

Table 3.7   Values of Student's t for use in Equation (3.48)

Subscript designates percent confidence level.

Degrees of freedom ν	t_{50}	t_{80}	t_{90}	t_{95}	t_{98}	t_{99}	$t_{99.9}$
1	1.000	3.078	6.314	12.706	31.821	63.657	636.619
2	0.816	1.886	2.920	4.303	6.965	9.925	31.598
3	0.765	1.638	2.353	3.182	4.541	5.841	12.941
4	0.741	1.533	2.132	2.776	3.747	4.604	8.610
5	0.727	1.476	2.015	2.571	3.365	4.032	6.859
6	0.718	1.440	1.943	2.447	3.143	3.707	5.959
7	0.711	1.415	1.895	2.365	2.998	3.499	5.405
8	0.706	1.397	1.860	2.306	2.896	3.355	5.041
9	0.703	1.383	1.833	2.262	2.821	3.250	4.781
10	0.700	1.372	1.812	2.228	2.764	3.169	4.587
11	0.697	1.363	1.796	2.201	2.718	3.106	4.437
12	0.695	1.356	1.782	2.179	2.681	3.055	4.318
13	0.694	1.350	1.771	2.160	2.650	3.012	4.221
14	0.692	1.345	1.761	2.145	2.624	2.977	4.140
15	0.691	1.341	1.753	2.131	2.602	2.947	4.073
16	0.690	1.337	1.746	2.120	2.583	2.921	4.015
17	0.689	1.333	1.740	2.110	2.567	2.898	3.965
18	0.688	1.330	1.734	2.101	2.552	2.878	3.922
19	0.688	1.328	1.729	2.093	2.539	2.861	3.883
20	0.687	1.325	1.725	2.086	2.528	2.845	3.850
21	0.686	1.323	1.721	2.080	2.518	2.831	3.819
22	0.686	1.321	1.717	2.074	2.508	2.819	3.792
23	0.685	1.319	1.714	2.069	2.500	2.807	3.767
24	0.685	1.318	1.711	2.064	2.492	2.797	3.745
25	0.684	1.316	1.708	2.060	2.485	2.787	3.725
26	0.684	1.315	1.706	2.056	2.479	2.779	3.707
27	0.684	1.314	1.703	2.052	2.473	2.771	3.690
28	0.683	1.313	1.701	2.048	2.467	2.763	3.674
29	0.683	1.311	1.699	2.045	2.462	2.756	3.659
30	0.683	1.310	1.697	2.042	2.457	2.750	3.646
40	0.681	1.303	1.684	2.021	2.423	2.704	3.551
60	0.679	1.296	1.671	2.000	2.390	2.660	3.460
120	0.677	1.289	1.658	1.980	2.358	2.617	3.373
∞	0.674	1.282	1.645	1.960	2.326	2.576	3.291

somewhat unreliable. A better method for estimating confidence intervals was developed by Student[2] by introducing the variable t such that

$$\Delta = \frac{t\sigma}{\sqrt{n}} \qquad \text{[3.48]}$$

where, now, t replaces the z variable previously used. It can be shown that

$$t = \frac{\bar{x} - X}{\sigma}\sqrt{n} \qquad \text{[3.49]}$$

where n = number of observations

$\bar{x}$ = mean of n observations

X = mean of normal population which the samples are taken from

Student then developed a distribution function $f(t)$ such that

$$f(t) = \frac{K_0}{\left(1 + \dfrac{t^2}{n-i}\right)^{n/2}}$$

$$f(t) = K_0\left(1 + \frac{t^2}{v}\right)(v+1)/2 \qquad \text{[3.50]}$$

where K_0 is a constant which depends on n and v is $(n-1)$ degrees of freedom. When $n \to \infty$, the distribution function approaches the normal distribution. Table 3.7 gives values of Student's t for different degrees of freedom and levels of confidence. t_{90} means a 90 percent confidence level. Note that for $v \to \infty$, t_{90} is 1.645, which agrees with Table 3.4.

CONFIDENCE LEVEL FROM t-DISTRIBUTION. Ten observations of a voltage are made with $\bar{e} = 15$ V and $\sigma = \pm 0.1$ V. Determine the 5 and 1 percent significance levels. | **Example 3.22**

Solution

For $n = 10$ we have $v = 10 - 1 = 9$. At the 5 percent significance level the probability is 95 percent and we find from Table 3.7

$$t = 2.262$$

From Eq. (3.48)

$$\Delta = \frac{(2.262)(0.1)}{\sqrt{10}} = 0.0715 \text{ V}$$

at the 1 percent significance level $P = 0.99$. $v = 9$, and

$$t = 3.250$$

| [2]Pen name of William S. Gosset (1876–1937), an Irish chemist.

so that

$$\Delta = \frac{(3.25)(0.1)}{10} = 0.1028 \text{ V}$$

We thus could state with a 95 percent confidence level that the voltage is 15 V $\pm$ 0.0715 V or with a 99 percent confidence level that it is 15 V $\pm$ 0.1028 V.

Example 3.23 | **ESTIMATE OF SAMPLE SIZE.**   For the steel bar in Example 3.11, obtain a new estimate for the number of measurements required using the t-distribution.

Solution

From Example 3.10 we have

$$\Delta = 0.2 \text{ mm}$$
$$\sigma = 0.5 \text{ mm}$$

and from Eq. (3.44)

$$\Delta = \frac{t\sigma}{\sqrt{n}}$$

or

$$t = 0.4\sqrt{n} \qquad \qquad \text{[a]}$$

At this point we note that t is a function of n through Table 3.7 so that Eq. (a) must be solved by iteration to obtain a value for n. Remembering that $\nu = n - 1 = 24$, the trials are for the 95 percent confidence level:

n	t_{95} (from Table 3.7)	t [Calculated, Eq. (a)]
25	2.064	2.000
26	2.060	2.040
27	2.056	2.078

Therefore, we shall require 27 measurements for the t-distribution in contrast to the 25 measurements required in Example 3.11.

Example 3.24 | **CONFIDENCE LEVEL.**   Ten measurements are made of the thickness of a metal plate which give 3.61, 3.62, 3.60, 3.63, 3.61, 3.62, 3.60, 3.62, 3.64, and 3.62 mm. Determine the mean value and the tolerance limits for a 90 percent confidence level.

Solution

The mean value is calculated from

$$x_m = \frac{1}{n}\sum^{i} x_i = \frac{1}{10}(36.17) = 3.617 \text{ mm}$$

The sample standard deviation is calculated from

$$\sigma = \left[\frac{\sum (x_i - x_m)^2}{n-1} \right]^{1/2}$$

$$= \left[\frac{1.41 \times 10^{-3}}{10-1} \right]^{1/2} = 0.0125 \text{ mm}$$

Entering Table 3.7 with $v = 10 - 1 = 9$, we obtain t_{90} for a 90 percent confidence level:

$$t_{90} = 1.833$$

Thus, we have from Eq. (3.44)

$$\Delta = \frac{t\sigma}{10} = \frac{(1.833)(0.0125)}{10}$$
$$= 0.00726 \text{ mm}$$

or $\qquad\qquad x = 3.617 \text{ mm} \pm 0.00726 \text{ mm} \quad$ (90 percent confidence)

CONFIDENCE LEVEL.   If the results of the measurements of Example 3.24 are stated as | **Example 3.25**

$$x_m = 3.617 \text{ mm} \pm 0.01 \text{ mm}$$

what confidence level should be assigned to this statement?

Solution

We still have $\sigma = 0.0125$ mm and from Eq. (3.44)

$$\Delta = \frac{t\sigma}{\sqrt{n}} = 0.01 = \frac{t(0.0125)}{10}$$

or $\qquad\qquad\qquad t = 2.53$

Entering Table 3.7 with $v = 10 - 1 = 9$, we find by interpolation

$$t = 2.53 = t_{96.4}$$

indicating a confidence level of 96.4 percent.

LOWER CONFIDENCE LEVEL.   Repeat Examples 3.24 and 3.25 for a confidence level | **Example 3.26**
of 90 percent (10 percent level of significance).

Solution

We still have

$$\Delta = 0.2 \text{ mm}$$
$$\sigma = 0.5 \text{ mm}$$

For 90 percent confidence level we obtain from Table 3.7

$$z = 1.65$$

and from

$$\Delta = \frac{z\sigma}{\sqrt{n}} = 0.2 = \frac{(1.65)(0.5)}{\sqrt{n}}$$

$$n = 17.01, \quad \text{rounded to } n = 18$$

For the t-distribution

$$\Delta = \frac{t\sigma}{\sqrt{n}}$$

and $$t = 0.4\sqrt{n}$$ **[a]**

and, again, an iterative procedure is required. This time we must use t_{90} from Table 3.7. The trials are, with $\nu = n - 1$:

n	t_{90} (from Table 3.7)	t [Calculated, Eq. (a)]
17	1.746	1.649
18	1.740	1.697
19	1.734	1.743

and 19 measurements would be required by the t-distribution.

Example 3.27 | **TRADE-OFF IN NUMBER OF MEASUREMENTS.**   In a certain pressure measurement a known precision of ±6 kPa can be obtained with a large number of measurements. This significance level of the pressure determination is directly related to rejection of a certain production part. The cost per part rejected is P and the cost per pressure measurement is C. Determine the relative relationship between P and C for levels of significance of 5 and 10 percent such that the mean pressure measured will be ±3 kPa.

Solution

For this problem the important parameter is the total cost of rejection and measurement for the two levels of significance. For the 5 percent level we have $z = 1.96$, $\sigma = 6$ kPa, and $\Delta = 3$ kPa applied to Eq. (3.19) to give

$$\Delta = \frac{z\sigma}{\sqrt{n}} = 3 = \frac{(1.96)(6)}{\sqrt{n}}$$

and $n = 15.37$, or rounded off to

$$n = 16 \text{ measurements for 5 percent level of significance}$$

For the 10 percent significance level we have $z = 1.65$, $\sigma = 6$ kPa, and $\Delta = 3$ kPa

so that $$3 = \frac{(1.65)(6)}{\sqrt{n}}$$

and $n = 10.89$, or rounded off to

$$n = 11 \text{ measurements for 10 percent level of significance}$$

The cost of the measurements is nC and the cost of rejection is Px (significance level expressed as a decimal).

The total cost T for each case is

$$T \text{ (5 percent level)} = 0.05P + 16C$$
$$T \text{ (10 percent level)} = 0.1P + 11C$$

If the total costs are to be the same for the 5 and 10 percent levels, we would have

$$0.05P + 16C = 0.1P + 11C$$

and
$$P = 100C$$

---

t-TEST COMPARISON OF DIFFERENT SAMPLES

The *t*-test may also be used to compare two samples to determine if significant variations exist. In this case t is calculated from

$$t = \frac{x_{m1} - x_{m2}}{\left[\dfrac{\sigma_1^2}{n_1} + \dfrac{\sigma_2^2}{n_2} \right]^{1/2}} \qquad \text{[3.51]}$$

where the subscripts 1 and 2 refer to the two sets of data. The degrees of freedom for the two samples are approximated by

$$\nu = \frac{\left[\sigma_1^2/n_1 + \sigma_2^2/n_2 \right]^2}{\dfrac{\left(\sigma_1^2/n_1 \right)^2}{n_1 - 1} + \dfrac{\left(\sigma_2/n_2 \right)^2}{n_2 - 1}} \qquad \text{[3.52]}$$

where ν is rounded down to an integer [13]. The procedure for comparing the samples is as follows:

1. The values of x_m, σ, and n are determined for each sample.
2. ν is estimated from Eq. (3.52).
3. t is calculated from Eq. (3.51).
4. A level of significance is selected for comparison of the samples. (This may be determined from 1 percent confidence level.)
5. A value of t is determined from Table 3.7 for the calculated value of ν in step 2 but a significance level of *one-half* the value in step 4.
6. If the value of t calculated in step 3 falls within the range of the value obtained in step 5, the two samples may be assumed to be statistically the same, within the confidence level selected in step 4.

COMPARISON OF TWO SAMPLES.   In a production process a metal part is measured in a sampling process with the following results: **Example 3.28**

$$x_{m_1} = 2.84 \text{ cm} \qquad \sigma_1 = 0.05 \text{ cm} \qquad n_1 = 12$$

A week later another set of measurements is made with the following results:

$$x_{m_2} = 2.86 \text{ cm} \qquad \sigma_2 = 0.03 \text{ cm} \qquad n_2 = 16$$

Do these tests indicate that the same results are obtained with a confidence level of 90 percent? (significance level = 10 percent)

Solution

Following the procedure outlined above, we first compute the degrees of freedom from Eq. (3.52):

$$\nu = \frac{[(0.05)^2/12 + (0.03)^2/16]^2}{\dfrac{[(0.05)^2/12]^2}{12 - 1} + \dfrac{[(0.03)^2/16]^2}{16 - 1}}$$

$$= 16.84$$

rounded down to 16.

We now enter Table 3.7 with $\nu = 16$ and a significance level of 5 percent (*one-half* of 10 percent) and obtain

$$t_{95} = 2.120 \qquad\qquad\qquad [a]$$

This figure is to be compared with that calculated from Eq. (3.47). We have

$$t = \frac{2.84 - 2.86}{\left[\dfrac{(0.05)^2}{12} + \dfrac{(0.03)^2}{16} \right]^{1/2}}$$

$$= -1.230$$

This value is less than that obtained in (*a*) above; so we may conclude that the two metal samples are the same within a confidence level of 90 percent.

3.16 GRAPHICAL ANALYSIS AND CURVE FITTING

Engineers are well known for their ability to plot many curves of experimental data and to extract all sorts of significant facts from these curves. The better one understands the physical phenomena involved in a certain experiment, the better one is able to extract a wide variety of information from graphical displays of experimental data. Because these physical phenomena may encompass all engineering science, we cannot discuss them here except to emphasize that the person who is usually most successful in analyzing experimental data is the one who understands the physical processes behind the data. Blind curve-plotting and cross-plotting usually generate an excess of displays, which are confusing not only to the management or supervisory personnel who must pass on the experiments, but sometimes even to the experimenter. To be blunt, the engineer should give considerable thought to the kind of information being looked for before even taking the graph paper out of the package, or activating the computer software.

Assuming that the engineer knows what is to be examined with graphical presentations, the plots may be carefully prepared and checked against appropriate theories. Frequently, a *correlation* of the experimental data is desired in terms of an analytical expression between variables that were measured in the experiment. When the data may be approximated by a straight line, the analytical relation is easy to obtain; but when almost any other functional variation is present, difficulties are usually encountered. This fact is easy to understand since a straight line is easily recognizable on a graph, whereas the functional form of a curve is rather doubtful. The curve could be a polynomial, exponential, or complicated logarithmic function and still present roughly the same appearance to the eye. It is most convenient, then, to try to plot the data in such a form that a straight line will be obtained for certain types of functional relationships. If the experimenter has a good idea of the type of function that will represent the data, then the type of plot is easily selected. It is frequently possible to estimate the functional form that the data will take on the basis of theoretical considerations and the results of previous experiments of a similar nature.

Table 3.8 summarizes several different types of functions and plotting methods that may be used to produce straight lines on the graph. The graphical measurements which may be made to determine the various constants are also shown. It may be remarked that the method of least squares may be applied to all these relations to obtain the best straight line to fit the experimental data. A number of computer software packages are available to accomplish the functional plots illustrated in Table 3.8. See, for example, Refs. [15], [16], [23], and [28].

Please note that when using logarithmic or semilog graph coordinates, it is unnecessary to make log calculations; the scaling of the coordinate automatically accomplishes this.

Incorporation of graphics in reports and presentations is discussed in Chap. 15.

3.17   CHOICE OF GRAPH FORMATS

The engineer has many graph formats available for presenting experimental data or calculation results. While bar charts, column charts, pie charts, and similar types of displays have some applications, by far the most frequently used display is the x-y graph with choices of coordinates to match the situation. This basic graph has several

Table 3.8   Methods of plotting various functions to obtain straight lines

Functional Relationship	Method of Plot	Graphical Determination of Parameters
$y = ax + b$	y vs. x on linear paper	Slope = a; b
$y = ax^b$	$\log y$ vs. $\log x$ on loglog paper	Slope = b; $\log a$; $\log x = 0$ or $x = 1.0$
$y = ae^{bx}$	$\log y$ vs. x on semilog paper	Slope = $b \log e$; $\log a$
$y = \dfrac{x}{a + bx}$	$\dfrac{1}{y}$ vs. $\dfrac{1}{x}$ on linear paper	Slope = a; Extrapolated; b
$y = a + bx + cx^2$	$\dfrac{y - y_1}{x - x_1}$ vs. x on linear paper	Slope = c; $b + cx_1$

(Continued)

Table 3.8 (Continued)

Functional Relationship	Method of Plot	Graphical Determination of Parameters
$y = \dfrac{x}{a + bx} + c$	$\dfrac{x - x_1}{y - y_1}$ vs. x on linear paper	
$y = ae^{bx+cx^2}$	$\log\left[\left(\dfrac{y}{y_1}\right)^{1/(x-x_1)}\right]$ vs. x on semilog paper	
$y = 1 - e^{-bx}$	$\log\left(\dfrac{1}{1-y}\right)$ vs. x on semilog paper	
$y = a + \dfrac{b}{x}$	y vs. $\dfrac{1}{x}$ on linear paper	
$y = a + b\sqrt{x}$	y vs. $\sqrt{x}$ on linear paper	

variations in format that we shall illustrate by plotting the simple table of x-y data shown below.

x	y
1	2
2	3.1
3	12
4	18
5	20
6	37
7	51
8	70
9	82
10	90

Six formats for plotting the data are shown in Fig. 3.9a through f. The choice of format depends on both the source and type of data as well as the eventual use to be made of the display. The following paragraphs discuss the six alternatives. The computer graphics were generated in Microsoft Excel.

$a.$ This display presents just the raw data points with a data marker for each point. It might be selected as an initial type of display before deciding on a more suitable alternative. It may be employed for either raw experimental data points or for points calculated from an analytical relationship. With computer graphics a wide selection of data marker styles is available.

$b.$ This display presents the points with the same data markers connected by a smooth curve drawn either by hand or by a computer graphics system; in this case, by computer. This display should be used with caution. If employed for presentation of experimental data, it implies that the smooth curve describes the physical phenomena represented by the data points. The engineer may want to avoid such an implication and choose not to use this format.

$c.$ This display is the same as (b) but with the data markers removed. It would almost *never* be employed for presentation of experimental data because the actual data points are not displayed. It also has the same disadvantage as (b) in the implication that the physical phenomena are represented by the smooth connecting curve. In contrast, this type of display is obviously quite suitable for presenting the results of calculations. The calculated points could be designated with data markers as in (b) or left off as in (c). The computer-generated curve offers the advantage of a smooth curve with a minimum number of calculated points.

$d.$ This display presents the data points connected with straight-line segments instead of a smooth curve, and avoids the implication that the physical situation behaves in a certain "smooth" fashion. The plot is typically employed for calibration curves where linear interpolation will be used between points, or when a numerical integration is to be performed based on the connecting straight-line segments. If used for presentation of experimental data, the implication is the same as in (b) and (c) that the physical system actually behaves as indicated, in this case with a somewhat jerky pattern.

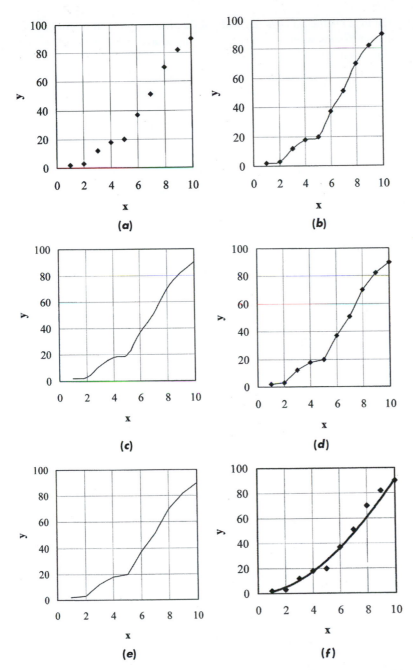

Figure 3.9 Choices of x-y graph formats. Points plotted (a) with data markers, but without connecting line segments, (b) with data markers joined by a smooth curve drawn by computer, (c) without data markers joined by smooth curve drawn by computer, (d) with data markers joined by straight-line segments, (e) without data markers joined by straight-line segments, (f) with data markers and correlation curve generated by computer.

e. The format in (e) is the same as (d) without the data markers. It might be used for calculation results where the engineer wants to avoid computer smoothing between the calculated points.

f. Finally, the format presented in (f) is one that is frequently selected to present experimental results where uncertainties in the measurements are expected to result in scatter of the data points. A smooth curve is drawn through the data points as the experimentalist's best estimate of the behavior of the phenomena under study. The smooth curve may be drawn by hand or generated through a least-squares process executed by the computer. A trendline equation may or may not be displayed along with the curve. When experimental uncertainties are expected to contribute significantly to the scatter of data, as they do in many cases, a full discussion of their nature should be offered in the accompanying narrative material. Examples 3.29 to 3.32 discuss the generation and display of correlation trendlines.

In some cases a display like that shown in (f) may be used to present calculated points along with a particular type of curve fit for the points which differs from the calculation equation. The cubic polynomial fit shown in case (i) of Example 3.30 illustrates such an application.

Example 3.29 | **CORRELATION OF DATA WITH POWER RELATION.**   The data for a series of experiments are shown in the table below.

x	y
70700	175
66600	168
51000	138
23000	81
30900	99
36242	109
52331	142
70000	168
84100	191
21810	76
33217	117
44000	144
56700	146
63000	157
87276	197
82200	201
83092	192
93800	208

In this case, the data were collected in two sets so the tabular listing is not in ascending values of x. Because of the nature of the physical problem the data are expected to correlate in terms of a power relation

$$y = ax^b \qquad \text{[a]}$$

A computer will be used to plot the data and obtain the values of the constants a and b. If the data are plotted *sequentially* on a point-to-point linear graph, the result shown in Fig. Example 3.29a is obtained. The jagged nature of the lines results from data scatter and the fact that the data are not tabulated with continuously increasing values of x. Obviously, such a graph is inappropriate.

Taking the logarithm of both sides of Eq. (*a*) gives

$$\log y = \log a + b \log x \qquad\qquad [b]$$

which suggests that a straight line will be obtained when the data are plotted on logarithmic coordinates. This arrangement is noted as the second entry of Table 3.8. When a loglog plot is executed, the result is as shown in Fig. Example 3.29b. In this case, the plots were generated in Microsoft Excel but may be accomplished with other software as well [15], [16], and [28].

A least-squares fit of the data to a power relation is given as the trendline and equation shown in Fig. Example 3.29b, along with the corresponding value of r^2 calculated from Eq. (3.40). The trendline and value of r^2 are also calculated by Excel but may be computed with other programs. The value of $r^2 = 0.9778$ indicates a good correlation.

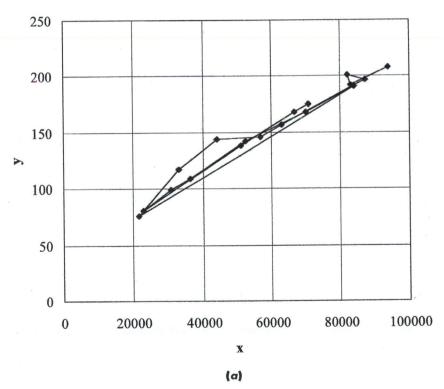

(a)

Figure Example 3.29   (a) Data plotted sequentially on linear coordinates. (b) Data plotted on loglog coordinates and computer least-squares fit.

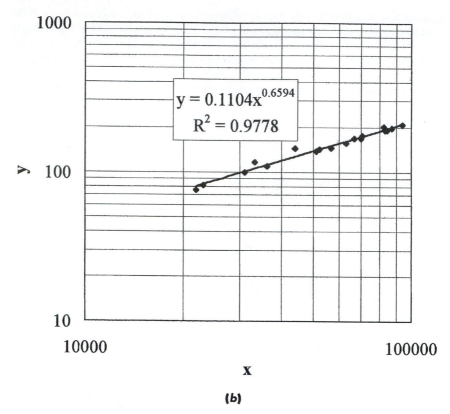

Figure Example 3.29   *(Continued)*

    Note that the line in Fig. Example 3.29*b* does indeed fit the data. A knowledgeable engineering person would obtain essentially the same result with a hand-drawn correlation line on graph paper. With the computer, it is not necessary to make the graphical determinations indicated in Table 3.8. Nevertheless, we emphasize once again that a computer-generated correlation should *never* be accepted without visual confirmation using a graphical display like that indicated above.

Example 3.30 | **ALTERNATIVE DISPLAYS AND CORRELATION TRENDLINES FOR EXPONENTIAL FUNCTION.**   This example illustrates different ways of graphing and obtaining least-squares correlations for data as applied to a calculated exponential function. First, the value of the function

$$y = 2.5e^{-0.2x}$$

is calculated for a number of values of x from 1 to 20 as shown in the accompanying table.

x	$y = 2.5 \exp(-0.2x)$
1	2.046826883
2	1.675800115
3	1.37202909
4	1.12332241
5	0.919698603
6	0.75298553
7	0.61649241
8	0.504741295
9	0.413247221
10	0.338338208
13	0.185683946
17	0.083433175
20	0.045789097

We are dealing with a known exponential function, so some of the correlations that will be examined are obviously designed for illustrative purposes only. In each case where a correlation trendline is presented the corresponding equation will be displayed on the graph along with a value of r^2 calculated from Eq. (3.40). The closer the value of r^2 to unity, the better the correlation. The graphs and correlation trendlines have been generated in Microsoft Excel but could be obtained from other software packages as well [15], [16], and [28]. The following comments apply to the figures noted.

a. The calculated values of the exponential relation are displayed with data markers, but without connecting line segments.

b. The calculated values of the exponential relation are plotted with data markers along with connecting line segments.

c. A smooth curve is plotted through the points with data markers omitted.

d. The data points are displayed without connecting line segments along with a least-squares *linear* fit to the data points. The linear relation obviously does not work, and is evidenced by a low value of r^2.

e. As suggested from the third entry of Table 3.8, the exponential relation should plot as a straight line on semilog coordinates. This figure gives such a display along with a least-squares fit to an exponential relation and the corresponding value of r^2 calculated from Eq. (3.40). A perfect fit is obtained, as should be expected from the exact calculations of the exponential function. It should be noted that the line drawn through the data markers is the correlation line, and *not* a line connecting the points. Note that $r^2 = 1.0$.

f. An exponential relation is again fitted to the data with a least-squares analysis but this time on a linear plot. Again, a perfect fit is obtained. We note from this chart that the least-squares calculation is independent of the type of coordinate system employed for the display.

g. This plot displays an attempt at a least-squares fit to a power relation like that in the second entry of Table 3.8, along with the plot using linear coordinates. Note the poor fit and low value of r^2.

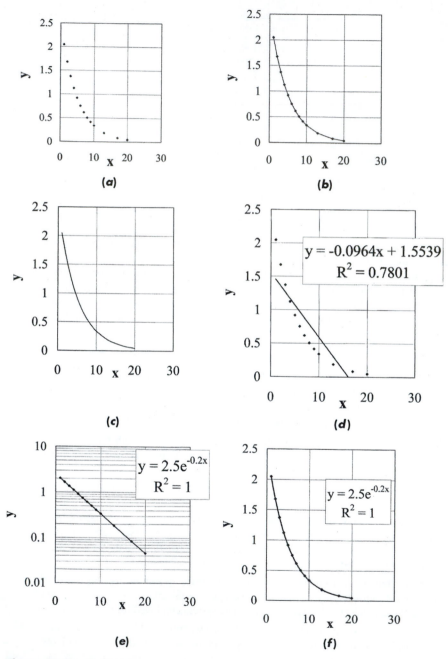

Figure Example 3.30

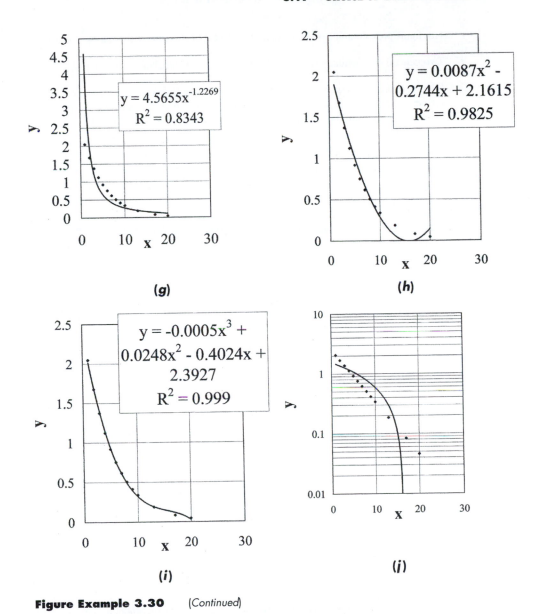

(g)

$$y = 4.5655x^{-1.2269}$$
$$R^2 = 0.8343$$

(h)

$$y = 0.0087x^2 - 0.2744x + 2.1615$$
$$R^2 = 0.9825$$

(i)

$$y = -0.0005x^3 + 0.0248x^2 - 0.4024x + 2.3927$$
$$R^2 = 0.999$$

(j)

Figure Example 3.30   (Continued)

h. This figure shows a second-order polynomial (quadratic) fit of the data on linear coordinates. The results are fairly good, except at the larger values of x. The least-squares analysis of this type of fit was described in Eqs. (3.33) to (3.35).

i. A cubic polynomial fit is performed in this figure with very good results. Note the almost perfect value of r^2. In this example we know the functional form of the data points, but with actual experimental results the functional form may be unknown. In such cases a polynomial fit is sometimes tried and frequently works very well.

j.   This figure presents another failed attempt at a linear fit, this time with semilog coordinates.

Two general conclusions may be made from the above calculations and displays:

1. A least-squares analysis of a set of data points is independent of the type of coordinate system used for the presentation, although the type of plot may suggest the functional form or correlation to be attempted. In this regard, the information of Table 3.8 can be quite helpful.

2. If one can anticipate the functional form of the data, the type of plot and presentation of a correlation trendline is simplified.

Example 3.31 | **EVOLUTION OF A CORRELATION USING COMPUTER GRAPHICS.**   To illustrate how a data correlation may evolve using computer graphics we consider the set of data shown below.

x	y	$y - 2$
1	2.2	0.2
3	2.6	0.6
5	2.7	0.7
10	2.65	0.65
30	3	1
50	3.1	1.1
100	3.15	1.15
500	3.7	1.7
1000	4.01	2.01
5000	4.96	2.96
10000	5.8	3.8
50000	7.7	5.7
100000	9.1	7.1

From the physical nature of the problem y is expected to behave according to

$$y = a + bx^c \qquad \text{[a]}$$

where a, b, and c are constants which must be determined from the experimental data. Normally, one would insist upon more data points than shown in the table, but we are considering an abbreviated set to keep the presentation simple.

We consider a sequence of graphics that may be used to correlate the data. Obviously, x takes on a wide range of values and the linear plot of y vs. x shown in Fig. Example 3.31a causes a compression of data markers for the lower range of x. Inspecting Eq. (a) we see that y will approach the value of the constant a for very small values of x. Thus, we should expect to be able to estimate the value of a by inspecting the behavior of the chart for small values of x; however, the compression in Fig. (a) makes that a very difficult task.

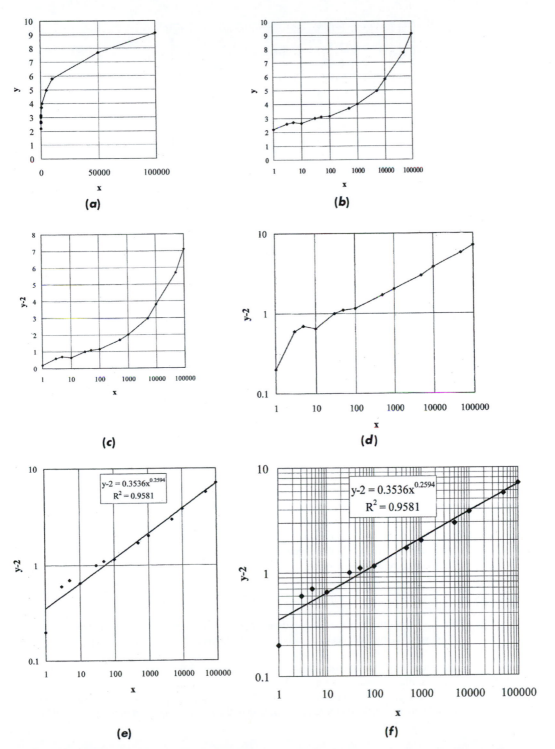

Figure Example 3.31 (a) to (f) Evolution of correlation. (g) Effect of elimination of first data point.

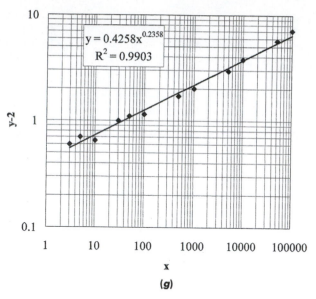

Figure Example 3.31   (Continued)

The situation is helped considerably by replotting the x axis with a logarithmic scale as shown in Fig. (b). Now, we see that y appears to be approaching a value of about 2 for very small values of x. We therefore add a "$y - 2$" column to the data table and replot the data as shown in Fig. (c). Equation (a) is now written as

$$y - 2 = bx^c \tag{b}$$

which should plot as a straight line when displayed as $\log(y - 2)$ vs. $\log x$. Such a display is shown in Fig. (d). We may suspect that the jagged or nonstraight line is the result of scatter in the experimental data, and thus a point-to-point graph is not appropriate. The point-to-point graph is dropped in Fig. (e), and a computer-generated trendline for a power relationship is calculated and displayed on the chart along with a value of r^2 calculated from Eq. (3.40). The final data correlation is therefore

$$y = 2 + 0.3536x^{0.2594} \tag{c}$$

Finally, the cosmetics of the presentation is improved with addition of major and minor gridlines along with enlarged data markers and a wider trendline. This results in the graph shown in (f).

The data point at $x = 1$, $y = 2.2$ appears out of line with the other points and suggests examination of the experimental setup which produced that point. Perhaps there was more experimental uncertainty for that point or a glitch in the measurements. If so, the point might be discarded, thereby improving the correlation. As mentioned earlier in the discussion, we would normally insist on more data points than the relatively small number presented here.

All of the graphical displays shown can be generated easily and displayed on a computer worksheet in a short period of time. If an examination of the experimental data indicates that the point at $x = 1$ can be eliminated, a new correlation can be calculated and displayed almost immediately. The result of such an elimination is shown in Fig. (g). Note the improved value of r^2.

CORRELATION TRENDLINES USING OFFSET POINTS. Several of the charts in Table 3.8 employ offsets from one of the data points, that is, $x - x_1$, $y - y_1$, etc., to linearize the presentation. The set (x_1, y_1) refers to *one* of the sets of data points. In this example we first consider the x-y data shown in the table below.

Example 3.32

x	y	$x - x_1$	$y - y_1$	$(y - y_1)/(x - x_1)$
1	3	−1	−4	4
2	7	0	0	
3	13	1	6	6
4	21	2	14	7
5	31	3	24	8
6	43	4	36	9
7	57	5	50	10
8	73	6	66	11
9	91	7	84	12

This is a contrived group of points which fits exactly to the relation

$$y = x^2 + x + 1 \qquad [a]$$

The data are plotted in Fig. Example 3.32a and a least-squares trendline fit is performed for a second-degree polynomial. The result is shown with a perfect value of $r^2 = 1$. The least-squares analysis is performed as described in Eqs. (3.33), (3.34), and (3.35) and executed here in Microsoft Excel.

Next, the data are replotted by selecting the second data point as the offset with

$$x_1 = 2 \qquad y_1 = 7$$

and $(y - y_1)/(x - x_1)$ versus x as suggested in Table 3.8 for a quadratic relation. The result is shown in Fig. (b) and the data plot as a straight line on this coordinate system. A linear least-squares fit is made to give

$$\frac{y - y_1}{x - x_1} = \frac{y - 7}{x - 2} = x + 3 \qquad [b]$$

Solving for y gives

$$y = x^2 + x + 1$$

the same relation as in Eq. (a).

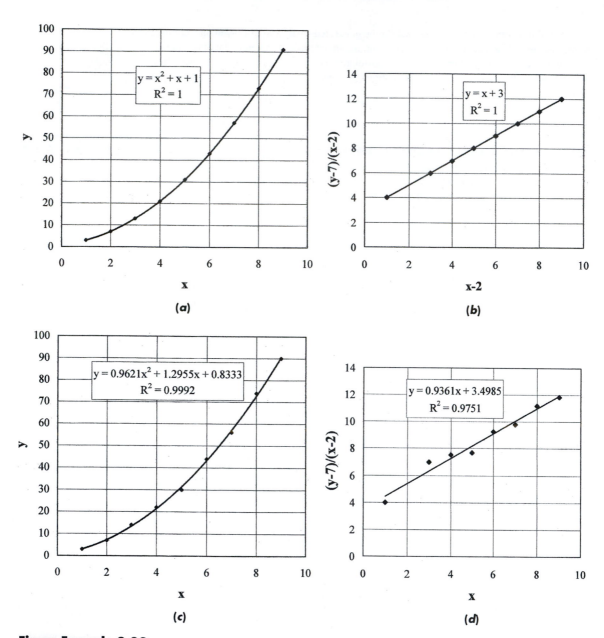

Figure Example 3.32

Now, we modify the data slightly as shown in the following table so that the points do not fit the relation of Eq. (a) exactly.

x	y	$x - x_1$	$y - y_1$	$(y - y_1)/(x - x_1)$
1	3	-1	-4	4
2	7	0	0	
3	14	1	7	7
4	22	2	15	7.5
5	30	3	23	7.666666667
6	44	4	37	9.25
7	56	5	49	9.8
8	74	6	67	11.16666667
9	90	7	83	11.85714286

These data are plotted in Fig. (c) and fitted with a second-degree polynomial trendline to give

$$y = 0.9621x^2 + 1.2955x + 0.8333 \qquad \textbf{[c]}$$

with a value of $r^2 = 0.9992$. As before, we replot the data with $x_1 = 2$ and $y_1 = 7$ as displayed in Fig. (d). A linear trendline fit is performed, giving the result

$$\frac{y - 7}{x - 2} = 0.9361x + 3.4985 \qquad \textbf{[d]}$$

with $r^2 = 0.9751$. Solving Eq. (d) for y,

$$y = 0.9361x^2 + 1.6263x + 3.003 \qquad \textbf{[e]}$$

or a somewhat different result from Eq. (c). The difference is the result of the two least-squares analyses; one for a second-degree polynomial, and the other on a linear basis. Equation (c) is more correct in that it is a direct fit to the quadratic relation.

Comment

The use of offset points for plotting data can be a convenient artifice for determining the functional relationship between data, just as loglog and semilog plots are useful for establishing power or exponential relations. Once the functional form is established, a direct least-squares regression may be the most reliable representation of the data. The computer graphics and least-squares fits are executed here in Microsoft Excel but could be generated with other software packages.

3.18   CAUSATION, CORRELATIONS, AND CURVE-FITS

Let us consider the first term, *causation*, or cause and effect. As a homely and almost absurd example, we might imagine that experiments are conducted in a large number of barnyards around the country and observations made of the behavior of roosters in the morning. Suppose that a large amount of data are collected that indicate roosters crow at about the time the sun rises. Suppose then that a statistical analysis is made of the data which indicate a very strong *correlation* between roosters crowing and the sun rising. Despite the strong correlation it would be grossly in error to conclude that the rooster *causes* the sun to rise.[3]

Not all experimental results would be as simple to evaluate as the rooster-sunrise hypothesis. Biomedical engineering applications are particularly interesting and sometimes difficult to interpret. In such applications graphical displays may not be as quantitative as in hardware-oriented experiments, and they are usually subject to more conjecture. We will give two biomedical examples to emphasize the concepts of correlation and causation further.

First, consider a set of observations and measurements that were conducted with Parkinson's disease (PD) patients [33]. These persons experience stiffness, involuntary motions, a characteristic tremor of hands, arms, and legs, speech problems, and in some cases deterioration in mental/cognitive capabilities. Over the years it has become possible to at least partially quantify these specific symptoms and to express the overall progress of PD. In the course of examining the specific symptoms and functional performance, investigators questioned whether some of the human performance measures were more indicative of overall progression of PD than others. After considerable study, investigators found that the behavior measures could be plotted as shown in Fig. 3.10. It was found that mental/cognitive performance was a primary factor in the overall evaluation of motor performance and that a threshold level exists which apparently governs the maximum level that can be achieved in the other human performance factors. In other words, the ability to achieve at some level of mental/cognitive function seen as a limiting resource is a *necessary* but not a sufficient condition in order to achieve at a minimum level of motor performance such as speed of movement, or coordination. Due to the nonlinear relationship of the data, the inverse is not true. High achievement of motor ability is not necessary for high mental/cognitive achievement. Assignment of any quantitative degree of causation other than a categorical one to this data presentation would be highly speculative. Still, the nonquantitative relationship can be of clinical use in evaluation of PD behavior and performance. *The important point is that the threshold effects show up quite clearly on a graphical display but would be difficult to discern from a corresponding tabular presentation.* The experimentalist involved with research efforts is encouraged to construct plots of a speculative nature in the hope that some new or unanticipated effect may be observed, even though they may not be quantifiable.

The second biomedical example we consider is concerned with human consumption of alcohol. There has been statistical evidence for some time that moderate consumption of alcohol results in a lower probability for heart attacks. But, as someone remarked, *statistics* does not *cause* or *prevent* heart attacks. In 1994, the results of extensive experimental evidence were presented in Ref. [34] that showed a measured increase in blood levels of t-plasminogen activator with an increase in consumption of alcoholic beverages. Such an increase has been shown to be associated with a reduction in the rate of myocardial infarction (heart attacks). In addition, convincing statistical evidence was given that excluded the possibility of smoking, blood

---

[3]Unless one wishes to go back in time for several centuries to some primitive culture where the crow of the rooster supposedly summons the God of Roosters to ask the Sun God to cause the sun to rise.

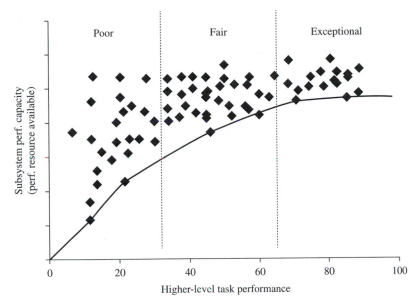

Figure 3.10   Typical scatter plot obtained when the measured performance capacity for a given subsystem drawn upon in a higher-level task is plotted against performance in the more complex task.

pressure, obesity, or other contributing factors influencing the final results. The important point here is that a statistical *correlation* was available for years, which many people believed reliable and therefore prompted appropriate action on their part. The follow-up with corroborating laboratory measurements linked to specific biochemical effects helped to remove remaining doubts. One could argue that a cause-and-effect result still has not been established, because it is not *known for sure* that increase in t-plasminogen activator levels reduces the risk of heart attacks. Such is the uncertainty in biomedical science, and the reader is left to self-speculate on the matter.

Sometimes it is difficult to distinguish between *correlations* and *curve-fitting*. Indeed, when one performs a least-squares analysis to obtain the best equation to fit a set of experimental data, a correlation is obtained that expresses the result in terms of the parameters that the experimentalist perceives (or knows) to be important. On the other hand, the curve-fitting or forced-fit of a specific type of function to provide for easy calculations, as illustrated in the polynomial expressions of Table 8.5, is strictly an artifice of convenience.

Suppose experiments are conducted to measure the heat transfer to boil water at atmospheric pressure in an open container. The actual boiling process is a complex phenomenon, but it would come as no surprise to find that for a constant pressure of 1 atm the boiling heat-transfer rate increases with an increase in the surface temperature in contact with the water. The heat flux per unit area could be plotted versus the temperature difference between the surface and boiling temperature of the water

(100°C at atmospheric pressure) and a curve-fit or correlation performed to obtain the specific function

$$\text{Heat flux} = f(\text{temperature difference})$$

The specific functional relation would be helpful to supply to other persons desiring to calculate boiling heat-transfer rates for water. The correlation is quite useful, although it does not explain the complex boiling phenomena involving bubble formation, collapse, and eventual dissipation in the body of the fluid. In this example, the terms *correlation* and *curve-fit* are almost synonymous, because the curve-fit is constructed from the original experimental data points. The term *causation* is only loosely applicable in this case because we certainly expect an increase in heat transfer with an increase in temperature difference, but other factors such as surface tension (which influences bubble size) and viscosity (which influences bubble motion) are also important.

In conclusion, the important point of this discussion is that the experimentalist should always be mindful of the difference between the terms *causation, correlation*, and *curve-fit* and not attach more meaning to a correlation of data than is justified. At the same time, one should not depreciate the utility that data correlations can afford, or possible simplification of calculations with use of curve-fits.

Example 3.33

MORE (COMPLEX) IS NOT ALWAYS BETTER   As a final note to our discussion of correlations we consider the set of data shown in Table Ex. 3.33(a), involving six data points. Let us construct least-squares fits to these data points using first- through fifth-order polynomials. The computer-generated correlations are shown in Fig. Ex. 3.33(a). The correlation equations and corresponding values of the correlation coefficient r^2 are shown, as well as plots of the correlation curves.

Clearly, the linear relation does not fit well, as illustrated by a visual inspection and the relatively low value of the correlation coefficient. The second-, third-, and fourth-degree

Table Example 3.33(a)

x	y	Linear	Fifth	Fourth	Third	Second
0	−1	−3.8571	−1	−0.8294	−0.9365	−0.4643
1	2	1.4858	2.0006	1.1467	1.4683	0.8072
2	3	6.8287	3.0024	4.706	4.4921	4.1145
3	11	12.1716	11.0012	9.2929	9.0793	9.4576
4	15	17.5145	14.984	15.8518	16.1743	16.8365
5	27	22.8574	26.925	26.8271	26.7215	26.2512
0.5		−1.18565	3.269653	−0.132569	0.24755	−0.083025
1.5		4.15725	1.216297	2.798331	2.8438	2.206375
2.5		9.50015	6.850391	6.847131	6.53125	6.531575
3.5		14.84305	13.78883	12.20823	12.2543	12.89258
4.5		20.18595	17.13773	20.57603	20.95735	21.28938

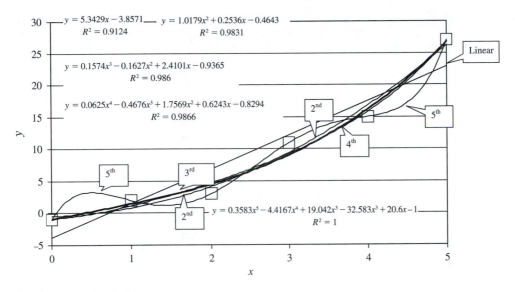

Figure Example 3.33(a)

curves are almost identical in their visual match and the value of the corresponding correlation coefficients. The fifth-order curve is problematical. It has a perfect correlation coefficient ($r^2 = 1.0$), passes through the data points exactly, but snakes through the other values with an oscillatory behavior. The values of the ordinate y calculated from the correlating equations at off-data points $x = 2.5, 3.5, 4.5$, etc., illustrate this behavior.

If one were to merely inspect the correlating equations and their corresponding correlation coefficients, it would be very easy to wrongfully select the fifth-order equation as the best relation. But, as we have mentioned several times before, *all correlations should be confirmed with a visual comparison with the data*. Thus, we reject the fifth-order correlation on the basis of this comparison. The linear relation is also easily discarded; but the second-, third-, and fourth-order relations appear equally satisfactory, in both a visual comparison and values of the correlation coefficient. Which one shall we choose? All selections should be so easy. Choose the simplest—the quadratic equation:

$$y = 1.0179x^2 + 0.2536x - 0.4643 \qquad \textbf{[a]}$$

Table Example 3.33(b)

x	y
0	0
1	1
2	4
3	9
4	16
5	25

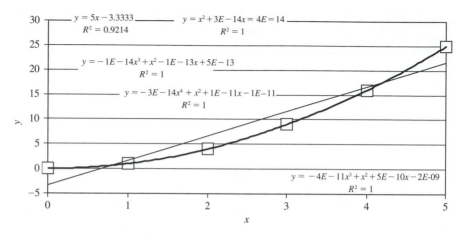

Figure Example 3.33(b)

Computer correlations are not always problematical. Sometimes they agree with the obvious observation. Consider the set of data shown in Table Ex. 3.33(b). A quick inspection of this set reveals the obvious result that the data follow the relation

$$y = x^2 \qquad\qquad [b]$$

exactly. When a least-squares computer correlation is attempted to fit first-, second-, third-, fourth-, and fifth-order formulas, the results are shown in Fig. Ex. 3.33(b). The linear relation fails again, but the other relations fit rather well. Note, however, that the coefficients of all but the x^2 term are very nearly equal to zero, and all the formulas reduce to Eq. (b). The coefficients differ slightly from zero as a result of round-off error in the computer calculations. In this simple example, the computer has given us the correct result, though in slightly disguised form.

3.19 GENERAL CONSIDERATIONS IN DATA ANALYSIS

Our discussions in this chapter have considered a variety of topics: statistical analysis, uncertainty analysis, curve plotting, least squares, among others. With these tools the reader is equipped to handle a variety of circumstances that may occur in experimental investigations. As a summary to this chapter let us now give an approximate outline of the manner in which one would go about analyzing a set of experimental data:

1. *Examine the data for consistency.* No matter how hard one tries, there will always be some data points that appear to be grossly in error. If we add heat to a container of water, the temperature must rise, and so if a particular data point indicates a *drop* in temperature for a heat *input,* that point might be eliminated. In other words, the data should follow commonsense consistency, and points that do not

appear proper should be eliminated. If very many data points fall in the category of "inconsistent," perhaps the entire experimental procedure should be investigated for gross mistakes or miscalculation.

2. *Perform a statistical analysis of data where appropriate.* A statistical analysis is only appropriate when measurements are repeated several times. If this is the case, make estimates of such parameters as standard deviation, and so forth. In those cases where the uncertainty of the data is to be prescribed by statistical analysis, a calculation should be performed using the *t*-distribution. This may be used to determine levels of confidence and levels of significance. The number of measurements to be performed may be determined for different levels of confidence.

3. *Estimate the uncertainties in the results.* We have discussed uncertainties at length. Hopefully, these calculations will have been performed in advance, and the investigator will already know the influence of different variables by the time the final results are obtained.

4. *Anticipate the results from theory.* Before trying to obtain correlations of the experimental data, the investigator should carefully review the theory appropriate to the subject and try to glean some information that will indicate the trends the results may take. Important dimensionless groups, pertinent functional relations, and other information may lead to a fruitful interpretation of the data. This step is particularly important in determining the graphical form(s) to be selected for presentation of data. If, for example, measurements are made of a first-order system subjected to a step input, we would expect exponential behavior in accordance with Eq. (2.11) and a semilog plot like that shown in the third or eighth entries of Table 3.8 should be selected for data presentation.

5. *Correlate the data.* The word "correlate" is subject to misinterpretation. In the context here we mean that the experimental investigator should make sense of the data in terms of physical theories or on the basis of previous experimental work in the field. Certainly, the results of the experiments should be analyzed to show how they conform to or differ from previous investigations or standards that may be employed for such measurements.

3.20 SUMMARY

By now the reader will have sensed the central theme of this chapter as that of uncertainty analysis and the use of this analysis to influence experiment design, instrument selection, and evaluation of the results of experiments. At this point we must reiterate statements we have made before. We still must recognize that uncertainty is *not* the same as error, even though some people interchange the terms. As we saw in Chap. 2, the determination of "error" is eventually related to a comparison with a standard. Even then, there is still "uncertainty" in the error because the "standard" has its own uncertainty.

In the chapters that follow we shall examine a large number of instruments and measurement devices and shall see how the concepts of error, uncertainty, and calibration apply.

3.21 REVIEW QUESTIONS

3.1. How does an error differ from an uncertainty?

3.2. What is a fixed error; random error?

3.3. Define standard deviation and variance.

3.4. In the normal error distribution, what does $P(x)$ represent?

3.5. What is meant by measure of precision?

3.6. What is Chauvenet's criterion and how is it applied?

3.7. What are some purposes of uncertainty analyses?

3.8. Why is an uncertainty analysis important in the preliminary stages of experiment planning?

3.9. How can an uncertainty analysis help to reduce overall experimental uncertainty?

3.10. What is meant by standard deviation of the mean?

3.11. What is a least-squares analysis?

3.12. What is the correlation coefficient?

3.13. What is meant by a regression analysis?

3.14. What is meant by level of significance; level of confidence?

3.15. How can statistical analysis be used to estimate experimental uncertainty?

3.16. How is Student's t-distribution used?

3.17. How is the chi-square test used?

3.18. How can statistical analysis be used to determine the number of measurements needed for a required level of confidence?

3.19. What is the coefficient of determination?

3.20. Why should one always make a graphical plot of data?

3.21. What does it mean when the correlation coefficient is 1.0?

3.22 PROBLEMS

Note: Problems marked with [C] may be worked with any appropriate computer software available to the reader. In some cases, graphing calculators with printout capabilities may also be employed.

3.1. The resistance of a resistor is measured 10 times, and the values determined are 100.0, 100.9, 99.3, 99.9, 100.1, 100.2, 99.9, 100.1, 100.0, and 100.5. Calculate the uncertainty in the resistance.

3.2. A certain resistor draws 110.2 V and 5.3 A. The uncertainties in the measurements are ± 0.2 V and ± 0.06 A, respectively. Calculate the power dissipated in the resistor and the uncertainty in the power.

3.3. A small plot of land has measured dimensions of 50.0 by 150.0 ft. The uncertainty in the 50-ft dimension is ± 0.01 ft. Calculate the uncertainty with which the 150-ft dimension must be measured to ensure that the total uncertainty in the area is not greater than 150 percent of that value it would have if the 150-ft dimension were exact.

3.4. Two resistors R_1 and R_2 are connected in series and parallel. The values of the resistances are

$$R_1 = 100.0 \pm 0.1 \; \Omega$$
$$R_2 = 50.0 \pm 0.03 \; \Omega$$

Calculate the uncertainty in the combined resistance for both the series and the parallel arrangements.

3.5. A resistance arrangement of 50 Ω is desired. Two resistances of 100.0 ± 0.1 Ω and two resistances of 25.0 ± 0.02 Ω are available. Which should be used, a series arrangement with the 25-Ω resistors or a parallel arrangement with the 100-Ω resistors? Calculate the uncertainty for each arrangement.

[C] 3.6. The following data are taken from a certain heat-transfer test. The expected correlation equation is $y = ax^b$. Plot the data in an appropriate manner and use the method of least squares to obtain the best correlation.

x	2040	2580	2980	3220	3870	1690	2130	2420	2900	3310	1020	1240	1360	1710	2070
y	33.2	32.0	42.7	57.8	126.0	17.4	21.4	27.8	52.1	43.1	18.8	19.2	15.1	12.9	78.5

Calculate the mean deviation of these data from the best correlation.

3.7. A horseshoes player stands 30 ft from the target. The results of the tosses are:

Toss	Deviation from Target, ft	Toss	Deviation from Target, ft
1	0	6	+2.4
2	+3	7	-2.6
3	-4.2	8	+3.5
4	0	9	+2.7
5	+1.5	10	0

On the basis of these data, would you say that this is a good player or a poor player? What advice would you give this player in regard to improving at the game?

3.8. Calculate the probability of drawing a full house (three of a kind and two of a kind) in the first 5 cards from a 52-card deck.

3.9. Calculate the probability of filling an inside straight with one draw from the remaining 48 cards of a 52-card deck.

3.10. A voltmeter is used to measure a known voltage of 100 V. Forty percent of the readings are within 0.5 V of the true value. Estimate the standard deviation for the meter. What is the probability of an error of 0.75 V?

3.11. In a certain mathematics course the instructor informs the class that grades will be distributed according to the following scale provided that the average class score is 75:

Grade	A	B	C	D	F
Score	90–100	80–90	70–80	60–70	Below 60

Estimate the percentage distribution of grades for 5, 10, and 15 percent failing. Assume that there are just as many As as Fs.

[C] 3.12. For the following data points y is expected to be a quadratic function of x. Obtain this quadratic function by means of a graphical plot and also by the method of least squares:

x	1	2	3	4	5
y	1.9	9.3	21.5	42.0	115.7

3.13. It is suspected that the rejection rate for a plastic-cup-molding machine is dependent on the temperature at which the cups are molded. A series of short tests is conducted to examine this hypothesis with the following results:

Temperature	Total Production	Number Rejected
T_1	150	12
T_2	75	8
T_3	120	10
T_4	200	13

On the basis of these data, do you agree with the hypothesis?

3.14. A capacitor discharges through a resistor according to the relation $E/E_0 = e^{-t/RC}$ where E_0 = voltage at time zero, R = resistance, C = capacitance. The value of the capacitance is to be measured by recording the time necessary for the voltage to drop to a value E_1. Assuming that the resistance is known accurately, derive an expression for the percent uncertainty in the capacitance as a function of the uncertainty in the measurements of E_1 and t.

3.15. In heat-exchanger applications a log mean temperature is defined by

$$\Delta T_m = \frac{(T_{h_1} - T_{c_1}) - (T_{h_2} - T_{c_2})}{\ln\left[(T_{h_1} - T_{c_1})/(T_{h_2} - T_{c_2})\right]}$$

where the four temperatures are measured at appropriate inlet and outlet conditions for the heat-exchanger fluids. Assuming that all four temperatures are measured with the same absolute uncertainty w_T, derive an expression for the percentage uncertainty in ΔT_m in terms of the four temperatures and the value of w_T. Recall that the percentage uncertainty is

$$\frac{w_{\Delta T_m}}{\Delta T_m} \times 100$$

[C] 3.16. A certain length measurement is made with the following results:

Reading	1	2	3	4	5	6	7	8	9	10
x, in	49.36	50.12	48.98	49.24	49.26	50.56	49.18	49.89	49.33	49.39

Calculate the standard deviation, the mean reading, and the uncertainty. Apply Chauvenet's criterion as needed.

3.17. Devise a method for plotting the gaussian normal error distribution such that a straight line will result. (*Ans.* $(1/\eta) \ln \left[\sqrt{2\pi} P(\eta)\right]$ versus η.) Show how such a plot may be labeled so that it can be used to estimate the fraction of points which lie below a certain value of η. Subsequently, show that this plot may be used to investigate the normality of a set of data points. Apply this reasoning to the data points of Example 3.7 and Probs. 3.6 and 3.7.

3.18. A citizens' traffic committee decides to conduct its own survey and analysis of the influence of drinking on car accidents. By some judicious estimates the committee determines that in their community 30 percent of the drivers on a Saturday evening between 10 P.M. and 2 A.M. have consumed some alcohol. During this same period there were 50 accidents, varying from minor scratched fenders to fatalities. In these 50 accidents 50 of the drivers had had something to drink (there are 100 drivers for 50 accidents). From these data, what conclusions do you draw about the influence of drinking on car accidents? Can you devise a better way to perform this analysis?

3.19. The grades for a certain class fall in the following ranges:

Number	10	30	50	40	10	8
Score	90–100	80–90	70–80	60–70	50–60	Below 50

The arithmetic mean grade is 68. Devise your own grade distribution for this class. Be sure to establish the criteria for the distribution.

3.20. A certain length measurement is performed 100 times. The arithmetic mean reading is 6.823 ft, and the standard deviation is 0.01 ft. How many readings fall within (a) ±0.005 ft, (b) ±0.02 ft, (c) ±0.05 ft, and (d) 0.001 ft of the mean value?

3.21. A series of calibration tests is conducted on a pressure gage. At a known pressure of 1000 psia it is found that 30 percent of the readings are within 1 psia of the true value. At a known pressure of 500 psia, 40 percent of the readings are within 1 psia. At a pressure of 200 psia, 45 percent of the readings are within 1 psia. What conclusions do you draw from these readings? Can you estimate a standard deviation for the pressure gage?

3.22. Two resistors are connected in series and have the following values:

$$R_1 = 10,000 \ \Omega \pm 5\% \qquad R_2 = 1 \ M\Omega \pm 10\%$$

Calculate the percent uncertainty for the series total resistance.

3.23. Apply Chauvenet's criterion to the data of Example 3.14 and then replot the data on probability paper, omitting any excluded points.

3.24. Plot the data of Example 3.7 on probability paper. Replot the data, taking into account the point eliminated in Example 3.13. Comment on the normality of these two sets of data.

3.25. Two groups of secretaries operate under the same manager. Both groups have the same number of people, use the same equipment, and turn out about the same amount of work. During one maintenance period group A had 10 service calls on the equipment, while group B had only 6 calls. From these data, would you conclude that group A was harder on the equipment?

[C] 3.26. A laboratory experiment is conducted to measure the viscosity of a certain oil. A series of tests gives the values as 0.040, 0.041, 0.041, 0.042, 0.039, 0.040, 0.043, 0.041, and 0.039 ft^2/s. Calculate the mean reading, the variance, and the standard deviation. Eliminate any data points as necessary.

[C] 3.27. The following data are expected to follow a linear relation of the form $y = ax + b$. Obtain the best linear relation in accordance with a least-squares analysis. Calculate the standard deviation of the data from the predicted straight-line relation:

x	0.9	2.3	3.3	4.5	5.7	6.7
y	1.1	1.6	2.6	3.2	4.0	5.0

[C] 3.28. The following data points are expected to follow a functional variation of $y = ax^b$. Obtain the values of a and b from a graphical analysis:

x	1.21	1.35	2.40	2.75	4.50	5.1	7.1	8.1
y	1.20	1.82	5.0	8.80	19.5	32.5	55.0	80.0

[C] 3.29. The following data points are expected to follow a functional variation of $y = ae^{bx}$. Obtain the values of a and b from a graphical analysis:

x	0	0.43	1.25	1.40	2.60	2.9	4.3
y	9.4	7.1	5.35	4.20	2.60	1.95	1.15

[C] 3.30. The following heat-transfer data points are expected to follow a functional form of $N = aR^b$. Obtain the values of a and b from a graphical analysis and also by the method of least squares:

R	12	20	30	40	100	300	400	1000	3000
N	2	2.5	3	3.3	5.3	10	11	17	30

What is the average deviation of the points from the correlating relationship?

[C] 3.31. In a student laboratory experiment a measurement is made of a certain resistance by different students. The values obtained were:

Reading	1	2	3	4	5	6	7	8	9	10	11	12
Resistance, kΩ	12.0	12.1	12.5	11.8	13.6	11.9	12.2	11.9	12.0	12.3	12.1	11.85

Calculate the standard deviation, the mean reading, and the uncertainty.

3.32. In a certain decade resistance box resistors are arranged so that four resistances may be connected in series to obtain a desired result. The first selector uses 10 resistances of 1000, 2000, ..., 9000, the second uses 10 of 100, 200, ..., 900, the third uses 10 of 20, ..., 90, and the fourth, 1, 2, ..., 9 Ω. Thus, the overall range is 0 to 9999 Ω. If all the resistors have an uncertainty of ±1.0 percent, calculate the percent uncertainties for total resistances of 9, 56, 148, 1252, and 9999 Ω.

3.33. Calculate the chances and probabilities that data following a normal distribution curve will fall within 0.2, 1.2, and 2.2 standard deviations of the mean value.

3.34. Suggest improvements in the measurement uncertainties for Example 3.5 which will result in reduction in the overall uncertainty of flow measurement to ±1.0 percent.

3.35. What uncertainty in the resistance for the first part of Example 3.3 is necessary to produce the same uncertainty in power determination as results from the current and voltage measurements?

3.36. Use the technique of Sec. 3.5 with Example 3.5.

3.37. Use the technique of Sec. 3.5 with Examples 3.4 and 3.3.

[C] **3.38.** Obtain the correlation coefficient for Prob. 3.27.

[C] **3.39.** Obtain the correlation coefficient for Prob. 3.28.

[C] **3.40.** Obtain the correlation coefficient for Probs. 3.29 and 3.30.

[C] **3.41.** Obtain the correlation coefficient for Probs. 3.6 and 3.12.

3.42. For the heat exchanger of Prob. 3.15 the temperatures are measured as $T_{h_1} = 100°C$, $T_{h_2} = 80°C$, $T_{c_1} = 75°C$, and $T_{c_2} = 55°C$. All temperatures have an uncertainty of $\pm 1°C$. Calculate the uncertainty in ΔT_m using the technique of Sec. 3.5.

3.43. Repeat Prob. 3.42 but with $T_{c_1} = 90°C$ and $T_{c_2} = 70°C$.

3.44. Four resistors having nominal values of 1, 1.5, 3, and 2.5 kΩ are connected in parallel. The uncertainties are ± 10 percent. A voltage of 100 V $\pm$ 1.0 V is impressed on the combination. Calculate the power drawn and its uncertainty. Use Sec. 3.5.

3.45. A radar speed-measurement device for state police is said to have an uncertainty of ± 4 percent when directed straight at an oncoming vehicle. When directed at some angle θ from the straight-on position, the device measures a component of the vehicle speed. The police officer can only obtain a value for the angle $\tilde{\theta}$ through a visual observation having an uncertainty of $\pm 10°$. Calculate the uncertainty of the speed measurement for θ values of 0, 10, 20, 30, and 45°. Use the techniques of both Secs. 3.4 and 3.5.

3.46. An automobile is to be tested for its acceleration performance and fuel economy. Plan this project taking into account the measurements which must be performed and expected uncertainties in these measurements. Assume that three different drivers will be used for the tests. Make plans for the number of runs which will be used to reduce the data. Also, prepare a detailed outline with regard to form and content of the report which will be used to present the results.

3.47. A thermocouple is used to measure the temperature of a known standard maintained at 100°C. After converting the electrical signal to temperature the readings are: 101.1, 99.8, 99.9, 100.2, 100.5, 99.6, 100.9, 99.7, 100.1, and 100.3. Using whatever criteria seem appropriate, make some statements about the calibration of the thermocouple.

3.48. Seven students are asked to make a measurement of the thickness of a steel block with a micrometer. The actual thickness of the block is known very accurately as 2.000 cm. The seven measurements are: 2.002, 2.001, 1.999, 1.997, 1.998, 2.003, and 2.003 cm. Comment on these measurements using whatever criteria you think appropriate.

3.49. A collection of 120 rock aggregate samples is taken and the volumes are measured for each. The mean volume is 6.8 cm^3 and the standard deviation is

0.7 cm^3. How many rocks would you expect to have volumes ranging from 6.5 to 7.2 cm^3?

[C] **3.50.** Plot the equation $y = 5e^{1.2x}$ on semilog paper. Arbitrarily assign fictitious data points on both sides of the line so that the line appears by eye as a reasonable representation. Then, using these points, perform a least-squares analysis to obtain the best fit to the points. What do you conclude from this comparison?

[C] **3.51.** The following data are presumed to follow the relation $y = ax^b$. Plot the values of x and y on loglog graph paper and draw a straight line through the points. Subsequently, obtain the values of a and b. Then, determine the values of a and b by the method of least squares. Compute the standard deviation for both cases. If a packaged computer routine for the least-squares analysis is available, use it:

x	y
4	105
5.3	155
11	320
21	580
30	1050
50	1900

3.52. The variables x and y are related by the quadratic equation

$$y = 2 - 0.3x + 0.01x^2$$

for $0 < x < 2$. Compute the percentage uncertainty in y for uncertainties in x of ± 1, 2, and 3 percent. Use both an analytical technique and the numerical technique discussed in Sec. 3.5.

3.53. For the relation given in Prob. 3.52, consider y as the primary variable with uncertainties of ± 1, 2, and 3 percent. On this basis, compute the resulting uncertainties in x. Use both the analytical and numerical techniques.

3.54. Reynolds numbers for pipe flow may be expressed as

$$\mathrm{Re} = \frac{4\dot{m}/\pi d}{\mu}$$

where

$\dot{m}$ = mass flow, kg/s
d = pipe diameter, m
μ = viscosity, kg/m · s

In a certain system the flow rate is 12 lbm/min, ± 0.5 percent, through a 0.5-in-diameter (± 0.005-in) pipe. The viscosity is 4.64×10^{-4} lbm/h · ft,

±1 percent. Calculate the value of the Reynolds number and its uncertainty. Use both the analytical and numerical techniques.

3.55. The specific heat of a gas at constant volume is measured by determining the temperature rise resulting from a known electrical heat input to a fixed mass and volume. Then,

$$P = El = mc_v \, \Delta T = mc_v \, (T_2 - T_1)$$

where the mass is calculated from the ideal-gas law and the volume, that is,

$$m = \frac{p_1 V}{RT_1}$$

Suppose the gas is air with $R = 287$ J/kg · K and $c_v = 0.714$ kJ/kg · °C, and the measurements are to be performed on a 1-liter volume (known accurately) starting at $p_1 = 150$ kPa and $T_1 = 30$°C. Determine suitable power and temperature requirements, assign some uncertainties to the measured variables, and estimate the uncertainty in the value of specific heat determined.

3.56. A model race car is placed on a tethered circular track having a diameter of 10 m ± 1 cm. The speed of the car is determined by measuring the time required for traveling each lap. A handheld stopwatch is used for the measurement, and the estimated uncertainty in *both* starting and stopping the watch is ±0.2 s. For a nominal speed of 100 mi/h, calculate the uncertainty in the speed measurement when made over 1, 2, 3, and 4 laps.

[C] 3.57. In a cooling experiment the system is presumed to behave as a first-order system following a relation like

$$y = Ce^{-at}$$

The following data points are collected:

y	t
0.9	0.1
0.8	0.5
0.4	0.9
0.3	1.2
0.2	1.7
0.1	2.3
0.01	4.6

Plot the data on an appropriate graph to obtain a straight line. Then perform a least-squares analysis to obtain the best values of C and a. Calculate the correlation coefficient for the least-squares fit.

[C] **3.58.** A certain experiment is presumed to behave according to the following equation:

$$y = a + cx^2$$

for $-1 < y < +1$ and $-1 < x < +1$. The following data points are collected:

y	x
0.01	−1.0
0.35	−0.8
0.65	−0.6
0.82	−0.4
−0.01	0
0.83	0.4
0.64	0.6
0.34	0.8
0.01	1.0

Plot the data on appropriate graph paper to obtain a straight line. Then, perform a least-squares analysis to obtain the best values of a and C. Calculate the correlation coefficient for the least-squares fit.

[C] **3.59.** Using the data of Prob. 3.58, perform a least-squares analysis for a general quadratic fit of the data: that is,

$$y = a + bx + cx^2$$

to obtain the values of a, b, and c. Compare with the results of Prob. 3.58.

3.60. Ten measurements are made of a certain resistance giving the following values in kΩ: 1.21, 1.24, 1.25, 1.21, 1.23, 1.22, 1.22, 1.21, 1.23, 1.24. Obtain the value of the mean reading and calculate the tolerance limits for confidence levels of 90 and 95 percent.

3.61. Nine students are asked to measure the length of a metal sample with a caliper. The results of the measurements expressed in millimeters are: 85.9, 86.0, 86.2, 85.8, 85.9, 85.8, 86.0, 85.9, and 86.1. Calculate the mean measurement and calculate the tolerance limits for 90 and 95 percent confidence levels.

3.62. Twelve pressure measurements are made of a certain source giving the following results in kPa: 125, 128, 129, 122, 126, 125, 125, 130, 126, 127, 124, and 123. Obtain the mean value and set the limits for 90 and 95 percent confidence levels.

3.63. Nine voltage measurements are made which give a mean value of 11 V with an unbiased standard deviation of ±0.03 V. Determine the 5 and 1 percent significance levels for these measurements.

3.64. A resistor is measured with a device which has a known precision of ± 0.1 kΩ when a large number of measurements is taken. How many measurements are necessary to ensure that the resistance is within ± 0.05 kΩ with a 5 percent level of significance? Make the calculation both with and without the t-distribution. Repeat for a 5 percent level of significance of ± 0.1 kΩ.

3.65. A power supply is stated to provide a constant voltage output of 50.0 V ± 0.2 V where the tolerance is stated to be "one sigma." Calculate the probability that the voltage will lie between 50.2 and 50.4 V.

3.66. As part of the quality control process, samples of parts are periodically taken for measurement to see if they conform to specifications. In one sample the diameter of a part is measured with the following results:

$$x_{m_1} = 3.56 \text{ mm}, \qquad \sigma_1 = 0.06 \text{ mm}, \qquad n_1 = 20$$

A second sample is taken with the following results:

$$x_{m_2} = 3.58 \text{ mm}, \qquad \sigma_2 = 0.03 \text{ mm}, \qquad n_2 = 23$$

Examine these data to determine if they yield the same results with a confidence level of 90 percent. Repeat for a confidence level of 95 percent.

3.67. The length of a production part is sampled twice with the following results:

$$x_{m_1} = 3.632 \text{ cm}, \qquad \sigma_1 = 0.06 \text{ cm}, \qquad n_1 = 17$$
$$x_{m_2} = 3.611 \text{ cm}, \qquad \sigma_2 = 0.02 \text{ cm}, \qquad n_2 = 24$$

Determine if the two samples yield the same results with a confidence level of 90 percent.

3.68. In a production process there is a trade-off between the cost of quality control sampling and rejection of the production parts. For a certain dimension measurement the determination can be made with a known precision of ± 0.1 mm with a large number of measurements. Suppose the parts are to be measured within ± 0.05 mm with levels of significance of 5 and 10 percent. Determine the relationship between the cost of measurement and the cost per part rejected such that the total cost will be the same for both levels of significance.

3.69. Repeat Prob. 3.68 using 1 and 5 percent levels of significance.

3.70. The density of air is to be determined by measuring its pressure and temperature for insertion in the ideal-gas equation of state; that is,

$$p = \rho RT$$

The value of R for air is 287.1 J/kg-K and may be assumed exact for this calculation. The temperature and pressure are measured as

$$T = 55 \pm 0.4°C$$
$$p = 125 \pm 0.5 \text{ kPa}$$

Determine the nominal value for the density in kg/m^3 and its uncertainty.

3.71. A length measurement is made with a metal scale having graduations of 1 mm. Thirteen measurements are made of the length of a part giving the results in centimeters of 8.55, 8.65, 8.7, 8.5, 8.5, 8.6, 8.65, 8.6, 8.65, 8.7, 8.55, 8.6, and 8.65. Examine these data by whatever means are appropriate and state conclusions regarding the accuracy or uncertainty of the measurement.

3.72. The variable y is expressed in terms of the variable x through the following relations:

$$y = 5x^2$$
$$y = 5x^3$$
$$y = 5x^4$$
$$y = 5x^2 + 3x + 2$$

A measurement of x is performed which yields $x = 3 \pm 0.1$. Calculate the magnitude and percent uncertainties in y for each of the above relations.

3.73. The variable y is given by the quadratic relation

$$y = 3x^2 - 2x + 5$$

To what precision must x be determined if y is to have a precision of 1 percent at a nominal value of $y = 13$?

[C] 3.74. Folklore has it that in the early days of production of automobiles the number of orders was determined by measuring the height of the stack of paper! Suppose the following data are obtained:

Height of Stack, mm	Number of Sheets in Stack
9.5	105
11.3	122
13.5	142
17.4	190
19.9	218

Perform a least-squares analysis to obtain the best linear relation between the height of the stack and the number of sheets in the stack. Determine the correlation coefficient for this relation.

[C] 3.75. The following data are expected to follow a second-degree polynomial (quadratic) relationship for y as a function of x. Plot the data in such a manner that the graphical display will be a straight line if a quadratic relation exists. Using available computer software, obtain a least-squares fit for this plot and the resulting $y = f(x)$ along with a value for r^2. Also plot the data on linear coordinates with y vs. x and obtain a least-squares fit for a quadratic relation for the data. Do the two relations agree? If not, why? For comparison purposes, also obtain a linear least-squares fit to the data, that is, $y = a + bx$.

x	y
0	1.9
1	3.6
2	5.5
3	9.8
4	14.5
5	19
6	25
7	35
8	40
9	53
10	60

[C] 3.76. The data for a certain experimental test are tabulated below. Examine the data with multiple graphs using available computer software to arrive at a suitable correlation for the data $y = f(x)$. Examine several functional forms and perform least-squares analyses to arrive at the final conclusion.

x	y
20000	60
25000	79
35000	90
45000	135
47000	110
50000	130
62000	160
65000	150
70000	180
75000	190
80000	191
90000	200
100000	210
105000	250
110000	240
120000	280
135000	300
140000	290
145000	330
150000	340

[C] 3.77. The tabulated values of x and y are expected to follow a functional form of

$$y = \frac{x}{c_0 + c_1 x} + c_2$$

Plot the following data in an appropriate way to obtain a straight line and determine the values of the constants c_0, c_1, and c_2. *Ans.* $y = x/(1 + 2x) + 3$.

x	y
0	3
1	3.25
2	3.4
3	3.4286
4	3.4444
5	3.5455
6	3.4615

[C] **3.78.** The following data are expected to follow some type of exponential relation. Plot the data in an appropriate way to determine if an exponential relation exists. Determine a least-squares fit to the exponential relation, as needed. Also obtain least-squares fits for a linear and quadratic relation to fit the data. What conclusions can be drawn? *Ans.* $y = 1 - e^{-0.1x}$.

x	y
0	0
1	0.09516
2	0.1813
3	0.2592
4	0.3297
5	0.3935
6	0.4512
7	0.5034
10	0.6321
15	0.7769
20	0.8647

[C] **3.79.** The tabulated data are expected to follow a functional form of $y = f[\exp(c_0 x + c_1 x^2)]$. Plot the data in an appropriate manner to display as a straight line. Obtain a least-squares fit to the line and use this fit to obtain values of the constants c_0 and c_1. *Ans.* $y = 2 \exp(0.1x + x^2)$.

x	y
0	2
0.1	2.0404
0.2	2.1237
0.3	2.255
0.4	2.4428
0.5	2.6997
0.6	3.0439
0.7	3.5103
1	6.0083
2	133.372
3	21876.04

[C] **3.80.** A turbine flowmeter has an output reading of flow in terms of the rotational speed of an internal turbine. The calibration data for a certain meter are given below:

Flow Coefficient	Frequency Parameter
1070	33
1087	64
1088	71
1090	95
1092	185
1093	280
1093	350
1094	450
1095	550
1096	650

Plot the flow coefficient vs. log (frequency parameter) and determine the range where the flow coefficient is constant within $\pm 1/2$ percent. Then, using available computer software, obtain a polynomial relationship between the two parameters that may be used for data analysis of the output of the flowmeter.

[C] **3.81.** A certain meter behaves according to the following table.

x(Input)	y(Output)
40000	0.957
50000	0.962
60000	0.966
80000	0.973
100000	0.977
150000	0.982
200000	0.984
300000	0.984
1E+06	0.984

Using available computer software, plot y vs. log x and obtain an appropriate relationship between y and x.

[C] **3.82.** The following calibration data are available for a certain temperature measurement device. Using available computer software, plot y versus x and obtain second-, third-, and fourth-order polynomial fits to the data.

x(Input)	y(Output)
−150	−4.648
−100	−3.379
−50	−1.819
−25	−0.94
0	0
25	0.993
50	2.036
75	3.132
100	4.279
150	6.704
200	9.288
300	14.862
400	20.872

3.83. A certain result function has the form of

$$R = (ax_3 + bx_1)x_2^{0.75}x_4^{0.25}$$

Using suitable variable substitutions and manipulations of Eqs. (3.2a) and (3.2b) obtain a relationship for the uncertainty w_R without resorting to partial derivatives.

3.84. Repeat Prob. 3.83 using direct application of Eq. (3.2) and the partial derivatives.

[C] 3.85. Two resistance-capacitance filters are illustrated in the first two entries of Table 4.2. Using available computer software, plot the ratio E_o/E_i for each filter as a function of the variable ωT using (a) linear coordinates, (b) semilog coordinates, and (c) loglog coordinates. Extend the plots far enough to approach the asymptotic values. Also plot the variable ϕ as a function of ωT for each of these two filters.

[C] 3.86. A certain instrument has the dynamic input-output response shown below

x(Input)	y(Output)
0	0
1	0.2642
2	0.594
3	0.8009
4	0.9084
5	0.9596
6	0.9826
10	0.9995

Plot y and x on appropriate coordinates and determine a functional relationship between the two variables. *Ans.* $y = 1 - (1 + x)e^{-x}$.

[C] **3.87.** The following input-output behavior is observed for an electronic circuit.

x(Input)	y(Output)
0	2
0.1	2.7629
0.2	3.793
0.3	5.3825
0.5	11.5092
0.8	41.8105
1	109.1963
2	44052.9
3	1.31E+08

Plot y and x on appropriate coordinates and determine a functional relationship between the two variables. *Ans.* $y = 2 \times \exp(3x + x^2)$.

3.88. In a certain dynamic measurement application the input(x) – output(y) relationship is given by

$$y = 1 + (1 + x)e^{-x}$$

With what precision must x be determined to yield a precision of ± 1 percent for y at a value of $x = 5$?

[C] **3.89.** The following test scores are recorded for a class:

22, 35, 99, 87, 77, 78, 89, 100, 68, 84, 86, 75, 69, 44, 56, 99, 95, 34, 73,

51, 100, 96, 21, 79, 87, 89, 69, 81, 100, 55, 89, 73, 75, 76, 59, 92

Determine the class average score and calculate the standard deviation. Following the method illustrated in Example 3.14, plot a histogram and cumulative frequency distribution versus score using 10-point increments in grades. Also plot a cumulative frequency distribution vs. score for a normal distribution having the same mean and standard deviation as the test scores. What do you conclude? Reconstruct the histograms using increments in standard deviation of your own selection. Make at least two selections.

[C] **3.90.** Numerical values of y as a function of the three variables x_1, x_2, and x_3 are shown in the accompanying table. Assuming a correlation of the form of the linear relation in Eq. (3.43), use available computer software to determine the values of the appropriate constants.

X_1	X_2	X_3	y
1	3.2	3	9.5
2	4.5	4	15.6
3	4	5.6	21
4	5.7	6	31
5	6	7	32
6	7	8	41
7	8	9	45
8	9	10	53

[C] **3.91.** The cooling performance of a certain air-conditioning unit, Q, is related to the outdoor temperature T_0 and the indoor temperature T_i through the values in the accompanying table:

Q(kBtu/h)	T_i(°F)	T_0(°F)
54.3	72	85
51.8	72	95
49.4	72	105
46.6	72	115
50.4	67	85
48	67	95
45.5	67	105
42.9	67	115
46.5	62	85
44.4	62	95
42.2	62	105
40	62	115
45.7	57	85
43.9	57	95
42	57	105
40	47	115

Using appropriate computer software determine a linear relation in the form of Eq. (3.43) to fit these data, and calculate the resulting values of S_{total} and $S_{regression}$.

[C] **3.92.** The experimental result y is a function of three parameters through the relation

$$y = x_1 x_2 / x_3$$

These variables have uncertainties of

$$W_{x1} = \pm 1.0$$
$$W_{x2} = \pm 0.5$$
$$W_{x3} = 0.2$$

The variables are measured over the ranges:

$$5 < x_1 < 100$$
$$5 < x_2 < 100$$
$$15 < x_3 < 100$$

Calculate the percentage uncertainty in y at the upper and lower limits of the measurement ranges.

3.93. An experimental result has the functional form of

$$y = x_1 + (x_2 x_3)^{1/2}$$

in terms of three measured variables, all of which have uncertainties of ± 1.0 over a Measurement range of 10 to 100 in each of the variables. Calculate the percentage uncertainty in y for the upper and lower limits of the measured variables.

[C] 3.94. Using the perturbation technique of Sec. 3.5, evaluate the derivatives $\partial R/\partial x_n$ for each of the variables in Prob. 3.93.

[C] 3.95. Using appropriate computer software obtain a curve-fit formula for the discharge coefficient for a $1 \times \frac{1}{2}$ in venturi as a function of throat Reynolds number Re_d. Obtain data points from the curve in Fig. 7.10.

3.96. Using the perturbation technique of Sec. 3.5, obtain values for the derivatives $\partial R/\partial x_n$ for each of the variables in Prob 3.92.

3.97. Using the perturbation method of Sec. 3.5, rework Example 3.4.

3.98. The polynomial relation of Eq. (8.12) [see Chap. 8] is used to fit the experimental data of Table 8.3a. Note that only a fifth-order polynomial is employed for Iron-Constantan thermocouples, while a ninth-order polynomial is employed for others. Using appropriate computer software attempt to fit a ninth-order polynomial to the Iron-Constantan data of Table 8.3a. Discuss the result.

3.99. Consider the radiation temperature error which results from nonblackbody conditions indicated in Fig. 8.27 [see Chap. 8]. What error would result from assuming a linear variation of the error for $0.4 < \varepsilon < 1.0$? Use a least-squares analysis to obtain the linear relation.

3.100. Suppose the electric discharge behavior of Fig. 2.3c is wrongfully modeled with a linear relation determined by a least-squares fit to the experimental decay curve. Using the definition of time constant as the time required to achieve a response of 63.2 percent, what error will result in determining the time constant from the assumed linear relation?

3.101. The clustering of experimental data can suggest various analytic expressions that may fit through the data. Occasionally, the data suggest a combination of curves to be used for the representation. Consider a set of data that appears to follow a straight-line relation from $x = 0$ up to a value of $x = x_1$. The slope of the line in this range appears to be m, and the y-intercept is b. Above $x = x_1$, the data appear to follow a second-order polynomial (parabola) shape up to a position where $x = x_2$ and $y = y_2$. Considering m, b, x_1, x_2, and y_2 as given values, determine the equation of the second-order polynomial. (Hint: The slope of the straight line and parabola are equal at $x = x_1$.)

3.102. Using the results of Prob. 3.101 plot the resultant straight-line polynomial combinations for $b = 0.25$, $x_1 = 0.5$, $x_2 = 1.0$, $y_2 = 1.0$, and slope m taking on the values 0.2, 0.5, and 1.0.

3.103. Suppose the straight-line polynomial combinations of Prob. 3.102 and the lines $y = y_2$, $x = 0$, and $x = 1.0$ form the boundaries of a set of data. Determine the area enclosed by these boundaries.

3.104. Suppose that an uncertainty analysis is made of the experimental data presented in Prob. 3.30, and it is found that the variable R has a much smaller uncertainty than the corresponding value for N. The method of least squares employed for the solution of Prob. 3.30 correctly assumes $N = f(R)$. Suppose further that an error is discovered in the uncertainty analysis such that the uncertainty in determination of R is much larger than that for N, thereby suggesting that the least-squares analysis should be conducted with an assumed correlation relation of

$$R = (N/a)^{1/b} = (1/a)^{1/b} N^{1/b}$$

Perform such a least-squares analysis and compare the values of a and b with those determined in Prob. 3.30. How much deviation is there between the two correlations at the upper and lower limits of the data, that is, at $R = 12$ and 3000?

3.105. The data of Prob. 3.28 have an uncertainty in x much smaller than that in y. Obtain a correlation that assumes the uncertainty in y is much smaller than that in x and compare the values of the two correlations at the limits of the data, that is, at $x = 1.21$ and 8.1.

3.106. Obtain a least-squares correlation for the data in Prob. 3.27 in a linear form of

$$x = (1/a)y - b/a$$

and compare with the result in Prob. 3.27.

3.107. Repeat Prob. 3.76 but express the correlation in the form $x = f(y)$.

3.23 REFERENCES

1. Kline, S. J., and F. A. McClintock: "Describing Uncertainties in Single-Sample Experiments," *Mech. Eng.*, p. 3, January 1953.

2. Mood, A. M., and F. A. Graybill: *Introduction to the Theory of Statistics,* 2d ed., McGraw-Hill, New York, 1963.

3. Schenck, H.: *Theories of Engineering Experimentation,* McGraw-Hill, New York, 1961.

4. Wilson, E. B.: *An Introduction to Scientific Research,* McGraw-Hill, New York, 1952.

5. Young, H. D.: *Statistical Treatment of Experimental Data,* McGraw-Hill, New York, 1962.

6. Hicks, C. R.: *Fundamental Concepts in the Design of Experiments,* Holt, Rinehart, and Winston, New York, 1964.

7. Moffat, R. J.: "Contributions to the Theory of Single-Sample Uncertainty Analysis," *J. Fluids Engr.,* vol. 104, p. 250, June 1982.

8. *Measurement Uncertainty,* ANSI/ASME Power Test Code 19.1-1985, American Society of Mechanical Engineers, New York, 1986.

9. Press, W. H., B. P. Flannery, S. A. Teukolsky, and W. T. Vettering: *Numerical Recipes, the Art of Scientific Computing,* Chap. 12, pp. 381–449, Cambridge

University Press, Cambridge Mass., 1986. Also FORTRAN Example Book and diskette.

10. Chapra, S. C., and R. P. Canale: *Numerical Methods for Engineers,* 3d ed., McGraw-Hill, New York, 1998.

11. Constantinides, A.: *Applied Numerical Methods with Personal Computers,* McGraw-Hill, New York, 1987.

12. Draper, N. R., and H. Smith: *Applied Regression Analysis,* 2d ed., John Wiley & Sons, New York, 1981.

13. Weiss, N. A., and M. J. Hassett: *Introductory Statistics,* 2d ed., Addison-Wesley, Reading, MA, 1987.

14. Moffat, R. J.: "Uncertainty Analysis in the Planning of an Experiment," *J. Fluids Engr.,* vol. 107, pp. 173–181, 1985.

15. ———: *Mathcad 8,* Mathsoft, Inc., Cambridge, MA, 1999.

16. ———: *TK Solver,* Universal Technical Systems Inc., Rockford, IL, 1999.

17. Wieder, S.: *Introduction to MathCAD for Scientists and Engineers,* McGraw-Hill, New York, 1992.

18. Viniotis, Yannis: *Probability and Random Processes,* McGraw-Hill, New York, 1998.

19. Rogers, David: *Procedural Elements of Computer Graphics,* 2d ed., McGraw-Hill, New York, 1998.

20. Palm, W.: *MATLAB for Engineering Applications,* McGraw-Hill, New York, 1999.

21. Heath, Michael T.: *Introduction to Scientific Computing,* McGraw-Hill, New York, 1997.

22. Neter, John, M. H. Kutner, W. Wasserman, and C. Nachtsheim: *Applied Linear Regression Models,* 3d ed., McGraw-Hill, New York, 1996.

23. Gottfried, B.: *Spreadsheet Tools for Engineers: Excel 97 Version,* McGraw-Hill, New York, 1998.

24. Keller, H., and J. G. Crandall: *Mastering MATHCAD, Version 7,* McGraw-Hill, New York, 1999.

25. Chapra, Steve, and R. Canale: *Introduction to Computing for Engineers,* 2d ed., McGraw-Hill, New York, 1994.

26. Buchanan, J. L., and P. R. Turner: *Numerical Method and Analysis,* McGraw-Hill, New York, 1992.

27. Barnes, J. Wesley: *Statistical Analysis for Engineers and Scientists: A Computer-Based Approach,* McGraw-Hill, New York, 1994.

28. Holman, J. P.: *What Every Engineer Should Know About EXCEL*, CRC Press, Taylor and Francis, Boca Raton, FL, 2006.

29. Ross, S. M.: "Pierce's Criterion for the Elimination of Suspect Experimental Data," *Journal of Engineering Technology*, pp. 38–41, Fall 2003.

30. Pierce, Benjamin: "Criterion for the Rejection of Doubtful Observations," *Astronomical Journal II*, vol. 45, pp. 161–163, 1852.

31. Gould, B. A.: "On Pierce's Criterion for the Rejection of Doubtful Observations, with Tables for Facilitating Its Application," *Astronomical Journal IV*, vol. 83, pp. 81–87, 1855.

32. Chauvenet, William: *A Manual of Spherical and Practical Astronomy*, 5th ed., vol. II, p. 558, Lippincott, Philadelphia, 1863, reprinted by Dover Publications, New York, 1960.

33. Kondraske, G. V., and R. M. Stewart: "New Methodology for Identifying Hierarchical Relationships Among Performance Measures: Concepts and Demonstration in Parkinson's Disease," *IEEE Eng Biol Soc*, pp. 5279–5282, in CD-ROM Proc. 31st Intl. Conf., Minneapolis, Sept. 2–6, 2009.

34. Ridker, Paul M., et al.: "Association of Moderate Alcohol Consumption and Plasma Concentration of Endogenous Tissue-Type Plasminogen Activator," *Jour. Amer. Med. Assn.*, vol. 272, no. 12, 1994.

BASIC ELECTRICAL MEASUREMENTS AND SENSING DEVICES

4.1 INTRODUCTION

Many measuring devices depend on some basic electrical principle for their operation, and nearly all data-gathering, transmission, and analysis systems depend on electronic devices. For example, the remote measurement and recording of temperature is ordinarily accomplished in the following way. A transducer is installed at the location of interest, and this device converts the temperature at any given time to an equivalent electric voltage. This voltage is then transmitted to a receiving station where it is displayed in an appropriate fashion. Electrical devices are involved at every stage of this process.

As a consequence of the pervasive nature of electronics in modern engineering, it is desirable to discuss some electrical devices currently employed and to emphasize their uses in the measurement process. We shall first consider the measurement of the basic electrical quantities of current and voltage. Next, we shall examine some simple circuits that may be used for modification and measurement of input signals. Here we shall be concerned specifically with signal amplification and the techniques used to minimize the effects of unwanted noise. Finally, the physical principles and operating characteristics of important electrical transducers will be studied and their applications surveyed.

4.2 FORCES OF ELECTROMAGNETIC ORIGIN

The operation of all electrical devices depends on two facts: charge exists and charged entities interact. In terms of basic physics, there are two kinds of charge, positive and negative; like charges repel and unlike charges attract. Such a crude statement suffices to explain simple experiments but is inadequate to allow an understanding of the operation of sophisticated devices. To provide such understanding, we must carefully model the interaction of charges. Once this is done, we shall be able to understand basic electrical phenomena.

Consider a point charge of q coulombs (C). If such a charge were alone in the universe, it would move in a straight line determined by its initial velocity. However, if this charge is moving in a space in which other charges exist, it will "feel" the presence of these other charges. In particular, we define an electric field intensity $\mathbf{E}$ which exists as a result of the presence of charges other than the point charge q under consideration. Also, we can define a magnetic flux density $\mathbf{B}$ which exists in the space as a result of the motion of these other charges.

A careful specification of the interaction between the point charge of q C and other charges is expressed by Lorentz' law. If the point charge is moving with a velocity $\mathbf{v}$, the resultant force on the charge is given by the equation

$$\mathbf{F} = q(\mathbf{E} + \mathbf{v} \times \mathbf{B}) \qquad \text{newtons} \qquad \textbf{[4.1]}$$

where $\mathbf{E}$ is in volts per meter, $\mathbf{v}$ is in meters per second, and $\mathbf{B}$ is in webers per square meter [20].

In order to apply this equation to a useful case, consider a current-carrying conductor placed in a magnetic field as shown in Fig. 4.1. No electric field is present in this case. The electric current i in the conductor is defined as the ratio of the charge dq passing a cross section of the conductor to the transit time dt.

$$i = \frac{dq}{dt} \qquad \textbf{[4.2]}$$

We write

$$i\,ds = \frac{dq}{dt}\,ds = v\,dq \qquad \textbf{[4.3]}$$

where ds is an element of length along the conductor covered by the moving charge in a time dt. The force exerted on the charge dq is given by Eq. (4.1) as

$$d\mathbf{F} = dq\,(\mathbf{v} \times \mathbf{B}) = i(d\mathbf{s} \times \mathbf{B}) \qquad \textbf{[4.4]}$$

Remember that the electric field $\mathbf{E}$ is 0. This last equation gives the incremental force $d\mathbf{F}$ exerted on a length $d\mathbf{s}$ of the conductor because of the interaction between the charge moving through the conductor and the magnetic flux density in which the

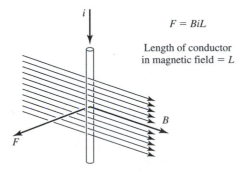

Figure 4.1   Current-carrying conductor in a magnetic field.

conductor is immersed. Carefully observe that this force is perpendicular to the plane defined by the conductor axis and the magnetic field **B**. To find the total force **F** acting on a length L of the conductor, we integrate:

$$\mathbf{F} = \int_0^L i(d\mathbf{s} \times \mathbf{B}) \qquad\qquad [4.5]$$

In order to simplify our results, we consider only the situation in which the conductor axis is perpendicular to **B**. Then,

$$d\mathbf{s} \times \mathbf{B} = ds\,B \sin 90° = ds\,B$$

and the total force F is given by the expression

$$F = BiL \qquad\qquad [4.6]$$

To integrate Eq. (4.5) and obtain Eq. (4.6), we assumed that the magnetic field **B** is constant along the length L of the conductor.

Equation (4.6) is very important because it provides a bridge between a basic electrical quantity i and a mechanical property F. In fact, this equation reduces the problem of measuring a current to the more familiar problem of measuring a force. Consider the apparatus shown in Fig. 4.2. With no current flowing through the conductor, the spring will be at its unstretched length. As current flows through the conductor, the spring will stretch and develop the force required to balance the electromagnetic force. The total distance x moved by the spring is found by equating the two forces.

$$Kx = BiL$$

where K is the spring constant. This equation can be solved for x,

$$x = \frac{BL}{K}i$$

and rearranged in the following way:

$$i = \left(\frac{K}{BL}\right)x \qquad\qquad [4.7]$$

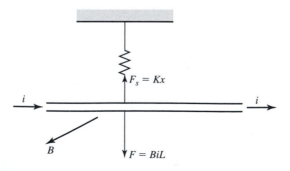

Figure 4.2   The primitive ammeter.

As long as the quantities B, L, and K are known, this last equation offers a direct method of determining the current i flowing in a conductor. We need only measure the total deflection x and then calculate the current. As will be shown, actual current meters are somewhat more complex geometrically, but they do use this basic principle; the current i generates a BiL force which is balanced by a restraining force caused by the stretching or compressing of a spring. The distance the spring moves from equilibrium is a direct measure of the current i.

We shall return to consideration of electrical measuring devices in Secs. 4.4 and 4.5. Before this, a few pages will be devoted to discussing the nature of electrical signals and the various quantities which are commonly used to characterize such signals. This discussion will allow a better appreciation of measuring instruments.

Consider the primitive current-measuring device shown in Fig. 4.2. If a time-varying current i is applied to the conductor, the position x of the conductor will also change with time. As long as the current changes slowly (so that the mechanical system is in equilibrium at each instant of time) Eq. (4.7) remains valid, and a plot of the current-time function could be obtained by recording the position at different times (by means of a motion picture camera, for example) and applying Eq. (4.7). Figure 4.3 is a possible result of this process. Such a plot is an analog representation of the current i as a function of time t. In this context the phrase "analog representation" means that the current is treated as a continuous variable. That is, we assume that we can measure and plot the magnitude of the current with an arbitrarily high degree of precision. In terms of the primitive meter this implies that quite sophisticated techniques must be used to measure the spring-stretching distance x. A laser interferometer could be used, for example. The key idea is that the current i can be treated as a continuous variable. It can take any numerical value. This is the crux of the phrase "analog representation."

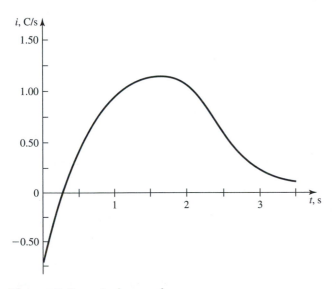

Figure 4.3   Analog waveform.

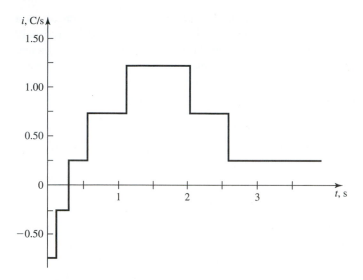

Figure 4.4   Digital waveform.

Such a representation is the most natural approach to physical quantities. We tend to think of them as continuous in their magnitudes. On the other hand, there are many situations, as we shall see, in which it is helpful to consider *digital representations* of physical quantities. A digital representation allows *only discrete* values. Figure 4.4 shows a digital representation of the analog waveform sketched in Fig. 4.3. In this example the current is considered in discrete "chunks" of 0.5 C/s.

Several features of digital representation are important. First, as the discrete "chunks" are made smaller the fit between the digital representation and the analog representation becomes closer. In some situations the digital representation is a more accurate rendering of the physical reality. For example, if we were forced to measure the deflection of the spring in the primitive instrument of Fig. 4.2 with a meter stick, we could only measure position to, perhaps, the nearest millimeter, and the current i could be found only in discrete pieces.

The concept of a digital representation may appear unfamiliar, but there are many instances where the representation and processing of physical signals in digital form is of significant value. A major portion of modern instruments process the signals in digital form [17, 21]. Now, we shall briefly examine commonly used measures of time-varying quantities.

4.3 WAVEFORM MEASURES

If a scalar physical quantity of interest is constant with respect to time, the specification of this quantity involves only one number. For example, Fig. 4.5 shows a constant current of 7 C/s; to specify this current, one needs only to state that a current of 7 A

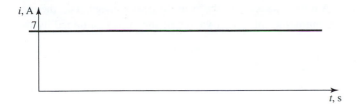

Figure 4.5   A 7-A direct current.

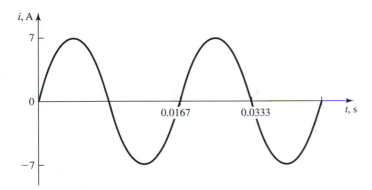

Figure 4.6   A sinusoidal current.

is flowing through the conductor.[1] There is no need in such a situation to display a plot of current versus time. Engineers describe such a physical situation by simply stating that a 7-A direct current is flowing. The symbol dc is shorthand for the two words *direct current*.

In the case of time-varying physical quantities the situation is more complex. Consider the current shown in Fig. 4.6 and given by the following equation:

$$i(t) = 7\sin(377t) \qquad A$$

Is it possible to describe such a waveform meaningfully by a single number? The answer to this question is yes, but in order to appreciate the correct single number, let us examine some incorrect choices. The first possible method is to characterize a time-varying periodic waveform by its average.

$$i_{av} = \frac{1}{T}\int_0^T i(t)\,dt$$

where T is the period of the wave. For the sample wave under consideration the average current which flows is zero, and this certainly does not constitute a meaningful

| [1] One ampere equals one coulomb per second.

measure of current $i(t)$. Another measure is the peak-to-peak current; i.e., the difference between the maximum and minimum values of the current attained by the waveform:

$$i_{p\text{-}p} = 7 - (-7) = 14 \text{ A}$$

This is a meaningful measure, but many other waveforms also display the same peak-to-peak current. An example is given in Fig. 4.7. Clearly, the current shown in Fig. 4.7 will be significantly different in its physical effects from the current shown in Fig. 4.6.

Suppose that the current $i(t)$ of interest is flowing through a resistor. The power dissipated in the resistor at each instant of time is given by the equation

$$p(t) = Ri^2(t)$$

and the average power dissipated in the resistor is given by the expression

$$P_{\text{av}} = \frac{1}{T}\int_0^T p(t)\, dt = R\left\{\frac{1}{T}\int_0^T i^2(t)\, dt\right\} \qquad \textbf{[4.8]}$$

Note that we can write

$$P_{\text{av}} = RI_{\text{eff}}^2$$

as long as we define I_{eff}^2 by the equation

$$I_{\text{eff}}^2 = \frac{1}{T}\int_0^T i^2(t)\, dt$$

The right-hand side of the last equation defines the root-mean-square (rms) value of the periodic waveform $i(t)$. We write

$$I_{\text{rms}} = \left\{\frac{1}{T}\int_0^T i^2(t)\, dt\right\}^{1/2} \qquad \textbf{[4.9]}$$

and this single number is a meaningful representation of the time-varying current $i(t)$ in that the average power dissipated in a resistor when this current flows is given by

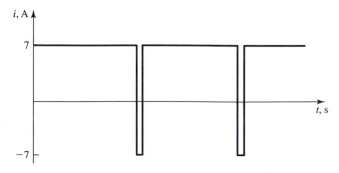

Figure 4.7   Another 14-A peak-to-peak current.

the expression

$$P_{\text{av}} = RI_{\text{rms}}^2$$

The rms (root-mean-square) value associated with a periodic time-varying waveform is a single number which meaningfully characterizes such a function [18]. It will be very useful in the following sections.

4.4 BASIC ANALOG METERS

Equation (4.6) provided the basic foundation upon which we constructed the primitive current-measuring meter in Fig. 4.2. In order to extend this concept to a more realistic model of actual meters, we construct a coil, as shown in Fig. 4.8, place it in a magnetic field, and measure the force exerted on the coil as a result of the electric current flowing in the coil. If the coil has N turns and the length of each turn in the magnetic field is L, the force on the coil is

$$F = NBiL \qquad\qquad \textbf{[4.10]}$$

The force is measured by observing the deflection of a spring. The above principles form the basis of the construction of the mirror galvanometer shown in Fig. 4.9. A permanent magnet is used to produce the magnetic field, while the telescope arrangement and expanded scale improve the readability of the instrument. The meter shown in Fig. 4.9 is designated as the *D'Arsonval moving-coil type*. The metal ribbon furnishes the torsional-spring restraining force in this case, while a filamentary suspension would be used for a more sensitive instrument.

Instead of the mirror and light-beam arrangement shown in Fig. 4.9, the D'Arsonval movement could be used as a pointer-type instrument, as shown in Fig. 4.10; however, such an instrument has a lower sensitivity than the mirror galvonometer because of the additional mass of the pointer and decreased readability resulting from the relatively shorter scale length.

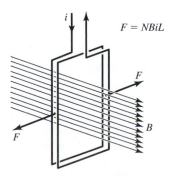

Figure 4.8   Current-carrying coil in a magnetic field.

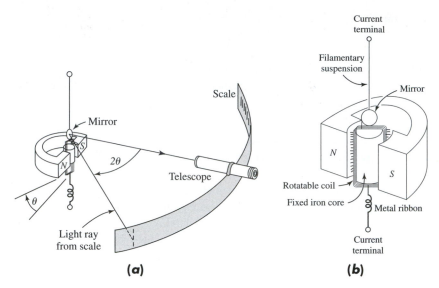

Figure 4.9   A typical galvanometer. (*a*) Optical system; (*b*) D'Arsonval movement.

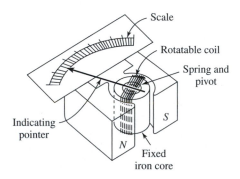

Figure 4.10   D'Arsonval movement used as a pointer-type instrument.

It is clear that D'Arsonval movement, in one form or another, may be used for the measurement of direct current. When this movement is connected to an alternating current, the meter will either vibrate or, if the frequency is sufficiently high, indicate zero. In either event, the D'Arsonval movement is not directly applicable to the measurement of alternating current.

Two common types of movements used for ac measurement are the iron-vane, or moving-iron, and electrodynamometer arrangements. In the iron-vane instrument, as shown in Fig. 4.11, the current is applied to a *fixed coil*. The iron vane is movable and connected to a restraining spring as shown. The displacement of the vane is then proportional to the inductive force exerted by the coil. The meter is subject to eddy-current losses in the iron vane and various hysteresis effects which limit its accuracy.

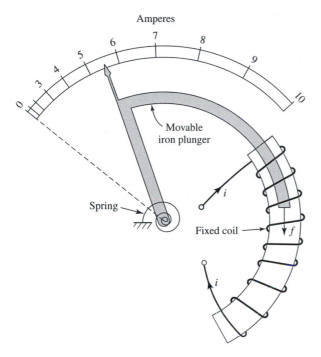

Figure 4.11   Principle of operation of the iron-vane or moving-iron instrument.

The features of the electrodynamometer movement are shown in Fig. 4.12. This movement is similar to the D'Arsonval movement, except that the permanent magnet is replaced by an electromagnet, which may be actuated by an alternating current. Consequently, the field in the electromagnet may be made to operate in synchronization with an alternating current in the moving coil. In order to use the electrodynamometer movement for ac measurements, it is necessary to connect the electromagnet and moving coil in series as shown in Fig. 4.13.

Both the iron-vane and the electrodynamometer movements are normally used for low-frequency applications with frequencies from 25 to 125 Hz. Special designs of the electrodynamometer movement may be used to extend its range to about 2000 Hz.

Both the iron-vane and the electrodynamometer instruments indicate the rms value to the alternating current, and the meter deflection varies with I_{rms}^2. The scale of the instrument is not necessarily based on a square law because the proportionality constant between the rms current and the meter deflection changes somewhat with the current.

An important feature of the electrodynamometer instrument is that it may be calibrated with direct current and that the calibration will hold for ac applications within the frequency range of the instrument. The iron-vane instrument is not as versatile because of the residual magnetism in the iron when direct current is used.

A rectifier arrangement may also be used for ac measurements. In this device an ac waveform is modified by some type of rectifier such that current is obtained with

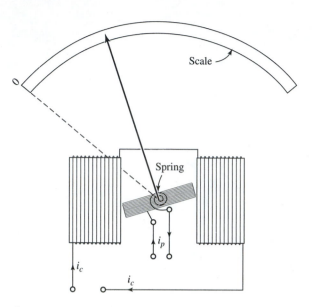

Figure 4.12   Basic features of electrodynamometer movement.

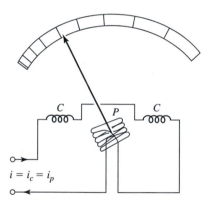

Figure 4.13   Electrodynamometer movement used as an ammeter.

a steady dc component. A dc instrument may be used to indicate the value of the ac current applied to the rectifier.

For measurements of high-frequency alternating currents, a thermocouple meter may be used. This type of meter is indicated in Fig. 4.14. The alternating current is passed through a heater element, and the temperature of the element is indicated by a thermocouple connected to an appropriator dc instrument.

The thermocouple indicates the rms value of the current because the average power dissipated in the heater is equal to $I_{rms}^2 R$. The instrument reading is independent

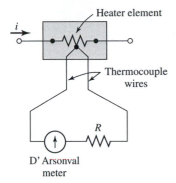

Figure 4.14 Schematic of a thermocouple meter.

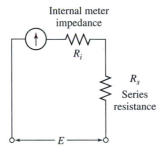

Figure 4.15 Direct current meter used as a voltmeter.

of waveform because of this relationship between the thermal emf generated in the thermocouple and the power dissipated in the heater. The thermal emf generated in the thermocouple varies approximately with the square of current, although slight deviations from the square law may be obtained because of change in heater resistance with temperature and other side effects. Alternating currents with frequencies up to 100 MHz may be measured with thermocouple meters. One may also measure high-frequency current by first rectifying the signal to direct current and then measuring the direct current.

A dc voltmeter may be constructed very easily by modifying the basic dc sensing device as shown in Fig. 4.15. In this arrangement a large resistor is placed in series with the movement; thus, when the instrument is connected to a voltage source, the current in the instrument is an indication of the voltage. The range of the voltmeter may be altered by changing the internal series resistor. The voltmeter is usually rated in terms of the input voltage for full-scale deflection or in terms of the ratio of internal resistance to the voltage for full-scale deflection. A series-resistor arrangement may also be used

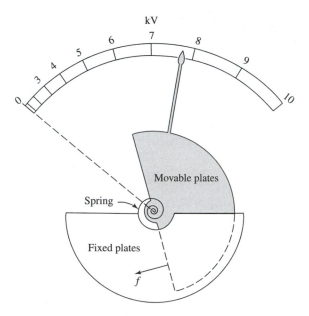

Figure 4.16   Electrostatic-voltmeter movement.

with the iron-vane and electrodynamometer instruments for measurements of rms values of ac voltages. For low frequencies up to about 125 Hz the electrodynamometer meter may be calibrated with dc voltage and calibration used for ac measurements.

Electrostatic forces may also be used to indicate electric potential difference. For this purpose two plates are arranged as shown in Fig. 4.16. One plate is fixed, and the other is mounted in jeweled bearings so that it may move freely. A spiral spring provides a restraining force on the two plates. If two complete disks were used instead of the sectoral plate arrangement, the net torque would be zero; however, with the arrangement shown, the fringe effects of the electric field produce a net force in the indicated direction which is proportional to the square of the rms voltage. As the movable plate changes position, the capacitance changes, and hence the proportionality between the stored energy and the voltage varies with the impressed voltage.

The electrostatic voltmeter may be used for either ac or dc voltage measurements, but potentials above 100 V are required in order to produce a sufficiently strong torque in the system. The meter may be calibrated with direct current and then used for measurement of rms values of ac voltages, regardless of the waveform. Electrostatic voltmeters are generally applicable up to frequencies of 50 MHz. It may be noted that the electrostatic voltmeter has an extremely high-input impedance for dc applications but a much lower ac impedance as a result of the capacitance reactance. The capacitance may be about 20 pF for a 5000-V meter.

In the preceding paragraphs we have discussed some of the more important analog devices that are used for the measurement of electric current and voltage. Next, we examine digital devices.

4.5   BASIC DIGITAL METERS

We have already discussed analog and digital representations of time-varying signals. Analog meters are reasonably simple in that they depend on balancing a spring force with a force generated by the interaction between a current and a magnetic field. However, the use of such analog meters necessarily is limited by the accuracy with which the position of the indicator with respect to the scale can be read. Parallax is also a problem, and many times two individuals can read the same meter and arrive at different values of the quantity being measured.

With a digital meter, the value of the measured quantity is shown directly as a series of digits. This means that parallax is no longer a problem and different individuals will read the same values. In addition, the inherent accuracy of digital meters is much greater than that of analog meters. For these reasons, we shall briefly discuss the digital voltmeter and related instruments.

The heart of digital meters is an oscillator or "clock" which is often a quartz crystal. When such a crystal is connected to proper electrical components, it establishes an output voltage which is almost sinusoidal with a fixed frequency. This frequency is controlled by the dimensions of the quartz crystal. A familiar example of an application of such devices is the digital watch. All digital meters have such a device, although not all have a quartz oscillator. In some, the timing function is carried out by means of integrated circuits. The clock output is usually "shaped" electronically into a series of pulses, one pulse for each cycle of clock oscillation.

The cornerstone of digital instrumentation is the digital voltmeter. Typical operation is characterized by the following: (1) the capability to generate an internal reference voltage and decrease it linearly from 10 V to 0 V at a rate of 0.1 V/s, (2) the ability to compare the reference voltage to the voltage being measured and to generate a signal when the two are equal, and (3) the ability to generate another signal when the reference voltage has reached 0 V.

For simplicity, assume that the clock generates one pulse each second. Let us see how voltage can be determined using this set of capabilities. All we need to do is start counting the clock pulses when the reference voltage equals the voltage being measured and stop the counting when the reference voltage reaches zero. If, for example, a total count of 23 is obtained in a given measurement, this means that 23 s elapsed between the "start count" and the "stop count" times. Since the reference voltage was decreasing at 0.1 V/s, the measured voltage is 2.3 V. Figure 4.17 illustrates these concepts.

A digital voltmeter can be converted into an ammeter by including a precision resistor within the instrument and measuring the voltage drop across this resistor caused by an unknown current. In a similar fashion, unknown resistances can be measured by incorporating an accurate current passing through the unknown resistor. In both of these cases the instrument design is such that the digital readout is the desired current or resistance value.

The simple example given above to illustrate digital meter operation serves its purpose but is somewhat misleading in that we have specified a very slow clock. In fact, the clocks used in available instrumentation emit pulses at much higher rates

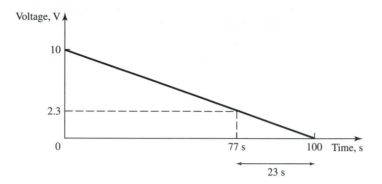

Figure 4.17   Digital voltage measurement.

than those specified above. The digital meter is, of course, naturally wedded to digital display. Since pulses are available in the system, these are used to drive the display circuits directly. In this way a set of light-emitting diodes, for example, can be used with little additional circuit complexity.

It is worth mentioning one significant additional difference between analog and digital instruments. The analog meters basically respond to, and measure, currents. The amount of torque produced on a meter movement is directly related to the amount of current flowing through the meter. The situation is complementary in the case of digital instruments. These devices respond to and measure voltages directly. This is a result of the physical behavior of the integrated circuits of which digital instruments are composed.

4.6   BASIC INPUT CIRCUITS

A block diagram of the general method of gathering, processing, and displaying physical information is shown in Fig. 4.18. The first block represents the transducer, which serves to convert the value of the physical property of interest into an electrical signal in some "faithful" fashion. A later section of this chapter concerns itself with such transducers, and we shall mention a few examples here. A microphone converts pressure into an equivalent electric voltage. A thermocouple converts temperature into an equivalent electric voltage. A solar cell converts incident light into an equivalent electric current, and so forth.

The electrical output of a transducer must be connected to some additional circuitry if we are to use it for anything. Care must be taken to ensure that this input circuit does not change the value of the transducer output.

The third block in Fig. 4.18 is labeled "a signal conditioning"; this refers to various techniques which are available to reduce the effects of noise. This term is used in a general sense to mean any electrical signals which may be present in addition to the output of the transducer. The next block is the transmission segment, which simply

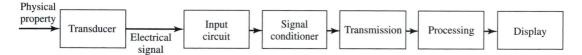

Figure 4.18   Block diagram of measurement and display.

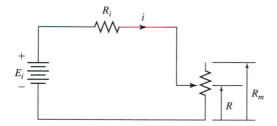

Figure 4.19   Current-sensitive input circuit.

indicates that we may in some instances need to make a measurement at one physical location and transmit the signal to another place for further processing, display, and/or storage.

The next-to-last step points out the need to process the signal after transmission. We shall not go into detail here, but, for example, if you were to listen to the output of an AM radio transmitter, you would hear nothing. That is because the frequency of such transmissions is much higher than the response capabilities of the human ear. A radio receiver provides the processing necessary to convert the signal to an audible frequency.

The display block in Fig. 4.18 indicates the need ultimately to display the electrical signals in some comprehensible form. Examples include cathode-ray screens, floppy disks, magnetic tape, lines of printer output, and xy plotters. In this section, we shall discuss a few representative types of input circuits.

Consider a gas sensor, the resistance of which changes as a function of the gas concentration surrounding the sensor. A simple type of input circuit uses the current flow through the sensor's resistance as an indication of the value of the resistance. Let the sensor be in series with a battery and represent the battery as a series combination of an ideal voltage source E_i and an internal resistance R_i, as shown in Fig. 4.19. A change in the gas concentration results in a change in the resistance, as indicated by the movable contact. The current is given by

$$i = \frac{E}{R + R_i} \qquad \text{[4.11]}$$

The maximum resistance of the transducer is R_m, and the current may be written in dimensionless form as

$$\frac{i}{E_i/R_i} = \frac{1}{(R/R_m)(R_m/R_i) + 1} \qquad \text{[4.12]}$$

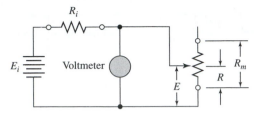

Figure 4.20   Voltage-sensitive input circuit. Change in resistance R is indicated through change in voltage indication.

It would be desirable to have the current output vary linearly with the resistance of the transducer. Unfortunately, this is not the case, as the last equation shows, although the output may be approximately linear for some ranges of operation.

The circuit shown above may be modified by using a voltmeter, as shown in Fig. 4.20. Let us assume that the internal impedance of the voltmeter is very large compared with the resistance in the circuit, so that we may neglect the current drawn by the meter. That is, the meter looks like an open circuit. The current flow is still given by

$$i = \frac{E_i}{R + R_i} \tag{4.13}$$

Let E be the voltage across the transducer, as indicated in Fig. 4.20. Then

$$\frac{E}{E_i} = \frac{iR}{i(R + R_i)} = \frac{(R/R_m)(R_m/R_i)}{1 + (R/R_m)(R_m/R_i)} \tag{4.14}$$

Now we have obtained a voltage indication as a measure of the resistance R, but a nonlinear output is still obtained. The advantage of the circuit in Fig. 4.20 over the one in Fig. 4.19 is that a voltage measurement is frequently easier to perform than a current measurement, as we have noted for digital meters. The voltage-sensitive circuit is called a *ballast circuit*.

We can define the sensitivity of the ballast circuit as the rate of change of the transducer voltage with respect to the transducer resistance R. Thus,

$$S = \frac{dE}{dR} = \frac{E_i R_i}{(R_i + R)^2} \tag{4.15}$$

We would like to design the circuit so that the sensitivity S is a maximum. The circuit-design variable which is at our disposal is the fixed resistance R_i so that we wish to maximize the sensitivity with respect to this variable. The maximizing condition

$$\frac{dS}{dR_i} = 0 = \frac{E_i(R - R_i)}{(R_i + R)^3} \tag{4.16}$$

is applied. Thus, for maximum sensitivity we should take $R_i = R$. But since R is a variable, we may select the value of R_i only for the range of R where the sensitivity is to be a maximum.

In both of the above circuits a current measurement has been used as an indicator of the value of the variable resistance of the transducer. In some instances it is more

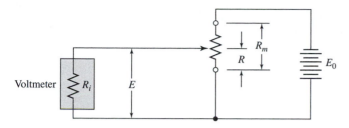

Figure 4.21   Simple voltage-divider circuit.

convenient to use a voltage-divider circuit as sketched in Fig. 4.21. In this arrangement a fixed voltage E_0 is impressed across the total transducer resistance R_m, while
the variable contact is connected to a voltmeter with internal resistance R_i. If the
impedance of the meter is sufficiently high, the indicated voltage E will be directly
proportional to the variable resistance R; that is,

$$\frac{E}{E_0} = \frac{R}{R_m} \qquad \text{for } R_i \gg R \qquad \textbf{[4.17]}$$

With a finite meter resistance, a current is drawn which affects the voltage measurement. Considering the internal resistance of the meter, the current drawn from the
voltage source is

$$i = \frac{E_0}{R_m - R + R_i R/(R + R_i)} \qquad \textbf{[4.18]}$$

The indicated voltage is therefore

$$E = E_0 - i(R_m - R)$$

or

$$\frac{E}{E_0} = \frac{R/R_m}{(R/R_m)(1 - R/R_m)(R_m/R_i) + 1} \qquad \textbf{[4.19]}$$

As a result of the loading action of the meter, the voltage does not vary in a linear
manner with the resistance R. If Eq. (4.19) is taken as the true relationship between
voltage and resistance, then an expression for the loading error may be written as

$$\text{Loading error} = \frac{(E/E_0)_{\text{true}} - (E/E_0)_{\text{ind}}}{(E/E_0)_{\text{true}}}$$

We wish to know how much the voltage ratio of Eq. (4.19) differs from the simple
linear relation of Eq. (4.17), which would be observed under open-circuit conditions
or for very large meter impedances. We thus use Eq. (4.19) for the true value and
Eq. (4.17) for an assumed linear indication and calculate the loading error as

$$\text{Fractional loading error} = -\left(\frac{R}{R_m}\right)\left(1 - \frac{R}{R_m}\right)\left(\frac{R_m}{R_i}\right) \qquad \textbf{[4.20a]}$$

We may note that the behavior of Eq. (4.20a) is such that the loading error becomes
zero at each end of the resistance scale, that is, for $R/R_m = 0$ and 1.0. Equation (4.20a)

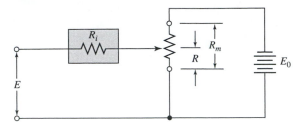

Figure 4.22   Simple voltage-balancing potentiometer circuit.

expresses the loading error as a fraction of the voltage reading obtained in Eq. (4.19). The raw error or deviation from a linear indication is

$$\text{Deviation from linear} = \frac{-(R/R_m)^2(1 - R/R_m)}{(R/R_m)(1 - R/R_m) + R_i/R_m} \qquad \textbf{[4.20b]}$$

The voltage-divider circuit shown in Fig. 4.21 has the disadvantage that the indicated voltage is affected by the loading of the meter. This difficulty may be alleviated by utilizing a voltage-balancing potentiometer circuit, as shown in Fig. 4.22. In this arrangement a known voltage E_0 is impressed on the resistor R_m, while the unknown voltage is impressed on the same resistor through the galvanometer with internal resistance R_i and the movable contact on the resistor R_m. At some position of the movable contact the galvanometer will indicate zero current, and the unknown voltage may be calculated from

$$\frac{E}{E_0} = \frac{R}{R_m} \qquad \textbf{[4.21]}$$

Notice that the internal resistance of the galvanometer does not affect the reading in this case; however, it does influence the sensitivity of the circuit. The voltage-balancing potentiometer circuit is used for precise measurements of small electric potentials, particularly those generated by thermocouples. In order to determine accurately the unknown voltage E, the supply voltage E_0 must be accurately known. A battery is ordinarily used for the supply voltage, but this represents an unreliable source because of aging characteristics. The aging problem is solved by using a standard-cell arrangement to standardize periodically the battery voltage, as shown schematically in Fig. 4.23. When switch S_2 is in position A and switch S_1 is closed, the circuit is the same as Fig. 4.22. Then, when switch S_1 is opened and switch S_2 is placed in position B, the standard cell E_s is connected in the circuit. The variable compensator resistance R_c is adjusted until the galvanometer indicates balance conditions. The protective resistor R_1 may then be bypassed by closing switch S_1 and a fine adjustment of the compensator resistor effected. The protective resistor R_1 is placed in the circuit to avoid excessive current drain on the standard cell and also to protect the galvanometer. Note that the standard-cell voltage is impressed on a fixed portion of the resistance R. Once the battery has been standardized, switch S_2 may be placed in the A position and the unknown voltage E measured. The battery may

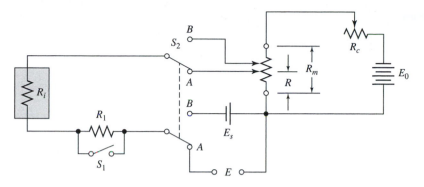

Figure 4.23   Potentiometer circuit incorporating features for standardization of battery voltage.

be standardized as often as necessary for the particular application. The standard cell would only be used for laboratory applications. Solid-state voltage standards would be used for most measurements.

VOLTAGE SENSITIVITY.   The output of a transducer with a total resistance of 150 Ω is to be measured with a voltage-sensitive circuit like that shown in Fig. 4.20. The sensitivity is to be a maximum at the midpoint of the transducer. Calculate the sensitivity at the 25 and 75 percent positions, assuming a voltage source E_i of 100 V.

Example 4.1

Solution

For maximum sensitivity at the midpoint of the range, we take

$$R_i = R = \tfrac{1}{2}R_m = 75\,\Omega$$

At the 25 percent position $R = (0.25)(150) = 37.5\,\Omega$, and the sensitivity is calculated from Eq. (4.15):

$$S = \frac{dE}{dR} = \frac{E_i R_i}{(R_i + R)^2} = \frac{(100)(75)}{(75 + 37.5)^2} = 0.592 \text{ V}/\Omega$$

At the 75 percent position the corresponding sensitivity is

$$S = \frac{(100)(75)}{(75 + 112.5)^2} = 0.213 \text{ V}/\Omega$$

LOADING ERROR OF VOLTAGE-SENSITIVE CIRCUIT.   The voltage-divider circuit is used to measure the output of the transducer in Example 4.1. A 100-V source is used ($E_0 = 100$ V), and the internal resistance of the meter R_i is 10,000 Ω. Calculate the loading error at the 25 and 75 percent positions on the transducer and the actual voltage readings which will be observed at these points.

Example 4.2

Solution

We have

$$\frac{R_m}{R_i} = \frac{150}{10,000} = 1.5 \times 10^{-2}$$

At the 25 percent position $R/R_m = 0.25$ and the fractional loading error is calculated from Eq. (4.20a) as

$$\text{Fractional loading error} = -(0.25)(1 - 0.25)$$
$$= 0.281\%$$

The deviation from linear is obtained from Eq. (4.20b)

$$\text{Deviation from linear} = \frac{-(0.25)^2(1 - 0.25)}{(0.25)(1 - 0.25) + 1/0.015}$$
$$= -7.01 \times 10^{-4}$$

The indicated voltage is therefore

$$E = (100)(0.25 - 7.01 \times 10^{-4}) = 24.93 \text{ V}$$

At the 75 percent position $R/R_m = 0.75$ and

$$\text{Fractional loading error} = -(0.75)(1 - 0.75)(0.015)$$
$$= 2.81 \times 10^{-3} = 0.281\%$$

The deviation from linear is

$$\text{Deviation from linear} = \frac{-(0.75)^2(1 - 0.75)}{(0.75)(1 - 0.75) + 1/0.015}$$
$$= -2.103 \times 10^{-3}$$

and the indicated voltage is

$$E = (100)(0.75 - 2.103 \times 10^{-3}) = 74.79 \text{ V}$$

Improvement in measurement and accuracy is provided by so-called bridge circuits, which are employed in a variety of applications for the measurement of resistance, inductance, and capacitance under both steady-state and transient conditions. The equivalent electric circuit of some transducers can be represented as an impedance, and, as a consequence, the capability of measuring such electrical quantities accurately is important. Many bridge circuits have been developed, but we shall limit our discussion to some of the more prominent types and their applications to various measurement and control problems.

The Wheatstone bridge is normally used for the comparison and measurement of resistances in the range of 1 Ω to 1 MΩ. A schematic of the bridge is given in Fig. 4.24. The cornerstone of the bridge consists of the four resistances (R_1, R_2, R_3, R_x), which are arranged in a diamond shape. R_2 and R_3 are normally known resistors, R_1 is a variable resistance, and R_x is the unknown resistance value associated with

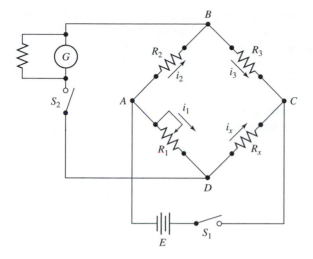

Figure 4.24   Schematic of basic Wheatstone bridge.

the transducer output. The voltage E is applied to the bridge by closing the switch S_1, and by adjusting the variable resistance R_1, the bridge may be balanced. This means simply that the potential difference between points B and D is zero. This balanced condition may be sensed by closing the switch S_2 and adjusting the value of R_1 until the sensing device indicates no current flow. When this occurs, the voltage drop across R_2 must equal the voltage drop across R_1 since this implies that the voltage difference between B and D must be zero. Using the currents defined in Fig. 4.24, we obtain

$$i_2 R_2 = i_1 R_1$$

Further,

$$i_2 = i_3 = \frac{E}{R_2 + R_3} \qquad \text{if balanced}$$

and

$$i_1 = i_x = \frac{E}{R_1 + R_x} \qquad \text{if balanced}$$

If the currents are eliminated from these relations, the result is

$$\frac{R_2}{R_3} = \frac{R_1}{R_x} \qquad\qquad \textbf{[4.22]}$$

or

$$R_x = \frac{R_1 R_3}{R_2} \qquad\qquad \textbf{[4.23]}$$

This last equation allows the calculation of the value of R_x in terms of the other three resistors in the bridge.

   If accurate measurements are to be made with a bridge circuit, the values of the resistors (R_1, R_2, R_3) must be precisely known, and the sensor must be sufficiently sensitive to detect small degrees of imbalance. When the unknown resistance R_x is connected to the circuit, care must be taken to use connecting leads which have a resistance that is small in comparison with the unknown.

The term "ratio arms" is frequently used in describing two known adjacent arms in a Wheatstone bridge. The galvanometer is usually connected to the junction of these two known resistors. In Fig. 4.24 R_2 and R_3 would normally be called the ratio arms.

In practice, the Wheatstone bridge is usually employed in an unbalanced mode and the galvanometer is replaced with a digital voltmeter of high internal impedance. The output voltage is a measure of the resistance, as we shall see in a later discussion.

Example 4.3 | **UNCERTAINTY IN WHEATSTONE BRIDGE.**   For the basic Wheatstone bridge in Fig. 4.24, determine the uncertainty in the measured resistance R_x as a result of an uncertainty of 1 percent in the known resistances. Repeat for 0.05 percent.

Solution

We use Eq. (3.2) to estimate the uncertainty. We have

$$\frac{\partial R_x}{\partial R_1} = \frac{R_3}{R_2} \qquad \frac{\partial R_x}{\partial R_2} = -\frac{R_1 R_3}{R_2^2} \qquad \frac{\partial R_x}{\partial R_3} = \frac{R_1}{R_2}$$

For a 1 percent uncertainty in the known resistances this gives

$$\frac{w_{R_x}}{R_x} = (0.01^2 + 0.01^2 + 0.01^2)^{1/2} = 0.01732$$

$$= 1.732\%$$

For a 0.05 percent uncertainty in the known resistances the corresponding uncertainty in R_x is 0.0866 percent.

The basic Wheatstone bridge circuit may also be employed for the measurement of ac impedances. The main problem is that two balance operations must be made to obtain a null, the first for the real part of the waveforms and the second for the imaginary part. There are some types of ac bridges that may be balanced with two independent adjustments. Several of these types are shown in Table 4.1, along with the balance conditions which may be used to determine the values of the unknown quantities.

Since an alternating current is involved, the null conditions may not be sensed by a galvanometer as in the case of the dc Wheatstone bridge; some type of ac instrument must be used. This could be a rectifier meter, oscilloscope, or digital multimeter.

It may be noted from Table 4.1 that for some types of ac bridge circuits the balanced condition is independent of frequency (types a, c, and e), while for others the frequency must be known in order to apply the balanced conditions (types b, d, and f). Thus, if the capitance and resistance values are known for a Wien bridge, the bridge could be used as a frequency-measuring device.

Table 4.1   Summary of bridge circuits

Circuit	Balance Relations	Name of Bridge and Remarks
(a)	$$C_x = \frac{C_3 R_2}{R_1}$$ $$R_x = \frac{R_3 R_1}{R_2}$$	Basic Wheatstone bridge. Greatest sensitivity when bridge arms are equal.
(b)	$$\frac{C_x}{C_3} = \frac{R_2}{R_1} - \frac{R_3}{R_x}$$ $$C_x C_3 = \frac{1}{\omega^2 R_3 R_x}$$ If $C_3 = C_x$ and $R_3 = R_x$ $$f = \frac{1}{\pi R_3 C_3}$$	Wien bridge. May be used for frequency measurement with indicated relations.
(c)	$$L_x = C_1 R_1 R_2$$ $$R_x = \frac{C_1 R_1}{C_2} - R_3$$	Owen bridge.

(Continued)

Table 4.1   (Continued)

Circuit	Balance Relations	Name of Bridge and Remarks
(d)	$\omega^2 LC = 1$ $R_x = \dfrac{R_3 R_1}{R_2}$	Resonance bridge. At balance conditions may be used for frequency measurement with $f = \dfrac{1}{2\pi\sqrt{LC}}$
(e)	$L_x = R_1 R_3 C$ $R_x = \dfrac{R_1 R_3}{R_2}$	Maxwell bridge.
(f)	$L_x = \dfrac{R_1 R_3 C}{1 + \omega^2 C^2 R_2^2}$ $R_x = \dfrac{\omega^2 C^2 R_1 R_2 R_3}{1 + \omega^2 C^2 R_2^2}$	Hay bridge.

As mentioned earlier, bridge circuits are useful for experimental measurements. The Wheatstone bridge is widely used for measuring the output resistance of various transducers, such as resistance thermometers, strain gages, and other devices that register the change in the physical variable as a change in output resistance. The ac bridges are used for inductance and capacitance measurements. Many transducers produce a change in either of these quantities upon a change in the physical quantity of interest. Bridge circuits are also useful in an unbalanced condition because a small change in the impedance in one of the bridge arms can produce a relatively large change in the detector signal, which can be used to control other circuits.

UNBALANCED BRIDGES

Bridge circuits may operate on either a *null* or a *deflection principle*. The null condition has been described above as where the galvanometer or sensing device reads zero at balance conditions. At any other condition the galvanometer reading will be *deflected* from the null condition by a certain amount, which depends on the degree of unbalance. Thus, the signal at the galvanometer or detector may be used as an indication of the unbalance of the bridge and may indicate the deviation of one of the arms from some specified balance condition. The use of the deflection bridge is particularly important for the measurements of dynamic signals where insufficient time is available for achieving balance conditions.

Consider the bridge circuit shown in Fig. 4.25. R_1, R_2, R_3, and R_4 are the four arms of the bridge; R_g is the galvanometer resistance; and i_b and i_g are the battery and galvanometer currents, respectively. R_b represents the resistance of the battery or power-supply circuit. When the bridge is only *slightly* unbalanced, it can be shown that the value of R_b does not appreciably influence the effective resistance of the bridge circuit as presented to the galvanometer (see Ref. [2]). As a result, the following relation for the galvanometer current may be derived:

$$i_g = \frac{E_g}{R + R_g}$$

[4.24]

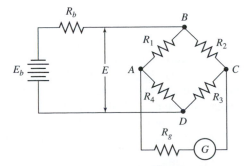

Figure 4.25   Schematic for analysis of unbalanced bridge.

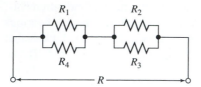

Figure 4.26   Equivalent circuit of bridge as presented to the galvanometer.

where R is the effective resistance of the bridge circuit presented to the galvanometer and is given by

$$R = \frac{R_1 R_4}{R_1 + R_4} + \frac{R_2 R_3}{R_2 + R_3}$$ **[4.25]**

This effective resistance R is indicated in Fig. 4.26. The voltage presented at the terminals of the galvanometer E_g is

$$E_g = E\left(\frac{R_1}{R_1 + R_4} - \frac{R_2}{R_2 + R_3}\right)$$ **[4.26]**

The voltage impressed on the bridge E depends on the battery or external circuit resistance R_b and the resistance of the total bridge circuit as presented to the battery circuit, which we shall designate as R_0. It can be shown that for small unbalances the resistance R_0 may be calculated by assuming that the galvanometer is not connected in the circuit. Thus,

$$R_0 = \frac{(R_1 + R_4)(R_2 + R_3)}{R_1 + R_4 + R_2 + R_3}$$ **[4.27]**

The voltage impressed on the bridge is then

$$E = E_b \frac{R_0}{R_0 + R_b}$$ **[4.28]**

CURRENT-SENSITIVE AND VOLTAGE-SENSITIVE BRIDGES

Because the deflection described above and indicated by Eq. (4.24) is based on the determination of the galvanometer or voltmeter current, the circuit is said to be current-sensitive. If the deflection measurement were made with an electronic voltmeter, oscilloscope, or other high-impedance device, the current flow in the detector circuit would be essentially zero since as $R_g \rightarrow \infty$, $i_g \rightarrow 0$ in Eq. (4.24). Even so, there is still an unbalanced condition at the galvanometer or detector terminals of the bridge. This condition is represented by the voltage between terminals A and C for the case of a high-impedance detector. Such an arrangement is called a *voltage-sensitive deflection-bridge circuit*. The voltage indication for such a bridge may be determined from Eq. (4.26).

For relatively large unbalances in a bridge circuit it may become necessary to shunt the galvanometer to prevent damage. Such an arrangement is shown in

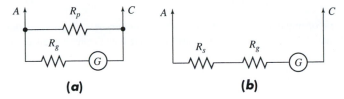

Figure 4.27 (a) Galvanometer shunt arrangement for use with bridge of Fig. 4.26; (b) galvanometer with series resistance for use with bridge in Fig. 4.26.

Fig. 4.27a. In this case the voltage across the galvanometer may still be given to a sufficiently close approximation by Eq. (4.26), but the current through the galvanometer must be computed by considering the shunt resistance R_p. The total resistance of the bridge and shunt combination *as seen by the galvanometer circuit* is given by

$$R_e = \frac{R_p R}{R_p + R} \qquad \text{[4.29]}$$

and the galvanometer current is now given by

$$i_g = \frac{E_g}{R_e + R_g} \qquad \text{[4.30]}$$

The galvanometer current may also be reduced by using a series resistor, as shown in Fig. 4.27b. For this case the galvanometer current becomes

$$i_g = \frac{E_g}{R + R_g + R_s} \qquad \text{[4.31]}$$

where R_s is the value of the series resistor.

OUTPUT MEASUREMENTS OF BRIDGES

For static measurements with a Wheatstone bridge the output will most likely be measured with a digital voltmeter having a very high input impedance. The output may be stored with various digital devices, including computers. For transient measurements an oscilloscope may be employed with or without digital storage facilities. Various levels of data acquisition systems are available for both indication and recording of the outputs.

DEFLECTION BRIDGE. The Wheatstone bridge circuit of Fig. 4.24 has ratio arms (R_2 and R_3) of 6000 and 600 Ω. A galvanometer with a resistance of 70 Ω and a sensitivity of 0.04 μA/mm is connected between B and D, and the adjustable resistance R_1 reads 340 Ω. The galvanometer deflection is 39 mm, and the battery voltage is 4 V. Assuming no internal battery resistance, calculate the value of the unknown resistance R. Repeat for R_2 and R_3 having values of 600 and 60 Ω, respectively. **Example 4.4**

Solution

In this instance the bridge is operated on the deflection principle. For purposes of analyzing the circuit we use Figs. 4.25 and 4.26. The galvanometer current is calculated from the deflection and sensitivity as

$$i_g = (39)(0.04 \times 10^{-6}) = 1.56 \ \mu A$$

In the circuit of Fig. 4.25 the resistances are

$$R_b = 0 \qquad R_1 = 340 \qquad R_2 = 6000 \qquad R_3 = 600 \qquad R_4 = R_x \qquad R_g = 70$$

We also have $E = 4.0 \ V$.

Combining Eqs. (4.24) to (4.26), we obtain

$$i_g = \frac{E[R_1/(R_1 + R_4) - R_2/(R_2 + R_3)]}{R_g + [R_1 R_4/(R_1 + R_4) + R_2 R_3/(R_2 + R_3)]}$$

Solving for R_4, we have

$$R_4 = \frac{E R_1 R_3 - i_g[R_g R_1 (R_2 + R_3) + R_2 R_3 R_1]}{i_g(1 + R_1 + R_g)(R_2 + R_3) + E R_2}$$

Using numerical values for the various quantities, we obtain

$$R_4 = 33.93 \ \Omega$$

Taking $R_2 = 600$ and $R_3 = 60$, we have

$$R_4 = 33.98 \ \Omega$$

4.7 AMPLIFIERS

Experimental measurements occur in many forms: for example, the voltage output of a bridge circuit, the frequency signal of a counting circuit, and voltage signals representative of a change in capacitance. In many cases the signals are comparatively weak and must be amplified before they can be used to "drive" an output device. In other instances there is a serious mismatch between the impedance of the measurement transducer and that of the output circuits so that some interface must be provided to allow effective impedance matching. An example, in a familar setting, of the former class of problems involves connecting a phonograph pickup cartridge directly to a loudspeaker. The cartridge is a transducer which converts the force produced by mechanical motions in the groove of a record directly into an electric voltage. This voltage could be applied directly to a loudspeaker, but it is so small that no audible sound will result. All sound systems provide a significant amount of amplification to increase this voltage before it is applied to the speakers. As a parenthetical comment, the buyer of such systems generally pays more when he or she insists that the voltage have a low distortion as it undergoes this amplification process.

In another example, the output of an audio CD player is usually insufficient to drive serious speaker systems and thus requires amplification.

Most amplifiers today are based on solid-state devices or integrated circuits. Vacuum tubes are employed in only a few specialized applications, but we shall be mainly concerned with the overall operating characteristics of amplifiers and the ways they can be used to accomplish a specific objective. We have already mentioned the effects of frequency and amplitude distortion in Sec. 2.9. Such effects can certainly be present in amplifiers if they are driven beyond the limits of their linear operation. We shall examine some techniques to minimize such distortion.

EFFECT OF FEEDBACK

Consider the amplifier schematic, shown in Fig. 4.28, having an input voltage E and an output voltage E_o with a gain of A. Without the feedback-attenuator loop we would have

$$A = \frac{E_o}{E_i} \tag{4.32}$$

Now, we add the feedback loop which attenuates the output voltage by a factor k and feeds it back into the input. The feedback voltage is $E_f = kE_o$, which is negative and subtracted from the input voltage, resulting in an input to the amplifier of $E = E_i - E_f$. Now, the gain of the amplifier and feedback system is

$$A_f = \frac{E_o}{E_i} = \frac{E_o}{E_o/A + kE_o} = \frac{A}{1 + kA} \tag{4.33}$$

If the amplifier gain is very large, that is, $kA \gg 1$, then the gain is very nearly

$$A_f \approx \frac{A}{kA} = \frac{1}{k} \tag{4.34}$$

Note that the gain of the overall system becomes constant so long as kA is sufficiently large. We may define the input and output impedances for the feedback system as

$$Z_{i,f} = \frac{E_i}{I_i} \quad \text{and} \quad Z_{o,f} = \frac{E_o}{I_o}$$

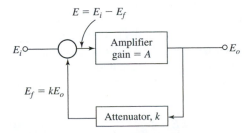

Figure 4.28 Amplifier with feedback loop.

respectively. The input impedance for the amplifier alone is

$$Z_i = \frac{E}{I_i} = \frac{E_o}{A Z_i}$$

From Eq. (4.33)

$$E_i = E_o \left(k + \frac{1}{A} \right)$$

Combining these relations, we obtain

$$Z_{i,f} = \frac{E_o(k + 1/A)}{E_o/A Z_i} = Z_i(1 + kA) \qquad \textbf{[4.35]}$$

Thus, we find that by the use of feedback we can increase the input impedance by a factor of $1 + kA$, or by a very large factor when the amplifier gain is large. The voltage appearing to drive the output impedance is AE and the output voltage is

$$E_o = AE - I_o Z_o$$

For $E_i = 0$ the input voltage to the amplifier will be $E = -kE_o$ and

$$E_o = -AkE_o - I_o Z_o$$

$$E = \frac{I_o Z_o}{kA + 1} \qquad \textbf{[4.36]}$$

The effective output impedance for the system is then

$$Z_{o,f} = \frac{E_o}{I_o} = \frac{Z_o}{1 + kA} \qquad \textbf{[4.37]}$$

So, the effect of feedback is to lower the output impedance by a factor of $1 + kA$. We have already mentioned that an amplifier will only be able to accept a limited range of input voltages before it can be driven into nonlinear behavior. The effect of feedback is to extend this range greatly because the feedback voltage lowers the voltage which is presented to the input of the amplifier.

Now, consider a typical frequency-response curve for an amplifier, as shown in Fig. 4.29. The curve is flat up to a certain frequency and then drops off; that is, the gain which can be delivered by the amplifier is reduced at higher frequencies. However, when negative feedback is employed, lower gain is experienced (by a factor of k), but this has the effect of extending the frequency response.

Negative feedback also has the effect of reducing amplifier noise which is present in the output signal. Consider the two cases shown in Fig. 4.30. In (a) we have the open-loop amplifier and the noise N is assumed to appear at the output terminal. In (b) we have the negative feedback configuration. As before, we can write $E = E_i - kE_o$ and the output voltage is

$$E_o = AE + N = A(E_i - kE_o) + N \qquad \textbf{[4.38]}$$

so that

$$E_o = \frac{AE_i + N}{1 + kA} \qquad \textbf{[4.39]}$$

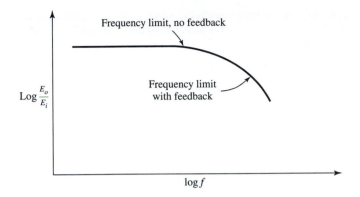

Figure 4.29   Typical amplifier frequency-response curve.

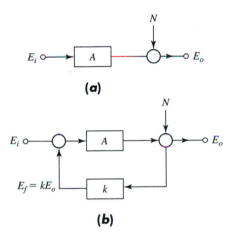

Figure 4.30   Effect of feedback on noise.

For $kA \gg 1$ this reduces to

$$E_o = \frac{E_i}{k} \pm \frac{N}{kA} \qquad\qquad \text{[4.40]}$$

where the $\pm$ indicates that the noise is random.

EFFECT OF FEEDBACK ON SIGNAL-TO-NOISE RATIO.   An amplifier has a gain of | **Example 4.5**
1000 and a signal-to-noise (S/N) ratio of 1000 in the open-loop configuration. Calculate the
S/N ratio when negative feedback with $k = 0.1$ is used.

Solution

For $E_i = 1$ the open-loop output would be $A E_i = E_o$.

$$E_o = 1000 \pm 1$$

that is, the noise is $1/1000$ of the output signal.

To produce this same output signal with the feedback arrangement, we would need an input signal of $E_i = 100$ because the gain is $A_f = 1/k = 10$. Then, from Eq. (4.40) the output would be (N is still 1.0)

$$E_o = \frac{100}{0.1} \pm \frac{1}{(0.1)(1000)}$$
$$= 1000 \pm 0.01$$

Now, the S/N ratio is

$$\text{S/N} = \frac{1000}{0.01} = 100,000$$

So, by the use of feedback we have increased the S/N ratio by a factor of 100.

4.8 DIFFERENTIAL AMPLIFIERS

A *differential* or *balanced* amplifier is a device that provides for two inputs and an output proportional to the *difference* in the two input voltages. For good operation it is necessary to have well-matched internal components and provisions for trimming the system so that a zero differential voltage input will produce a zero output. The differential amplifier is particularly useful for amplification and measurement of small signals subjected to stray electric fields (typically line voltage at 60 Hz and 115 V). Such stray fields can induce substantial voltages in the input lines unless stringent efforts are made to provide shielding. However, the application of equal voltages to the inputs or equal ac voltages of the same phase will cancel out. This feature is called *common mode rejection*. The descriptive property of the amplifier is called the *common mode rejection ratio* (CMRR), defined as the common mode signal at input divided by the common mode signal at output.

$$\text{CMRR} = \frac{\text{CM}_i}{\text{CM}_o} \qquad \textbf{[4.41]}$$

Values greater than 10^6 (120 dB) are not uncommon.

4.9 OPERATIONAL AMPLIFIERS

An operational amplifier (op-amp) is a dc differential amplifier incorporating many solid-state elements in a compact package and shown schematically in Fig. 4.31. It operates from dc to some upper frequency limit of the order of 1.0 MHz. The (+) input is called the *noninverting* input because the output from this source is in phase with the input. The *inverting* input (−) has the opposite behavior; that is, the output resulting from that source is 180° out of phase with the input. Op-amps using npn transistors will have input impedances of greater than 1 MΩ, while those constructed with field-effect transistors may have input impedances of the order of 10^{12} Ω. Because of these high impedances the input current for op-amps is essentially zero.

The output impedance of op-amps is very low, typically less than $1.0\ \Omega$, and the open-loop gain is very high, of the order of 10^6 (120 dB) [15].

In drawing the schematic diagrams for op-amps it is common practice to leave off the power supply and null adjustment inputs. Of course, one recognizes that these inputs must always be provided. With this brief description of the general characteristics of operational amplifiers, we now examine some specific amplifier-feedback configurations which are useful in experimental measurements. Figure 4.32 gives the schematics for the examples to be discussed.

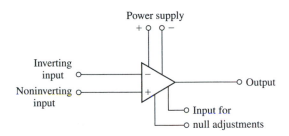

Figure 4.31   Schematic for operational amplifier.

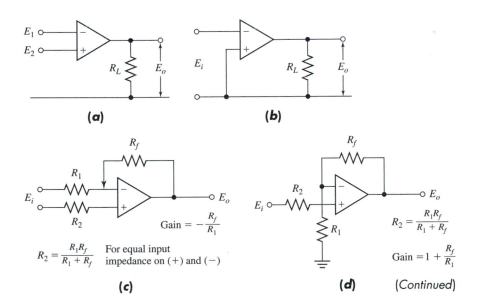

(a)

(b)

$$R_2 = \frac{R_1 R_f}{R_1 + R_f} \quad \text{For equal input}\atop \text{impedance on } (+) \text{ and } (-)$$

$$\text{Gain} = -\frac{R_f}{R_1}$$

(c)

$$R_2 = \frac{R_1 R_f}{R_1 + R_f}$$

$$\text{Gain} = 1 + \frac{R_f}{R_1}$$

(d)   (*Continued*)

Figure 4.32   Operational amplifier configurations. (*a*) Open-loop, differential input; (*b*) open-loop, single input; (*c*) inverting, differential input; (*d*) noninverting; (*e*) noninverting voltage follower; (*f*) voltage follower with gain greater than unity; (*g*) voltage follower with resistance in feedback; (*h*) inverting summer or adder; (*i*) noninverting adder; (*j*) circuit for high-input impedance and common mode rejection; (*k*) integrator; (*l*) differentiator; (*m*) current-to-voltage converter; (*n*) charge amplifier.

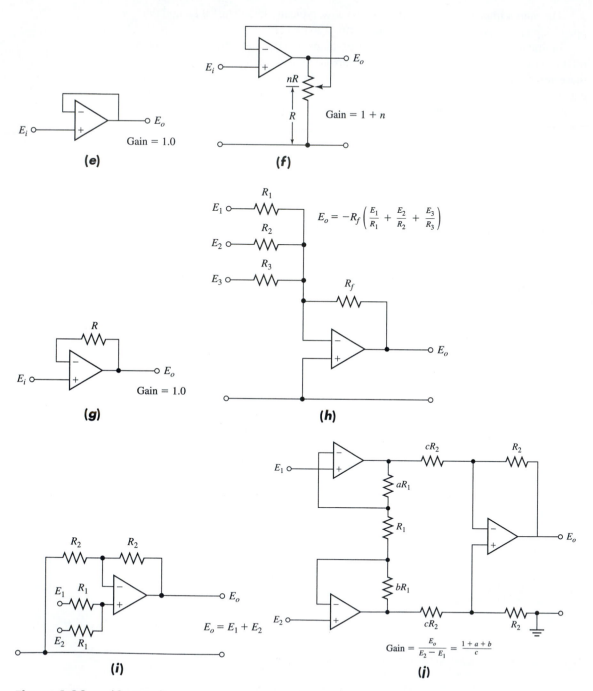

Figure 4.32   (Continued)

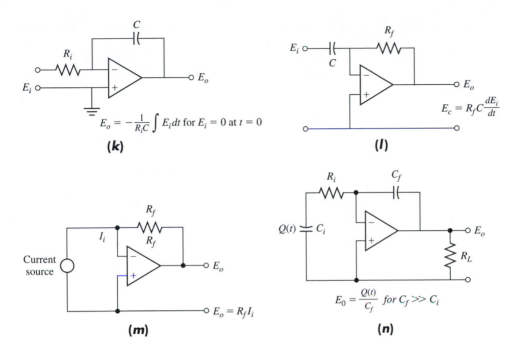

$$E_o = -\frac{1}{R_i C} \int E_i dt \text{ for } E_i = 0 \text{ at } t = 0$$

(k)

$$E_c = R_f C \frac{dE_i}{dt}$$

(l)

$$E_o = R_f I_i$$

(m)

$$E_0 = \frac{Q(t)}{C_f} \text{ for } C_f \gg C_i$$

(n)

Figure 4.32 *(Continued)*

The open-loop configurations in (*a*) and (*b*) are not often used because even a very small input will drive the output to full capacity of the power supply. Because the input impedance is high, the current presented to the input of the inverting terminal in (*c*) will be nearly zero. Designating the voltage present at the amplifier input as E_s requires that

$$\frac{E_i - E_s}{R_1} + \frac{E_o - E_s}{R_f} = 0$$

But $E_o = -AE_s$, or $E_s = -E_o/A$ so that

$$\frac{E_o}{R_f} = -\frac{E_i}{R_i} - \frac{E_o}{A}\left(\frac{1}{R_f} + \frac{1}{R_1}\right) \qquad \textbf{[4.42]}$$

Because the gain is very large, the last term drops out and the gain of the negative feedback arrangement becomes

$$\frac{E_o}{E_i} = -\frac{R_f}{R_1} \qquad \textbf{[4.43]}$$

The very high gain has the effect of causing E_s to be very nearly zero. A similar deviation could be used to obtain the gain for the noninverting configuration in Fig. 4.32*d*. The voltage follower in (*e*) has a gain of unity and can serve as a power amplifier to convert a high-impedance input into a low-impedance output. The potentiometer arrangement in (*f*) can be used to produce a gain greater than unity through the split between nR and R. Addition of an arbitrary resistor as in (g) still produces unity gain with the voltage follower.

Still further use of the zero-input current condition could be to derive the relations shown for the inverting and noninverting adder configurations shown in (h) and (i). A configuration for amplifying a high-impedance input with excellent common mode rejection is shown in (j).

In many experimental applications a transducer will be employed that gives a signal proportional to a certain rate, such as a fluid flowmeter, which indicates the flow rate of the fluid through the device. In addition to the rate of flow, the experimentalist may also wish to know the total quantity of fluid flowing over a given length of time. In other words, we need to integrate the input signal over a given time. The configuration in (k) accomplishes this objective. For very high gain we again have nearly zero-input voltage and sum the input currents to the amplifier as

$$\frac{E_i}{R_i} + C\frac{dE_o}{dt} = 0$$

Thus,
$$E_o = -\frac{1}{R_iC}\int E_i\, dt \tag{4.44}$$

Normally, a switching process is installed across the capacitor to make the voltage zero at the time the integration is to begin.

In a similar way, the op-amp can be used as a differentiator, as shown in Fig. 4.32l. If one has a constant *current* input as in (m), we would have

$$I_i - \frac{E_o}{R_f} = 0 \tag{4.45}$$

and we have a current-to-voltage convertor with $E_o = R_f I_i$.

With some transducers the input signal is electric charge. Examples are piezoelectric transducers which convert an impressed-force input to charge and condenser microphones used to measure acoustic phenomena. The configuration in Fig. 4.32n is used for such measurements. The charge amplifier is also employed to modify the output signals from charge-coupled devices (CCDs) in digital imaging systems. It then becomes part of the integrated circuit for the image-sensing device. See Appendix B for further information.

The particular amplifier configuration to be employed depends on the application, and one must carefully observe the operating characteristics of the sensors or transducers connected to the amplifier inputs. Impedance matching is quite important. If the source impedance is large, the amplifier circuit might cause sufficient loading so that most of the signal power is lost. In this case the input noise level in relation to the reduced signal may become excessive and reduce the overall accuracy of measurement. One way to avoid this problem is to employ first a voltage follower to reduce the impedance (the output impedance of an op-amp is low) and then follow with whatever other amplification and signal processing are required.

4.10 TRANSFORMERS

Transformers are also used to match impedance in many experimental situations. An ideal n-turn transformer is shown in Fig. 4.33 along with the associated four terminal variables, two voltages, and two currents. The equations relating the voltages and currents are

$$v_2 = nv_1$$
$$i_2 = \frac{i}{n}i_1$$

[4.46]

Dividing the first by the second results in the following:

$$\frac{v_2}{i_2} = n^2\frac{v_1}{i_1}$$

and as long as the voltages and currents are single-frequency sinusoids, we may state that the output impedance Z_2 is related to the input impedance Z_1 by the equation

$$Z_2 = n^2 Z_1$$

[4.47]

This equation states that the output impedance of an n-turn ideal transformer is n^2 greater than the input impedance. As a concrete example, consider an experiment in which we need to apply the 1000-Hz output voltage of an operational amplifier to a loudspeaker to produce an audible sound. The op-amp has an output impedance of 50 Ω at this frequency, and the loudspeaker has an equivalent impedance of 3 Ω. For efficient power transfer we wish the loudspeaker impedance to "look" as much as possible like 50 Ω. This can be accomplished by using a transformer, as shown in Fig. 4.34. Here

$$Z_1 = 3 \ \Omega$$

and by the definition of an ideal transformer

$$Z_2 = n^2 Z_1$$

or

$$50 = n^2(3)$$

and taking the closest integer value for n, we obtain

$$n = 4$$

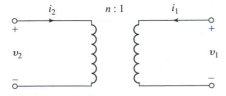

Figure 4.33 Ideal transformer.

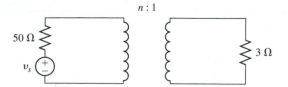

Figure 4.34   Impedance matching via transformer.

The insertion of a four-turn transformer between the output of the operational amplifier and the loudspeaker will ensure a close-to-optimum transfer of energy from the amplifier to the speaker.

This completes our discussion of the basics of operational amplifiers, their uses in measurement procedures, and impedance matching. We next turn to a condensed examination of power supplies since they are needed if one is to use operational amplifiers and other integrated circuits. They are also necessary for driving various types of detectors and transducers.

4.11   POWER SUPPLIES

There are two commonly used sources of dc energy: batteries and electronic modules which convert 60-Hz electric energy (available from power lines) to direct current. These modules are called *power supplies*.

Batteries are constantly being improved in terms of both their cost per unit energy and their weight and volume per unit energy. Despite this progress, they are still relatively expensive and bulky; they are preferred only in situations where portability is desired.

The specific details underlying the operation of dc power supplies are of minor interest here, but some general features will be pointed out. In many cases combinations of solid-state diodes are used to convert the pure 60-Hz sine wave available at an electric outlet to the positive, "rectified" wave shown in Fig. 4.35. This voltage contains a dc component as well as a series of sinusoids with frequencies which are integer multiples of 60 Hz. The "ripple" caused by these harmonics is undesirable and is reduced by further filtering, as will be discussed in the next section. The output of

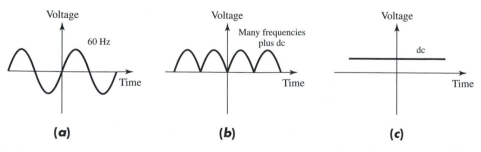

Figure 4.35   Rectification.

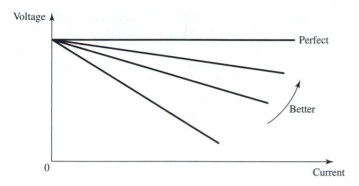

Figure 4.36   Power-supply voltage regulation.

a properly designed power supply is similar to that shown in Fig. 4.35 and is nearly constant. There is, to be sure, a nonzero amount of residual ripple, and this is normally specified by the manufacturer in percentage terms.

It is also important that the output voltage of a dc power supply remain constant as the current supplied to the circuits it is driving increases. If we represent the power supply as an ideal constant voltage source in series with some characteristic resistance, the goal of the power-supply designer is to reduce the value of this resistance as much as possible. Figure 4.36 illustrates various possible voltage-versus-current characteristics for power supplies. The curve labeled "perfect" indicates a practically unattainable ideal in that the voltage supplied is independent of the current drawn. The degree to which the output voltage is independent of the current supplied is termed *voltage regulation*. There are available integrated circuits which, using operational amplifiers, do an excellent job of maintaining a constant output voltage. For example, one such circuit will maintain an output voltage to within 0.05 V of 5 V for output currents ranging from 5 to 500 mA. Such regulation is more than adequate for most electronic circuits.

4.12   SIGNAL CONDITIONING

Noise is present in all physical situations in which measurements are attempted or information is conveyed, and noise is, in fact, a rather subtle ramification of the second law of thermodynamics. Noise is fundamentally another manifestation of the attempts of the physical universe to attain a state of randomness. Proper experimental design and procedures can greatly reduce its effects in many situations, as we shall now discuss.

A significant help to the experimenter is any a priori knowledge which may be available concerning the frequency ranges in which the desired signal exists. For example, if atmospheric temperatures are being measured, it is reasonable to expect that such temperatures will not change significantly over times shorter than several minutes. As a consequence, the meaningful output voltage of a thermocouple includes only the low-frequency part. Higher-frequency components are noise due to random fluctuations and can be eliminated without loss of experimental information, if, indeed, such elimination is possible. Another example would be the investigation by an electric utility company of the audible noise produced by an ultra-high-voltage transmission line. Because the human ear responds to acoustic signals only in the range of frequencies between 20 Hz and 18 kHz, the processing of sensor outputs can be safely limited to this frequency range.

This discussion naturally leads to an examination of the possibility of filtering electrical signals. This term means that we desire circuits which transmit or pass only certain bands or ranges of frequencies of an input signal. The unwanted parts of the signal can be characterized as noise, but in addition, there is also noise present in the frequency band of interest. Filtering will not solve all the problems, but it does provide a significant degree of experimental improvement. If a relatively narrow range of frequencies is important, it is usually possible to operate with much simpler electronic circuits for amplification and processing purposes than when a broad frequency range must be examined.

Various arrangements may be used for filter circuits, but they all fall into three categories: (1) lowpass, (2) highpass, and (3) bandpass circuits. Some filters are composed only of passive elements, while others involve amplification to eliminate losses and are termed *active filters* (discussed in Chap. 14). A lowpass filter permits the transmission of signals with frequencies below a certain cutoff value with little or no attenuation, while a highpass filter allows the transmission of signals with frequencies above a cutoff value. The bandpass filter permits the transmission of signals with frequencies in a certain range or band while attenuating signals with frequencies both above and below the limits of this band. The approximate performance for the three types of filters is shown in Fig. 4.37. The cutoff frequency is designated by f_c for both the highpass and lowpass filters, while the limits of frequency transmission for the bandpass filter are given by the symbols f_1 and f_2. Note carefully that the filters do not provide discontinuities in attenuation at the transition frequencies; that is, there is transmission of signals with frequencies above or below a given cutoff value, although the signal attenuation becomes more pronounced as the frequency moves away from the cutoff point.

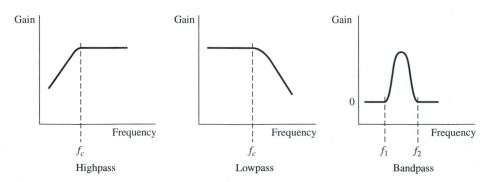

Figure 4.37   Approximate performance curves for three types of filters.

Table 4.2 Some simple *RL* and *RC* filter sections

Diagram	Type	Time Constant	Formula and Asymptotic Approximation
	A lowpass *RC*	$T = RC$	$\dfrac{E_o}{E_i} = \dfrac{1}{\sqrt{1+\omega^2 T^2}} \approx \dfrac{1}{\omega T}$ $\phi_A = -\tan^{-1}(R\omega C)$
	B highpass *RC*	$T = RC$	$\dfrac{E_o}{E_i} = \dfrac{1}{\sqrt{1+\dfrac{1}{\omega^2 T^2}}} \approx \omega T$ $\phi_B = \tan^{-1}\dfrac{1}{R\omega C}$
	C lowpass *RL*	$T = \dfrac{L}{R}$	$\dfrac{E_o}{E_i} = \dfrac{1}{\sqrt{1+\omega^2 T^2}} \approx \dfrac{1}{\omega T}$ $\phi_C = -\tan^{-1}\dfrac{\omega L}{R}$
	D highpass *RL*	$T = \dfrac{L}{R}$	$\dfrac{E_o}{E_i} = \dfrac{1}{\sqrt{1+\dfrac{1}{\omega^2 T^2}}} \approx \omega T$ $\phi_D = \tan^{-1}\dfrac{R}{\omega L}$

A summary of several types of passive filters is given in Fig. 4.38 on the next page. The various filter sections may be used separately or in combination to produce sharper cutoff performance. In the *m*-derived sections the quantity *m* is defined by

$$m = \sqrt{1 - \left(\frac{f_c}{f_\infty}\right)^2} \qquad \textbf{[4.48]}$$

for a lowpass filter and

$$m = \sqrt{1 - \left(\frac{f_\infty}{f_c}\right)^2} \qquad \textbf{[4.49]}$$

for a highpass filter, where f_c is the desired cutoff frequency and f_∞ is a frequency having high attenuation. When only one *m*-derived section is used, a value of $m = 0.6$ is recommended. When a filter is designed for a particular application, it is usually important that the impedance of the filter circuit be matched to the connecting circuit. Consideration of the matching of impedances in such circuits is beyond the scope of our discussions.

Table 4.2 presents some basic *RL* and *RC* filter sections. In a number of experimental situations these simple two-element filters offer significant improvement

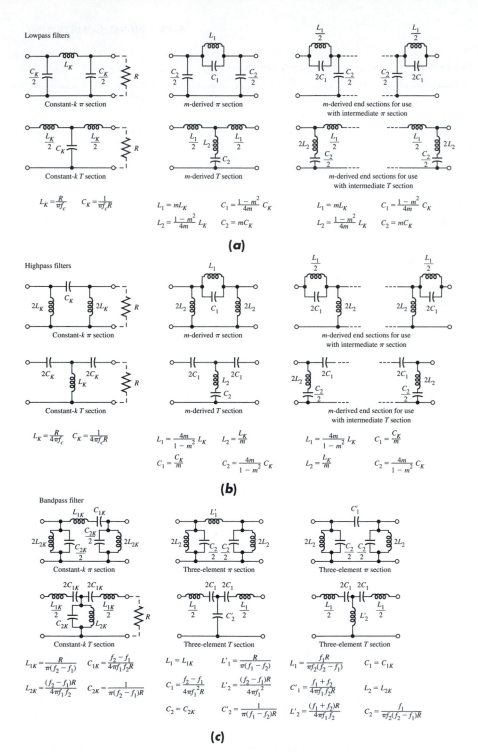

Figure 4.38 Basic filter sections and design formulas. *R* is in ohms, *C* is in farads, *L* is in henrys, and *f* is in hertz.

in noise reduction. The filters in this table do, of course, attenuate signal levels because they contain resistances. The designer often has to compensate for this resistive attenuation by including some additional amplifying circuits.

Sharper roll-off characteristics may be achieved using multiple filter sections like those shown in Table 4.2 placed in a series arrangement. In the lowpass RC filter with n sections in series the output-input relation is

$$\frac{E_o}{E_i} = \frac{1}{[1 + (\omega^2 T^2)^n]^{1/2}} \qquad \textbf{[4.50]}$$

with a phase shift of

$$\phi_{\text{total}} = \sum_{i=1}^{n} \phi_i(\omega) \qquad \textbf{[4.51]}$$

where the ϕ_i are the phase shifts of the sections. A series of n highpass RC filters produces the characteristic of

$$\frac{E_o}{E_i} = \frac{1}{\left[1 + \left(\dfrac{1}{\omega^2 T^2}\right)^n\right]^{1/2}} \qquad \textbf{[4.52]}$$

The filter circuits shown in Fig. 4.38 are commonly employed and are reasonably efficient passive-element types. It is possible to construct filters in RC and RL arrangements instead of the LC designs given in Fig. 4.38; however, the circuits which employ resistive elements are generally not as efficient because they remove energy from the system. Some simple RL and RC filter sections are shown in Table 4.2, along with the appropriate equation describing their input-output voltage characteristics. The phase-lag angle ϕ is also given.

The topic of *active filters* has been significantly influenced by progress in converting signals from analog to digital form before performing the filtering operations. Numerous references are available on filter design, and the interested reader may wish to consult Refs. [10] to [14] for more information.

DEGREE OF AMPLIFICATION AND ATTENUATION

A measurement of the degree of amplification or attenuation provided by a circuit is given by its gain or amplification ratio. Gain is defined as

$$\frac{\text{Output}}{\text{Input}}$$

The output and input quantities may be voltage, current, or power, depending on the application. Gain is a dimensionless quantity, but engineers speak of decibels of gain or loss in terms of the following definition involving a logarithmic ratio:

$$\text{Decibels} = 10 \log \frac{P_2}{P_1} \qquad \textbf{[4.53]}$$

where P_1 and P_2 are the input and output powers, respectively. The voltage or current gain may be defined in a similar manner.

Since power may be expressed as E^2/R or I^2R, for a resistive load we would have

$$\frac{P_2}{P_1} = \frac{E_2^2}{E_1^2} = \frac{I_2^2}{I_1^2}$$

Expressed in terms of voltages or current, the decibel notation would then be

$$\text{Decibels} = 20 \log \frac{E_2}{E_1} \qquad\qquad \textbf{[4.54]}$$

$$\text{Decibels} = 20 \log \frac{I_2}{I_1} \qquad\qquad \textbf{[4.55]}$$

For passive filters the ratio of output power to input power is less than 1, and, as a consequence, the gain in decibels is negative. To characterize such circuits with a positive quantity, engineers define an *insertion loss* by the following equation:

$$10 \log \frac{P_1}{P_2} \qquad\qquad \textbf{[4.56]}$$

One should note that the number of decibels of gain is simply the negative of the decibels of insertion loss.

Example 4.6 | **VOLTAGE AMPLIFICATION.**   A 1.0-mV signal is applied to an amplifier such that an output of 1.0 V is produced. The 1.0-V signal is then applied to a second amplification stage to produce an output of 25 V. For the three voltage points indicated, calculate the voltage in decibel notation referenced to (*a*) 1.0 mV and (*b*) 1.0 V. Also, calculate the overall voltage amplification in decibels.

Solution

The schematic for the amplifiers is shown in Fig. Example 4.6.

Figure Example 4.6

(*a*) For $E_0 = 1.0$ mV we have

$$20 \log \frac{E_1}{E} = \text{dB}_1$$

$$20 \log \frac{1.0 \times 10^{-3}}{1.0 \times 10^{-3}} = \text{dB}_1 \qquad \text{dB}_1 = 0$$

$$20 \log \frac{1.0}{1.0 \times 10^{-3}} = \text{dB}_2 \qquad \text{dB}_2 = 60$$

$$20 \log \frac{25}{1.0 \times 10^{-3}} = \text{dB}_3 \qquad \text{dB}_3 = 87.92$$

(b) For $E_0 = 1.0$ V we have

$$20 \log \frac{1.0 \times 10^{-3}}{1.0} = dB_1 \qquad dB_1 = -60$$

$$20 \log \frac{1.0}{1.0} = dB_2 \qquad dB_2 = 0$$

$$20 \log \frac{25}{1.0} = dB_3 \qquad dB_3 = 27.92$$

The overall amplification is calculated from

$$20 \log \frac{E_3}{E_1} = 20 \log \frac{25}{1.0 \times 10^{-3}} = 87.92 \, dB$$

Notice that this result could have been obtained from the decibel calculations in (a) and (b) for either reference level. In other words,

$$\text{Gain}_{13}(dB) = dB_3 - dB_1$$

when all are expressed on the same reference basis.

RC LOWPASS FILTER.   A simple RC circuit is to be used as a lowpass filter. It is desired that the output voltage be attenuated 3 dB at 100 Hz. Calculate the required value of time constant $T = RC$. | **Example 4.7**

Solution

Network A of Table 4.2 is the desired arrangement. We wish to have

$$\frac{E_o}{E_i} = -3 \, dB$$

so that

$$-3 = 20 \log \frac{E_o}{E_i}$$

or

$$\frac{E_o}{E_i} = 0.708$$

Thus,

$$\frac{E_o}{E_i} = \frac{1}{(1 + \omega^2 T^2)^{1/2}} = 0.708$$

where $\omega^2 = (2\pi f)^2 = [2\pi(100)]^2$.

We find that $T = 1.59 \times 10^{-3}$ s. The circuit might be constructed using a 1.0-μF capacitor and a 1.59-kΩ resistor.

RC HIGHPASS FILTER.   A highpass RC filter is to be designed for an attenuation of -3 dB at 150 Hz. Calculate the required value of the time constant and the attenuation of the filter at 100, 200, and 400 Hz. What is the limiting insertion loss at higher frequencies? | **Example 4.8**

Solution

Network B of Table 4.2 is the desired arrangement. As in Example 4.7, we find for 3-dB attenuation

$$\frac{E_o}{E_i} = 0.708 \qquad\qquad [a]$$

and

$$0.708 = \frac{1}{[1 + 1/(\omega^2 T^2)]^{1/2}} \qquad [b]$$

where

$$\omega^2 = (2\pi f)^2 = [2\pi(150)]^2 = 8.883 \times 10^5$$

Solving Eq. (b), we obtain

$$T = 1.064 \times 10^{-3} \text{ s}$$

and the circuit could be constructed with a 1.0-μF capacitor and a 1.06-kΩ resistor.
    At 100 Hz

$$\omega^2 T^2 = [2\pi(100)]^2 (1.064 \times 10^{-3})^2 = 0.4469$$

and at 200 Hz

$$\omega^2 T^2 = [2\pi(200)]^2 (1.064 \times 10^{-3})^2 = 1.788$$

From Eq. (b) the attenuation for 100 Hz is

$$\frac{E_o}{E_i} = \frac{1}{\left[1 + \dfrac{1}{0.4469}\right]^{1/2}} = 0.5558$$

At 200 Hz we have

$$\frac{E_o}{E_i} = \frac{1}{\left[1 + \dfrac{1}{1.788}\right]^{1/2}} = 0.8008$$

and at 400 Hz the result is

$$\frac{E_o}{E_i} = 0.9367$$

The corresponding decibel attenuations are obtained from Eq. (4.54):

$$\text{dB (100 Hz)} = 20\log (0.5558) = -5.1$$
$$\text{dB (200 Hz)} = 20\log (0.8008) = -1.93$$
$$\text{dB (400 Hz)} = 20\log (0.9367) = -0.57$$

At higher frequencies the attenuation approaches zero. At very low frequencies the attenuation approaches ωT, so for a low frequency of say, 10 Hz, we obtain $\omega^2 T^2 = 0.4469 \times 10^{-2}$, $\omega T = 0.06685$, or an attenuation of -23.5 dB.

Example 4.9 | **TWO-STAGE RC HIGHPASS FILTER.**   The single-stage highpass filter of Example 4.8 is to be replaced with two sections in series, still keeping the attenuation of -3 dB at 150 Hz. Determine the performance at 100, 200, and 400 Hz for this arrangement.

Solution

For this problem we have $n = 2$ stages and $E_o/E_i = 0.708$ from Example 4.8 so that we may use Eq. (4.52) to obtain

$$0.0708 = \frac{1}{\left[1 + \left(\dfrac{1}{\omega^2 T^2}\right)^2\right]^{1/2}} \qquad [a]$$

at $\omega^2 = 8.883 \times 10^5$ (150 Hz). We may solve to obtain

$$T^2 = 1.1284 \times 10^{-6} \text{ s}^2$$

Then at 100 Hz

$$\omega^2 T^2 = 0.4455$$

at 200 Hz

$$\omega^2 T^2 = 1.782$$

at 400 Hz

$$\omega^2 T^2 = 7.128$$

From Eq. (4.52) the corresponding attenuations are at 100 Hz

$$\frac{E_o}{E_i} = 0.4069 = -7.81 \text{ dB}$$

at 200 Hz

$$\frac{E_o}{E_i} = 0.8721 = -1.19 \text{ dB}$$

at 400 Hz

$$\frac{E_o}{E_i} = 0.9903 = -0.085 \text{ dB}$$

At a much lower frequency of 10 Hz the attenuation would be -47 dB.

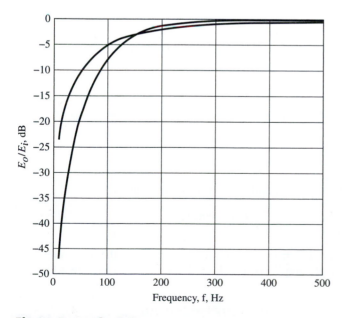

Figure Example 4.9

Comment

The use of series filter sections can produce a much sharper roll-off of the frequency response curve. The response curves for the single- and dual-stage filters are shown in Fig. Example 4.9. Note that the two response curves cross at the design point of −3 dB and 150 Hz.

OTHER NOISE-REDUCTION METHODS

Passive filters, as discussed and characterized above, offer one possible strategy for combating the effects of noise in experimental measurements. Remember that we must have some knowledge of where we expect our signals to appear in the frequency spectrum if filtering is to be of any value. We conclude this section by discussing a few other methods which are useful for minimizing noise.

For these techniques to be effective it is necessary that the signals of interest be repetitive. This may seem to be an exceptionally stringent requirement, but clever experimental procedures can result in its satisfaction in many situations. For example, to measure the vibration of a wall which results from an impulse of force, the proper experimental procedure is to make the force repetitive. In this way the signal of interest (the wall's vibration) will also be repetitive. In addition, any dc signal can always be converted into a repetitive form by "chopping" it by electrical or mechanical means. As long as such steps have been taken to ensure that the signal to be measured is repetitive, either of the following techniques may be used to reduce the effects of noise.

The first is based on the observation that the signal will be repetitive but the associated noise will not be. Consider taking the output of the transducer for two successive signals, adding them, dividing the resultant sum by 2, and then displaying the output. What will have happened? The signal of interest will be present, but two different noise components will have been added together and then divided by 2. Since the noise will be different in each of the two successive times, it follows, as long as the noise is uncorrelated, that the amount of noise present in the resultant signal will decrease. This can be more clearly seen by considering the following two sums which represent the first and second "pieces" of output from the transducer.

$$v_s(t) + n_1(t)$$
$$v_s(t) + n_2(t)$$

After these are added together and divided by 2, the result is

$$v_s(t) + \frac{n_1(t) + n_2(t)}{2}$$

and this sum is the desired signal v_s as well as an average of the noise present in the two "pieces." After N samples of the desired signal plus noise are added and averaged, the result is

$$v_s(t) + \frac{n_1(t) + n_2(t) + \cdots + n_N(t)}{N} \qquad \textbf{[4.57]}$$

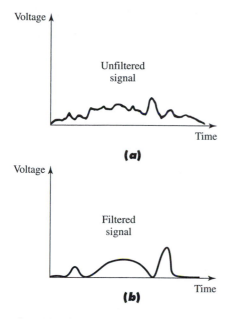

Voltage

Unfiltered
signal

Time

(a)

Voltage

Filtered
signal

Time

(b)

Figure 4.39   CAT filtering.

This shows the output of such a noise-reduction system to be composed of the desired signal $v_s(t)$ as well as an average of various samples of the noise. One may wonder what assurances can be offered that the average of noise samples will be smaller than a given segment $n(t)$. A detailed answer to that question involves a careful study of stochastic processes and is not necessary for our purposes. In physical terms we can be sure that this averaging process will reduce the effective level of noise as long as the origins of the noise are truly unrelated to the processes giving rise to the signal of interest v_s. If such is the case, the noise and signal are said to be uncorrelated, and averaging will improve the ratio of signal to noise in the system and, as a consequence, improve the quality and accuracy of our measurements.

The averaging process clearly requires storing of "pieces" of the transducer output in computer memory, followed by appropriate analysis techniques. Figure 4.39 indicates a sketch of the possible improvement that may be accomplished with such operation.

PHASE-SENSITIVE DETECTOR OR LOCK-IN AMPLIFIER

We conclude this section with a discussion of the lock-in amplifier. This instrument uses a principle different from that used by a CAT to reduce noise. An electronic circuit known as a *phase-sensitive detector* is the key to understanding the lock-in amplifier. Such a detector has two input terminals, and, assuming that the two input signals are sinusoids with the same frequency, it produces an output which depends on the cosine of the phase angle between the two signals. For the detector sketched

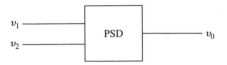

Figure 4.40   Phase-sensitive detector.

in Fig. 4.40 the input voltages are given by

$$v_1(t) = V_1 \sin(\omega t)$$
$$v_2(t) = V_2 \sin(\omega t + \phi)$$

and the output voltage is

$$v_o = K V_1 V_2 \cos \phi \qquad \text{[4.58]}$$

where K is a constant.

With an understanding of this element, the basic behavior of a lock-in amplifier can be described quite simply. Recall that we have a repetitive signal of interest accompanied by some inevitable noise. This entire transducer output is passed through a narrow bandpass filter which has its pass frequency located at the frequency of repetition of the input signal. The output of this filter constitutes one of the inputs to the phase-sensitive detector. The other necessary input is a sinusoid with the same frequency and a phase which is maintained constant with respect to the chopper used to modulate the transducer's output. This means that the desired signal (or, at least, that part of it which survived the filtering operation) and the reference signal have a fixed phase (ϕ_f) maintained between them. Any noise which is present in the filter output will not maintain such a fixed phase, and the output of the phase-sensitive detector can be written as

$$v_o(t) = K V_s V_R \cos \phi_f + K V_N V_R \cos \phi(t) \qquad \text{[4.59]}$$

The system is sketched in Fig. 4.41.

The phase between the reference voltage and the noise will change as time progresses, and the cosine of this angle will take on both positive and negative values. If the noise is truly uncorrelated, the time average of the second term in the last equation will be zero. The last part of a lock-in amplifier is a suitable integrator which serves to time-average. The output of the system depends only on the amplitude and phase of the signal of interest v_s.

$$\overline{v_o(t)} = \frac{1}{T} \int_0^T v_o(t)\, dt = K V_s V_R \cos \phi_f \qquad \text{[4.60]}$$

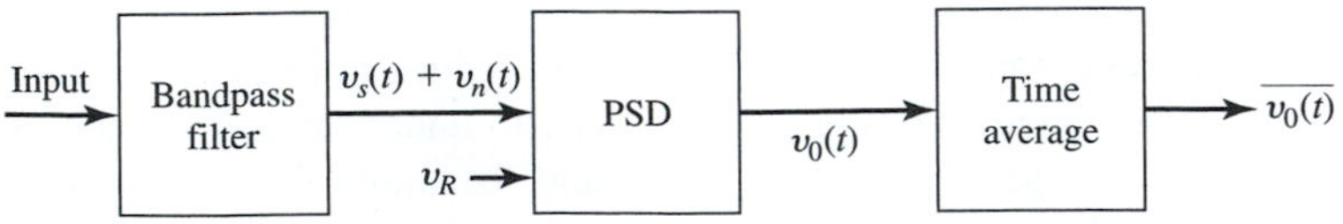

Figure 4.41   Lock-in amplifier.

Well-designed lock-in amplifiers result in significant S/N ratio improvements. In fact, the degree of improvement made possible by their careful design and use is nearly incredible. When we are faced with what appears to be a nearly hopeless situation in terms of the desired signal being buried in noise, this is a method which should be considered. Manufacturers of these instruments provide very helpful documentation and application assistance.

4.13   THE ELECTRONIC VOLTMETER

The electronic voltmeter (EVM) employs either solid-state analog or digital elements for operation. The EVM is one of the most useful laboratory devices for the measurement of voltage. It may be used for both ac and dc measurements and is particularly valuable because of its high-input-impedance characteristics, which make it applicable to the measurement of voltages in electronic circuits.

A block-diagram schematic of a simple EVM is shown in Fig. 4.42. The input voltage is connected through appropriate terminals to the function switch. If a dc voltage is to be measured, the signal is fed directly to the range selector switch operating as a voltage-divider circuit where the signal is reduced to a suitable range for the succeeding amplifier circuit. The voltage signal from the voltage divider is then applied to the amplifier stage, whose voltage output is used to drive a conventional D'Arsonval movement or digital display for readout purposes. The output from the amplifier stage could also be connected to a recorder, or oscilloscope, for the measurement of transient voltages.

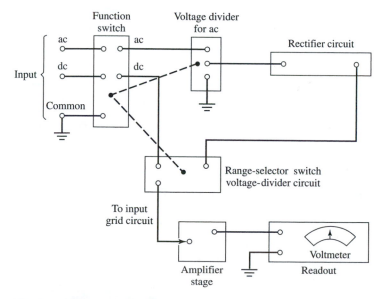

Figure 4.42   Block-diagram schematic of an electronic voltmeter.

For an ac voltage measurement the signal is fed through a voltage divider and then to a rectifier circuit, which produces a dc voltage proportional to the input ac signal. The dc voltage is then impressed on the voltage-divider network of the range-selector switch. The remainder of the circuit is the same as in the case of the dc measurement.

Various modifications of the above arrangement may be used, depending on the range of voltages to be measured. For very low-voltage ac signals an amplification stage may be added to the input instead of the voltage-divider arrangement. For very low dc signals a chopper device might be used at the input to produce an ac signal, which is more easily amplified to the voltage levels that may be handled with conventional circuitry.

The important point is that the input impedance of the EVM is very high, usually greater than 100 MΩ, so that the measured circuit is not loaded appreciably and indicated voltage more closely represents the true voltage to be measured. Example 4.10 illustrates the influence that the meter can exert on a circuit and reduction in error of measurement that can result when an EVM is used. It may be noted that a correction may be made for the meter impedance when the impedance of the measured circuit is known.

4.14 DIGITAL VOLTMETERS

As mentioned earlier in this chapter, a wide variety of voltmeters are now available that provide a digital output instead of the pointer-scale arrangement. A wide choice of input-signal modifiers, ac-dc converters, resistance dc converters, amplifiers, and more, is available so that, in a real sense, the digital voltmeter now enables the experimentalist to make precision measurements over a broad range of variables. Not surprisingly, the cost of digital voltmeters is directly related to their accuracy and versatility.

| **Example 4.10** | **ERROR IN VOLTAGE MEASUREMENT.**   The voltage at points A and B is to be measured. A constant 100 V is impressed on the circuit as shown. Two meters are available for the measurement: a small volt-ohmmeter with an internal impedance of 100,000 Ω and a range of 100 V, and an EVM with an input impedance of 17 MΩ. |

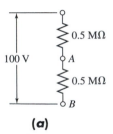

(a)

Figure Example 4.10(a)

Compare the error in measurement with each of these devices.

Solution

The true voltage, by inspection, is 50 V. With the volt-ohmmeter connected in the circuit, there results

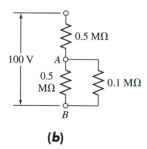

(b)

Figure Example 4.10(b)

and the voltage at A and B is

$$E_{AB} = 100\frac{1/[(1/0.5)+(1/0.1)]}{1/[(1/0.5)+(1/0.1)]+0.5}$$

$$= 100\left(\frac{0.0833}{0.833+0.5}\right) = 14.3 \text{ V}$$

or an error of -71 percent.

With the EVM connected in the circuit, there results

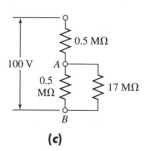

(c)

Figure Example 4.10(c)

and the voltage at A and B is

$$E_{AB} = 100\frac{1/[(1/0.5)+(1/17)]}{1/[(1/0.5+1/17)]+0.5}$$

$$= 100\left(\frac{0.4857}{0.4857+0.5}\right) = 49.27 \text{ V}$$

or an error of -1.46 percent.

4.15   THE OSCILLOSCOPE

We have seen that the EVM offers the advantage that it can be used to measure voltage without a substantial "loading" of the transducer or other source supplying the signal. The cathode-ray oscilloscope (CRO) is similar to the EVM in that it also has a high-input impedance and is a voltage-measuring device. In addition, the CRO is capable of displaying voltage signals as functions of time.

The heart of any oscilloscope is the cathode-ray tube (CRT), which is shown schematically in Fig. 4.43. Electrons are released from the hot cathode and accelerated toward the screen by the use of a positively charged anode. An appropriate grid arrangement then governs the focus of the electron beam on the screen. The exact position of the spot on the screen is controlled by the use of the horizontal and vertical deflection plates. A voltage applied on one set of plates produces the x deflection, while a voltage on the other set produces the y deflection. Thus, with appropriate voltages on the two sets of plates, the electron beam may be made to fall on any particular spot on the screen of the tube. The screen is coated with a phosphorescent material, which emits light when struck by the electron beam. If the deflection of the beam against a known voltage input is calibrated, the oscilloscope may serve as a voltmeter. Since voltages of the order of several hundred volts are usually required to produce beam deflections across the entire diameter of the screen, CRT is not directly applicable for many low-level voltage measurements, and amplification must be provided to bring the input signal up to the operating conditions for the CRT.

A schematic of the CRO is shown in Fig. 4.44. The main features are the CRT, as described above, the horizontal and vertical amplifiers, and the sweep and synchronization circuits. The sweep generator produces a sawtooth wave which may be used to provide a periodic horizontal deflection of the electron beam, in accordance with some desired frequency. This sweep then provides a time base for transient voltage measurements by means of the vertical deflection. Oscilloscopes provide internal circuits to vary the horizontal sweep frequency over a rather wide range as well as external connections for introducing other sweep frequencies. Internal switching is also provided, which enables the operator of the scope to "lock" the sweep frequency onto the frequency impressed on the vertical input. Provisions are also made for

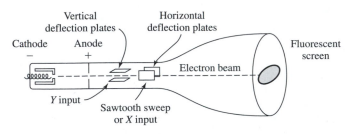

Figure 4.43   Schematic diagram of a cathode-ray tube (CRT).

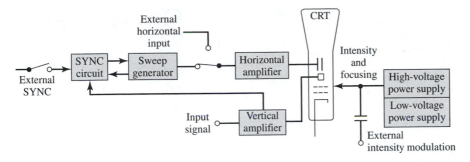

Figure 4.44   Block-diagram schematic of an oscilloscope.

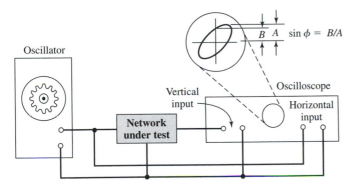

Figure 4.45   Schematic illustrating use of oscilloscope for phase measurements.

external modulation of the intensity of the electron beam. This is sometimes called the *z-axis input*. This modulation may be used to cause the trace to appear on the screen during certain portions of a waveform and to disappear during other portions. It may also be used to produce traces of a specified time duration on the screen of the CRT so that a time base is obtained along with the waveform under study.

A *dual-beam oscilloscope* provides for amplification and display of two signals at the same time, thereby permitting direct comparison of the signals on the CRT screen.

The CRO may be used to measure phase shift in an approximate fashion, as shown in Fig. 4.45. An oscillator is connected to the input of the circuit under test. The output of the circuit is connected to the CRO vertical input, whereas the oscillator signal is connected directly to the horizontal input. The phase-shift angle ϕ may be determined from the relation

$$\phi = \sin^{-1} \frac{B}{A} \qquad \textbf{[4.61]}$$

where B and A are measured, as shown in Fig. 4.45. For zero phase shift the ellipse will become a straight line with a slope of $45°$ to the right; for $90°$ phase shift it will become a circle; and for $180°$ phase shift it will become a straight line with a slope of $45°$ to the left.

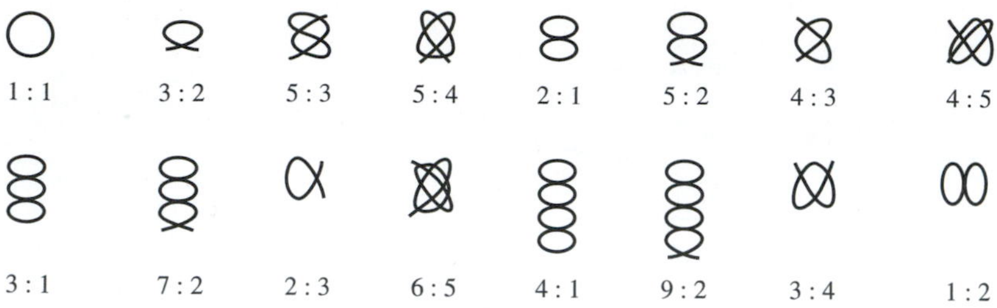

1 : 1   3 : 2   5 : 3   5 : 4   2 : 1   5 : 2   4 : 3   4 : 5

3 : 1   7 : 2   2 : 3   6 : 5   4 : 1   9 : 2   3 : 4   1 : 2

Figure 4.46   Lissajous diagrams for various frequency ratios as indicated.

The CRO offers a convenient means of comparing signal frequencies through the use of Lissajous diagrams. Two frequencies are impressed on the CRO inputs, one on the horizontal input and one on the vertical input. One of these frequencies may be a known frequency as obtained from a variable frequency oscillator or signal generator. If the two input frequencies are the same, the patterns that are displayed on the CRT screen are called Lissajous diagrams. There is a distinct relationship that governs the shape of these diagrams in accordance with the input frequencies:

$$\frac{\text{Vertical input frequency}}{\text{Horizontal input frequency}} = \frac{\text{Number of vertical maxima on Lissajous diagram}}{\text{Number of horizontal maxima on Lissajous diagram}} \qquad \textbf{[4.62]}$$

Some typical shapes for the Lissajous diagrams are shown in Fig. 4.46. It may be noted that these shapes can vary somewhat depending on the phase relation between the input signals.

DIGITAL OSCILLOSCOPES

We have seen in Sec. 4.2 that analog signals may be converted to digital signals through a sampling process. The digital oscilloscope operates according to this principle. Instead of displaying the analog signal directly, it first performs an analog-to-digital conversion and then stores the digital signals in a buffer memory. The signal may then be displayed on the CRT screen as points. With 12-bit conversion, 4096 vertical data locations are available.

Because the digital signal is stored, it may be recalled and reexamined on an expanded scale. In addition, the signal may be stored on inexpensive auxiliary devices for later study and manipulation with a computer.

SAMPLING OSCILLOSCOPES

Each oscilloscope is characterized by its manufacturer by a set of performance specifications. One of the most important is the maximum response frequency. This number indicates the highest-frequency sinusoid which the instrument will be able to follow

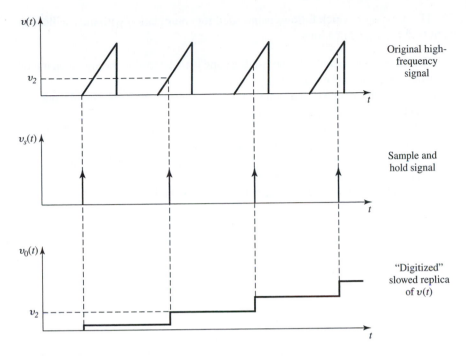

Figure 4.47   Waveforms in a sampling oscilloscope.

and display accurately. In terms of internal components, this maximum frequency is primarily determined by the frequency response of the amplifiers used in the oscilloscope.

A technique for examining higher frequencies, as long as they are repetitive, is known as sampling and is the electronic equivalent of the stroboscope. Figure 4.47 illustrates the operation of a sampling oscilloscope. The first waveform shown is repetitive and has a frequency which exceeds the maximum response rate of the oscilloscope amplifiers. In order to "slow" this waveform a series of electronic "snapshots" are taken and displayed, as shown in the second waveform. This "sampling" occurs at slightly later times in each cycle of the original waveform. As a consequence, the oscilloscope is able to display a digital approximation of the original signal.

4.16   OSCILLOSCOPE SELECTION

Advances in digital oscilloscope technology have been staggering and offer so many features as to be confusing even to an expert. In this section we shall outline a procedure one might use for oscilloscope selection in the form of questions which should be asked and some of the choices which may be available. This discussion is not to imply that analog scopes do not have a valid place in the laboratory because they have the advantage of immediate display of the signal as it occurs. Indeed, some oscilloscopes are available which combine analog plus digital features.

The discussion which follows is intended for coordination with the oscilloscope manufacturer's specifications.

1. The first step in selection of an oscilloscope is to determine the signal characteristics which are to be studied.
   a. The repetition rate of the signal should be determined: Is the signal a one-shot occurrence or a repetitive waveform?
   b. The frequency range of the signal should be determined along with the rise-time characteristics which must be measured. The scope selection should be based on the shortest rise time which is expected. The bandwidth is then approximately 0.35 per rise time.
   c. The vertical sensitivity required should be specified by the lowest voltage measurement which will be performed. Obviously, the smaller the input voltage the scope can accommodate, the higher the cost.
   d. The dynamic range of the input signal should be established. This is commonly expressed as zero to some upper voltage limit.
   e. Set the requirements for dc offset. If a measurement is made of a signal with a dynamic range of 1 V at a level of 40 V, then dc offset of 40 V should be available so that the full dynamic range can be displayed full-size on the screen.
   f. The horizontal scan should be specified to take into account the largest time which needs to be viewed and the smallest increment of time for viewing. These variables will determine the timing resolution required.
2. In addition to the above specification of the signal characteristics to be studied, one must also specify the overall accuracy which is required for the measurements. Is only a minimal visual observation required or must one make very precise waveform measurements? One may also specify multiple signal displays at the same time. It is possible to have a relatively simple two-channel scope or ones with over 100 input channels. Finally, there should be a specification of the storage requirements and whether digital storage, processing, and hard-copy output are desirable.
3. Once the signal characteristics and features are selected, one may then begin to select from among the technologies which are available. We outline some of these technologies in the paragraphs below:
   a. *Real-time digital sampling oscilloscope (DSO).* In this scope the digitizer samples the entire input waveform in one pass with a single trigger. This type of scope is appropriate for one-shot signal applications.
   b. *Random equivalent time DSO.* Equivalent sampling operates on a repetitive signal to obtain a very large number of samples of the signal. In random equivalent time sampling several samples are taken for each trigger over a large number of trigger events. Because of the sampling process, the bandwidth of the scope can far exceed its sampling rate. Bandwidths as high as 1 GHz are available.
   c. *Analog and digital sequential sampling.* In this application a repetitive signal is required as in the random sampling above, except that only one sample

is acquired for each trigger, with a constant delay after the trigger. For each subsequent sample the trigger to sample interval is increased by a fixed time. The fixed-time interval can be made as short as femtoseconds ($1 \text{ fs} = 10^{-15}$ s). In these scopes the input is sampled prior to signal conditioning (amplification), thus permitting very high bandwidths up to 14 GHz.

d. *Analog scopes*. The major advantage of an analog scope is that it gives a direct representation of the signal and the fastest update possible. This results from the fact that only beam retrace and trigger times are required between sweeps and thus the scope operates thousands of times faster than the transferral of data in and out of memory required in digital technology.

e. *CRT storage*. In a CRT storage oscilloscope the events are stored on the face of the CRT screen itself. This storage technique is ideal for capturing transient events.

f. *Triggering capability*. Several choices of triggering are available: auto-level-triggering for "hands-off" operation; peak-to-peak auto triggering with automatic-level limits; vertical mode triggering for stable viewing of two or more signals unrelated in time; single-sweep operation to capture a transient pulse for CRT photography; high- and low-frequency reject coupling for stable triggering on noisy signals; and time-, level-, and event-qualified triggering to capture a signal that is too high, too low, too wide, too narrow, too soon, too late, missing, or extra.

To summarize the above three sections, we say that the scope will be selected on the basis of bandwidth and rise time, sampling rate, horizontal resolution and record rate, horizontal magnification, update rate to get ready for new triggering, visual writing rate, the possibility of dual-time bases, vertical sensitivity, vertical accuracy and resolution, and the triggering capability. Once these specifications are met (which are quite formidable), we may then select one or more of the following special features:

1. *Automatic setup*. With this feature, a single button optimizes settings to acquire and display a signal. The proper sweep speed, vertical deflection, trigger level, position, and intensity are automatically calculated for a usable display.

2. *Store/recall*. This feature allows the operator to store oscilloscope settings in memory for later recall. This is very useful when certain operations are to be repeated. It is akin to driver's seat memory on luxury automobiles.

3. *On-board countertime and digital multimeter (DMM)*. This feature may be built into the scope to allow one to make frequency, period, width, rise-fall time, and propagation delay determinations at the touch of a button. All of this can be accomplished at the same time the operator is viewing the signal.

4. *On-board waveform calculations*. Depending on the scope selected (and cost), this feature can provide elaborate waveform calculations, including differentiation, integration, interpolation, smoothing, averaging, waveform passfail, standard waveform math, and fast Fourier transforms.

5. *Record capabilities*. In addition, the various parameters may be recorded and stored on floppy disks for later data processing or hard-copy printout.

From the above discussion it is obvious that the oscilloscope measurement capabilities available to the experimentalist are almost boundless. Because of the rapid development in this area of technology, an individual will be well advised to consult appropriate manufacturer's representatives for up-to-date information.

4.17 OUTPUT RECORDERS

The output of an electronic circuit or meter may be recorded in a number of ways. The storage oscilloscope is useful for recording dynamic signals for later analysis. Any number of connecting devices make it possible to store output in a desktop or laptop computer, portable magnetic hard drives, recording CDs, digital tape recorders, and video tape recorders for visual displays. Low-frequency signals may be recorded on strip-chart recorders that employ self-balancing potentiometric circuits and mechanisms. These devices are only applicable up to about 5 Hz but are very useful for recording thermocouple outputs and other low-voltage signals, or in applications where an immediate display is available for an extended time period of measurement.

The relatively low costs of digital storage of information suggest that this mode of output recording will be even more prevalent in the future. Some relative values for storage capability for digital media are digital flash memory "sticks" (approximately 13×50 mm); 256 GB, writable CD; 700 MB, writable DVD; 9 GB, and digital hard drives in excess of 3 TB.

4.18 COUNTERS—TIME AND FREQUENCY MEASUREMENTS

The engineer is called upon to perform counting-rate and frequency measurements over an extremely broad range of time intervals. A determination of revolutions per minute (rpm) on a slow-speed diesel engine might involve a simple mechanical revolutions counter and a hand-operated stopwatch. In this case the accuracy of the measurement would depend on the human-response time in starting and stopping the watch (about 0.2 s). The same measurement performed on an automobile engine might utilize an electric transducer which generates a pulse for each revolution of the engine. The pulses could be fed to some type of counting device, which would establish the number of pulses produced for a given increment of time. The accuracy of measurement would again depend on the accuracy of the specification of the time interval. In both of these simple cases a circular frequency measurement is being performed through the combination of a counting measurement and time-interval measurement. Most frequency measurements are performed by some type of counting operation.

A large variety of electronic counters is available commercially. These instruments have internal circuitry which enables them to be used for measurement of frequency, period, or time intervals over a very wide range. These instruments usually contain four sets of internal circuits:

1. Input-signal conditioning circuits, which transform the input signal into a series of pulses for counting.
2. The time base, which provides precise time increments during which the pulses are counted.
3. A signal "gate," which starts and stops the counting device.
4. A counter and display, which counts the pulses and provides a digital readout.

4.19   TRANSDUCERS

A large number of devices transform values of physical variables into equivalent electrical signals. Such devices are called *transducers* and have been mentioned frequently in the preceding sections of this chapter. We shall now discuss some of the more widely used transducers and their principles of operation. Table 4.3 presents a compact summary of transducer characteristics, and the reader should consult this table throughout the following discussions to maintain an overall perspective. It may be noted that the subsequent paragraphs will be primarily concerned with principles of operation and will not cover much of the detailed information contained in Table 4.3. Thus, the written discussion and tabular presentation are complementary and should be used jointly.

The discussion that follows in this chapter is concerned primarily with electric effects of transducers. Other transducers of a more specialized nature (thermocouples, strain gages, pressure transducers, and nuclear radiation detectors) will be discussed in subsequent chapters.

4.20   THE VARIABLE-RESISTANCE TRANSDUCER

The variable-resistance transducer is a very common device which may be constructed in the form of a moving contact on a slide-wire or a moving contact that moves through an angular displacement on a solid conductor like a piece of graphite. The device may also be called a *resistance potentiometer* or *rheostat* and is available commercially in many sizes, designs, and ranges. Costs can range from a few cents for a simple potentiometer used as a volume control in a radio circuit to hundreds of dollars for a precision device used for accurate laboratory work.

The variable-resistance transducer fundamentally is a device for converting either linear or angular displacement into an electric signal; however, through mechanical methods it is possible to convert force and pressure to a displacement so that the device may also be useful in force and pressure measurements.

Table 4.3   Summary of transducer characteristics

Type of Transducer and Principle of Operation	Type of Input	Input Range or Level	Input-Impedance Characteristics	Input Sensitivity	Error and Noise Characteristics
Variable resistance: Movement of contact on slide-wire; also called resistance potentiometers	Linear displacement or angular displacement	Minimum level as low as 0.1% of total resistance	Varies widely, depending on the total resistance characteristics and physical size	Commercial potentiometers can have sensitivity of less than 0.002 in, or 0.2° in an angular measurement	Deviation from nonlinearity of the order of 0.5% of total resistance. Noise is usually negligible of the order of $10\,\mu$V at the contact. Noise increases with "chatter" of contact
Differential transformer: See fig. and text discussion; linear variable differential transformer most widely used; converts displacement to voltage	Linear displacement	Total range from ± 0.005 to ± 3 in	Depends on size; forces from 0.1 to 0.3 g usually required	0.5% of total input range	Deviation from linearity about 0.5%; generally accurate to ± 1%
Capacitive transducer: Variable distance between plates registered as a change in capacitance	Displacement or change in dielectric constant between plates; also change in area of plates	Very broad; from 10^{-8} cm to several meters	The input force requirements are very small, of the order of a few dynes	Highly variable; can obtain sensitivities of the order of 1 pF/0.0001 in for airgap measurements of displacement	Errors may result from careless mechanical construction, humidity variations, noise, and stray capacitance in cable connections
Piezoelectric effect: Force impressed on crystals with asymmetrical charge distributions produces a potential difference at the surface of the crystal	Force or stress	Varies widely with crystal material; see sensitivity	Input force requirements are relatively large compared with other transducers	Varies with material: Quartz 0.05 V · m/N Rochelle salt 0.15 V · m/N Barium titanate 0.007 V · m/N	Subject to hysteresis and temperature effects
Photoelectric effect: Light striking metal cathode causes liberation of electrons which may be attracted to anode to produce electric current	Light	Wavelength range depends on glass-tube enclosure; photoemissive materials respond between 0.2 and 0.8 μm	Not applicable	Vacuum tubes: 0.002–0.1 μA/μW Gas-filled tubes: 0.01–0.15 μA/μW	Depends on plate voltage but of the order of 10^{-8} A at room temperature
Photoconductive transducer: Light striking a semiconductor material, such as selenium, metallic sulfides, or germanium, produces a decrease in resistance of the material	Light	Very broad; from thermal radiation through the x-ray region	Not applicable	About 300 μA/μW at maximum sensitivity of the device	Very low noise, usually less than associated circuit
Photovoltaic cell: Light falling upon a semiconducting material in contact with a metal plate produces a potential	Light	Depends on material: selenium 0.2–0.7 μm CuO 0.5–1.4 μm germanium 1.0–1.7 μm	Not applicable	1 mA/lm or 10^{-7} W/cm^2 · lm	Low noise
Ionization transducer: Displacement converted to voltage through a capacitance change	Displacement, 0.1–10 MHz excitation frequency	Less than 1 mm to several inches	Small force required	1–10 V/mm	Can be accurate to microinches
Magnetometer search coil: Changing magnetic field impressed on coil generates an emf proportional to the time rate of change of the field	Changing magnetic field	10^{-3} oersted to highest values obtainable	Not applicable	Depends on coil dimensions but can be of the order of 10^{-5} oersted	Accuracies of 0.05% have been obtained
Hall-effect transducer: Magnetic field impressed on a plate carrying an electric current generates a potential difference in a direction perpendicular to both the current and the magnetic field	Magnetic field	1–20,000 gauss	Not applicable	Depends on plate thickness and current; of the order of -1×10^{-8} V · cm/A · G for bismuth	Can be calibrated within 1%

Frequency Response	Temperature Effects	Type of Output	Output Range or Level	Output-Impedance Characteristics	Application	Remarks
Generally not above 3 Hz for commercial potentiometers	0.002–0.15% °C^{-1} due to a change in resistance; also, some thermoelectric effects depending on types of contacts used	Voltage or current depending on connecting circuit	Wide	Variable	Used for measurement of displacement	Simple, inexpensive, easy to use, many types available commercially
Frequency of applied voltage must be 10 times desired response; mechanical limitations also	Small influence of temperature may be reduced by using a thermistor circuit	Voltage proportional to input displacement	0.4–4.0 mV/ 0.001 in/V input depending on frequency; lower frequency produces lower output	Mainly resistive; low to medium impedance, as low as 20 Ω, depends on size	Used for measurement of displacement	Simple, rugged, inexpensive, high output, requires simple accessory equipment. Care must be taken to eliminate stray magnetic fields
Depends strongly on mechanical construction but may go to 50,000 Hz	Not strong if design allows for effects	Capacitance	Usually between 10^{-3} and 10^3 pF change in capacitance over output range	Usually 10^3–10^7 Ω	Displacement, area liquid level, pressure, sound-level measurements, and others. Particularly useful where small forces are available for driving the transducer	High-output impedance may require careful construction of output circuitry
Depends on external circuitry and mechanical mounting: 20–20 kHz easily obtained; no response to steady-state forces	Wide variation in crystals properties with temperature	Voltage proportional to input force	Wide, depends on crystal size and material; see sensitivity; can have output of several volts	High, of the order of 10^3 MΩ	Measurement of force, pressure, sound level (microphone use)	Simple, inexpensive, rugged
Linear 0–500 Hz; current response drops off 15% at 10,000 Hz	Generally not operable above 75–100°C	Current	Of the order of 2 μA	High, of the order of 10 MΩ	Very useful for counting purposes	Inexpensive, high output
Rise time varies widely with material and incident radiation from 50 μs to minutes	Response to longer wavelengths increases with reduction in temperature	Current drawn in the external circuit	Depends on incident intensity; see input sensitivity	High, varies from 1–10^3 MΩ; in commercial devices	Widely used for radiant measurements at all wavelengths	Fairly expensive; calls for precise circuitry to utilize its full potential
Rise time of the order of 1 μs, response into the megahertz range	Variations of 10% over 40°C, range depending on external resistance load	Voltage	100–250 mV in normal room light; selenium cells up to 500 V at high illumination	3,000–10,000 Ω capacitance of the order of 0.05 μF/cm^2 in selenium cells	Widely used for exposure meters, selenium cells, responds to x-rays	Inexpensive, nonlinear behavior, some aging effects
0–3000 Hz	Small	Voltage	Depends on excitation circuit; see input sensitivity	High, of the order of 1 MΩ	Can be used where accurate measurement of displacement is needed	Relatively insensitive to frequency of excitation circuitry
0 to radiofrequency	Small	Voltage	Depends on coil size: see input sensitivity	Depends on coil size	Measurement of magnetic field	Simple, although accurate, coil dimensions must be maintained for high accuracy
High	Large but can be calibrated	Voltage	Millivolts and microvolts	Low, of the order of 100 Ω for a bismuth detector	Measurement of magnetic field	Each transducer usually must be calibrated because of nonuniformity in semiconductors used for construction

4.21   THE DIFFERENTIAL TRANSFORMER (LVDT)

A schematic diagram of the differential transformer is shown in Fig. 4.48 Three coils are placed in a linear arrangement as shown with a magnetic core which may move freely inside the coils. The construction of the device is indicated in Fig. 4.49. An alternating input voltage is impressed in the center coil, and the output voltage from the two end coils depends on the magnetic coupling between the core and the coils. This coupling, in turn, is dependent on the position of the core. Thus, the output voltage of the device is an indication of the displacement of the core. As long as the core remains near the center of the coil arrangement, the output is very nearly linear, as indicated in Fig. 4.50. The linear range of commercial differential transformers is clearly specified, and the devices are seldom operated outside this range. When operating in the linear range, the device is called a *linear variable differential transformer* (LVDT). Near the null position a slight nonlinearity condition is encountered, as illustrated in Fig. 4.51. It will be noted that Fig. 4.50 considers the phase relationship of the output voltage, while the "V" graph in Fig. 4.51 indicates the absolute magnitude of the output. There is a 180° phase shift from one side of the null position to the other.

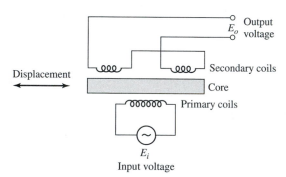

Figure 4.48   Schematic diagram of a differential transformer.

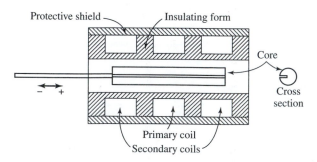

Figure 4.49   Construction of a commercial linear variable differential transformer (LVDT). (*Courtesy of Schaevitz Engineering Company.*)

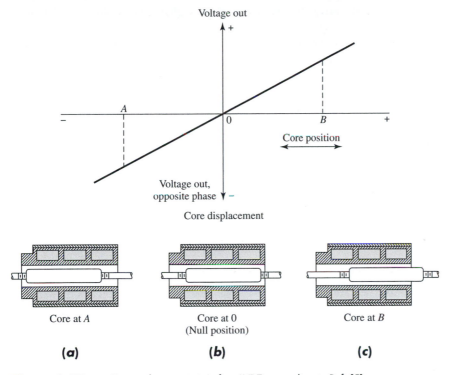

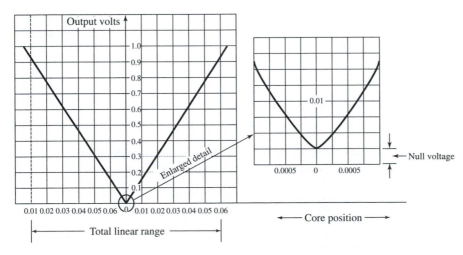

Figure 4.50   Output characteristics of an LVDT, according to Ref. [5].

Figure 4.51   V graph for an LVDT showing slight nonlinear behavior in the null region, according to Ref. [5].

The frequency response of LVDTs is primarily limited by the inertia characteristics of the device. In general, the frequency of the applied voltage should be 10 times the desired frequency response.

Commercial LVDTs are available in a broad range of sizes and are widely used for displacement measurements in a variety of applications. Force and pressure measurements may also be made after a mechanical conversion. Table 4.3 indicates the general characteristics of the LVDT. The interested reader should consult Ref. [5] for more detailed information.

4.22 CAPACITIVE TRANSDUCERS

Consider the capacitive transducer shown in Fig. 4.52. The capacitance (in picofarads) of this arrangement is given by

$$C = 0.225\epsilon \frac{A}{d} \qquad \textbf{[4.63]}$$

where
d = distance between the plates, in or cm
A = overlapping area, in² or cm²
ϵ = dielectric constant ($\epsilon = 1$ for air; $\epsilon = 3$ for plastics)

The constant 0.225 is 0.0885 when the area is in square centimeters and the separation distance is in centimeters.

This plate arrangement may be used to measure a change in the distance d through a change in capacitance. A change in capacitance may also be registered through a change in the overlapping area A resulting from a relative movement of the plates in a lateral direction or a change in the dielectric constant of the material between the plates. The capacitance may be measured with bridge circuits. The output impedance of a capacitor is given by

$$Z = \frac{1}{2\pi f C} \qquad \textbf{[4.64]}$$

where
Z = impedance, Ω
f = frequency, Hz
C = capacitance, F

In general, the output impedance of a capacitive transducer is high; this fact may call for careful design of the output circuitry to avoid loading.

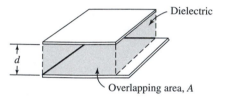

Figure 4.52   Schematic of a capacitive transducer.

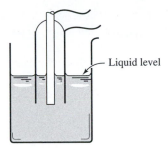

Figure 4.53   Use of a capacitive transducer for liquid-level measurement.

The capacitive transducer may be used for displacement measurements through a variation of either the spacing distance d or the plate area. It is commonly used for liquid-level measurements, as indicated in Fig. 4.53. Two electrodes are arranged as shown, and the dielectric constant varies between the electrodes according to the liquid level. Thus, the capacitance between the electrodes is a direct indication of the liquid level. A charge amplifier may be used to increase the signal level before transmission to readout circuits.

SENSITIVITY OF CAPACITIVE TRANSDUCER.   A capacitive transducer is constructed of two 1-in^2 plates separated by a 0.01-in distance in air. Calculate the displacement sensitivity of such an arrangement. The dielectric constant for air is 1.0006. | **Example 4.11**

Solution

The sensitivity is found by differentiating Eq. (4.64).

$$S = \frac{\partial C}{\partial d} = -\frac{0.225\epsilon A}{d^2}$$

Thus, $$S = -\frac{(0.225)(1.0006)(1)}{(0.01)^2} = -2.25 \times 10^3 \text{ pF/in}$$

UNCERTAINTY FOR CAPACITIVE TRANSDUCER.   For the capacitive transducer in Example 4.11 the allowable uncertainty in the spacing measurement is $w_d = \pm 0.0001$ in, while the estimated uncertainty in the plate area is ± 0.005 in^2. Calculate the tolerable uncertainty in the capacitance measurement in order to achieve the allowable uncertainty in the spacing measurement. | **Example 4.12**

Solution

Solving Eq. (4.64) for d, we have

$$d = 0.225\frac{\epsilon A}{C} \qquad \text{[a]}$$

Making use of Eq. (3.2), we obtain

$$\frac{w_d}{d} = \left[\left(\frac{w_c}{C}\right)^2 + \left(\frac{w_A}{A}\right)^2\right]^{1/2} \qquad \text{[b]}$$

We have

$$\frac{w_d}{d} = \frac{0.0001}{0.01} = 0.01 \qquad \frac{w_A}{A} = \frac{0.005}{1.0} = 0.005$$

so that

$$\frac{w_c}{C} = 0.00866 = 0.866\%$$

The nominal value of C is

$$C = \frac{(0.255)(1.0006)(1.0)}{0.01} = 22.513 \, \text{pF}$$

so that the tolerable uncertainty in C is

$$w_C = (22.513)(0.00866) = \pm 0.195 \, \text{pF}$$

4.23 PIEZOELECTRIC TRANSDUCERS

Consider the arrangement shown in Fig. 4.54. A piezoelectric crystal is placed between two plate electrodes. When a force is applied to the plates, a stress will be produced in the crystal and a corresponding deformation. With certain crystals this deformation will produce a potential difference at the surface of the crystal, and the effect is called the *piezoelectric effect*. The induced charge on the crystal is proportional to the impressed force and is given by

$$Q = dF \qquad\qquad \textbf{[4.65]}$$

where Q is in coulombs, F is in newtons, and the proportionality constant d is called the *piezoelectric constant*. The output voltage of the crystal is given by

$$E = gtp \qquad\qquad \textbf{[4.66]}$$

where t is the crystal thickness in meters, p is the impressed pressure in newtons per square meter, and g is called voltage sensitivity and is given by

$$g = \frac{d}{\epsilon} \qquad\qquad \textbf{[4.67]}$$

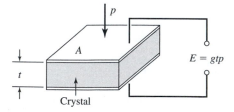

Figure 4.54   The piezoelectric effect.

Table 4.4   Piezoelectric constants

Material	Orientation	Charge Sensitivity d, $\frac{C/m^2}{N/m^2}$	Voltage Sensitivity g, $\frac{V/m}{N/m^2}$
Quartz	X cut; length along Y length longitudinal	2.25×10^{-12}	0.055
	X cut; thickness longitudinal	-2.04	-0.050
	Y cut; thickness shear	4.4	-0.108
Rochelle salt	X cut 45°; length longitudinal	435.0	0.098
	Y cut 45°; length longitudinal	-78.4	-0.29
Ammonium dihydrogen phosphate	Z cut 0°; face shear	48.0	0.354
	Z cut 45°; length longitudinal	24.0	0.177
Commercial barium titanate	Parallel to polarization	86–130	0.011
	Perpendicular to polarization	-56	0.005
Lead zirconate titanate	Parallel to polarization	190–580	0.02–0.03
Lead metaniobate	Parallel	80	0.036

Values of the piezoelectric constant and voltage sensitivity for several common piezoelectric materials are given in Table 4.4.

The voltage output depends on the direction in which the crystal slab is cut in respect to the crystal axes. In Table 4.4 an X (or Y) cut means that a perpendicular to the largest face of the cut is in the direction of the x axis (or y axis) of the crystal.

Piezoelectric crystals may also be subjected to various types of shear stresses instead of the simple compression stress shown in Fig. 4.54, but the output voltage is a complicated function of the exact crystal orientation. Piezoelectric crystals are used as pressure transducers for dynamic measurements. General information on the piezoelectric effect is given in Refs. [1], [23], and [24].

OUTPUT OF A PIEZOELECTRIC PRESSURE TRANSDUCER.   A quartz piezoelectric | **Example 4.13**

crystal having a thickness of 2 mm and a voltage sensitivity of 0.055 V · m/N is subjected to a pressure of 200 psi. Calculate the voltage output.

Solution

We calculate the voltage output with Eq. (4.66):

$$p = (200)(6.895 \times 10^3) = 1.38 \times 10^6 \text{ N/m}^2$$
$$t = 2 \times 10^{-3} \text{ m}$$

Thus, $E = (0.055)(2 \times 10^{-3})(1.38 \times 10^6) = 151.8 \text{ V}$

4.24 PHOTOELECTRIC EFFECTS

A photoelectric transducer converts a light beam into a usable electric signal. Consider the circuit shown in Fig. 4.55. Light strikes the photoemissive cathode and releases electrons, which are attracted toward the anode, thereby producing an electric current in the external circuit. The cathode and anode are enclosed in a glass or quartz envelope, which is either evacuated or filled with an inert gas. The photoelectric sensitivity is defined by

$$I = S\Phi \tag{4.68}$$

where
I = photoelectric current
Φ = illumination of the cathode
S = sensitivity

The sensitivity is usually expressed in units of amperes per watt or amperes per lumen.

Photoelectric-tube response to different wavelengths of light is influenced by two factors: (1) the transmission characteristics of the glass-tube envelope and (2) the photoemissive characteristics of the cathode material. Photoemissive materials are available which will respond to light over a range of 0.2 to 0.8 μm. Most glasses transmit light in the upper portion of this range, but many do not transmit below about 0.4 μm. Quartz, however, transmits down to 0.2 μm. Various noise effects are present in photoelectric tubes, and the interested reader should consult the discussion by Lion [2] for more information, as well as Ref. [25].

Photoelectric tubes are quite useful for measurement of light intensity. Inexpensive devices can be utilized for counting purposes through periodic interruption of a light source.

4.25 PHOTOCONDUCTIVE TRANSDUCERS

The principle of the photoconductive transducer is shown in Fig. 4.56. A voltage is impressed on the semiconductor material as shown. When light strikes the semiconductor material, there is a decrease in the resistance, thereby producing an increase

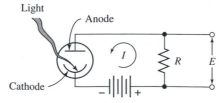

Figure 4.55 The photoelectric effect.

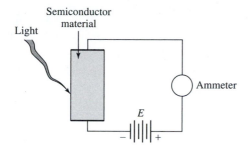

Light — Semiconductor material — Ammeter — E

Figure 4.56 Schematic of a photoconductive transducer.

in the current indicated by the meter. A variety of substances are used for photoconductive materials, and a rather detailed discussion of the pertinent literature on the subject is given in Refs. [2], [6], [23], and [25].

Photoconductive transducers enjoy a wide range of applications and are useful for measurement of radiation at all wavelengths. It must be noted, however, that extreme experimental difficulties may be encountered when operating with long-wavelength radiation, and some of them are discussed in Chap. 12 concerning measurement of thermal radiation.

The responsivity R_v of a detector is defined as

$$R_v = \frac{\text{rms output voltage}}{\text{rms power incident upon the detector}} \quad \textbf{[4.69]}$$

The *noise-equivalent power* (NEP) is defined as the minimum-radiation input that will produce a S/N ratio of unity. The *detectivity D* is defined as

$$D = \frac{R_v}{\text{rms noise-voltage output of cell}} \quad \textbf{[4.70]}$$

The detectivity is the reciprocal of NEP. A normalized detectivity D^* is defined as

$$D^* = (A \, \Delta f)^{1/2} D \quad \textbf{[4.71]}$$

where A is the area of the detector and Δf is a noise-equivalent bandwidth. The units of D^* are usually cm $\cdot$ Hz$^{1/2}$/W, and the term is used in describing the performance of detectors so that the particular surface area and bandwidth will not affect the results. Figures 4.57 and 4.58 illustrate the performance of several photoconductive detectors over a range of wavelengths. In these figures the wavelength is expressed in micrometers, where 1 μm = 10^{-6} m. The symbols represent lead sulfide (PbS), lead selenide (PbSe), lead telluride (PbTe), indium antimonide (InSb), and gold- and antimony-doped germanium (Ge:Au, Sb). For these figures D^* is a monochromatic detectivity for an incident radiation, which is chopped at 900 Hz and a 1-Hz bandwidth.

The lead-sulfide cell is very widely used for detection of thermal radiation in the wavelength band of 1 to 3 μm. By cooling the detector more favorable response at higher wavelengths can be achieved up to about 4 or 5 μm. For measurements at

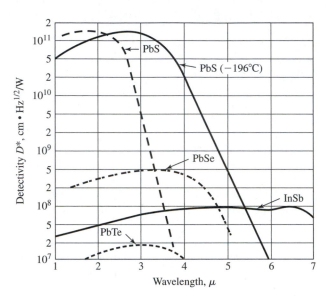

Figure 4.57   Absolute spectral response of typical detectors at room temperature, according to Ref. [6].

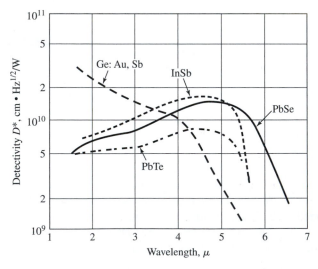

Figure 4.58   Absolute spectral response of typical detector cooled to liquid nitrogen temperature, (−195°C) according to Ref. [7].

longer wavelengths the indium-antimonide detector is preferred, but it has a lower detectivity than lead sulfide. Some of the applications of these cells are discussed in Chap. 12.

DETECTIVITY OF PHOTOCONDUCTIVE TRANSDUCER. Calculate the incident radiation at 2 μm that is necessary to produce a signal-to-noise (S/N) ratio of 40 dB with a lead-sulfide detector at room temperature, having an area of 1 mm^2.

| **Example 4.14**

Solution

We first insert Eqs. (4.69) and (4.70) in Eq. (4.71) to express the detectivity in terms of voltages and incident power. Thus,

$$D^* = (A \, \Delta f)^{1/2} \frac{E_o}{E_{noise}} \frac{1}{P_{incident}} \qquad [a]$$

From Fig. 4.57, $D^* = 1.5 \times 10^{11}$ cm $\cdot$ Hz$^{1/2}$/W for $\Delta f = 1$ Hz. For a S/N ratio of 40 dB we have

$$40 = 20 \log \frac{E_o}{E_{noise}}$$

so that

$$\frac{E_o}{E_{noise}} = 100$$

Using $A = 10^{-2}$ cm^2, Eq. (a) yields

$$1.5 \times 10^{11} = \frac{(10^{-2})^{1/2}(1)(100)}{P_{incident}}$$

Thus,

$$P_{incident} = 6.7 \times 10^{-11} \text{ W}$$

4.26 PHOTOVOLTAIC CELLS

The photovoltaic-cell principle is illustrated in Fig. 4.59. The sandwich construction consists of a metal base plate, a semiconductor material, and a thin transparent

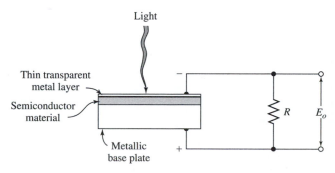

Figure 4.59 Schematic diagram of a photovoltaic cell.

metallic layer. This transparent layer may be in the form of a sprayed, conducting lacquer. When light strikes the barrier between the transparent metal layer and the semiconductor material, a voltage is generated as shown. The output of the device is strongly dependent on the load resistance R. The open-circuit voltage approximates a logarithmic function, but more linear behavior may be approximated by decreasing the load resistance.

Perhaps the most widely used application of the photovoltaic cell is the light exposure meter in photographic work. The logarithmic behavior of the cell is a decided advantage in such applications because of its sensitivity over a broad range of light intensities. Further information concerning photovoltaic and photoconductive detectors is available in Refs. [23], [24], and [25]. A discussion of CCD (charge-coupled-device) and CMOS (complementary-metal-oxide-semiconductor) sensors employed in digital photography systems is given in Appendix B.

4.27   IONIZATION TRANSDUCERS

A schematic of the ionization transducer is shown in Fig. 4.60. The tube contains a gas at low pressure while the RF generator impresses a field on this gas. As a result of the RF field, a glow discharge is created in the gas, and the two electrodes 1 and 2 detect a potential difference in the gas plasma. The potential difference is dependent on the electrode spacing and the capacitive coupling between the RF plates and the gas. When the tube is located at the central position between the plates, the potentials on the electrodes are the same; but when the tube is displaced from this central position, a dc potential difference will be created. Thus, the ionization transducer is a useful device for measuring displacement. Some of the basic operating characteristics are given in Table 4.3, and a detailed description of the output characteristics is given by Lion [2].

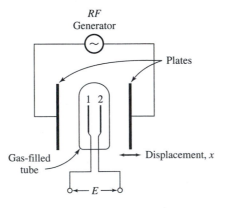

Figure 4.60   Schematic diagram of an ionization-displacement transducer.

4.28   MAGNETOMETER SEARCH COIL

A schematic of the magnetometer search coil is shown in Fig. 4.61. A flat coil with N turns is placed in the magnetic field as shown. The length of the coil is L, and the cross-section area is A. The magnetic field strength H and the magnetic flux density B are in the direction shown, where

$$B = \mu H \qquad\qquad \textbf{[4.72]}$$

and μ is the magnetic permeability. The voltage ouput of the coil E is given by

$$E = NA \cos \alpha \frac{dB}{dt} \qquad\qquad \textbf{[4.73]}$$

where α is the angle formed between the direction of the magnetic field and a line drawn perpendicular to the plane of the coil. The total flux through the loop is

$$\phi = A \cos \alpha B \qquad\qquad \textbf{[4.74]}$$

so that
$$E = N\frac{d\phi}{dt} \qquad\qquad \textbf{[4.75]}$$

Note that the voltage output of the device is dependent on the rate of change of the magnetic field and that a stationary coil placed in a steady magnetic field will produce a zero-voltage ouput. The search coil is thus a transducer that transforms a magnetic field signal into a voltage.

In order to perform a measurement of a steady magnetic field it is necessary to provide some movement of the search coil. A typical method is to use a rotating coil, as shown in Fig. 4.62. The rms value of the output voltage for such a device is

$$E_{\text{rms}} = \frac{1}{\sqrt{2}} NAB\omega \qquad\qquad \textbf{[4.76]}$$

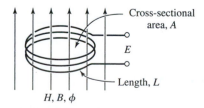

Figure 4.61   Schematic of a magnetometer search coil.

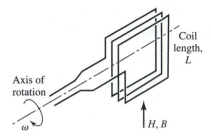

Figure 4.62   Use of a rotating search coil for measurement of steady-state magnetic fields.

where ω is the angular velocity of rotation. Oscillating coils are also used. The accuracy of the search coil device depends on the accuracy with which the dimensions of the coil are known. The coil should be small enough so that the magnetic field is constant over its area.

In the above equations the magnetic flux density is expressed in webers per square meter, the area is in square meters, the time is in seconds, the magnetic flux is in webers, the magnetic field strength (magnetic intensity) is in amperes per meter, and the magnetic permeability for free space is $4\pi \times 10^{-7}$ H/m. An alternative set of units uses B in gauss, H in oersteds, A in centimeters squared, and μ in abhenrys per centimeter. The magnetic permeability for free space in this instance is unity.

Example 4.15 | **ROTATING SEARCH COIL.**   A rotating search coil has 10 turns with a cross-sectional area of 5 cm². It rotates at a constant speed of 100 rpm. The output voltage is 40 mV. Calculate the magnetic field strength.

Solution

According to Eq. (4.76)

$$B = \frac{\sqrt{2}E_{rms}}{NA\omega} = \frac{\sqrt{2}(0.04)}{(10)(5 \times 10^{-4})[(100)(2\pi)/60]}$$

$$= 1.08 \text{ Wb/m}^2$$

$$H = \frac{B}{\mu} = \frac{1.08}{4\pi \times 10^{-7}}$$

$$= 8.6 \times 10^5 \text{ A/m}$$

4.29   HALL-EFFECT TRANSDUCERS

The principle of the Hall effect is indicated in Fig. 4.63. A semiconductor plate of thickness t is connected as shown so that an external current I passes through the material. When a magnetic field is impressed on a plate in a direction perpendicular to the surface of the plate, there will be a potential E_H generated as shown. This potential is called the Hall voltage and is given by

$$E_H = K_H \frac{IB}{t} \qquad \textbf{[4.77]}$$

where I is in amperes, B is in gauss, and t is in centimeters. The proportionality constant is called the *Hall coefficient* and has the units of volt-centimeters per ampere-gauss. Typical values of K_H for several materials are given in Table 4.5.

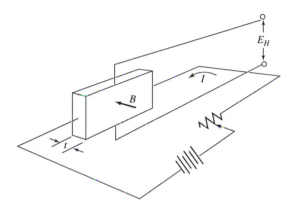

Figure 4.63   The Hall effect.

Table 4.5   Hall coefficients for different materials[1]

Material	Field Strength, G	Temp., °C	$\dfrac{K_H,}{A \cdot G}$
AS	4,000–8,000	20	4.52×10^{-11}
C	4,000–11,000	Room	-1.73×10^{-10}
Bi	1,130	20	-1×10^{-8}
Cu	8,000–22,000	20	-5.2×10^{-13}
Fe	17,000	22	1.1×10^{-11}
n-Ge	100–8,000	25	-8.0×10^{-5}
Si	20,000	23	4.1×10^{-8}
Sn	4,000	Room	-2.0×10^{-14}
Te	3,000–9,000	20	5.3×10^{-7}

[1]According to Lion [2].

Example 4.16 | **HALL-EFFECT VOLTAGE OUTPUT.** A Hall-effect transducer is used for the measurement of a magnetic field of 5000 G. A 2-mm slab of bismuth is used with a current of 3 A. Calculate the voltage output of the device.

Solution

We use Eq. (4.77) and the data of Table 4.5:

$$\begin{aligned}
E_H &= K_H \frac{IR}{t} \\
&= \frac{(-1 \times 10^{-8})(3)(5000)}{(2 \times 10^{-1})} \\
&= -7.5 \times 10^{-4} \text{ V}
\end{aligned}$$

4.30   DIGITAL DISPLACEMENT TRANSDUCERS

A digital displacement transducer can be used for both angular and linear measurements. In Fig. 4.64 an angular measurement device is shown. As the wheel rotates, light from the source is alternately transmitted and stopped, thereby submitting a digital signal to the photodetector. The signal is then amplified and sent to a counter. The number of counts is proportional to angular displacement. The *frequency* of the signal is proportional to angular velocity. Sensitivity of the device may be improved by increasing the number of cutouts.

A linear transducer which operates on a reflection principle is shown in Fig. 4.65. Small reflecting strips are installed on the motion device. Light from the source is then alternately reflected and absorbed with linear motion, thereby presenting a digital signal to the photodetector. Readout is the same as with the angular instrument. Calibration with a known displacement standard must be performed.

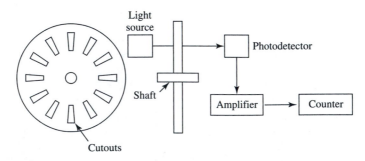

Figure 4.64   Digital transducer for angular displacement.

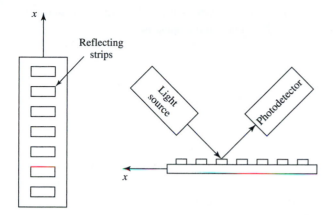

Figure 4.65   Digital transducer for linear displacement.

4.31   COMPARISON OF ANALOG AND DIGITAL INSTRUMENTS

As most instrument circuits and readout devices employ digital techniques one might automatically assume that everything should be converted to digital. But analog instruments have their place too. For example, when balancing a circuit to obtain a null condition, an analog instrument may be easier to use. When a panel of instruments is used to indicate the operating condition of a complicated plant or process, analog instruments may be preferable because a trained operator can visually sense the position of all the indicators more quickly than with several digital readouts.

Most physical measurements, such as those of resistance, voltage, force, and displacement, occur in the form of some analog signal. Certainly, if direct computer processing of the data is to be performed, analog-to-digital conversion must be performed, and a wide variety of commercial equipment is available for such conversion.

4.32   SUMMARY

This chapter began by considering the basic physical phenomena which underlie electrical instruments and circuits. We then examined the meaningful measures which can be used to characterize time-varying waveforms. The basic instrumentation, both analog and digital, used to measure these waveforms was discussed.

This set the stage for an examination of the general experimental setting in which measurements are made with the assistance of electronic systems. In general, the value of the particular physical property of interest is converted to an electrical signal by some suitable transducer, and the output of this transducer is fed into an input circuit. This circuit is intended to couple efficiently the transducer's signal to the additional equipment which is necessary to reduce the effects of noise upon the signal's distinguishability, transmit the signal, and display its form.

4.33 **REVIEW QUESTIONS**

4.1. What is the "bridge" between electric current and mechanical force?

4.2. State the differences between analog and digital representations of signals.

4.3. Why is the root-mean-square value of a periodic waveform useful?

4.4. How does an electrodynamometer differ from a D'Arsonval meter? What advantages does it have over the D'Arsonval meter?

4.5. How do measurements of alternating currents differ from measurements of direct currents?

4.6. How are very high-frequency currents measured?

4.7. What are the applications for an electrostatic voltmeter?

4.8. How can an ammeter be converted to a voltmeter?

4.9. What are the advantages of digital instruments?

4.10. How is the current-sensitive input circuit dependent on internal meter impedance?

4.11. What is meant by a ballast circuit? How does it differ from a voltage-divider circuit?

4.12. What are the advantages of a high-meter impedance when it is used with a voltage-divider circuit?

4.13. What is meant by loading error?

4.14. Why does the internal resistance of the galvanometer not influence the reading in a potentiometer circuit?

4.15. Differentiate between a bridge operated on the null principle and on the deflection principle.

4.16. What is meant by a voltage-sensitive deflection-bridge circuit?

4.17. How can an operational amplifier be used to average two inputs?

4.18. What purpose does a voltage follower serve?

4.19. How do transformers aid in matching impedances?

4.20. Why are dc power supplies needed?

4.21. How does the decibel notation for voltage level differ from that for power level?

4.22. What is the function of the phase-sensitive detector in a lock-in amplifier?

4.23. Why is an EVM useful for electrical measurements?

4.24. Why are sampling oscilloscopes used?

4.25. What are the advantages of digital oscilloscopes and digital recorders?

4.34 PROBLEMS

4.1. Expand Eq. (4.12) in series form and indicate the relation which may be used to obtain a linear approximation for the current as a function of the transducer resistance R. Also, show the error which results from this approximation.

4.2. A charged particle ($q = 0.5$ C) is moving with a velocity of 10 m/s in a magnetic field of 10 Wb/m^2.

(a) What is the magnitude of the largest force the charge can experience under these conditions?

(b) What is a physical geometry in which the particle will experience no force?

4.3. The basic meter of Fig. 4.2 is designed so that

$$B = 1 \, \text{Wb/m}^2$$
$$L = 0.1 \, \text{m}$$
$$K = 1 \, \text{N/m}$$

(a) What deflection x results when a direct current of 4 A flows through the meter?

(b) Find an expression for $x(t)$ if $i(t) = 2 \cos t$ A.

4.4. Calculate the rms values of the following periodic currents:

(a) $i(t) = 10 \cos(t)$ A

(b) $i(t) = 10 \cos(377t)$ A

(c) $i(t) = t \qquad\qquad 0 \le t \le 1$
$ = t - 1 \qquad 1 \le t \le 2$
$ = t - 2 \qquad 2 \le t \le 3 \qquad$ etc.

4.5. Expand Eq. (4.14) in series form and indicate the relation which may be used to obtain a linear approximation for the current as a function of the transducer resistance R. Also, show the error which results from this approximation.

4.6. Derive an expression for the sensitivity S of the circuit of Fig. 4.19 where

$$S = \frac{di}{dR}$$

Find the condition for maximum sensitivity.

4.7. Obtain a linear approximation for the sensitivity of the ballast circuit of Fig. 4.20. Under what conditions would this relation apply? Estimate the error in this approximation.

4.8. Show that the output voltage v_o of the following circuit is proportional to the derivative of the input voltage v_i.

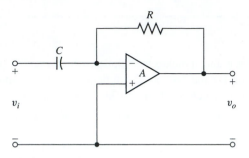

Figure Problem 4.8

If $C = 10^{-5}$ F, what value of R is required if

$$v_o = \frac{dv_i}{dt}$$

Assume the operational amplifier is ideal.

4.9. A diode has a current-voltage relation of the form

$$i = \alpha e^{\beta v}$$

where α and β are constants.

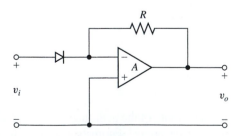

Find the relation between input and output voltages for the following circuit.

Figure Problem 4.9

4.10. The transformer in the following circuit has a turns ratio of 10. A sinusoidal voltage of 10 V rms is applied at terminals $11'$.

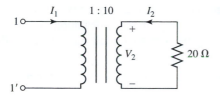

Figure Problem 4.10

What is I_2?

4.11. A Wheatstone bridge circuit has resistance arms of 400, 40, 602, and 6000 Ω taken sequentially around the bridge. The galvanometer has a resistance of 110 Ω and is connected between the junction of the 40- and 602-Ω resistors to the junction of the 400- and 6000- resistors. The battery has an emf of 3 V and negligible internal resistance. Calculate the voltage across the galvanometer and the galvanometer current.

4.12. It is known that a certain resistor has a resistance of approximately 800 Ω. This resistor is placed in a Wheatstone bridge, the other three arms of which have resistances of exactly 800 Ω. A 4-V battery with negligible internal resistance is used in the circuit. The galvanometer resistance is 100 Ω, and the indicated galvanometer current is 0.08 μA. Calculate the resistance of the unknown resistor.

4.13. Two galvanometers are available for use with a Wheatstone bridge having equal ratio arms of 100 Ω. One galvanometer has a resistance of 100 Ω and a sensitivity of 0.05 μA/mm, whereas the other has a resistance of 200 Ω and a sensitivity of 0.01 μA/mm. A 4-V battery is used in the circuit, and it has negligible internal resistance. An unknown resistance of approximately 500 Ω is to be measured with the bridge. Calculate the deflection of each galvanometer. State assumptions necessary to make this calculation.

4.14. Two known ratio arms of a Wheatstone bridge are 4000 and 400 Ω. The bridge is to be used to measure a resistance of 100 Ω. Two galvanometers are available: one with a resistance of 50 Ω and a sensitivity of 0.05 μA/mm, and one with a resistance of 500 Ω and a sensitivity of 0.2 μA/mm. Which galvanometer would you prefer to use? Assume that the galvanometer is connected from the junction of the ratio arms to the opposite corner of the bridge.

4.15. A Wheatstone bridge is constructed with ratio arms of 60 and 600 Ω. A 4-V battery with negligible internal resistance is used and connected from the junction of the ratio arms to the opposite corner. A galvanometer having a resistance of 50 Ω and a sensitivity of 0.05 μA/mm is connected between the other corners. When the adjustable arm reads 200 Ω, the galvanometer deflection is 30 mm. What is the value of the unknown resistance?

4.16. The four arms of a Wheatstone bridge have resistances of 500, 1000, 600, and 290 Ω taken in sequence around the bridge. The battery of 3 V connects between the 1000- and 600-Ω resistors and 500- and 290-Ω resistors. A galvanometer with a resistance of 50 Ω and a sensitivity of 0.05 μA/mm is connected across the other two terminals. The galvanometer is shunted by a resistance of 30 Ω. Calculate the galvanometer deflection. Repeat the calculation for a series resistance of 30 Ω connected in the galvanometer circuit instead of the shunt arrangement.

4.17. Suppose an *end resistor* is added to the circuit of Fig. 4.21 so that there results

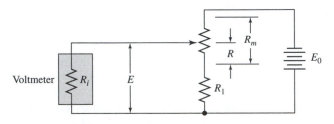

Figure Problem 4.17

where R_1 is the end resistor. Derive an expression for the voltage output and loading error of such an arrangement. What advantage does it offer over the circuit in Fig. 4.21? Suppose the end resistor were attached to the other end of the variable resistance. What would be the advantage in this circumstance? What would be the advantage if an end resistor were placed on each end of the variable resistor?

4.18. Design a bandpass filter to operate between the limits of 500 and 2000 Hz, with a load resistance of 16 Ω.

4.19. Design a lowpass filter with a cutoff frequency of 500 Hz with a load resistance of 1000 Ω.

4.20. Design a highpass filter with a cutoff frequency of 1000 Hz with a load resistance of 1000 Ω.

4.21. A voltage of 500 V is impressed on a 150-kΩ resistor. The impedance of the voltage source is 10 kΩ. Two meters are used to measure the voltage across the 150-kΩ resistor: a volt-ohmmeter with an internal impedance of 1000 Ω/V and an EVM with an impedance of 11 MΩ. Calculate the voltage indicated by each of these devices.

4.22. Plot the gain of the filter circuit of Example 4.7.

4.23. Five 1-in^2 plates are arranged as shown. The plate spacings are 0.01 in. The arrangement is to be used for a displacement transducer by observing the change in capacitance with the distance x.

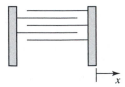

Figure Problem 4.23

Calculate the sensitivity of the device in picofarads per inch. Assume that the plates are separated by air.

4.24. Calculate the voltage-displacement sensitivity for the LVDT whose characteristics are shown in Fig. 4.51.

4.25. The piezoelectric crystal of Example 4.13 is used for a pressure measurement at a nominal 100 psi. The uncertainty in the voltage measurement is ± 0.0003 in. Estimate the uncertainty in the pressure measurement.

4.26. A rotating search coil like that shown in Fig. 4.62 has a nominal area of 1 cm^2 with 50 turns of small-diameter wire. The rotational speed is nominally 180 rpm. Calculate the voltage output when the coil is placed in a magnetic field of 1 Wb/m^2.

4.27. An amplifier has a stated power bandwidth of $+0, -1$ dB over a frequency range of 10 Hz to 30 kHz. If the 0-dB level is 35 W, what is the power that would occur at the lower limit of this specification?

4.28. An amplifier is stated to produce a 35-W output into an 8-Ω load when driven with an input voltage of 2.2 mV. The hum and noise is stated as -65 dB referenced to 10 mV at the input. What is the voltage output of the noise when operating into the 8-Ω load?

4.29. Suppose the amplifier of Prob. 4.28 requires an input of 180 mV to produce the 35-W output and has a hum and noise level of -75 dB referenced to 0.25 V at the input. What is the voltage output of the noise under these conditions? Calculate the relative decibel level of this noise and the noise calculated in Prob. 4.28.

4.30. An emf of 100 V is connected across a load of 10,000 Ω. In order to measure this voltage a meter having an internal impedance of 100 kΩ is connected across the load. Calculate the percent error that will result from the loading by the meter.

4.31. In Prob. 4.30, calculate the number of decibels below the true voltage indicated by the meter.

4.32. A certain transducer has an internal resistance of 10 kΩ. The output of the device is to be measured with a voltage-divider circuit using a voltage source of 50 V and a meter with an impedance of 100 kΩ. Calculate the maximum loading error between the ranges of 10 and 90 percent of the measured variables.

4.33. A simple ballast circuit is used to measure the output of a pressure pickup. The circuit is designed so that the internal resistance is six times the total transducer resistance. A source of 100 V is used to energize the circuit. Calculate the voltage output at 25, 50, 60, and 80 percent of full load on the transducer. What error would result if the voltage output were assumed to vary linearly with the load?

4.34. A simple Wheatstone bridge has arms of $R_1 = 121$ Ω, $R_2 = 119$ Ω, and $R_3 = 121$ Ω. What is the value of R_4 for balance? If $R_4 = 122$ Ω and the bridge is driven with a source at 100 V, what is the open-circuit output voltage?

4.35. A lead-sulfide detector is used to measure a radiation signal at 3.5 μm. What reduction in incident power, measured in decibels, is afforded when the detector is cooled to $-196°$C, assuming a constant S/N ratio?

4.36. An indium antimonide detector is used to sense radiant energy at 5 μm. What incident flux on a 4-mm^2 detector at room temperature is necessary to produce an S/N ratio of 45 dB?

4.37. A current-sensitivity input circuit is used to measure the output of a transducer having a total resistance of 500 Ω. A 12-V source is used, and the internal meter impedance is 7000 Ω. What is the current indication for 25, 50, 75, and 100 percent of full scale? What error would result at each of these readings if the current output were assumed to vary linearly with the transducer resistance?

4.38. A simple voltage-divider circuit is used to measure the output of a transducer whose total resistance is 1000 $\Omega \pm 1$ percent. The internal resistance of the voltmeter is 10,000 $\Omega \pm 5$ percent, and the meter is calibrated to read within $\frac{1}{2}$ percent of the true voltage at a full scale of 30 V. The source voltage is 24 V $\pm$ 0.05 V. Calculate the percent full-scale reading of the transducer for voltage readings of 6, 12, 18, and 24 V. Estimate the uncertainty for each of these readings.

4.39. Suppose the transducer of Prob. 4.38 is connected to an end resistor of 500 Ω, as shown in Prob. 4.17. Calculate the percent full-scale reading of the transducer for voltage readings of 12, 15, 18, and 24 V. Estimate the uncertainty for each of these readings, assuming the end resistor is accurate within 1 percent.

4.40. An amplifier has a power bandwidth of $+0.5, -1.0$ dB over a frequency range of 20 to 100 kHz. The nominal power output at 0 dB is 0.1 W. Calculate the power levels at the upper and lower limits of the specification for each end of the frequency range.

4.41. A voltage divider circuit measures the output of a transducer having a resistance of 800 $\Omega \pm 2$ percent. The internal resistance of the voltage measuring device is 15,000 $\Omega \pm 10$ percent. Calibration is performed so that it reads within 0.5 percent at a full-scale reading. The source voltage is 30 V $\pm$ 0.1 V. Calculate the percent full-scale reading of the transducer and uncertainty for readings of 5, 10, 15 V.

4.42. The noise level of a certain amplifier is stated to be 80 dB below the power output of 100 W, referenced to an input level of 1.0 mV. Calculate the noise-level input.

4.43. A velocity measurement is to be performed by simultaneously photographing a moving object and a "clock" composed of a rotating 10-cm-diameter disk driven by a synchronous motor at 1800 rpm. The rotational speed of the motor is accurate within ± 0.1 percent and the disk has markings in 1° increments. Estimate the accuracy with which the velocity of an object moving at 20 m/s could be measured with this arrangement if the displacement can be measured with an uncertainty of ± 1 mm.

4.44. An operational amplifier has an open-loop gain of 99 dB and an input impedance of 5 MΩ. Specify suitable resistances for an inverting amplifier having a gain of 25.

4.45. Repeat Prob. 4.44 for a noninverting amplifier.

4.46. Repeat Prob. 4.44 for a differential amplifier.

4.47. Plot the output response of a highpass RC filter versus ΩT. For what values of RC will the attenuation be less than 2 percent for frequencies of 20, 200, and 2000 Hz?

4.48. Plot the output response of a lowpass RC filter versus ΩT. For what values of RC will the attenuation be less than 2 percent for frequencies of 20, 200, and 2000 Hz?

4.49. An operational amplifier is to be used to sum three input voltages such that the output voltage is

$$E_o = E_1 + 2E_2 + 3E_3$$

Select the appropriate resistances and draw the circuit to accomplish this objective.

4.50. A voltage follower with a gain of 15 is to be designed using the op-amp of Prob. 4.44. Specify the necessary resistances.

4.51. Suppose an op-amp is to be used to integrate a signal over time of 1 ms, 0.5 s, 1.0 s, and 10 s. Design circuits to accomplish these objectives.

4.52. Using available computer software, plot the values of E_o/E_i vs. ωT for the RC lowpass filter of Table 4.2. What is the insertion loss in dB for this filter as a function of ωT?

4.53. Repeat Prob. 4.52 for the RC highpass filter of Table 4.2.

4.54. Using available computer software, plot the voltage ratio E_o/E_i vs. ωT for three identical lowpass RC filter sections in series. Also plot the insertion loss in dB.

4.55. A simple RC filter is to be designed to pass frequencies below a nominal value of 40 Hz. If the attenuation at 40 Hz is to be 5 dB, determine suitable values of R and C for the filter.

4.56. A simple RC filter is to be designed to pass frequencies above a nominal value of 40 Hz. If the attenuation at 40 Hz is to be 5 dB, determine suitable values of R and C for the filter.

4.57. Suppose the two filters of Probs. 4.55 and 4.56 are connected in series. Plot the output E_o/E_i for such an arrangement.

4.58. Estimate the linear voltage sensitivity for the LVDT having the characteristics shown in Fig. 4.51 over the displacement range from 0 to 0.06. How much does the sensitivity vary from this value in the null region?

4.59. A lead-sulfide detector is to be used to detect radiation at a wavelength of 1.0 μm. A signal-to-noise ratio of 30 dB is desired for the detection process and the detector area is 1 mm by 1 mm. What is the minimum incident radiation in watts that will be necessary for this detection process?

4.60. Negative feedback is added to an amplifier to increase the signal-to-noise ratio. In the open-loop configuration the amplifier has a gain of 2000 and an S/N ratio of 2000. A negative feedback loop with $k = 0.05$ is added to the amplifier. Calculate the initial S/N ratio in dB and that which results from addition of the feedback.

4.61. A noninverting configuration for an operational amplifier is arranged so that the feedback resistance is 1 MΩ while the noninverting resistance input is 100 kΩ. Calculate the gain.

4.62. Repeat Prob. 4.61 for an inverting, differential input configuration of the operational amplifier circuit.

4.63. Design a lowpass filter with a cutoff frequency of 40 Hz with a load resistance of 8 Ω.

4.64. Design a highpass filter with a cutoff frequency of 450 Hz and a load resistance of 8 Ω.

4.65. A lead-sulfide detector is employed to measure a light source at a wavelength of 3.0 μm. What reduction in incident power is possible when the detector is cooled to liquid nitrogen temperature of 76 K? Assume that the S/N ratio remains constant.

4.66. A power amplifier has a stated bandwidth of ± 1.0 dB from 10 to 50 kHz. The nominal power output 0 dB point is specified at 100 W. Calculate the power (W) at the limits of the ± 1.0 dB specification for each limit of the frequency specification.

4.67. The noise level of the amplifier in Prob. 4.66 is specihed as 90 dB below the rated output for a input level of 5 mV. Calculate the noise level at input.

4.68. A LVDT has the operating characteristics shown in Fig. 4.51. Calculate the voltage sensitivity over the displacement range from 0 to 0.04.

4.69. A lead-sulfide detector has an area of 2 mm by 2 mm. It is exposed to a radiation source of 2.0 μm. If a minimum S/N ratio of 25 dB is to be achieved, what minimum incident radiation is required? Express in watts.

4.35   REFERENCES

1. Cady, W. G.: *Piezoelectricity,* McGraw-Hill, New York, 1946.

2. Lion, K. S.: *Instrumentation in Scientific Research,* McGraw-Hill, New York, 1959.

3. ———: "Mechanical-electric Transducer," *Rev. Sci. Instr.,* vol. 27, no. 2, pp. 222–225, 1956.

4. Sweeney, R. J.: *Measurement Techniques in Mechanical Engineering,* John Wiley & Sons, New York, 1953.

5. Herceg, E. E.: *Schaevitz Handbook of Measurement and Control,* Schaevitz Engineering, Pennsauken, NJ.

6. "International Conference on Photoconductivity," *J. Phys. Chem. Solids,* vol. 22, pp. 1–409, December 1961.

7. Beyen, W., P. Bratt, W. Engeler, L. Johnson, H. Levinstein, and A. MacRaw: "Infrared Detectors Today and Tomorrow," *Proc. IRIS,* vol. 4, no. 1, March 1959.

8. Bartholomew, D.: *Electrical Measurements and Instrumentation,* Allyn & Bacon, Boston, 1963.

9. Doebelin, E. O.: *Measurement Systems: Analysis and Design,* 4th ed., McGraw-Hill, New York, 1990.

10. Wait, J. V., L. P. Huelsman, and G. A. Korn: *Introduction to Operational Amplifier Theory and Applications,* 2d ed., McGraw-Hill, New York, 1992.

11. Rutkowski, G. B.: *Handbook of Integrated-Circuit Operational Amplifiers,* Prentice-Hall, Englewood Cliffs, NJ, 1975.

12. Huelsman, L. P.: *Active and Passive Analog Filter Design: An Introduction,* McGraw-Hill, New York, 1993.

13. Williams, A. B., *Electronic Filter Design Handbook,* McGraw-Hill, New York, 1981.

14. Antoniou, A.: *Digital Filters, Analysis and Design,* 2d ed., McGraw-Hill, New York, 1993.

15. Stadler, Wolfram: *Analytical Robotics and Mechatronics,* McGraw-Hill, New York, 1995.

16. Mitra, S. K.: *Digital Signal Processing: A Computer Based Approach,* McGraw-Hill, New York, 1998.

17. Lindner, D. K.: *Introduction to Signals and Systems,* McGraw-Hill, New York, 1999.

18. Histand, M. B., and D. G. Alciatore: *Introduction to Mechatronics and Measurement Systems,* McGraw-Hill, New York, 1999.

19. Hayt, W. J.: *Engineering Electromagnetics,* 5th ed., McGraw-Hill, New York, 1989.

20. Gopalan, K. G.: *Introduction to Digital Electronic Circuits,* McGraw-Hill, New York, 1996.

21. Helfrick, A. D., and W. O. Cooper: *Modern Electronic Instrumentation and Measurement Techniques,* Prentice-Hall, Englewood Cliffs, NJ, 1990.

22. De Micheli, G.: *Synthesis and Optimization of Digital Circuits,* McGraw-Hill, New York, 1994.

23. Navon, D. R.: *Electronic Materials and Devices,* Houghton Mifflin, Boston, 1975.

24. McGee, T. D.: *Principles and Methods of Temperature Measurements,* Wiley, New York, 1988.

25. Northrop, R. B.: *Introduction to Instrumentation and Measurements,* CRC Press, Boca Raton, FL, 1997.

5

DISPLACEMENT AND AREA MEASUREMENTS

5.1 INTRODUCTION

Many of the transducers discussed in Chap. 4 represent excellent devices for measurement of displacement. In this chapter we wish to examine the general subject of dimensional and displacement measurements and indicate some of the techniques and instruments that may be utilized for such purposes, making use, where possible, of the information in the preceding sections.

Dimensional measurements are categorized as determinations of the size of an object, while a *displacement measurement* implies the measurement of the movement of a point from one position to another. An area measurement on a standard geometric figure is a combination of appropriate dimensional measurements through a correct analytical relationship. The determination of areas of irregular geometric shapes usually involves a mechanical, graphical, or numerical integration.

Displacement measurements may be made under both steady and transient conditions. Transient measurements fall under the general class of subjects discussed in Chap. 11. The present chapter is concerned only with static measurements.

5.2 DIMENSIONAL MEASUREMENTS

The standard units of length were discussed in Chap. 2. All dimensional measurements are eventually related to these standards. Simple dimensional measurements with an accuracy of ± 0.01 in (0.25 mm) may be made with graduated metal machinist scales or wood scales which have accurate engraved markings. For large dimensional measurements metal tapes are used to advantage. The primary errors in such measurement devices, other than readability errors, are usually the result of thermal expansion or contraction of the scale. On long metal tapes used for surveying purposes this can represent a substantial error, especially when used under extreme temperature conditions.

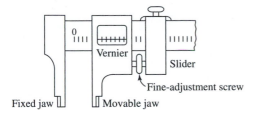

Figure 5.1 Vernier caliper.

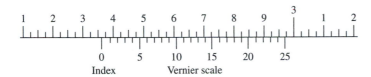

Figure 5.2 Expanded view of vernier scale.

It may be noted, however, that thermal expansion effects represent *fixed* errors and may easily be corrected when the measurement temperature is known.

Vernier calipers represent a convenient modification of the metal scale to improve the readability of the device. The caliper construction is shown in Fig. 5.1, and an expanded view of the vernier scale is shown in Fig. 5.2. The caliper is placed on the object to be measured and the fine adjustment is rotated until the jaws fit tightly against the workpiece. The increments along the primary scale are 0.025 in. The vernier scale shown is used to read to 0.001 in (0.025 mm) so that it has 25 equal increments (0.001 is $\frac{1}{25}$ of 0.025) and a total length of $\frac{24}{25}$ times the length of the primary scale graduations. Consequently, the vernier scale does not line up exactly with the primary scale, and the ratio of the last coincident number on the vernier to the total vernier length will equal the fraction of a whole primary scale division indicated by the index position. In the example shown in Fig. 5.2 the reading would be $2.350 + (\frac{14}{25})(0.025) = 2.364$ in.

The micrometer calipers shown in Fig. 5.3 represent a more precise measurement device than the vernier calipers. Instead of the vernier scale arrangement, a calibrated screw thread and circumferential scale divisions are used to indicate the fractional part of the primary scale divisions. In order to obtain the maximum effectiveness of the micrometer care must be exerted to ensure that a consistent pressure is maintained on the workpiece. The spring-loaded ratchet device on the handle enables the operator to maintain such a condition. When properly used, the micrometer can be employed for the measurement of dimensions within 0.0001 in (0.0025 mm).

Dial indicators are devices that perform a mechanical amplification of the displacement of a pointer or follower in order to measure displacements within about 0.001 in. The construction of such indicators provides a gear rack, which is connected to a displacement-sensing shaft. This rack engages a pinion which in turn is used to provide a gear-train amplification of the movement. The output reading is made on a circular dial.

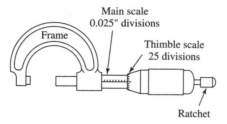

Main scale
0.025" divisions

Frame

Thimble scale
25 divisions

Ratchet

Figure 5.3   Micrometer calipers.

Example 5.1 | **ERROR DUE TO THERMAL EXPANSION.**   A 30-m (at 15°C) steel tape is used for surveying work in the summer such that the tape temperature in the sun is 45°C. A measurement indicates 24.567 ± 0.001 m. The linear thermal coefficient of expansion is $11.65 \times 10^{-6}/°C$ at 15°C. Calculate the true distance measurement.

Solution

The indicated tape length would be the true value if the measurement were taken at 15°C. At the elevated temperature the tape has expanded and consequently reads too small a distance. The actual length of the 30-m tape at 45°C is

$$L(1 + \alpha \, \Delta T) = [1 + (11.65 \times 10^{-6})(45 - 15)](30) = 30.010485 \text{ m}$$

Such a true length would be indicated as 30 m. The true reading for the above situation is thus

$$(24.567)[1 + (11.65 \times 10^{-6})(45 - 15)] = 24.576 \text{ m}$$

5.3   GAGE BLOCKS

Gage blocks represent industrial dimension standards. They are small steel blocks about $\frac{3}{8} \times 1\frac{3}{8}$ in with highly polished parallel surfaces. The thickness of the blocks is specified in accordance with the following tolerances:

Grade of Block	Tolerance, μ in[1]
AA	2
A	4
B	8

[1] Tolerances are for blocks less than 1 in thick; for greater thickness the same tolerances are per inch.

Gage blocks are available in a range of thicknesses that make it possible to stack them in a manner such that with a set of 81 blocks any dimension between 0.100 and

8.000 in can be obtained in increments of 0.0001 in. The blocks are stacked through a process of *wringing*. With surfaces thoroughly clean, the metal surfaces are brought together in a sliding fashion while a steady pressure is exerted. The surfaces are sufficiently flat so that when the wringing process is correctly executed, they will adhere as a result of molecular attraction. The adhesive force may be as great as 30 times atmospheric pressure.

Because of their high accuracy, gage blocks are frequently used for calibration of other dimensional measurement devices. For very precise measurements they may be used for direct dimensional comparison tests with a machined item. A discussion of the methods of producing gage-block standards is given in Ref. [3]. The literature of manufacturers of gage blocks furnishes an excellent source of information on the measurement techniques which are employed in practice.

5.4 OPTICAL METHODS

An optical method for measuring dimensions very accurately is based on the principle of light interference. The instrument based on this principle is called an *interferometer* and is used for the calibration of gage blocks and other dimensional standards. Other optical instruments in wide use are various types of microscopes and telescopes, including the conventional surveyor's transit, which is employed for measurement of large distances.

Consider the two sets of light beams shown in Fig. 5.4. In Fig. 5.4*a* the two beams are in phase so that the brightness at point P is augmented when they intersect. In Fig. 5.4*b* the beams are out of phase by half a wavelength so that a cancellation is observed, and the light waves are said to *interfere* with each other. This is the essence of the interference principle. The effect of the cancellation is brought about by allowing two light waves from a single source to travel along paths of different lengths. When the difference in the distance is an integral multiple of wavelengths,

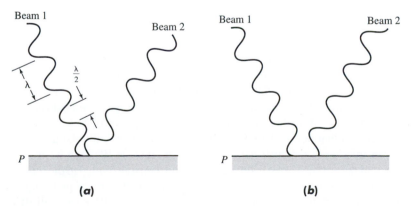

(a)   **(b)**

Figure 5.4   Interference principle. (*a*) Beams in phase; (*b*) beams out of phase.

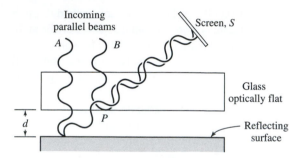

Figure 5.5   Application of interference principle.

there will be a reinforcement of the waves, while there will be a cancellation when the difference in the distances is an odd multiple of half-wavelengths.

Now, let us apply the interference principle to dimensional measurements. Consider the two parallel plates shown in Fig. 5.5. One plate is a transparent, strain-free glass accurately polished flat within a few microinches. The other plate has a reflecting metal surface. The glass plate is called an *optical flat*. Parallel light beams *A* and *B* are projected on the plates from a suitable collimating source. The separation distance between the plates *d* is assumed to be quite small. The reflected beam *A* intersects the incoming beam *B* at point *P*. Since the reflected beam has traveled farther than beam *B* by a distance of $2d$, it will create an interference at point *P* if this incremental distance is an odd multiple of $\lambda/2$. If the distance $2d$ is an even multiple of $\lambda/2$, the reflected beam will augment beam *B*. Thus, for $2d = \lambda/2$, $3\lambda/2$, etc., the screen *S* will detect no reflected light. Now, consider the same two plates, but let them be tilted slightly so that the distance between the plates is a variable. Now, if one views the reflected light beams, alternate light and dark regions will appear on the screen, indicating the variation in the plate spacing. The dark lines or regions are called *fringes,* and the change in the separation distance between the positions of two fringes corresponds to

$$\Delta(2d) = \frac{\lambda}{2} \qquad \textbf{[5.1]}$$

The interference principle offers a convenient means for measuring small surface defects and for calibrating gage blocks. The use of a tilted optical flat as in Fig. 5.5 is an awkward method of utilizing the principle, however. For practical purposes the interferometer, as indicated schematically in Fig. 5.6, is employed. Monochromatic light from the source is collimated by the lens *L* onto the splitter plate S_2, which is a half-silvered mirror that reflects half of the light toward the optically flat mirror *M* and allows transmission of the other half toward the workpiece *W*. Both beams are reflected back and recombined at the splitter plate S_2 and then transmitted to the screen. Fringes may appear on the screen resulting from differences in the optical path lengths of the two beams. If the instrument is properly constructed, these differences will arise from dimensional variations of the workpiece. The interferometer is primarily used for

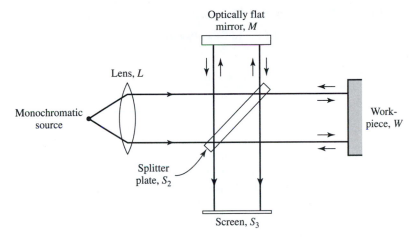

Figure 5.6   Schematic of interferometer.

Table 5.1   Monochromatic light sources

Source	Wavelength, μm	Half-wavelength Fringe Interval, μm
Helium	0.589	0.295
Krypton 86	0.606	0.303
Mercury 198	0.546	0.273
Sodium	0.598	0.299

calibration of gage blocks and other applications where extremely precise absolute dimensional measurements are required. For detailed information on experimental techniques used in interferometry the reader should consult Refs. [3] and [5]. The use of the interferometer for fluid-flow measurements will be discussed in Chap. 7.

As shown in Eq. (5.1), the wavelength of the monochromatic light source will influence the fringe spacing. Table 5.1 lists the wavelengths of some common light sources and the corresponding half-wavelength fringe interval.

INTERFERENCE MEASUREMENT.   A mercury light source employs a green filter such that the wavelength is 5460 Å. This light is colliminated and directed onto two tilted surfaces like those shown in Fig. 5.5. At one end the surfaces are in precise contact. Between the point of contact and a distance of 3000 in five interference fringes are observed. Calculate the separation distance between the two surfaces and the tilt angle at this position. | **Example 5.2**

Solution

The five fringe lines correspond to $\lambda/2, 3\lambda/2, \ldots, 9\lambda/2$; that is, for the fifth fringe line

$$2d = \frac{9\lambda}{2}$$

We have $\lambda = 5460 \times 10^{-8}$ cm $= 2.15 \times 10^{-5}$ in so that

$$d = \tfrac{9}{4}(2.15 \times 10^{-5}) = 48.4 \, \mu\text{in}$$

The tilt angle is

$$\phi = \tan^{-1} \frac{48.4 \times 10^{-6}}{3.000} = \frac{48.4 \times 10^{-6}}{3.000} = 16.1 \times 10^{-6} \text{ rad}$$

Mechanical displacement may also be measured with the aid of the electric transducers discussed in Chap. 4. The LVDT, for example, can be used to sense displacements as small as 1 μin. Use of LVDT devices for displacement measurements is described in Ref. [17]. Resistance transducers are primarily of value for measurement of fairly large displacements because of their poor resolution. Capacitance and piezoelectric transducers, on the other hand, provide high resolution and are suitable for dynamic measurements.

5.5 PNEUMATIC DISPLACEMENT GAGE

Consider the system shown in Fig. 5.7. Air is supplied at a constant pressure p_1. The flow through the orifice and through the outlet of diameter d_2 is governed by the separation distance x between the outlet and the workpiece. The change in flow with x will be indicated by a change in the pressure downstream from the orifice p_2. Thus, a measurement of this pressure may be taken as an indication of the separation distance x. For purposes of analysis we assume incompressible flow. (See Sec. 7.3 for a discussion of the validity of this assumption.) The volumetric flow through an orifice may be represented by

$$Q = CA\sqrt{\Delta p} \qquad\qquad \textbf{[5.2]}$$

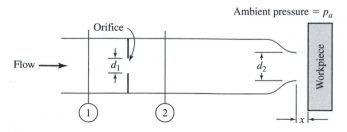

Figure 5.7   Pneumatic displacement device.

where C = discharge coefficient

  A = flow area of the orifice

  Δp = pressure differential across the orifice

There are *two* orifices in the situation depicted in Fig. 5.7, the obvious one and the orifice formed by the flow restriction between the outlet and the workpiece. We shall designate the area of the first orifice A_1 and that of the second A_2. Then, Eq. (5.2) becomes

$$Q = C_1 A_1 \sqrt{p_1 - p_2} = C_2 A_2 \sqrt{p_2 - p_a} \qquad \textbf{[5.3]}$$

where p_a is the ambient pressure and is assumed constant. Equation (5.3) may be rearranged to give

$$r = \frac{p_2 - p_a}{p_1 - p_a} = \frac{1}{1 + (A_2/A_1)^2} \qquad \textbf{[5.4]}$$

where it is assumed that the discharge coefficients C_1 and C_2 are equal. We may now observe that

$$A_1 = \frac{\pi d_1^2}{4} \qquad \textbf{[5.5]}$$

$$A_2 = \pi d_2 x \qquad \textbf{[5.6]}$$

Thus, we see the relation between the pressure ratio r and the workpiece displacement x. It has been shown experimentally [1] that the relation between r and the area ratio A_2/A_1 is very nearly linear for $0.4 < r < 0.9$ and that

$$r = 1.10 - 0.50 \frac{A_2}{A_1} \qquad \textbf{[5.7]}$$

for this range. Introducing Eqs. (5.5) and (5.6), we have

$$r = \frac{p_2 - p_a}{p_1 - p_a} = 1.10 - 2.00 \left(\frac{d_2}{d_1^2} \right) x \qquad \text{for } 0.4 < r < 0.9 \qquad \textbf{[5.8]}$$

The pneumatic displacement gage is mainly used for small displacement measurements.

UNCERTAINTY IN PNEUMATIC DISPLACEMENT GAGE.   A pneumatic displacement gage like the one shown in Fig. 5.7 has d_1 = 0.030 in and d_2 = 0.062 in. The supply pressure is 10.0 psig, and the differential pressure $p_2 - p_a$ is measured with a water manometer which may be read with an uncertainty of 0.05 in H_2O. Calculate the displacement range for which Eq. (5.8) applies and the uncertainty in this measurement, assuming that the supply pressure remains constant.   **Example 5.3**

Solution

We have

$$\frac{d_2}{d_1^2} = \frac{0.062}{(0.030)^2} = 68.8$$

When $r = 0.4$, we have from Eq. (5.8)

$$x = \frac{1.10 - 0.4}{(2.00)(68.8)} = 0.0509 \text{ in } (0.129 \text{ cm})$$

When $r = 0.9$, $x = 0.0145$ in (0.0368 cm).

Utilizing Eq. (3.2) as applied to Eq. (5.8), we have

$$w_r = \left[\left(\frac{\partial r}{\partial x} \right)^2 w_x^2 \right]^{1/2} = \pm \left| \frac{\partial r}{\partial x} \right| w_x$$

Furthermore,

$$w_r = \frac{w_{\Delta p}}{p_1 - p_a} \qquad \frac{\partial r}{\partial x} = -(2.00)(68.8) = -137.6$$

The uncertainty in the measurement of $p_2 - p_a$ is

$$w_{\Delta p} = (0.05)(0.0361) = 1.805 \times 10^{-3} \text{ psig } (12.44 \text{ N/m}^2)$$

Thus, the uncertainty in x is given by

$$w_x = \frac{1.805 \times 10^{-3}}{(137.6)(10.00)} = 1.313 \times 10^{-6} \text{ in} = 1.313 \ \mu\text{in } (0.033 \ \mu\text{m})$$

Comment

From this example we see that the pneumatic gage can be quite sensitive, even with modest pressure-measurement facilities at hand.

In addition to its application as a steady-state–displacement-measurement device, the pneumatic gage may be employed as a dynamic sensor in conjunction with properly designed fluidic circuits. By periodically interrupting the discharge, the device may serve as a periodic signal generator for fluidic circuits. Different interruption techniques may be employed to generate square-, triangle-, or sine-wave signals [7]. The device may also be employed for fluidic reading of coded information on plates or cards that pass under the outlet jet [8 and 9]. The rapidity with which such readings may be made depends on the dynamic response of the gage and its associated connecting lines and pressure transducers. Studies [11] have shown that the device can produce good frequency response for signals up to 500 Hz.

5.6   AREA MEASUREMENTS

There are many applications that require a measurement of a plane area. Graphical determinations of the area of the survey plots from maps, the integration of a function to determine the area under a curve, and analyses of experimental data plots all may rely on a measurement of a plane area. There are also many applications for the measurement of surface areas, but such measurements are considerably more difficult to perform.

5.7   THE PLANIMETER, A DEVICE OF HISTORICAL INTEREST

The planimeter is a mechanical integrating device that may be used for measurement of plane areas. We consider it here as an illustration of a novel mechanical device to perform area measurements. It is seldom used today. Consider the schematic representation shown in Fig. 5.8. The point O is fixed, while the tracing point T is moved around the periphery of the figure whose area is to be determined. The wheel W is mounted on the arm BT so that it is free to rotate when the arm undergoes an angular displacement. The wheel has engraved graduations and a vernier scale so that its exact number of revolutions may be determined as the tracing point moves around the curve. The planimeter and area are placed on a flat, relatively smooth surface so that the wheel W will only slide when the arm BT undergoes an axial translational movement. Thus, the wheel registers zero angular displacement when an axial translational movement of arm BT is experienced. Let the length of the tracing arm BT be L and the distance from point B to the wheel be a. The diameter of the wheel is D. The distance OB is taken as R. Now, suppose the arm BT is rotated an angle $d\theta$ and the arm OB through an angle $d\phi$ as a result of movement of the tracing point. The area swept out by the arms BT and OB is

$$dA = \tfrac{1}{2}L^2\,d\theta + LR\cos\beta\,d\phi + \tfrac{1}{2}R^2\,d\phi \qquad \textbf{[5.9]}$$

where β is the angle between the two arms. Similarly, the distance traveled by the rim of the wheel owing to rotation is

$$ds = a\,d\theta + R\cos\beta\,d\phi \qquad \textbf{[5.10]}$$

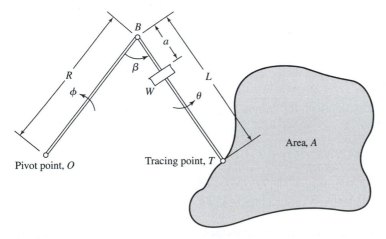

Figure 5.8   Schematic of a polar planimeter.

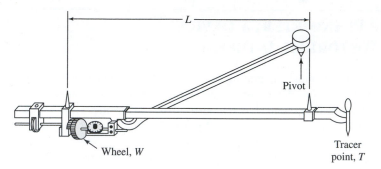

Figure 5.9   Construction of a polar planimeter.

We may now integrate these expressions and obtain

$$A = \frac{1}{2}L^2 \int d\theta + LR \int \cos \beta \, d\phi + \frac{1}{2}R^2 \int d\phi \qquad \textbf{[5.11]}$$

$$s = a \int d\theta + \int R \cos \beta \, d\phi \qquad \textbf{[5.12]}$$

Thus,

$$\int R \cos \beta \, d\phi = s - a \, d\theta$$

and

$$A = \left(\frac{1}{2}L^2 - aL\right) \int d\theta + Ls + \frac{1}{2}R^2 \int d\phi \qquad \textbf{[5.13]}$$

If the pole is outside the area as shown in Fig. 5.8, we have both $\int d\theta = 0$ and $\int d\phi = 0$, and the area is obtained as

$$A = Ls \qquad \text{(pole outside area)}$$

When the pole is inside the area, both $\int d\phi$ and $\int d\theta$ are equal to 2π because both arms make complete rotations. The area is then

$$A = Ls + (R^2 + L^2 - 2aL)\pi \qquad \text{(pole inside area)} \qquad \textbf{[5.14]}$$

The last term in Eq. (5.14) represents the area of the *zero circle,* which is the area the tracing point would sweep out when the pivot point is inside the area and the wheel reading is zero.

The instrument described above is called a *polar planimeter,* and its construction is indicated in Fig. 5.9.

5.8   GRAPHICAL AND NUMERICAL METHODS FOR AREA MEASUREMENT

A very simple method of plane-area measurement is to place the figure on coordinate paper and to count the number of squares enclosed by the figure. An appropriate

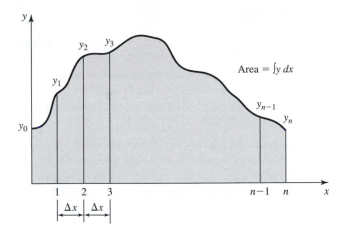

Figure 5.10   Plane-area determination.

scale factor is then applied to determine the area. Numerical integration is commonly applied to determine the area under an irregular curve. Perhaps the two most common methods are the trapezoidal rule and Simpson's rule. Consider the area shown in Fig. 5.10. The area under the curve is

$$A = \int y\,dx \tag{5.15}$$

For relatively small increments of Δx the area is given approximately by the rectangular rule, which calculates the area element ΔA_i as

$$\Delta A_i = y_i\,\Delta x = y_i(x_{i+1} - x_i)$$

so that

$$A = \sum \Delta A_i = \sum y_i(x_{i+1} - x_i)$$

Equal or nonequal increments in Δx may be employed with this rule. The rectangular rule will underestimate the area in regions where the curve has $dy/dx > 0$ and overestimate when $dy/dx < 0$. The main advantage of the rule is its simplicity and ease of calculation.

If the figure is divided into equal increments Δx along the x axis, the trapezoidal rule gives for the area

$$A = \left(\frac{y_0 + y_n}{2} + \sum_{i=1}^{n-1} y_i \right) \Delta x \tag{5.16}$$

If nonuniform increments in x are involved, the elemental areas A_i for the trapezoidal rule are

$$A_i = y_m(x_{i+1} - x_i) = \frac{\Delta x(y_{i+1} + y_i)}{2} \tag{5.17}$$

and the integral for the area under the curve is

$$A = \int dA = \sum A_i \qquad \textbf{[5.18]}$$

When the area is divided in an *even* number of increments, Simpson's rule gives

$$A = \frac{\Delta x}{3}\left\{ y_0 + y_n + \sum_{i=1}^{n-1} y_i[3 + (-1)^{i+1}] \right\} \qquad \textbf{[5.19]}$$

The above equations give the area under the curve $y = f(x)$ from $y = 0$ to the values on the curve. If the area of an enclosed figure is the desired objective, the result may be obtained from the same formulas by executing the summation; first through the positive values of Δx, followed by a subtraction of the summation executed through the negative values of Δx. For nonuniform values of Δx an application of the trapezoidal rule of Eq. (5.17) through the entire sequence of points will automatically take account of positive and negative elemental areas.

The trapezoidal rule is obtained by joining the ordinates of the curve with straight lines, while the result given by Eq. (5.19) is obtained by joining three points at a time with a parabola. Both the above equations are special cases of a general class of equations called the *Newton-Cotes integration formulas*. A derivation of these formulas is based on an approximation of the actual curve with a polynomial which agrees with it at $n + 1$ equally spaced points. The general form of the integration formulas is derived in Refs. [2, 15] and may be written in the following way:

$$\int_{x_1}^{x_2} f(x)\, dx = \Delta x \sum_{k=0}^{n} C_k f(x_k) \qquad \textbf{[5.20]}$$

where

$$\Delta x = \frac{x_2 - x_1}{n} \qquad \textbf{[5.21]}$$

$$x_k = x_1 + k\,\Delta x \qquad \textbf{[5.22]}$$

and the coefficients C_k are given by

$$C_k = \int_0^n \frac{s(s-1)\cdots(s-k+1)(s-k-1)\cdots(s-n)}{k(k-1)\cdots(k-k+1)(k-k-1)\cdots(k-n)}\, ds \qquad \textbf{[5.23]}$$

The variable s is defined according to

$$s = \frac{x - x_1}{n} \qquad \textbf{[5.24]}$$

Note that the Newton-Cotes formulas use n increments in the independent variable and that this requires a matching of a polynomial at $n + 1$ points in the region that is to be integrated. A discussion of the errors involved in the above formulas is given in Ref. [15]. Because of the ready availability of computing power the simple trapezoidal or Simpson's rule with a large number of small increments may be preferred over the more complex Newton-Cotes formulas.

A discussion of the various integration formulas is given in Ref. [15], including computer routines and application examples.

AREA OF SURVEY PLOT BY NUMERICAL INTEGRATION.   A surveyor determines   **Example 5.4**
the coordinates of an irregular hexagon-shaped parcel of land as shown in the accompanying
Fig. Example 5.4a and listed in the table as x_i and y_i. Notice that the coordinates are listed in
sequence with the final entry the same as the initial entry, indicating closure of the figure.

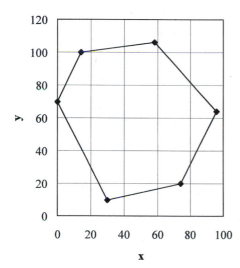

Figure Example 5.4a

$x(i)$	$y(i)$	$y(i) + y(i+1)$	$x(i+1) - x(i)$	$A(i)$	Sum $A(i)$
0	70				
14	100	170	14	1190	1190
58	106	206	44	4532	5722
96	64	170	38	3230	8952
74	20	84	−22	−924	8028
30	10	30	−44	−660	7368
0	70	80	−30	−1200	6168

The area of the plot will be determined by an application of the trapezoidal rule for
numerical integration. The increments in x are not uniform, so the integration must be carried
out with the summation of Eq. (5.17). The values of Δx and the elemental areas A_i are also
listed in the table along with a running sum of the area elements.
   The first three area elements are positive because the corresponding values of Δx are
positive. Similarly, the last three area elements are negative because their values of Δx are
negative. The sum of the first three area elements, 8952, represents the area under the first
three points and the x axis ($y = 0$), while the final entry, 6168, is the area under the first three
points, less the area under the last three points, or the enclosed area of the figure. Note that an
application of Eq. (5.17) for the sequence of points enclosing the figure automatically takes
account of the positive and negative area elements.
   Since the area in question is a polygon, that is, a figure formed with straight-line segments
for each side, the trapezoidal rule will give an *exact* answer for the area. If some type of

higher-order polynomial connecting line segments were used to effect the integration, the figure might look something like that shown in Fig. Example 5.4b. In this case, the outcome of the numerical integration would be uncertain. Simpson's rule, as expressed in Eq. (5.19), could not be applied because the increments in x are not uniform. Moreover, the parabola fit of Simpson's rule would not match with the straight-line segments joining the data points. Again, we note that the trapezoidal rule gives an exact answer in the example presented here.

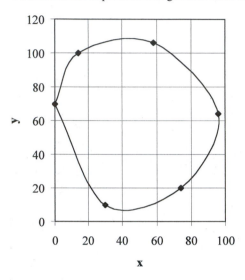

Figure Example 5.4b

Example 5.5

UNCERTAINTY OF SURVEY AREA DETERMINATION.   Determine the uncertainty in the determination of the survey area of Example 5.4 assuming uncertainties in measurements of x and y of ± 0.02. Assume the numerical integration calculation is exact.

Solution

The elemental areas are expressed in terms of the measured quantities as

$$A_i = \frac{(y_{i+1} + y_i)(x_{i+1} - x_i)}{2}$$

Performing the partial differentiations with respect to each of the four measured quantities and inserting in Eq. (3.2) gives for the uncertainty in the elemental areas

$$w_{Ai} = \frac{w_{xi}[(x_{i+1} - x_i)^2 + (y_{i+1} + y_i)^2]^{1/2}}{2^{1/2}} \qquad \textbf{[a]}$$

where it is understood that the uncertainties in x and y are equal.

We have

$$A = \sum A_i$$

and $\partial A / \partial A_i = 1$ for all the A_i. Applying Eq. (3.2) again gives

$$w_A = \left[\sum (w_{Ai})^2 \right]^{1/2} \qquad \textbf{[b]}$$

Performing the various calculations produces the table below.

A_i	Sum A_i	w_{Ai}	$w_{A\text{(total)}}$	w_A/A
1190	1190	2.412301805	2.412301805	0.002027144
4532	5722	2.978993118	3.833223187	0.00066991
3230	8952	2.463493454	4.556577663	0.000509001
−924	8028	1.228006515	4.719152466	0.000587837
−660	7368	0.753126815	4.778870159	0.000648598
−1200	6168	1.208304597	4.929259579	0.000799167

The uncertainties listed here are calculated with $w_{xi} = w_{yi} = 0.02$. From Eq. (*a*) we see that the uncertainties in each elemental area are a linear function of w_{xi}, so a value of $w_{xi} = 0.01$ would produce area uncertainties just half the values in the table.

5.9   SURFACE AREAS

Consider the general three-dimensional surface shown in Fig. 5.11. The surface is described by the function

$$z = f(x, y)$$

and the surface area is given in Ref. [4] as

$$A = \iint \left[\left(\frac{\partial z}{\partial x} \right)^2 + \left(\frac{\partial z}{\partial y} \right)^2 + 1 \right] dx\, dy \qquad \textbf{[5.25]}$$

If the function z is known and well behaved, the integral in Eq. (5.25) may be evaluated directly. Let us consider the case where the function is not given but specific values

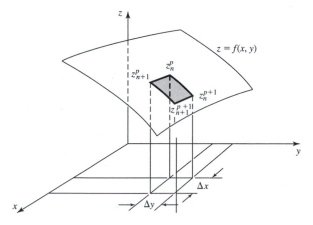

Figure 5.11   Surface-area determination.

of z are known for incremental changes in x and y. The increments in x and y are denoted by Δx and Δy, while the value of z is denoted by z_n^p, where the subscript n refers to the x increments and the superscript p refers to the y increment. We thus have the approximations

$$\frac{\partial z}{\partial x} \approx \frac{z_{n+1}^p - z_n^p}{\Delta x}$$

$$\frac{\partial z}{\partial y} \approx \frac{z_n^{p+1} - z_n^p}{\Delta y}$$

The integral Eq. (5.25) is now replaced by the double sum

$$A = \sum_p \sum_n \left[\left(\frac{z_{n+1}^p - z_n^p}{\Delta x} \right)^2 + \left(\frac{z_n^{p+1} - z_n^p}{\Delta y} \right)^2 + 1 \right]^{1/2} \Delta x \, \Delta y \qquad \textbf{[5.26]}$$

The surface area may be determined by performing this numerical summation.

Standard computer routines and calculator programs are available for performing the numerical computations indicated above. See, for example, Refs. [12] to [16].

5.10   Problems

5.1. A 12-in steel scale is graduated in increments of 0.01 in and is accurate when used at a temperature of 60°F. Calculate the error in an 11-in measurement when the ambient temperature is 100°F. Should the person using the scale be concerned about this error? Why?

5.2. Calculate the temperature error in a 76-ft measurement with a steel surveyor's tape at -10°F when the tape is accurate at 60°F.

5.3. Show that the spacing distance d in Fig. 5.5 can be represented by

$$d = \frac{2n - 1}{4} \lambda$$

where n is the number of fringe lines.

5.4. A pneumatic displacement gage is designed according to the arrangement in Fig. 5.7. An air supply pressure of 20 psig is available, and displacements are to be measured over a range of 0.050 in. The orifice diameter is 0.025 in. Calculate the maximum displacement which may be measured in the linear range of operation and the outlet tube diameter d_2.

5.5. Plot the equation

$$y = 3 + 4x - 6x^2 + 2.8x^3 - 0.13x^4$$

for the range $0 < x < 5$. Determine the area under the curve by counting squares and also by numerical integration using the trapezoidal relation and Simpson's rule. Calculate the error in each of these three cases by comparing the results with those obtained analytically.

5.6. Use the trapezoidal rule and Simpson's method to perform the integration.

$$A = \int_0^\pi \sin x \, dx$$

Use 4, 8, and 12 increments of x and calculate the error for each case.

5.7. Consider the sphere given by

$$x^2 + y^2 + z^2 = 25$$

Using the summation of Eq. (5.26), calculate the surface area bounded by $x = \pm 1$ and $y = \pm 1$. Use $\Delta x = \Delta y = 0.5$. Determine the error in the calculation by comparing it with the true value as calculated from Eq. (5.25).

5.8. By suitable numerical integration determine the surface area and volume of a right circular cone having a height of 12.5 cm and a base diameter of 15 cm. Compare the result with that obtained by an exact calculation.

5.9. A steel tape is used to measure a distance of 20 m at 10°C. Assuming that the tape will measure the true distance when the temperature is 20°C, what would be the indicated distance at 10°C? What would be the indicated distance at 40°C?

5.10. A mercury light source and green filter (5460 Å) are used with an interferometer to determine the distance between an optically flat glass plate and a metal plate, as shown in Fig. 5.5. For this arrangement nine interference fringes are observed. Calculate the separation distance between the two surfaces.

5.11. A pneumatic displacement gage is to be used to measure displacements between 0.1 and 0.2 mm. The output of the device is a hypodermic needle having a diameter of 0.4 mm. The measurement is to be performed with an uncertainty of 25 nm. Specify values of the orifice diameter, upstream pressure, and allowable uncertainty in the pressure measurement to accomplish this objective.

5.12. The pneumatic displacement gage of Example 5.3 is operated under the same conditions as given, but it is discovered that the supply pressure has a random fluctuation of ± 0.07 psig during the measurements. Calculate the uncertainty in the dimensional measurement under these new conditions, assuming that the uncertainty in $p_2 - p_a$ remains at 0.05 in H_2O.

5.13. The function y is given as

$$y = 1 + 3 + 4x^2$$

Perform the integration $\int y \, dx$ between $x = 0$ and $x = 4$ using the trapezoidal rule and Simpson's rule. Use four increments in x and calculate the error for each case. How many more increments would have to be taken for the trapezoidal rule to equal the accuracy of Simpson's rule?

5.14. Draw some kind of irregular figure on graph paper having small subdivisions. Determine the area by (*a*) counting squares, (*b*) the trapezoidal rule, and (*c*) Simpson's rule. Repeat by drawing a 10-cm-diameter circle using a sharp pencil and accurately measured radius from a metal scale. Discuss the "uncertainties" in these two experiments.

5.15. Plot the function $y = 3xe^{0.4x}$. Determine the area under the curve between $x = 1$ and $x = 3$ using several methods. Estimate the uncertainties.

5.16. Assemble a group of at least five people. Have each person measure two objects using a vernier caliper and a micrometer, with each person taking at least five readings. Analyze the data on the basis of the information in Chap. 3. What conclusions do you draw?

5.17. Determine values of the Newton-Cotes coefficients for cubic and quartic polynomials.

Answer:

For four points

$$A = \tfrac{3}{8} \Delta x(y_0 + 3y_1 + 3y_2 + y_3)$$

For five points

$$A = \frac{2 \Delta x}{45}(7y_0 + 32y_1 + 12y_2 + 32y_3 + 7y_4)$$

5.18. Plot the equation

$$y = x^2 e^{-0.2x}$$

on suitable graph paper for $0 < x < 1$. Select increments of x and determine the area under the curve using the trapezoidal and Simpson's rule. Also, calculate the true area.

5.19. For the curve $y = \sin x$, perform the integration $\int y \, dx$ from $x = 0$ to π using progressively smaller increments of x and both the trapezoid and Simpson's rule. Use a computer for the calculation and determine the number of increments for which the area is in error by 1 percent for both rules.

5.20. Using the function

$$y = 3e^{-2x}$$

obtain the integral $\int_0^2 y \, dx$ using the trapezoidal and Simpson's rule with increments of 0.2 and 0.5 for Δx. Also, calculate the error in these determinations.

5.21. Obtain the integral $\int_{0.1}^1 y \, dx$ of

$$y = \ln x$$

using increments of 0.1 and 0.2 in Δx. Calculate the error of the determination.

5.22. Obtain a numerical integration $\int_0^\infty y \, dx$ of the function

$$y = e^{-x}$$

using increments in x sufficiently small to obtain an accuracy of 99 percent.

5.23. A steel tape is used to measure a distance of 40 m at a temperature of 0°C. The tape indicates a true length at 15°C. What will be the indication at 0°C?

5.24. Apply the rectangular rule to the area determination of Example 5.4. What percent error results from use of this method?

5.25. An irregular polygon-shaped area has the following coordinates for the vertices of the polygon:

x	y
0	50
10	72
20	84
30	93
40	60
30	33
20	41
10	45
0	50

Determine the area of the enclosed figure using rectangular, trapezoidal, and Simpson's methods of numerical integration. Calculate the percent error for each determination.

5.26. Calculate the percent and absolute temperature error in a 30-m measurement with a steel surveyor's tape at 50°C when the tape is accurate at 15°C.

5.27. Divide a circle into two sets of even numbers of increments in the abscissa coordinate. Calculate the enclosed area using the rectangular, trapezoidal, and Simpson's rules. Determine the percent error for each calculation.

5.28. A rectangular area can be represented by the coordinates:

x	y
0	0.5
0	1
0.5	1
1	1
1.5	1
2	1
2	0.5
2	0
1.5	0
1	0
0.5	0
0	0
0	0.5

Determine the area using the rectangular and trapezoidal rules. Would it be possible to use these data as is, or in an abbreviated form, to employ Simpson's rule? If so, what would be the result?

5.29. The following coordinates represent a set of data:

x	y
0	1
1	1
2	4
3	7
4	13
5	21
6	31

Determine the integral $\int y\,dx$ using rectangular, trapezoidal, and Simpson's rules. What is the accuracy of each method? (*Hint:* Use the methods of Chap. 3 to determine the functional form that fits the data, and use this functional form to determine the exact value of the integral.)

5.30. Planck's blackbody radiation formula is

$$E_{b\lambda} = C_1 \lambda^{-5} / [\exp(C_2/\lambda T) - 1]$$

where $C_1 = 3.743 \times 10^8$, $C_2 = 14387$, λ is in μm and T in K.

Perform a numerical integration of this formula from 0 to 10 μm for temperatures of 1000 and 2000 K. Use any computer software that is available. Repeat for λ from 0 to 20 μm.

5.11   REFERENCES

1. Graneek, M., and J. C. Evans: "A Pneumatic Calibrator of High Sensitivity," *Engineer,* p. 62, July 13, 1951.

2. Hildebrand, F. B.: *Introduction to Numerical Analysis,* McGraw-Hill, New York, 1956.

3. Peters, C. G., and W. B. Emerson: "Interference Methods for Producing and Calibrating End Standards," *J. Res. Nat. Bur. Std.,* vol. 44, p. 427, 1950.

4. Sokolnikoff, I. S., and R. M. Redheffer: *Mathematics of Physics and Modern Engineering,* McGraw-Hill, New York, 1958.

5. "Metrology of Gauge Blocks," *Natl. Bur. Std. (U.S.), Circ.* 581, April 1957.

6. Fish, V. T., and G. M. Lance: "An Accelerometer for Fluidic Control Systems," ASME Paper No. 67-WA/FE-29.

7. Goldstein, S.: "A New Type of Pneumatic Triangle-Wave Oscillator," ASME Paper No. 66-WA/AUT-4.

8. Rosenbaum, H. M., and J. S. Gant: "A Pneumatic Tape Reader," 2d Cranfield Fluidics Conference, E2-13, January 1967.

9. Gant, G. C.: "A Fluidic Digital Displacement Indicator," 2d Cranfield Fluidics Conference, E1-1, January 1967.

10. Eckerlin, H. M., and D. A. Small: "A Method of Air Gauge Circuit Analysis," *Adv. Fluidics,* ASME, 1967.

11. Moses, H. L., D. A. Small, and G. A. Cotta: "Response of a Fluidic Air Gauge," *J. Basic Eng.,* pp. 475–478, September 1969.

12. Johnson, A. T.: "Microcomputer Programs for Instrumentation and Data Analysis," *Int. J. Appl. Eng. Ed.,* vol. 3, no. 2, p. 149, 1987.

13. Press, W. H., B. P. Flannery, S. A. Teukolsky, and W. T. Vettering: *Numerical Recipes, the Art of Scientific Computing,* Chap. 12, pp. 381–449, Cambridge University Press, Cambridge, 1986. Also FORTRAN Example Book and diskette.

14. Canale, R. P., and S. C. Chapra: *Electronic Tool Kit,* McGraw-Hill, New York, 1988.

15. Chapra, S. C., and R. P. Canale: *Numerical Methods for Engineers,* 3d ed., McGraw-Hill, New York, 1998.

16. Constantinides, A.: *Applied Numerical Methods with Personal Computers,* McGraw-Hill, New York, 1987.

17. Herceg, E. E.: *Schaevitz Handbook of Measurement and Control,* latest edition, Schaevitz Engineering, Pennsauken, N.J.

18. Buchanan, J. L., and P. R. Turner: *Numerical Method and Analysis,* McGraw-Hill, New York, 1992.

19. Chapra, Steve, and R. Canale: *Introduction to Computing for Engineers,* 2d ed., McGraw-Hill, New York, 1994.

chapter

6

PRESSURE MEASUREMENT

6.1 INTRODUCTION

Pressure is represented as a force per unit area. As such, it has the same units as stress and may, in a general sense, be considered as a type of stress. For our purposes we shall designate the force per unit area exerted by a fluid on a containing wall as the pressure. The forces that arise as a result of strains in solids are designated as stresses and discussed in Chap. 10. Thus, our discussion of pressure measurement is one restricted to fluid systems. *Absolute pressure* refers to the absolute value of the force per unit area exerted on the containing wall by a fluid. *Gage pressure* represents the difference between the absolute pressure and the local atmospheric pressure. *Vacuum* represents the amount by which the atmospheric pressure exceeds the absolute pressure. From these definitions we see that the absolute pressure may not be negative and the vacuum may not be greater than the local atmospheric pressure. The three terms are illustrated graphically in Fig. 6.1. It is worthwhile to mention that local fluid pressure may be dependent on many variables; parameters such as elevation, flow velocity, fluid density, and temperature are of frequent importance.

In the English system of units pressure is usually expressed in pounds per square inch absolute (psia). Gage pressure carrying the same unit is designated with the symbol psig. The standard SI unit for pressure is the newton per square meter (N/m^2) or pascal (Pa). Pressure is frequently expressed in terms of the height of a column of fluid (viz., mercury) which it will support at a temperature of 20°C. At standard atmospheric pressure this height is 760 mm of mercury having a density of 13.5951 g/cm^3. Some common units of pressure are

$$1 \text{ atmosphere (atm)} = 14.696 \text{ pounds per square inch absolute}$$
$$= 1.01325 \times 10^5 \text{ newtons per square meter (Pa)}$$
$$= 2116 \text{ pounds-force per square foot (lbf/ft}^2\text{)}$$
$$1 \text{ N/m}^2 \equiv 1 \text{ pascal (Pa)}$$
$$1 \text{ atmosphere (atm)} = 760 \text{ millimeters of mercury (mmHg)}$$
$$1 \text{ bar} = 10^5 \text{ newtons per square meter (100 kPa)}$$

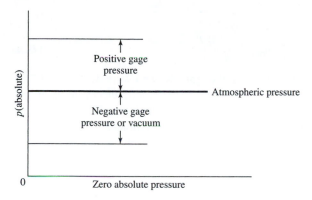

Figure 6.1   Relationship between pressure terms.

$$1 \text{ bar} = 10^5 \text{ Pa}$$
$$1 \text{ microbar} = 1 \text{ dyne per square centimeter}$$
$$= 2.089 \text{ pounds-force per square foot}$$
$$= 0.1 \text{ newton per square meter (0.1 Pa)}$$
$$1 \text{ millimeter of mercury (mmHg)} = 1333.22 \text{ microbar}$$
$$= 133.322 \text{ newtons per square meter}$$
$$\text{(133.3 Pa)}$$
$$1 \text{ micrometer} = 10^{-6} \text{ meters of mercury } (\mu\text{m, microns})$$
$$= 10^{-3} \text{ millimeters of mercury (mmHg)}$$
$$= 0.133322 \text{ newtons per square meter}$$
$$\text{(0.133 Pa)}$$
$$1 \text{ torr} \equiv 1 \text{ millimeter of mercury (mmHg)}$$
$$1 \text{ inch of mercury} = 70.73 \text{ pounds-force per square foot}$$
$$1 \text{ inch of water} = 5.203 \text{ pounds-force per square foot}$$
$$1 \text{ pound per square inch absolute} = 6894.76 \text{ newtons per square meter}$$
$$\text{(6.894 kPa)}$$
$$= 0.070307 \text{ kilograms-force per square}$$
$$\text{centimeter (kgf/cm}^2\text{), [kilopounds}$$
$$\text{per square centimeter (kp/cm}^2\text{)]}$$

Fluid pressure results from a momentum exchange between the molecules of the fluid and a containing wall. The total momentum exchange is dependent on the total number of molecules striking the wall per unit time and the average velocity of the molecules. For an ideal gas it may be shown that the pressure is given by

$$p = \tfrac{1}{3}nmv_{\text{rms}}^2 \qquad\qquad \textbf{[6.1]}$$

where n = molecular density, molecules/unit volume
         m = molecular mass
         v_{rms} = root-mean-square molecular velocity

It may also be shown that

$$v_{\text{rms}} = \sqrt{\frac{3kT}{m}} \qquad \textbf{[6.2]}$$

where T is the absolute temperature of the gas, K, and k is 1.3803×10^{-23} J/ molecule · K (Boltzmann's constant). Equation (6.1) is a kinetic-theory interpretation of the ideal-gas law. An expression for the pressure in a liquid would not be so simple.

The *mean free path* is defined as the average distance a molecule travels between collisions. For an ideal gas whose molecules act approximately like billiard balls

$$\lambda = \frac{\sqrt{2}}{8\pi r^2 n} \qquad \textbf{[6.3]}$$

where r is the effective radius of the molecule and λ is the mean free path. It is clear that the mean free path increases with a decrease in the gas density. At standard atmospheric pressure and temperature the mean free path is quite small, of the order of 10^{-5} cm. At a pressure of 1 μm, however, the mean free path would be of the order of 1 cm. At very low pressures the mean free path may be significantly greater than a characteristic dimension of the containing vessel. For air the relation for mean free path reduces to

$$\lambda = 8.64 \times 10^{-7} \frac{T}{p} \text{ ft} \qquad T \text{ in } °R \text{ and } p \text{ in lbf/ft}^2 \qquad \textbf{[6.3a]}$$

$$\lambda = 2.27 \times 10^{-5} \frac{T}{p} \text{ m} \qquad T \text{ in K and } p \text{ in Pa} \qquad \textbf{[6.3b]}$$

A variety of devices are available for pressure measurement, as we shall see in the following sections. Static, that is, steady-state, pressure is not difficult to measure with good accuracy. Dynamic measurements, however, are much more preplexing because they are influenced strongly by the characteristics of the fluid being studied as well as the construction of the measurement device. In many instances a pressure instrument that gives very accurate results for a static measurement may be entirely unsatisfactory for dynamic measurements. We shall discuss some of the factors that are important for good dynamic response in conjunction with the exposition associated with the different types of pressure-measurement devices.

Example 6.1 | **MEAN FREE PATH.**   Determine the mean free path for air at 20°C and pressures of 1 atm, 1 torr, 1 μm, and 0.01 μm.

Solution

The calculation is performed using Eq. (6.3b) with $T = 20°C = 293$ K and the following pressures in pascals:

$$1 \text{ atm} = 1.0132 \times 10^5 \text{ Pa}$$
$$1 \text{ torr} = 133.32 \text{ Pa}$$
$$1 \ \mu\text{m} = 0.13332 \text{ Pa}$$
$$0.01 \ \mu\text{m} = 1.332 \times 10^{-3} \text{ Pa}$$

Inserting the values in Eq. (6.3b) gives

$$\lambda(1 \text{ atm}) = 6.564 \times 10^{-8} \text{ m}$$
$$\lambda(1 \text{ torr}) = 4.989 \times 10^{-5} \text{ m}$$
$$\lambda(1 \text{ } \mu\text{m}) = 0.04989 \text{ m}$$
$$\lambda(0.01 \text{ } \mu\text{m}) = 4.989 \text{ m}$$

6.2 DYNAMIC RESPONSE CONSIDERATIONS

The transient response of pressure-measuring instruments is dependent on two factors: (1) the response of the transducer element that senses the pressure and (2) the response of the pressure-transmitting fluid and the connecting tubing, etc. This latter factor is frequently the one that determines the overall frequency response of a pressure-measurement system, and, eventually, direct calibration must be relied upon for determining this response. An estimate of the behavior may be obtained with the following analysis. Consider the system shown in Fig. 6.2. The fluctuating pressure has a frequency of ω and an amplitude of p_0 and is impressed on the tube of length L and radius r. At the end of this tube is a chamber of volume V where the connection to the pressure-sensitive transducer is made. The mass of fluid vibrates under the influence of fluid friction in the tube, which tends to dampen the motion. If the conventional formula for laminar friction resistance in the tube flow is used to represent this friction, the resulting expression for the pressure-amplitude ratio is

$$\left| \frac{p}{p_0} \right| = \frac{1}{\{[1 - (\omega/\omega_n)^2]^2 + 4h^2(\omega/\omega_n)^2\}^{1/2}} \qquad \textbf{[6.4]}$$

In this equation p is the amplitude of the pressure signal impressed on the transducer. The natural frequency ω_n is given by

$$\omega_n = \sqrt{\frac{3\pi r^2 c^2}{4LV}} \qquad \textbf{[6.5]}$$

and the damping ratio h is

$$h = \frac{2\mu}{\rho c r^3} \sqrt{\frac{3LV}{\pi}} \qquad \textbf{[6.6]}$$

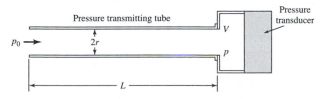

Figure 6.2   Schematic of pressure-transmitting system.

In the above formulas c represents the velocity of sound in the fluid, μ is the dynamic viscosity of the fluid, and ρ is the fluid density. The phase angle for the pressure signal is

$$\phi = \tan^{-1} \frac{-2h(\omega/\omega_n)}{1 - (\omega/\omega_n)^2} \qquad \textbf{[6.7]}$$

The velocity of sound for air may be calculated from

$$c = 49.1 T^{1/2} \text{ fts} \qquad \text{with } T \text{ in } °R$$

$$c = 20.04 T^{1/2} \text{ m/s} \qquad \text{with } T \text{ in } K$$

When the tube diameter is very small, as in a capillary, it is possible to produce a very large damping ratio so that Eq. (6.4) will reduce to the following for frequencies below the natural frequencies:

$$\left| \frac{p}{p_0} \right| = \frac{1}{[1 + 4h^2(\omega/\omega_n)^2]^{1/2}} \qquad \textbf{[6.8]}$$

If the transmitting fluid is a gas, the entire system can act as a Helmholtz resonator with a resonant frequency of

$$\omega_n = \left[\frac{\pi r^2 c^2}{V\left(L + \frac{1}{2}\sqrt{\pi^2 r^2}\right)} \right]^{1/2} \qquad \textbf{[6.9]}$$

More complete information on the dynamic response of pressure-measurement systems is given in Refs. [1], [7], and [11].

From both Eqs. (6.4) and (6.8) it is evident that a capillary tube may be used for effective damping of pressure signals. The tube is then said to act as an acoustical filter. The similarity of this system to that described in Sec. 2.7 and the discussion of the seismic instrument in Sec. 11.3 are to be noted. It should be noted also that the actual dynamic response for tube systems is strongly frequency-dependent, and the preceding formulas must be accepted with possible modification in high-frequency ranges. Tijdeman [18] and others [14 to 17] discuss experimental and analytical solutions which are available for such problems. In a dynamic pressure-measurement application one must also consider the frequency response of the pressure transducer and its movement in the overall measurement system. In general, one should try to design the system so that the natural frequency of the transducer is substantially greater than the signal frequency to be measured. Dynamic pressure measurements are particularly applicable to sound-level determinations, as discussed in Chap. 11. The calibration of pressure transducers for dynamic measurement applications is rather involved; it is discussed in detail in Ref. [19].

Example 6.2 | **NATURAL FREQUENCY FOR TUBE.**   A small tube, 0.5 mm in diameter, is connected to a pressure transducer through a volume of 3.5 cm³. The tube has a length of 7.5 cm. Air at 1 atm and 20°C is the pressure-transmitting fluid. Calculate the natural frequency for this system and the damping ratio.

Solution

We shall use Eq. (6.9) for this calculation. For air

$$\rho = \frac{p}{RT} = \frac{1.0132 \times 10^5}{(287)(293)} = 1.205 \text{ kg/m}^3$$

$$c = (20.04)(293)^{1/2} = 343 \text{ m/s}$$

$$\mu = 1.91 \times 10^{-5} \text{ kg/m} \cdot \text{s (Table A.6, appendix)}$$

Thus,

$$\omega_n = \left\{ \frac{\pi(0.25 \times 10^{-3})^2(343)^2}{(3.5 \times 10^{-6})[0.075 + (0.5)(\pi)(0.25 \times 10^{-3})^{1/2}]} \right\}^{1/2}$$

$$= 296 \text{ Hz}$$

The damping ratio is calculated with Eq. (6.6):

$$h = \frac{(2)(1.91 \times 10^{-5})}{(1.205)(343)(0.25 \times 10^{-3})^3} \left[\frac{(3)(0.075)(3.5 \times 10^{-6})}{\pi} \right]^{1/2}$$

$$= 2.96$$

ATTENUATION IN TUBE.   Calculate the attenuation of a 100-Hz pressure signal in the system of Example 6.2. | **Example 6.3**

Solution

For this calculation we employ Eq. (6.4). We have

$$\frac{\omega}{\omega_n} = \frac{100}{296} = 0.338$$

so that

$$\left| \frac{p}{p_0} \right| = \frac{1}{\{[1 - (0.338)^2]^2 + 4(2.96)^2(0.338)^2\}^{1/2}} = 0.457$$

6.3   MECHANICAL PRESSURE-MEASUREMENT DEVICES

Mechanical devices offer the simplest means for pressure measurement. In this section we shall examine the principles of some of the more important arrangements.

The fluid manometer is a widely used device for measurement of fluid pressures under steady-state and laboratory conditions. Consider first the *U-tube manometer* shown in Fig. 6.3. The difference in pressure between the unknown pressure p and the atmosphere is determined as a function of the differential height h. The density of the fluid transmitting the pressure p is ρ_f, and the density of the manometer fluid

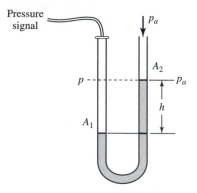

Figure 6.3   U-tube manometer.

is designated as ρ_m. A pressure balance of the two columns dictates that

$$p_a + \frac{g}{g_c} h \rho_m = p + \frac{g}{g_c} h \rho_f \qquad [6.10]$$

or

$$p - p_a = \frac{g}{g_c} h (\rho_m - \rho_f) \qquad [6.11]$$

Equation (6.11) gives the basic principle of the U-tube manometer. It is to be noted that the distance h is measured parallel to the gravitational force and that the differential pressure $p - p_a$ is measured at the location designated by the dashed line. If the location of the pressure source is at a different elevation from this point, there could be an appreciable error in the pressure determination, depending on the density of the transmitting fluid.

The *sensitivity* of the U-tube manometer may be defined as

$$\text{Sensitivity} = h/(p - p_a) = h/\triangle p = 1/(g/g_c)(\rho_m - \rho_f)$$

or for a manometer with $\rho_m \gg \rho_f$,

$$\text{Sensitivity} = 1/\rho_m (g/g_c)$$

A *well-type manometer* operates in the same manner as the U-tube manometer, except that the construction is as shown in Fig. 6.4. In this case the pressure balance of Eq. (6.10) still yields

$$p - p_a = \frac{g}{g_c} h (\rho_m - \rho_f)$$

This equation is seldom used, however, because the height h is not the fluid displacement which is normally measured. Typically, the well-type manometer is filled to a certain level at zero-pressure differential conditions. A measurement is then made of the displacement of the small column from this zero level. Designating this displacement by h', we have

$$h' A_2 = (h - h') A_1 \qquad [6.12]$$

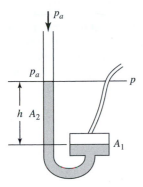

Figure 6.4   Well-type manometer.

since the volume displacements are the same on both sides of the manometer. Inserting Eq. (6.12) in (6.10) gives

$$p - p_a = \frac{g}{g_c} h' \left(\frac{A_2}{A_1} + 1 \right) (\rho_m - \rho_f) \qquad \textbf{[6.13]}$$

Commercial well-type manometers have the scale for the manometer column graduated so that the user need not apply the area correction factor to the indicated displacement h'. Thus, for an area ratio of $A_2/A_1 = 0.03$ a true reading of 10.0 in for h' would be indicated as 10.3 as a result of the special scale graduation. The indicated value is then substituted for h in Eq. (6.11).

Manometers may be oriented in an inclined position to lengthen the scale and to improve readability, or special optical sightglasses and vernier scales may be employed to provide more accurate location and indication of the manometer-fluid height than could be obtained with the naked eye. When mercury is the manometer fluid, variable-reluctance pickups may be used to sense accurately the fluid height. Special metal floats may also afford such a convenience with less dense fluids which are nonconductive.

For a manometer inclined at an angle θ with the horizontal the sensitivity becomes

$$\text{Sensitivity} = L/\Delta p = 1/(\rho_m - \rho_f) \sin \theta (g/g_c)$$

where L is the measured fluid displacement along the incline and $h = L \sin \theta$

Now consider a two-fluid manometer consisting of two reservoirs of diameter D connected by a U-tube of smaller diameter d. Reservoir 1 contains a fluid with density ρ_1, while reservoir 2 contains a fluid having a density ρ_2. The assembly is subjected to a pressure differential Δp. The sensitivity may be derived as

$$\text{Sensitivity} = h/\Delta p = \{[(d/D)^2(\rho_2 + \rho_1) + (\rho_2 - \rho_1)](g/g_c)\}^{-1}$$

where

$$h = h_{\text{initial interface}} - h_{\text{final interface}}$$

When $D \gg d$ and the difference in the two fluid densities is small, a very large sensitivity can result. This situation forms the basis for the *micromanometer* [24].

| **Example 6.4** | **U-TUBE MANOMETER.**   A U-tube manometer employs a special oil having a specific gravity of 0.82 for the manometer fluid. One side of the manometer is open to local atmospheric pressure of 29.3 inHg and the difference in column heights is measured as 20 cm $\pm$ 1.0 mm when exposed to an air source at 25°C. Standard acceleration of gravity is present. Calculate the pressure of the air source in pascals and its uncertainty. |

Solution

The manometer fluid has a density of 82 percent of that of water at 25°C; so,

$$\rho_m = 0.82\rho_w = (0.82)(996 \text{ kg/m}^3) = 816.7 \text{ kg/m}^3$$

The local atmospheric pressure is

$$p_a = 29.3 \text{ inHg} = 9.922 \times 10^4 \text{ Pa}$$

The "fluid" in this problem is the air which has a density at the above pressure and 25°C (298 K) of

$$\rho_f = \rho_a = \frac{p}{RT} = \frac{9.922 \times 10^4}{(287)(298)} = 1.16 \text{ kg/m}^3$$

For this problem the density is negligible compared to that of the manometer fluid, but we shall include it anyway. From Eq. (6.11)

$$p - p_a = \frac{g}{g_c} h(\rho_m - \rho_f)$$

$$= \frac{9.807}{1.0}(0.2)(816.7 - 1.16)$$

$$= 1600 \text{ Pa}$$

or
$$p = 1600 + 9.922 \times 10^4 = 1.0082 \times 10^5 \text{ Pa}$$

Comment

The uncertainty of the column height measurement is $1.0/200 = 0.5$ percent. If all other terms are exact, the uncertainty in the pressure measurement would be 0.5 percent of 1600 or 8 Pa. It is unlikely that the local atmospheric pressure and temperature or specific gravity of the fluid would be known exactly, so they too would make a contribution to the uncertainty. But no information is given to determine their influence.

When a well-type manometer is arranged as in Fig. 6.5, it is commonly called a barometer. The top of the column contains saturated mercury vapor at the local temperature. This saturation pressure is negligible in comparison to atmospheric pressure. The well is exposed to atmospheric pressure. The height h is thus a measure of the absolute atmospheric pressure. When $p_a = 14.696$ psia (1 atm), the height of a column of mercury at 68°F (20°C) would be 760 mm.

The column has a graduated scale fixed in position which requires that the instrument be zeroed for each reading because the level of the well will vary with height of the mercury in the column. The zeroing is accomplished with a screw adjustment which sets the level of the well at a reference position.

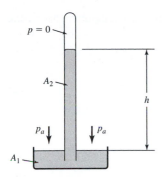

Figure 6.5   Manometer used as a barometer.

At this point we should note that many pressure-measurement devices indicate *gage* pressure, or the difference between the absolute and local atmospheric pressures. To obtain the absolute pressure, one must, of course, calculate

$$p \text{ (absolute)} = p \text{ (gage)} + p \text{ (atm)}$$

The local atmospheric pressure *must be obtained from a local measurement* near the place where the gage pressure is measured. Such a measurement might be performed with a mercury barometer. We must note that local atmospheric pressure can vary between the inside and outside of a building because of the ventilation systems.

A common mistake made by novice experimentalists is to take the local atmospheric pressure as the value given by the local weather bureau. Such practice can produce major errors because the value stated by the weather bureau is corrected to sea level, using the altitude at the weather station. In Denver, Colorado, for example, at an altitude of about 5000 ft, when the weather station reports a barometric pressure of 760 mm (29.92 in) of mercury, the true barometric pressure is only 632 mm, or 17 percent less than the 760-mm value. For altitudes between 0 and 36,000 ft the standard atmosphere is expressed by

$$p = p_0 \left(1 - \frac{BZ}{T_0} \right)^{5.26} \qquad \textbf{[6.14]}$$

where p_0 = standard atmospheric pressure at sea level

Z = altitude, m or ft

$T_0 = 518.69°R = 288.16 \text{ K} = 15°C$

$B = 0.003566°R/\text{ft} = 0.00650 \text{ K/m}$

INFLUENCE OF BAROMETER READING ON VACUUM MEASUREMENT.   A pressure measurement is made in Denver, Colorado (elevation 5000 ft), indicating a *vacuum* of 75 kPa. The weather bureau reports a barometer reading of 29.92 inHg. The absolute pressure is | **Example 6.5**

to be calculated from this information. What percent error would result if the above barometric pressure were taken at face value?

Solution

The absolute pressure is given by

$$p_{absolute} = p_{atm} - p_{vacuum} \qquad\qquad [a]$$

If the barometer report is taken at face value,

$$p_{atm} = (29.92)(25.4) = 760 \text{ mmHg} = 101.32 \text{ kPa}$$

and the absolute pressure is

$$p_{absolute} = 101.32 - 75 = 26.32 \text{ kPa} \qquad\qquad [b]$$

Assuming the correction for altitude is given by Eq. (6.14), the true atmospheric pressure at the weather bureau is

$$p_{atm} = (760)[1 - (0.003566)(5000)/518.69]^{5.26} = 632.3 \text{ mmHg} = 84.29 \text{ kPa}$$

Assuming the local atmospheric pressure where the measurement is taken has this same value, the true absolute pressure is therefore

$$P_{absolute} = 84.29 - 75 = 9.29 \text{ kPa} \qquad\qquad [c]$$

The percent error between the values in Eqs. (b) and (c) is

$$\% \text{ error} = \frac{26.32 - 9.29}{9.29} \times 100 = +183 \text{ percent}$$

Obviously, the *local* barometric pressure must be used instead of the value reported by the weather bureau.

6.4 DEAD-WEIGHT TESTER

The dead-weight tester is a device used for balancing a fluid pressure with a known weight. Typically, it is a device used for static calibration of pressure gages and is seldom employed for an actual pressure measurement. Our discussion will be concerned only with the use of the dead-weight tester as a calibration device.

Consider the schematic in Fig. 6.6. The apparatus is set up for calibration of the pressure gage G. The chamber and cylinder of the tester are filled with a clean oil by first moving the plunger to its most forward position and then slowly withdrawing it while the oil is poured in through the opening for the piston. The gage to be tested is installed and the piston inserted in the cylinder. The pressure exerted on the fluid by the piston is now transmitted to the gage when the valve is opened. This pressure may be varied by adding weights to the piston or by using different piston-cylinder combinations of varying areas. The viscous friction between the piston and the cylinder in the axial direction may be substantially reduced by rotating the piston-weight assembly while the measurement is taken. As the pressure is increased,

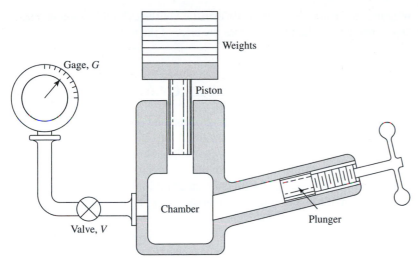

Figure 6.6   Schematic of a dead-weight tester.

it may be necessary to advance the plunger to account for the compression of the oil and any entrapped gases in the apparatus. High-pressure–dead-weight testers have a special lever system which is used to apply large forces to the piston.

The accuracies of dead-weight testers are limited by two factors: (1) the friction between the cylinder and the piston and (2) the uncertainty in the area of the piston. The friction is reduced by rotation of the piston and use of long enough surfaces to ensure negligible flow of oil through the annular space between the piston and the cylinder. The area upon which the weight force acts is not the area of the piston or the area of the cylinder; it is some effective area between these two which depends on the clearance spacing and the viscosity of the oil. The smaller the clearance, the more closely the effective area will approximate the cross-sectional area of the piston. It can be shown[1] that the percentage error due to the clearance varies according to

$$\text{Percent error} \sim \frac{(\rho \, \Delta p)^{1/2} b^3}{\mu D L} \qquad \textbf{[6.15]}$$

where ρ = density of the oil
Δp = pressure differential on the cylinder
b = clearance spacing
μ = viscosity
D = piston diameter
L = piston length

| [1]See, e.g., Ref. [10], p. 105.

At high pressures there can be an elastic deformation of the cylinder which increases the clearance spacing and thereby increases the error of the tester.

6.5   BOURDON-TUBE PRESSURE GAGE

Bourdon-tube pressure gages enjoy a wide range of application where consistent, inexpensive measurements of static pressure are desired. They are commercially available in many sizes (1- to 16-in diameter) and accuracies. The Heise gage[2] is an extremely accurate bourdon-tube gage with an accuracy of 0.1 percent of full-scale reading; it is frequently employed as a secondary pressure standard in laboratory work.

The construction of a bourdon-tube gage is shown in Fig. 6.7. The bourdon tube itself is usually an elliptical cross-sectional tube having a C-shape configuration. When the pressure is applied to the inside of the tube, an elastic deformation results, which, ideally, is proportional to the pressure. The degree of linearity depends on the quality of the gage. The end of the gage is connected to a spring-loaded linkage, which amplifies the displacement and transforms it to an angular rotation of the pointer. The linkage is constructed so that the mechanism may be adjusted for optimum linearity and minimum hysteresis, as well as to compensate for wear which may develop over

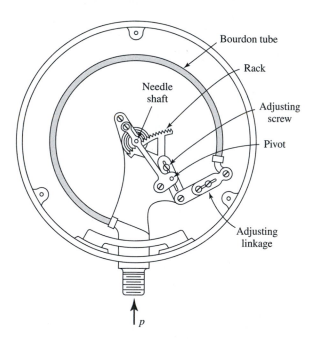

Figure 6.7   Schematic of a bourdon-tube pressure gage.

[2]Manufactured by Heise Gage Company, Newton, CT.

a period of time. Electrical-resistance strain gages (Sec. 10.7) may also be installed on the bourdon tube to sense the elastic deformation. A proprietary design of digital pressure transducers based upon piezoresistance strain gage response claims an accuracy of 0.02 percent of full scale and is available in full-scale ranges from 10 to 10,000 psia.[3]

6.6 DIAPHRAGM AND BELLOWS GAGES

Diaphragm and bellows gages represent similar types of elastic deformation devices useful for many pressure-measurement applications. Consider first the flat diaphragm subjected to the differential pressure $p_1 - p_2$, as shown in Fig. 6.8. The diaphragm will be deflected in accordance with this pressure differential and the deflection sensed by an appropriate displacement transducer. Electrical-resistance strain gages may also be installed on the diaphragm, as shown in Fig. 6.9. The output of these gages is a function of the local strain, which, in turn, may be related to the diaphragm deflection and pressure differential. Both semiconductor and foil grad strain gages are employed in practice. Accuracies of ± 0.5 percent of full scale are typical. The deflection generally follows a linear variation with Δp when the deflection is less than one-third the diaphragm thickness. Figure 6.10 compares the deflection characteristics of three diaphragm arrangements as given by Roark [9]. Note that the first two diaphragms have uniform pressure loading over the entire surface of the disk, while the third type has a load which is applied at the center boss. In all three cases it is assumed that the outer edge of the disk is rigidly fixed and supported. To facilitate linear response over a larger range of deflections than that imposed by the one-third-thickness restriction, the diaphragm may be constructed from a corrugated disk, as shown in Fig. 6.11. This type of diaphragm is most suitable for those applications where a mechanical device is used for sensing the deflection of the diaphragm. Larger deflections are usually necessary with a mechanical amplification device than for electric transducers. A good summary of the properties of corrugated diaphragms is given in Ref. [12].

The bellows gage is depicted schematically in Fig. 6.12. A differential pressure force causes a displacement of the bellows, which may be converted to an electrical signal or undergo a mechanical amplification to permit display of the output on an indicator dial. The bellows gage is generally unsuitable for transient measurements because of the larger relative motion and mass involved. The diaphragm gage, on the other hand, which may be quite stiff, involves rather small displacements and is suitable for high-frequency pressure measurements.

The deflection of a diaphragm under pressure may be sensed by a capacitance variation, as shown in Fig. 6.13. Such pressure pickups are well suited for dynamic measurements since the natural frequency of diaphragms can be rather high. The capacitance pickup, however, involves low sensitivity, and special care must be exerted in the construction of readout circuitry. A schematic diagram of a LVDT-diaphragm

---

[3] Heise Model DXD, Ashcroft Inc., Stratford, CT.

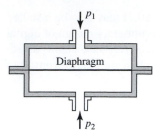

Figure 6.8   Schematic of a diaphragm gage.

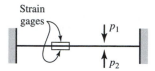

Figure 6.9   Diaphragm gage using electrical-resistance strain gages.

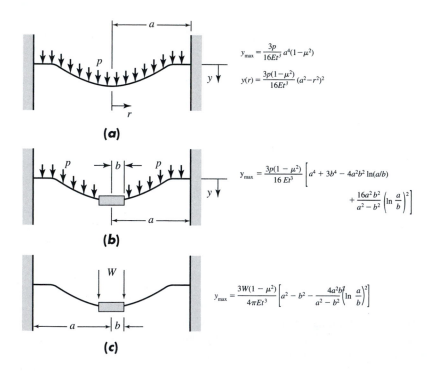

$$y_{max} = \frac{3p}{16Et^3} a^4(1-\mu^2)$$

$$y(r) = \frac{3p(1-\mu^2)}{16Et^3}(a^2-r^2)^2$$

(a)

$$y_{max} = \frac{3p(1-\mu^2)}{16\,Et^3}\left[a^4 + 3b^4 - 4a^2b^2\ln(a/b) \right.$$
$$\left. + \frac{16a^2b^2}{a^2-b^2}\left(\ln\frac{a}{b}\right)^2 \right]$$

(b)

$$y_{max} = \frac{3W(1-\mu^2)}{4\pi Et^3}\left[a^2 - b^2 - \frac{4a^2b^2}{a^2-b^2}\left(\ln\frac{a}{b}\right)^2 \right]$$

(c)

Figure 6.10   Deflection characteristics of three diaphragm arrangements, according to Ref. [9]. (a) Edges fixed, uniform load over entire surface; (b) outer edge fixed and supported, inner edge fixed, uniform load over entire actual surface; (c) outer edge fixed and supported, inner edge fixed, uniform load along inner edge.

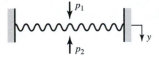

Figure 6.11   Corrugated-disk diaphragm.

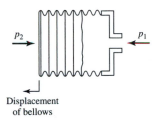

Displacement
of bellows

Figure 6.12   Schematic of a bellows pressure gage.

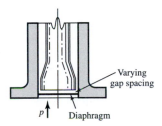

Figure 6.13   Capacitance pressure gage.

differential pressure gage is shown in Fig. 6.14. Commercial models of this type of gage permit measurement of pressures as low as 0.00035 psi (0.25 Pa).

The natural frequency of a circular diaphragm fixed at its perimeter is given by Hetenyi [5] as

$$f = \frac{10.21}{a^2} \sqrt{\frac{g_c E t^2}{12(1 - \mu^2)\rho}} \text{ Hz} \qquad \text{[6.16]}$$

where E = modulus of elasticity, psi or Pa

t = thickness, in or m

a = radius of diaphragm, in or m

ρ = density of material, lbm/in^3 or kg/m^3

g_c = dimensional conversion constant

= 385.9 lbm · in/lbf · s^2 or 1.0 kg · m/N · s^2

μ = Poisson's ratio

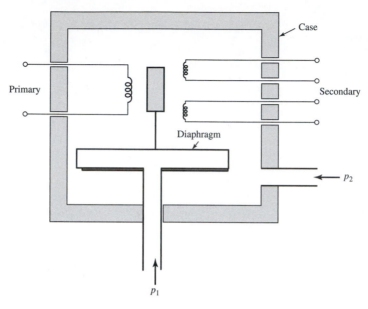

Figure 6.14   Schematic of diaphragm LVDT combination used as differential pressure gage.

Equation (6.16) may be simplified to the following relation for steel diaphragms:

$$f = 1.934 \times 10^6 \frac{t}{\pi a^2} \qquad (t \text{ and } a \text{ in in.}) \qquad \textbf{[6.17a]}$$

$$= 4.912 \times 10^4 \frac{t}{\pi a^2} \qquad (t \text{ and } a \text{ in m}) \qquad \textbf{[6.17b]}$$

Piezoresistive or semiconductor pressure transducers usually consist of a silicon diaphragm with a semiconductor strain gage bonded to the diaphragm to measure deflection.

Reference [21] gives the construction details of a number of commercial pressure transducers and techniques for thermal compensation.

Example 6.6 | **NATURAL FREQUENCY OF A DIAPHRAGM GAGE.**   A diaphragm pressure gage is to be constructed of spring steel ($E = 200$ GN/m^2, $\mu = 0.3$) 5.0 cm in diameter and is to be designed to measure a maximum pressure of 1.4 MPa. Calculate the thickness of the gage required so that the maximum deflection is one-third this thickness. Calculate the natural frequency of this diaphragm.

Solution

Using the relation from Fig. 6.10, we have

$$\frac{1}{3}t = \frac{3\,\Delta p}{16Et^3}a^4(1-\mu^2)$$

$$t^4 = \frac{(0)(1.4\times 10^6)(0.025)^4[1-(0.3)^2]}{(16)(2\times 10^{11})}$$

$$t = 1.09 \text{ mm}$$

We may calculate the natural frequency from Eq. (6.16)

$$f = \frac{10.21}{(0.025)^2}\left[\frac{(1.0)(2\times 10^{11})(0.00109)^2}{(12)[1-(0.3)^2](7800)}\right]^{1/2}$$

$$= 27{,}285 \text{ Hz}$$

6.7   THE BRIDGMAN GAGE[4]

It is known that resistance of fine wires changes with the pressure according to a linear relationship.

$$R = R_1(1 + b\,\Delta p) \qquad\qquad \textbf{[6.18]}$$

R_1 is the resistance at 1 atm, b is the pressure coefficient of resistance, and Δp is the gage pressure. The effect may be used for measurement of pressures as high as 100,000 atm [4]. A pressure transducer based on this principle is called a *Bridgman gage*. A typical gage employs a fine wire of Manganin (84% Cu, 12% Mn, 4% Ni) wound in a coil and enclosed in a suitable pressure container. The pressure coefficient of resistance for this material is about 1.7×10^{-7} psi^{-1} (2.5×10^{-11} Pa^{-1}). The total resistance of the wire is about 100 Ω, and conventional bridge circuits are employed for measuring the change in resistance. Such gages are subject to aging over a period of time so that frequent calibration is required; however, when properly calibrated, the gage can be used for high-pressure measurement with an accuracy of 0.1 percent. The transient response of the gage is exceedingly good. The resistance wire itself can respond to variations in the megahertz range. Of course, the overall frequency response of the pressure-measurement system would be limited to much lower values because of the acoustic response of the transmitting fluid. Many of the problems associated with high-pressure measurement are discussed more fully in Refs. [2], [6], and [13].

---

[4]P. W. Bridgman, *Proc. Natl. Acad. Sci., U.S.*, vol. 3, p. 10, 1917.

6.8   LOW-PRESSURE MEASUREMENT

The science of low-pressure measurement is a rather specialized field which requires considerable care on the part of the experimentalist. The purpose of our discussion is to call attention to the more prominent types of vacuum instruments and to describe the physical principles upon which they operate. For readers requiring more specialized information we refer them to the excellent monograph by Dushman and Lafferty [3]. The reader may also consult this reference for information on the various techniques for producing and maintaining a vacuum.

For moderate vacuum measurements the bourdon gage, manometers, and various diaphragm gages may be employed. Our discussion in this section, however, is concerned with the measurement of low pressures which are not usually accessible to the conventional gages. In this sense we are primarily interested in absolute pressures below 1 torr (1 mmHg, 133 Pa).

6.9   THE MCLEOD GAGE[5]

The McLeod gage is a modified mercury manometer which is constructed as shown in Fig. 6.15. The movable reservoir is lowered until the mercury column drops below the opening O. The bulb B and capillary C are then at the same pressure as the vacuum source p. The reservoir is subsequently raised until the mercury fills the bulb and rises in the capillary to a point where the level in the reference capillary is located at the zero point. The volume of the capillary per unit length is denoted by a so that the volume of the gas in the capillary is

$$V_c = ay \qquad \text{[6.19]}$$

where y is the length of the capillary occupied by the gas.

We designate the volume of the capillary, bulb, and tube down to the opening as V_B. If we assume isothermal compression of the gas in the capillary, we have

$$p_c = p \frac{V_B}{V_c} \qquad \text{[6.20]}$$

Now, the pressure indicated by the capillary is

$$p_c - p = y \qquad \text{[6.21]}$$

where we are expressing the pressure in terms of the height of the mercury column. Combining Eqs. (6.19) to (6.21) gives

$$p = \frac{ay^2}{V_B - ay} = \frac{yV_c}{V_B - ay} \qquad \text{[6.22]}$$

[5]H. McLeod, *Phil. Mag.*, vol. 48, p. 110, 1874.

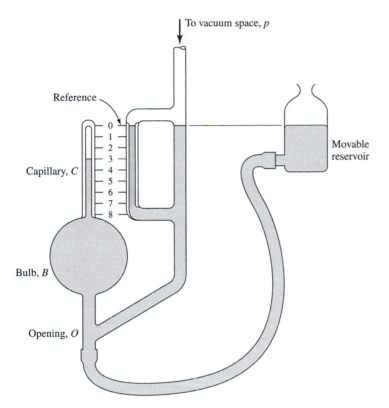

Figure 6.15   The McLeod gage.

For most cases $ay \ll V_B$ and

$$p = \frac{ay^2}{V_B} \qquad \textbf{[6.23]}$$

Commercial McLeod gages have the capillary calibrated directly in micrometers. The McLeod gage is sensitive to condensed vapors that may be present in the sample because they can condense upon compression and invalidate Eq. (6.20). For dry gases the gage is applicable from 10^{-2} to 10^{2} μm (0.0013 to 13.3 Pa).

ERROR IN McLEOD GAGE.   A McLeod gage has $V_B = 100 \text{ cm}^3$ and a capillary diameter of 1 mm. Calculate the pressure indicated by a reading of 3.00 cm. What error would result if Eq. (6.23) were used instead of Eq. (6.22)? | **Example 6.7**

Solution

We have

$$V_c = \frac{\pi(1)^2}{4}(30.0) = 23.6 \text{ mm}^3$$

$$V_B = 10^5 \text{ mm}^3$$

From Eq. (6.22)

$$p = \frac{(23.6)(30.0)}{10^5 - 23.6} = 0.0071 \text{ torr} = 7.1 \; \mu\text{m} \; (0.94 \text{ Pa})$$

The fractional error in using Eq. (6.23) would be

$$\text{Error} = \frac{ay}{V_B} = 2.36 \times 10^{-4}$$

or a negligibly small value.

6.10   PIRANI THERMAL-CONDUCTIVITY GAGE[6]

At low pressures the effective thermal conductivity of gases decreases with pressure. The Pirani gage is a device that measures the pressure through the change in thermal conductance of the gas. The gage is constructed as shown in Fig. 6.16. An electrically heated filament is placed inside the vacuum space. The heat loss from the filament is dependent on the thermal conductivity of the gas and the filament temperature. The lower the pressure, the lower the thermal conductivity and, consequently, the higher the filament temperature for a given electric-energy input. The temperature of the filament could be measured by a thermocouple, but in the Pirani-type gage the measurement is made by observing the variation in resistance of the filament material (tungsten, platinum, etc.). The resistance measurement may be performed with an appropriate bridge circuit. The heat loss from the filament is also a function of the ambient temperature, and, in practice, *two* gages are connected in series, as shown in Fig. 6.17, to compensate for possible variations in the ambient conditions. The measurement gage is evacuated, and both it and the sealed gage are exposed to the same environment conditions. The bridge circuit is then adjusted (through resistance R_2) to produce a null condition. When the test gage is now exposed to the particular pressure conditions, the deflection of the bridge from the null position will be compensated for changes in environment temperature.

Pirani gages require an empirical calibration and are not generally suitable for use at pressures much below 1 μm. The upper limit is about 1 torr (133 Pa), giving an overall range of about 0.1 to 100 Pa. For higher pressures the thermal conductance changes very little with pressure. It must be noted that the heat loss from the filament is also a function of the conduction losses to the filament supports and radiation losses to the surroundings. The lower limit of applicability of the gage is the point where these effects overshadow the condition into the gas. The transient response of the Pirani gage is poor. The time necessary for the establishment of thermal equilibrium may be of the order of several minutes at low pressures.

---

[6]M. Pirani, *Verhandl. deut. physik. Ges.*, vol. 8, p. 686, 1906.

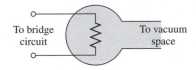

Figure 6.16   Schematic of Pirani gage.

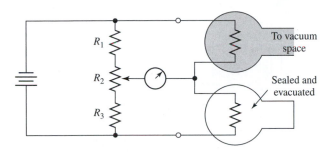

Figure 6.17   Pirani-gage arrangement to compensate for change in ambient temperature.

6.11   THE KNUDSEN GAGE[7]

Consider the arrangement shown in Fig. 6.18. Two vanes V along with the mirror M are mounted on the thin-filament suspension. Near these vanes are two heated plates P, each of which is maintained at a temperature T. The separation distance between the vanes and plates is *less than* the mean free path of the surrounding gas. Heaters are installed so that the temperature of the plates is higher than that of the surrounding gas. The vanes are at the temperature of the gas T_g. The molecules striking the vanes from the hot plates have a higher velocity than those leaving the vanes because of the difference in temperature. Thus, there is a net momentum imparted to the vanes which may be measured by observing the angular displacement of the mirror, similar to the technique used in a lightbeam galvanometer. The total momentum exchange with the vanes is a function of molecular density, which, in turn, is related to the pressure and temperature of the gas. An expression for the gas pressure may thus be derived in terms of the temperatures and the measured force. For small temperature differences $T - T_g$ it may be shown that this relation is [3]

$$p = 4F \frac{T_g}{T - T_g} \qquad \textbf{[6.24]}$$

where the pressure is in dynes per square centimeter when the force is in dynes. The temperatures are in degrees kelvin.

| [7]M. Knudsen, *Ann. Physik*, vol. 32, p. 809, 1910.

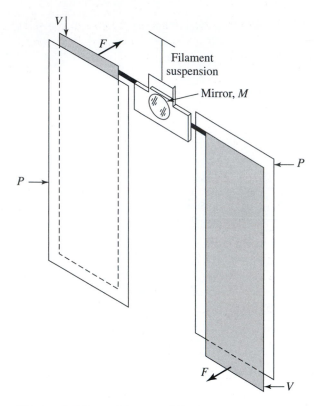

Figure 6.18   Schematic of Knudsen gage.

The Knudsen gage furnishes an absolute measurement of the pressure which is independent of the molecular weight of the gas. It is suitable for use between 10^{-5} and 10 μm (10^{-6} to 1 Pa) and may be used as a calibration device for other gages in this region.

6.12   THE IONIZATION GAGE

Consider the arrangement shown in Fig. 6.19, which is similar to the ordinary triode vacuum tube. The heated cathode emits electrons, which are accelerated by the positively charged grid. As the electrons move toward the grid, they produce ionization of the gas molecules through collisions. The plate is maintained at a negative potential so that the positive ions are collected there, producing the plate current i_p. The electrons and negative ions are collected by the grid, producing the grid current i_g. It is found that the pressure of the gas is proportional to the ratio of plate current to grid current.

$$p = \frac{1}{S}\frac{i_p}{i_g}$$ **[6.25]**

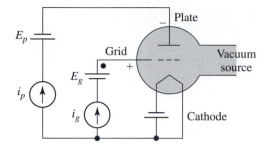

Figure 6.19   Schematic of ionization gage.

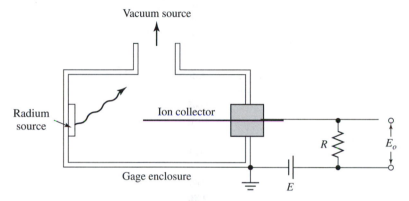

Figure 6.20   Schematic of Alphatron gage.

where the proportionality constant S is called the "sensitivity" of the gage. A typical value for nitrogen is $S = 20\ \text{torr}^{-1}$ (2.67 kPa^{-1}), but the exact value must be determined by calibration of the particular gage. The value of S is a function of the tube geometry and the type of gas.

Conventional ionization gages are suitable for measurements between 1.0 and $10^{-5}\ \mu\text{m}$ (0.13 to 1.3×10^{-6} Pa), and the current output is usually linear in this range. At higher pressures there is the danger of burning out the cathode. Special types of ionization gages are suitable for measurements of pressures as low as 10^{-2} torr (0.13 nPa). Very precise experimental techniques are required, however, in order to perform measurements at these high vacuums. The interested reader should consult Ref. [3] for additional information.

6.13   THE ALPHATRON[8]

The Alphatron is a radioactive ionization gage, shown schematically in Fig. 6.20. A small radium source serves as an alpha-particle emitter. These particles ionize the gas inside the gage enclosure, and the degree of ionization is determined by measuring

| [8]National Research Corp., Cambridge, MA.

the voltage output E_o. The degree of ionization is a direct linear function of pressure for a rather wide range of pressures, from 10^{-3} to 10^3 torr (0.1 to 10^5 Pa). The output characteristics, however, are different for each type of gas used. The lower pressure limit of the gage is determined by the length of the mean free path of the alpha particles as compared with the enclosure dimensions. At very low pressures the mean free path becomes so large that very few collisions are probable in the gage, and hence the ionization level is very small. The Alphatron has the advantages that it may be used at atmospheric pressure as well as high vacuum and that there is no heated filament to contend with as in the conventional ionization gage. Consequently, there is no problem of accidentally burning out a filament because of an inadvertent exposure of the gage to high (above 10^{-1} torr or 13 Pa) pressures.

6.14   SUMMARY

Figure 6.21 gives a convenient summary of the pressure ranges for which the gages discussed are normally employed in practice.

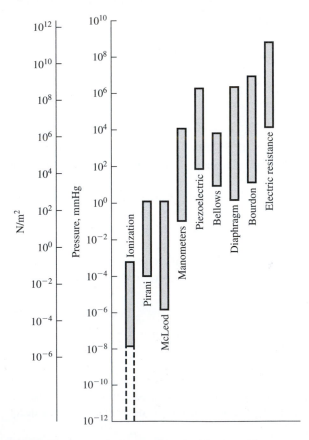

Figure 6.21   Summary of applicable range of pressure gages.

Development of solid-state pressure transducers employing semiconductors, quartz resonators, and integrated circuits is described in the survey article of Ref. [22]. In some cases accuracies of 0.001 percent have been achieved.

6.15 REVIEW QUESTIONS

6.1. Distinguish among gage pressure, absolute pressure, and vacuum.

6.2. To transmit a high-frequency pressure signal, one should select
(*a*) A short, small-diameter tube
(*b*) A short, large-diameter tube
(*c*) A long, small-diameter tube
(*d*) A long, large-diameter tube

6.3. What are the advantages of the manometer pressure-measurement device?

6.4. What is the advantage of a well-type manometer?

6.5. What are some advantages of the bourdon-tube, diaphragm, and bellows gages?

6.6. Describe the principle of operation of a McLeod gage.

6.7. Describe the Pirani gage.

6.8. When is the Knudsen gage used?

6.9. Describe the ionization gage. How does it differ from the Pirani gage? What disadvantages does it have?

6.16 PROBLEMS

6.1. A mercury barometer is constructed like that shown in Fig. 6.5. The column is a glass tube 0.250 in ID and 0.375 in OD, and the well is a glass dish 1.50 in ID. Calculate the percent error which would result if an area correction factor were not used.

6.2. Derive an expression for the radius of a simple diaphragm as shown in Fig. 6.10 using the following restrictions:
(*a*) The maximum deflection is one-third the thickness.
(*b*) The maximum deflection must be 100 times as great as the uncertainty in the deflection measurement w_y.
Assume that the maximum pressure differential Δp is given as well as w_y.

6.3. Determine the factor to convert pressure in inches of water to pounds per square foot.

6.4. The effective radius of an air molecule is about 1.85×10^{-8} cm. Calculate the mean free path at 70°F and the following pressures: 1 atm, 1 torr, 1 μm, 1 inH_2O, and $10^{-3} \mu$m.

6.5. Reduce Eq. (6.3) to an expression for mean free path in terms of pressure in micrometers, temperature in degrees kelvin, molecular weight of the gas, and effective molecular diameter.

6.6. A dynamic pressure measurement is to be made with an apparatus similar to that shown in Fig. 6.2. The appropriate dimensions are

$$L = 2.50 \text{ in}$$

$$r = 0.005 \text{ in}$$

$$V = 0.10 \text{ in}^3$$

The fluid is air at 70°F and 14.7 psia. Plot the pressure-amplitude ratio versus ω/ω_n according to both Eqs. (6.4) and (6.8).

6.7. Plot the error in Eq. (6.8) versus $r^2 L/V$.

6.8. Calculate the resonant frequency of the system in Prob. 6.6, assuming that it acts as a Helmholtz resonator.

6.9. A well-type manometer uses a special bromide fluid having a specific gravity of 2.95. The well has a diameter of 3.00 in and the tube has a diameter of 0.200 in. The manometer is to be used to measure a differential pressure in a water-flow system. The scale placed alongside the tube has no correction factor for the area ratio of the manometer. Calculate the value of a factor that may be multiplied by the manometer reading in inches to find the pressure differential in pounds per square inch.

6.10. A vacuum gage is to use an LVDT-diaphragm combination like that shown in Fig. 6.14. The LVDT has a sensitivity of 2.5 nm, and the diaphragm is to be constructed of steel ($E = 2 \times 10^{11}$ Pa, $\mu = 0.3$) with a diameter of 15 cm. Calculate the diaphragm thickness in accordance with the restriction that the maximum deflection does not exceed one-third this thickness. What is the lowest pressure which may be sensed by this instrument?

6.11. Calculate the natural frequency of the diaphragm in Prob. 6.10.

6.12. A Bridgman gage uses a coil of Manganin wire having a nominal resistance of 100 Ω at atmospheric pressure. The gage is to be used to measure a pressure of 1000 psig with an uncertainty of 0.1 percent. What is the allowable uncertainty in the resistance measurement?

6.13. Suppose the Bridgman gage of Prob. 6.12 is connected to the bridge circuit of Fig. 4.25 so that the gage is R_1 and all resistances are equal to 100 Ω at a pressure of 1 atm. The battery voltage is 4.0 V, and the detector is a high-impedance voltage-measuring device. The bridge is assumed to be in balance at $p = 1$ atm. Calculate the voltage output of the bridge at $p = 1000$ psig.

6.14. Rework Example 6.2 assuming the diameter of the capillary to be 0.2 mm.

6.15. A Knudsen gage is to be designed to operate at a maximum pressure of 1.0 μm. For this application the spacing of the vane and plate is to be less than 0.3 mean free path at this pressure. Calculate the force on the vanes at pressures 1.0 and

0.01 μm when the gas temperature is 20°C and the temperature difference is 50.0 K.

6.16. A capacitance-diaphragm pressure gage as shown in Fig. 6.13 is to be used to measure pressure differentials as high as 1000 psi at frequencies as high as 15,000 Hz. The diameter of the diaphragm is not to exceed 0.500 in. Calculate the thickness and diameter of the diaphragm to accomplish this (the natural frequency should be at least 30,000 Hz). Choose a suitable gap spacing, and estimate the capacitance-pressure sensitivity of the device. Assume the dielectric constant is that of air.

6.17. A bourdon-tube pressure gage having an internal volume of 1.0 in^3 is used for measuring pressure in a fluctuating air system having frequencies as high as 100 Hz. Design an acoustical filter which will attenuate all frequencies above 20 Hz by 99 percent. Plot the frequency response of this filter.

6.18. An acoustic filter is to be designed to attenuate sharp pressure transients in air above 50 Hz. The volume of air contained in the pressure-transducer cavity is 0.6 in^3, and a capillary tube connects the cavity to the pressure source. If the 50-Hz frequency is to be attenuated by 50 percent, determine the capillary length and diameter for system natural frequencies of (*a*) 50 Hz, (*b*) 100 Hz, and (*c*) 500 Hz.

6.19. A U-tube manometer uses tubes of 0.250 and 0.500 in diameters for the two legs. When subjected to a certain pressure, the difference in height of the two fluid columns is 10.0 inHg. What would have been the reading if both tubes were the same diameter? The measurement is performed on air.

6.20. A pressure signal is fed through a line having an inside diameter of 1.5 mm and a length of 1.5 m. The line is connected to a pressure transducer having a volume of approximately 5 cm^3. Air at 690 kPa and 90°C is the transmitting fluid. Calculate the natural frequency and damping ratio for this system.

6.21. The manometer of Prob. 6.19 uses a fluid having a specific gravity of 1.85. The sensing fluid is water. What is the pressure difference when the difference in heights of the columns is 5.0 in? Assume that both legs of the manometer are filled with water.

6.22. A diaphragm-pressure gage is constructed of spring steel to measure a pressure differential of 1000 psi. The diameter of the diaphragm is 0.5 in. Calculate the diaphragm thickness so that the maximum deflection is one-third the thickness. What is the natural frequency of this diaphragm?

6.23. A Bridgman gage is to be used to measure a pressure of 10,000 psi using a Manganin element having a resistance of 100 Ω at atmospheric pressure. Calculate the resistance of the gage under high-pressure conditions. If the gage is one leg of a bridge whose other legs all have values of exactly 100 Ω, calculate the voltage output of the bridge for a constant-voltage source of 24 V.

6.24. A McLeod gage is available which has a volume V_B of 150 cm^3 and a capillary diameter of 0.3 mm. Calculate the gage reading for a pressure of 30 μm.

6.25. What is the approximate range of mean free paths for air over the range of pressures for which the Knudsen gage is applicable?

6.26. A special high-pressure U-tube manometer is constructed to measure pressure differential in air at 13.8 MPa and 20°C. When an oil having a specific gravity of 0.83 is used as the fluid, calculate the differential pressure in pounds per square inch absolute that would be indicated by a 135-mm reading.

6.27. A diaphragm like that shown in Fig. 6.10c has $a = 1.0$ in, $b = 0.125$ in, and $t = 0.048$ in and is constructed of spring steel. It is subjected to a total loading of 600 lbf. Calculate the deflection.

6.28. A U-tube manometer is used to measure a differential air pressure with a fluid having a specific gravity of 0.8. The air is at 400 kPa and 10°C. Calculate the differential pressure for difference in heights of the manometer legs of 12 cm. Express in units of both psia and pascals.

6.29. Calculate the mean free path for air at standard conditions of 1 atm and 20°C.

6.30. A diaphragm pressure gage is constructed as in Fig. 6.10b, with $a = 2.5$ cm, $b = 0.3$ cm, and $t = 0.122$ cm. The material is spring steel. What pressure, in pascals, will be necessary to cause a deflection of 0.04 cm?

6.31. Suppose the diaphragm of Prob. 6.30 is constructed as in Fig. 6.10a. What pressure would cause the same deflection? What would be the natural frequency of this diaphragm?

6.32. Prepare a matrix table for pressure conversions among the following units: psia, Pa, mmHg, inH$_2$O, atm, and kp/cm^2.

6.33. In Laramie, Wyoming, the weather bureau reports the barometric pressure as 29.8 in of mercury. At the University of Wyoming (in Laramie) a group of students measures the air pressure in a 100-liter tank as 10 kPa *gage* pressure. The temperature of the air is 20°C. If the altitude for Laramie is 7200 ft, calculate the mass of air in the tank. What error would result if the local atmospheric pressure were taken as the value quoted by the weather bureau?

6.34. On the same day, at the same time, another group of students at the University of Wyoming (as in Prob. 6.33) makes a vacuum measurement indicating 10 psig *vacuum*. Calculate the percent error in determination of *absolute* pressure if the atmospheric pressure were taken as the weather bureau value.

6.35. A U-tube manometer contains a fluid having a specific gravity of 1.75 and is used to measure a differential pressure in water. What will be the differential pressure, in pascals, for a reading of 10.5 cm?

6.36. The same manometer as in Prob. 6.35 is used to measure the same differential pressure in air at 1 atm and 20°C. What would be the reading under these conditions?

6.37. A well-type manometer has the measurement leg inclined at 30° from the horizontal. The diameter of the measurement column is 5 mm and the diameter of the well is 5 cm. An oil having a specific gravity of 0.85 is used as the fluid. A differential pressure in air at 1 atm and 20°C is made which produces a displacement in the measurement column of 15 cm from the zero level. What is the differential pressure in pascals?

6.38. A U-tube manometer uses mercury as the manometer fluid to measure a differential pressure in water at 80°F. Both sides of the manometer have diameters of 5 mm. What differential pressure, in pascals, will result in a column height measurement of 13 cm?

6.39. Suppose the manometer in Prob. 6.38 had unequal diameters of 5 and 10 mm. What would be the differential pressure, in pascals, for a column height measurement of 13 cm in this circumstance?

6.40. What pressure differential would be indicated for the manometer in Prob. 6.39 if the 13-cm measurement is the height of the small column from the *zero level* instead of the difference in heights of the two columns?

6.41. A diaphragm gage constructed as in Fig. 6.10*a* is to be fabricated of spring steel and used to measure a differential pressure of 10 kPa in air at 20°C. Assuming that the maximum deflection is not to exceed one-third the thickness, calculate the value of the thickness for a diaphragm diameter of 2.5 cm. Also, calculate the natural frequency of this diaphragm.

6.42. A special U-tube manometer is used to make a differential pressure measurement in air at 20°C and 65 atm. The manometer fluid has a specific gravity of 0.85. The differential column height is 15.3 cm ± 1.0 mm. Calculate the pressure differential in pascals and its uncertainty.

6.43. An air-pressure signal at 5 atm and 50°C is fed through a 1.0-mm-diameter line having a length of 0.8 m, and is connected to a transducer volume of 3 cm³. Calculate the natural frequency and damping ratio of this system. By how much will a pressure signal having a frequency one-half the natural frequency be attenuated?

6.44. A McLeod gage has a capillary diameter of 0.2 mm and a volume V_B of 125 cm³. What gage reading will result from an absolute pressure of 20 μm?

6.45. A vacuum chamber stands outdoors at the airport in a Texas airbase in the summer. The temperature of the chamber is 120°F. A vacuum gage indicates 13 psi of vacuum and the weather station at the airport reports the barometric pressure as 29.83 inHg. The volume of the tank is 3.0 m³ and the elevation of the airport is 600 ft above sea level. Calculate the mass of air in the tank. What percentage error would result if the barometric pressure were not corrected for elevation?

6.46. A microphone operates on the principle of a diaphragm gage with a capacitance pickup. Suppose a 140-dB sound-pressure source which produces a maximum pressure fluctuation of 0.029 psia in air is to be imposed on the diaphragm at a frequency of 5000 Hz. The diameter of the diaphragm is to be 1.0 cm. Determine the thickness for the diaphragm such that the natural frequency is 10,000 Hz. Assume spring-steel construction and a thickness of one-third the maximum displacement.

6.47. Calculate the standard barometric pressure at the top of a 14,000-ft mountain.

6.48. A well-type manometer uses mercury for measuring a differential pressure in water at 90°F. The measuring column has a diameter of 4.0 mm and the well diameter is 5 cm. Calculate the differential pressure for a column height reading of 25 cm from the zero level.

6.49. A Bridgman gage is to be employed for measurement of a pressure of 700 atm using a Manganin element which has a resistance of 90 Ω at 1 atm. Calculate the resistance of the gage under the high-pressure condition.

6.50. A tube having a diameter of 1.2 mm and length of 10 cm is connected to a pressure transducer which has a volume of 1.5 cm^3. Calculate the natural frequency and damping ratio for this system when operating with air at 500 kPa and 50°C.

6.51. Two U-tube manometers are connected in series using mercury as the manometer fluids. The tube connecting the manometers is filled with water. A differential air pressure is imposed on the system such that the sum of the differential column heights in both manometers is 30 cm. Calculate the differential air pressure for a temperature of 20°C.

6.52. Determine the following conversion factors:

1 inH$_2$O = _____ psi

1 inH$_2$O = _____ Pa

1 mmHg = _____ psi

1 mmHg = _____ kgf/cm^2

6.53. A measurement is made of air pressure in a tank. The gage indicates 825 kPa while the local barometer reading is 750 mmHg. What is the absolute pressure in the tank?

6.54. An experiment is conducted which requires the calculation of the mass of helium in a 5-liter tank maintained at −10°C. The pressure in the tank is measured as 80 kPa vacuum and the location of the measurement is at an altitude of 600 ft above sea level. Unfortunately the local barometric pressure was not taken so the local weather bureau information is used for that day, which is reported as 29.85 inHg. Calculate the mass of helium in the tank taking the weather barometer reading at face value. Repeat the calculation with the reading corrected for elevation. Comment on the results of these calculations.

6.55. A U-tube-type manometer uses mercury as the sensing fluid for measuring differential pressure across an orifice which has water as the flow medium. The difference in mercury column heights is measured as 45.2 ± 0.1 cm. Calculate the difference in pressure in psi if the apparatus temperature is 20°C.

6.56. What error would result in Prob. 6.52 if the water were neglected in the calculation?

6.57. A manometer inclined at an angle of 20° with the horizontal employs a fluid having a specific gravity of 0.82 and is used to measure a differential pressure of 2 in H$_2$O in air at 50 psi and 70°F. What displacement of fluid along the length of the manometer tube will be registered?

6.58. A diaphragm like that shown in Fig. 6.10a is to be designed to measure a differential pressure of 10 kPa in air at 1 atm and 20°C. Determine suitable dimensions for the diaphragm to accomplish this objective.

6.59. Calculate the natural frequency for the diaphragm selected in Prob. 6.58.

6.60. A precision bourdon tube is stated to have an absolute accuracy of ±0.1 percent of full-scale reading of 1 MPa. What is the percent accuracy (uncertainty) when the gage is employed for a differential pressure of 40 kPa in air at 3 atm and 20°C?

6.61. An assembly of two reservoirs and a small-diameter tube inclined at 30° with the horizontal operates as a micromanometer using water and an oil of specific gravity 0.85 as the two fluids. Calculate the sensitivity of the device expressed in mm/Pa.

6.62. Suppose the manometer in Prob. 6.61 is used to measure the dynamic pressure in an airflow system defined by

$$\Delta p = \rho u^2 / 2g_c$$

where ρ is the density of the flowing fluid (air) and u is the velocity. What velocity would be indicated by a deflection of 1.0 mm?

6.63. Repeat Probs. 6.61 and 6.62 if the two fluids are water and a special oil having a specific gravity of 0.9.

6.64. A U-tube manometer contains a special bromine fluid having a specific gravity of 2.95, and is used to measure a differential pressure in a water system. Calculate the differential pressure in psi for a manometer reading of 15 cm.

6.65. The same manometer and fluid as in Prob. 6.64 is used to measure the same differential pressure, but in air at 0.5 atm and 35°C. What would be the reading (height of column) in this situation?

6.66. A diaphragm gage is to be constructed of spring steel and configured as illustrated in Fig. 6.10a. The design is to accommodate a maximum differential pressure of 200 kPa with air at 30°C. The diameter of the diaphragm is specified as 4.0 cm. Calculate the thickness of the diaphragm and its natural frequency.

6.67. A high-pressure U-tube manometer is to be employed for measurement of a differential pressure in air at conditions of 75 atm and 35°C. Water is used as the manometer fluid. Calculate the pressure differential indicated by a water column height of 20 cm.

6.68. A 15-cm-long tube has an inside diameter of 1.0 mm and is connected to a transducer which has a volume of 1.2 mL. If the system is to operate with air at 2 atm and 20°C, calculate the natural frequency and damping ratio of the tube-transducer system.

6.69. A novice experimentalist performs some vacuum measurements using a sensitive instrument that records the difference between the local barometric pressure and the absolute pressure in the vacuum system, that is, the so-called negative gage pressure. At a location where the altitude from sea level is approximately 1520 m the experimentalist discovers that the calculated absolute pressure of the system being measured becomes *negative* at sufficiently high vacuum readings, which of course is impossible. You are asked to solve the problem. At approximately what vacuum, or negative gage pressure reading, would you expect the anomaly to occur?

6.17   REFERENCES

1. Arons, A. B., and R. H. Cole: "Design and Use of Piezo-Electric Gages for Measurement of Large Transient Pressures," *Rev. Sci. Instr.,* vol. 21, pp. 31–38, 1950.

2. Bridgman, P. W.: *The Physics of High Pressure,* Macmillan, New York, 1931.

3. Dushman, S., and J. M. Lafferty: *Scientific Foundations of Vacuum Technique,* 2d ed., Wiley, New York, 1962.

4. Hall, H. T.: "Some High Pressure–High Temperature Apparatus Design Considerations," *Rev. Sci. Instr.,* vol. 29, p. 267, 1958.

5. Hetenyi, M. (ed.): *Handbook of Experimental Stress Analysis,* Wiley, New York, 1950.

6. Howe, W. H.: "The Present Status of High Pressure Measurements," *ISA J.,* vol. 2, pp. 77, 109, 1955.

7. Iberall, A. S.: "Attenuation of Oscillatory Pressures in Instrument Lines," *Trans. ASME,* vol. 72, p. 689, 1950.

8. Neubert, H. K. P.: *Instrument Transducers,* Oxford University Press, Fair Lawn, NJ, 1963.

9. Young, W. C.: *Roark's Formulas for Stress and Strain,* 6th ed., McGraw-Hill, New York, 1998.

10. Sweeney, R. J.: *Measurement Techniques in Mechanical Engineering,* Wiley, New York, 1953.

11. Taback, I.: "The Response of Pressure Measuring Systems to Oscillating Pressure," *NACA Tech. Note* 1819, February 1949.

12. Wildhack, W. A., R. F. Dresslea, and E. C. Lloyd: "Investigation of the Properties of Corrugated Diaphragms," *Trans. ASME,* vol. 79, pp. 65–82, 1957.

13. Giardini, A. A. (ed.): *High Pressure Measurements,* Butterworth & Co., London, 1963.

14. Karam, J. T., and M. E. Franke: "The Frequency Response of Pneumatic Lines," *J. Basic Eng., Trans. ASME,* ser. D. vol. 90, pp. 371–378, 1967.

15. Nichols, N. B.: "The Linear Properties of Pneumatic Transmission Lines," *Trans. Instr. Soc. Am.,* vol. 1, pp. 5–14, 1962.

16. Watts, G. P.: "The Response of Pressure Transmission Lines," Preprint No. 13.3-1.65, 20th Annual ISA Conf. and Exhibit, Los Angeles, Calif., October 4–7, 1965.

17. Bergh, H., and H. Tijdeman: "Theoretical and Experimental Results for the Dynamic Response of Pressure Measuring Systems," *NLR Tech. Rept.* F. 238, 1965.

18. Tijdeman, H.: "Remarks on the Frequency Response of Pneumatic Lines," *J. Basic Eng.,* pp. 325–328, June 1969.

19. Schweppe, J. L., et al.: "Methods for the Dynamic Calibration of Pressure Transducers," *Natl. Bur. Std. (U.S.), Monograph* 67, 1963.

20. Funk, J. E., D. J. Wood, and S. P. Chao: "The Transient Response of Orifices and Very Short Lines," ASME paper 71-WA/FE-14, December 1971.

21. Omega Engineering, Inc.: *Pressure, Strain, and Force Measurement Handbook,* Omega Engineering, Stamford, CT, 1991.

22. ——: "Inside Pressure Measurement," *Mech. Eng.,* vol. 109, no. 5, pp. 41–56, May 1987.

23. Benedict, R. P.: *Fundamentals of Temperature, Pressure and Flow Measurement,* 3d ed., Wiley, New York, 1984.

24. Brombacher, W.G.: "Survey of Micromanometers," NBS Monograph 114, Washington DC, 1970.

chapter

7

FLOW MEASUREMENT

7.1   INTRODUCTION

The measurement of fluid flow is important in applications ranging from measurements of blood-flow rates in a human artery to the measurement of the flow of liquid oxygen in a rocket. Many research projects and industrial processes depend on a measurement of fluid flow to furnish important data for analysis. In some cases extreme precision is called for in the flow measurement, while in other instances only crude measurements are necessary. The selection of the proper instrument for a particular application is governed by many variables, including cost. For many industrial operations the accuracy of a fluid-flow measurement is directly related to profit. A simple example is the gasoline pump at the neighborhood service station; another example is the water meter at home. It is easy to see how a small error in flow measurement on a large natural gas or oil pipeline could make a difference of thousands of dollars over a period of time. Thus, the laboratory scientist is not the only person who is concerned with accurate flow measurement; the engineer in industry is also vitally interested because of the impact flow measurements may have on the profit-and-loss statement of the company.

Flow-rate-measurement devices frequently require accurate pressure and temperature measurements in order to calculate the output of the instrument. Chapters 6 and 8 consider these associated measurement topics in detail, and the reader should consult the appropriate sections from time to time to relate specific pressure and temperature measurement devices to the material in the present chapter. We may remark at this time, however, that the overall accuracy of many of the most widely used flow-measurement devices is governed primarily by the accuracy of some pressure or temperature measurement. Commercial organizations[1] offer flowmeter calibrations traceable to NIST standards.

---

[1] Flow Dynamics, Scottsdale, AZ.

Flow rate is expressed in both volume and mass units of varying sizes. Some commonly used terms are

1 gallon per minute (gpm)
   $= 231$ cubic inches per minute (in^3/min)
   $= 63.09$ cubic centimeters per second (cm^3/s)
1 liter
   $= 0.26417$ gallon $= 1000$ cubic centimeters
1 cubic foot per minute (cfm, or ft^3/min)
   $= 0.028317$ cubic meter per minute
   $= 471.95$ cubic centimeters per second
1 standard cubic foot per minute of air at 20°C, 1 atm
   $= 0.07513$ pound-mass per minute
   $= 0.54579$ gram per second

We should alert the reader to the fact that commercial gas-flow meters typically specify flow ratings in volume flow rate at standard conditions of 1 atm and 20°C. The units employed are standard cubic feet per minute (scfm) and standard cubic centimeters per minute (sccm).

Our objective in this chapter is to present a broad discussion of flow measurements and to indicate the principles of operation of a number of devices that are commonly used. We shall also give the calculation methods that are connected with some of these devices and discuss some methods of flow visualization. In concluding the chapter a tabular comparison of the various methods will be presented, pointing out their range of applicability and expected accuracies.

7.2   POSITIVE-DISPLACEMENT METHODS

The flow rate of a nonvolatile liquid like water may be measured through a direct-weighing technique. The time necessary to collect a quantity of the liquid in a tank is measured, and an accurate measurement is then made of the weight of liquid collected. The average flow rate is thus calculated very easily. Improved accuracy may be obtained by using longer or more precise time intervals or more precise weight measurements. The direct-weighing technique is frequently employed for calibration of water and other liquid flowmeters, and thus may be taken as a standard calibration technique. Obviously, it is not suited for transient flow measurements.

Positive-displacement flowmeters are generally used for those applications where consistently high accuracy is desired under steady-flow conditions. A typical positive-displacement device is the home water meter shown schematically in Fig. 7.1. This meter operates on the nutating-disk principle. Water enters the left side of the meter and strikes the disk, which is eccentrically mounted. In order for the fluid to move through the meter the disk must "wobble" or nutate about the vertical axis since both the top and bottom of the disk remain in contact with the mounting chamber. A partition separates the inlet and outlet chambers of the disk. As the disk

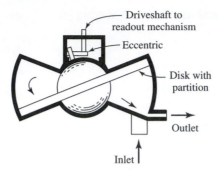

Figure 7.1   Schematic of a nutating-disk meter.

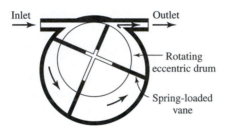

Figure 7.2   Schematic of rotary-vane flowmeter.

nutates, it gives direct indication of the volume of liquid which has passed through the meter. The indication of the volumetric flow is given through a gearing and register arrangement which is connected to the nutating disk. The nutating-disk meter may give reliable flow measurements within 1 percent, over an extended period of time.

Another type of positive-displacement device is the rotary-vane meter shown in Fig. 7.2. The vanes are spring-loaded so that they continuously maintain contact with the casing of the meter. A fixed quantity of fluid is trapped in each section as the eccentric drum rotates, and this fluid eventually finds its way out the exit. An appropriate register is connected to the shaft of the eccentric drum to record the volume of the displaced fluid. The uncertainties of rotary-vane meters are of the order of 0.5 percent, and the meters are relatively insensitive to viscosity since the vanes always maintain good contact with the inside of the casing.

The lobed-impeller meter shown in Fig. 7.3 may be used for either gas- or liquid-flow measurements. The impellers and case are carefully machined so that accurate fit is maintained. In this way the incoming fluid is always trapped between the two rotors and is conveyed to the outlet as a result of their rotation. The number of revolutions of the rotors is an indication of the volumetric flow rate.

Remote sensing of all the positive-displacement meters may be accomplished with rotational transducers or sensors and with appropriate electronic counters.

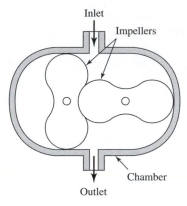

Inlet

Impellers

Chamber

Outlet

Figure 7.3   Schematic of lobed-impeller flowmeter.

UNCERTAINTY IN FLOW CAUSED BY UNCERTAINTIES IN TEMPERATURE AND | **Example 7.1**
PRESSURE.   A lobed-impeller flowmeter is used for measurement of the flow of nitrogen
at 20 psia and 100°F. The meter has been calibrated so that it indicates the volumetric flow with
an accuracy of ± one-half of 1 percent from 1000 to 3000 cfm. The uncertainties in the gas
pressure and temperature measurements are ±0.025 psi and ±1.0°F, respectively. Calculate
the uncertainty in a mass flow measurement at the given pressure and temperature conditions.

Solution

The mass flow is given by

$$\dot{m} = \rho Q$$

where the density of nitrogen is given by

$$\rho = \frac{p}{R_{N_2} T}$$

Using Eq. (3.2), we obtain the following equation for the uncertainty in the mass flow:

$$\frac{w_{\dot{m}}}{\dot{m}} = \left[\left(\frac{w_Q}{Q} \right) + \left(\frac{w_p}{p} \right)^2 + \left(\frac{w_T}{T} \right)^2 \right]^{1/2}$$

Using the given data, we obtain

$$\frac{w_{\dot{m}}}{\dot{m}} = \left[(0.005)^2 + \left(\frac{0.025}{20} \right)^2 + \left(\frac{1}{560} \right)^2 \right]^{1/2} = 5.05 \times 10^{-3}$$

or 0.505 percent. Thus, the uncertainties in the pressure and temperature measurements do not
appreciably influence the overall uncertainty in the mass flow measurements.

7.3 FLOW-OBSTRUCTION METHODS

Several types of flowmeters fall under the category of obstruction devices. Such devices are sometimes called *head meters* because a head-loss or pressure-drop measurement is taken as an indication of the flow rate. They are also called *differential pressure meters*. Let us first consider some of the general relations for obstruction meters. We shall then examine the applicability of these relations to specific devices.

Consider the one-dimensional flow system shown in Fig. 7.4. The continuity relation for this situation is

$$\dot{m} = \rho_1 A_1 u_1 = \rho_2 A_2 u_2 \qquad \text{[7.1]}$$

where u is the velocity. If the flow is adiabatic and frictionless and the fluid is incompressible, the familiar Bernoulli equation may be written

$$\frac{p_1}{\rho_1} + \frac{u_1^2}{2g_c} = \frac{p_2}{\rho_2} + \frac{u_2^2}{2g_c} \qquad \text{[7.2]}$$

where now $\rho_1 = \rho_2$. Solving Eqs. (7.1) and (7.2) simultaneously gives for the pressure drop

$$p_1 - p_2 = \frac{u_2^2 \rho}{2g_c}\left[1 - \left(\frac{A_2}{A_1}\right)^2\right] \qquad \text{[7.3]}$$

and the volumetric flow rate may be written

$$Q = A_2 u_2 = \frac{A_2}{\sqrt{1 - (A_2/A_1)^2}}\sqrt{\frac{2g_c}{\rho}(p_1 - p_2)} \qquad \text{[7.4]}$$

where $Q = \text{ft}^3/\text{s or m}^3/\text{s}$
$A = \text{ft}^2 \text{ or m}^2$
$\rho = \text{lbm/ft}^3 \text{ or kg/m}^3$
$p = \text{lbf/ft}^2 \text{ or N/m}^2$
$g_c = 32.17 \text{ lbm} \cdot \text{ft/lbf} \cdot \text{s or } 1.0 \text{ kg} \cdot \text{m/N} \cdot \text{s}^2$

Thus, we see that a channel like the one shown in Fig. 7.4 could be used for a flow measurement by simply measuring the pressure drop $(p_1 - p_2)$ and calculating the flow from Eq. (7.4). No such channel, however, is frictionless, and some losses are always present in the flow. The volumetric flow rate calculated from Eq. (7.4) is the ideal value, and it is usually related to the actual flow rate and an empirical

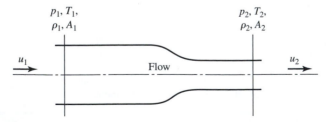

Figure 7.4   General one-dimensional flow system.

discharge coefficient C by the following relation:

$$\frac{Q_{\text{actual}}}{Q_{\text{ideal}}} = C \qquad\qquad [7.5]$$

The discharge coefficient is not a constant and may depend strongly on the flow Reynolds number and the channel geometry.

When the flow of an ideal gas is considered, the following equation of state applies:

$$p = \rho R T \qquad\qquad [7.6]$$

where T is the absolute temperature and R is the gas constant for the particular gas, which can be expressed in terms of the universal gas constant $\Re$ and the molecular weight by

$$R = \frac{\Re}{M}$$

The value of $\Re$ is 8314 kJ/kg · mol · K or 1545 ft · lbf/lbm · mol · °R. For reversible adiabatic flow the steady-flow energy equation for an ideal gas is

$$c_p T_1 + \frac{u_1^2}{2g_c} = c_p T_2 + \frac{u_2^2}{2g_c} \qquad\qquad [7.7]$$

where c_p is the specific heat at constant pressure and is assumed constant for an ideal gas. When Eqs. (7.1), (7.6), and (7.7) are combined, there results

$$\dot{m}^2 = 2g_c A_2^2 \frac{\gamma}{\gamma - 1} \frac{p_1^2}{RT_1} \left[\left(\frac{p_2}{p_1}\right)^{2/\gamma} - \left(\frac{p_2}{p_1}\right)^{(\gamma+1)/\gamma} \right] \qquad\qquad [7.8]$$

where the velocity of approach, that is, the velocity at section 1 of Fig. 7.4, is assumed to be very small. This relationship may be simplified to

$$\dot{m} = \sqrt{\frac{2g_c}{RT_1}} A_2 \left[p_2 \Delta p - \left(\frac{1.5}{\gamma} - 1\right)(\Delta p)^2 + \cdots \right]^{1/2} \qquad\qquad [7.9]$$

with $\Delta p = p_1 - p_2$ and $\gamma = c_p/c_v$ is the ratio of specific heats for the gas. Equation (7.9) is valid for $\Delta p < p_1/4$. When $\Delta p < p_1/10$, a further simplification may be made to give

$$\dot{m} = A_2 \sqrt{\frac{2g_c p_2}{RT_1}(p_1 - p_2)} \qquad\qquad [7.10]$$

where $\dot{m}$ = mass flow rate, lbm/s or kg/s
   A = area, ft^2 or m^2
   g_c = 32.17 lbm · ft/lbf · s^2 or 1.0 kg · m/N · s^2
   p = pressure, lbf/ft^2 or N/m^2(Pa)
   R = gas constant, lbf · ft/lbm · °R or N · m/kg · K
   T = absolute temperature, °R or K

Note that Eq. (7.10) reduces to Eq. (7.4) when the relation for density from Eq. (7.6) is substituted. Thus, for small values of Δp compared with p_1 the flow of a compressible fluid may be approximated by the flow of an incompressible fluid.

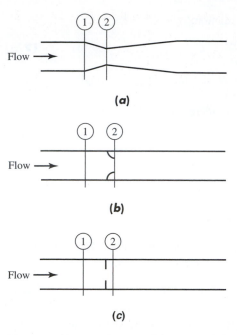

Figure 7.5   Schematic of three typical obstruction meters. (a) Venturi; (b) flow nozzle; (c) orifice.

Three typical obstruction meters are shown in Fig. 7.5. The venturi offers the advantages of high accuracy and small pressure drop, while the orifice is considerably lower in cost. Both the flow nozzle and the orifice have a relatively high permanent pressure drop. Flow-rate calculations for all three devices are made on the basis of Eq. (7.4) with appropriate empirical constants defined as follows:

$$M = \text{velocity of approach factor} = \frac{1}{\sqrt{1 - (A_2/A_1)^2}} \qquad \textbf{[7.11]}$$

$$K = \text{flow coefficient} = CM \qquad \textbf{[7.12]}$$

$$\beta = \text{diameter ratio} = \frac{d}{D} = \sqrt{\frac{A_2}{A_1}} \qquad \textbf{[7.13]}$$

When flow measurements of a compressible fluid are made, an additional parameter, the *expansion factor Y*, is used. For venturis and nozzles this factor is given by

$$Y_a = \left[\left(\frac{p_2}{p_1} \right)^{2/\gamma} \frac{\gamma}{\gamma - 1} \frac{1 - (p_2/p_1)^{(\gamma-1)/\gamma}}{1 - (p_2/p_1)} \frac{1 - (A_2/A_1)^2}{1 - (A_2/A_1)^2 (p_2/p_1)^{2/\gamma}} \right]^{1/2}$$

$$\textbf{[7.14]}$$

while for orifices an empirical expression for Y is given as

$$Y_1 = 1 - \left[0.41 + 0.35 \left(\frac{A_2}{A_1} \right)^2 \right] \frac{p_1 - p_2}{\gamma p_1} \qquad \textbf{[7.15]}$$

when either flange taps or vena contracta taps are used. For orifices with pipe taps the following relation applies:

$$Y_2 = 1 - [0.333 + 1.145(\beta^2 + 0.7\beta^5 + 12\beta^{13})]\frac{p_1 - p_2}{\gamma p_1}$$ [7.16]

The empirical expansion factors given by Eqs. (7.15) and (7.16) are accurate within ± 0.5 percent for $0.8 < p_2/p_1 < 1.0$. Plots of the expansion factors Y_a and Y_1 are given in Figs. 7.14 and 7.15, respectively.

We thus have the following semiempirical equations, which are conventionally applied to venturis, nozzles, or orifices:

VENTURIS, INCOMPRESSIBLE FLOW:

$$Q_{actual} = CMA_2\sqrt{\frac{2g_c}{\rho}}\sqrt{p_1 - p_2}$$ [7.17]

NOZZLES AND ORIFICES, INCOMPRESSIBLE FLOW:

$$Q_{actual} = KA_2\sqrt{\frac{2g_c}{\rho}}\sqrt{p_1 - p_2}$$ [7.18]

The use of the flow coefficient instead of the product CM is merely a matter of convention. When compressible fluids are used, the above equations are modified by the factor Y and the fluid density is evaluated at inlet conditions. We then have

VENTURIS, COMPRESSIBLE FLOW:

$$\dot{m}_{actual} = YCMA_2\sqrt{2g_c\rho_1(p_1 - p_2)}$$ [7.19]

NOZZLES AND ORIFICES, COMPRESSIBLE FLOW:

$$\dot{m}_{actual} = YKA_2\sqrt{2g_c\rho_1(p_1 - p_2)}$$ [7.20]

In Eqs. (7.17) to (7.20) the appropriate units are

Q = volume flow rate, ft^3/s or m^3/s
A = area, ft^2 or m^2
g_c = 32.17 lbm · ft/lbf · s^2 or 1.0 kg · m/N · s^2
ρ = density, lbm/ft^3 or kg/m^3
p = pressure, lbf/ft^2 or N/m^2

Detailed tabulations of the various coefficients have been made in Ref. [1], some of which are presented in Figs. 7.9 through 7.15. Examples 7.2 and 7.3 illustrate the use of these charts for practical calculations.

INSERTION (PERMANENT) PRESSURE LOSSES

Because of the fluid friction and turbulence resulting from the insertion of the obstruction meters in the flow channel, there is a permanent pressure loss which must be overcome by the flow system. Approximate values of these losses as a fraction of the measured pressure differential are shown in Table 7.1. The venturi is clearly superior to the orifice or flow nozzle in this regard.

Table 7.1

| | $\Delta p_{loss}/(p_1 - p_2)$ | | |
β	Square-edged Orifice	Flow Nozzle	Venturi $\alpha_2 = 7°$
0.4	0.86	0.8	0.1
0.5	0.78	0.7	0.1
0.6	0.67	0.55	0.1

7.4 PRACTICAL CONSIDERATIONS FOR OBSTRUCTION METERS

The construction of obstruction meters has been standardized by the American Society of Mechanical Engineers [1 and 2]. The recommended proportions of venturi tubes are shown in Fig. 7.6. Note that the pressure taps are connected to manifolds which surround the upstream and throat portions of the tube. These manifolds receive a sampling of the pressure all around the sections so that a good average value is obtained. The discharge coefficients for such venturi tubes are shown in Fig. 7.9, with the tolerance limits indicated by the dashed lines. In general, the discharge coefficient

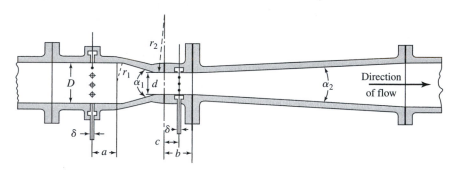

D = Pipe diameter inlet and outlet
d = Throat diameter as required
a = 0.25D to 0.75D for 4″ ≤ D ≤ 6″, 0.25D to 0.50D for 6″ ≤ D ≤ 32″
b = d
c = $d/2$
δ = 3/16 in to 1/2 in according to D. Annular pressure chamber
  with at least four piezometer vents
r_2 = 3.5d to 3.75d
r_1 = 0 to 1.375D
α_1 = 21∞ ± 2∞
α_2 = 5∞ to 15∞

Figure 7.6   Recommended proportions of venturi tubes, according to Ref. [1].

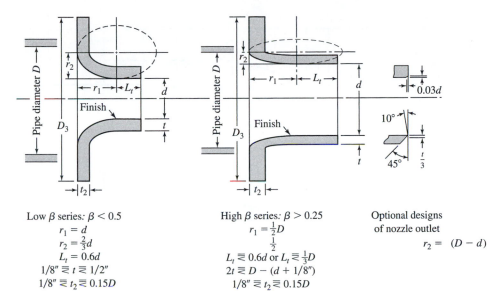

Low β series: $\beta < 0.5$
$r_1 = d$
$r_2 = \frac{2}{3}d$
$L_t = 0.6d$
$1/8'' \gtreqless t \gtreqless 1/2''$
$1/8'' \gtreqless t_2 \gtreqless 0.15D$

High β series: $\beta > 0.25$
$r_1 = \frac{1}{2}D$
$\frac{1}{2}$
$L_t \gtreqless 0.6d$ or $L_t \gtreqless \frac{1}{3}D$
$2t \gtreqless D - (d + 1/8'')$
$1/8'' \gtreqless t_2 \gtreqless 0.15D$

Optional designs
of nozzle outlet
$r_2 = (D - d)$

Figure 7.7   Recommended proportions of the ASME long-radius flow nozzle, according to Ref. [1].

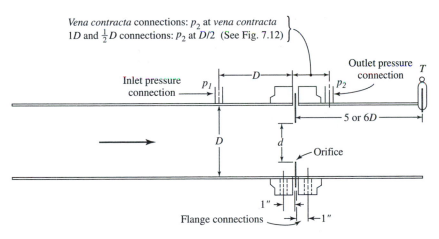

Figure 7.8   Recommended location of pressure taps for use with concentric, thin-plate, square-edged orifices, according to Ref. [1].

is smaller for pipes less than 2 in diameter, and the approximate behavior is indicated in Fig. 7.10. More precise values of the discharge coefficient for a venturi may be obtained by direct calibration, in which cases accuracies of ±0.5 percent may be obtained fairly easily. The recommended dimensions for ASME flow nozzles are shown in Fig. 7.7, and the discharge coefficients are shown in Fig. 7.11.

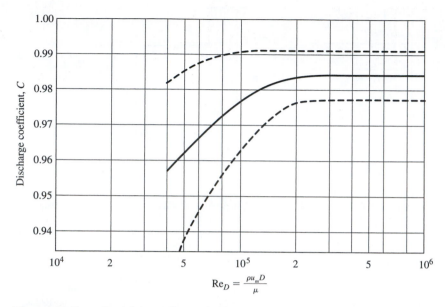

Figure 7.9   Discharge coefficients for the venturi tube shown in Fig. 7.6, according to Ref. [1]. Values are applicable for $0.25 < \beta < 0.75$ and $D > 2$ in.

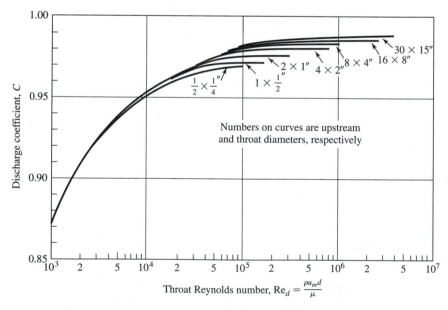

Figure 7.10   Approximate venturi coefficients for various throat diameters, according to Ref. [15].

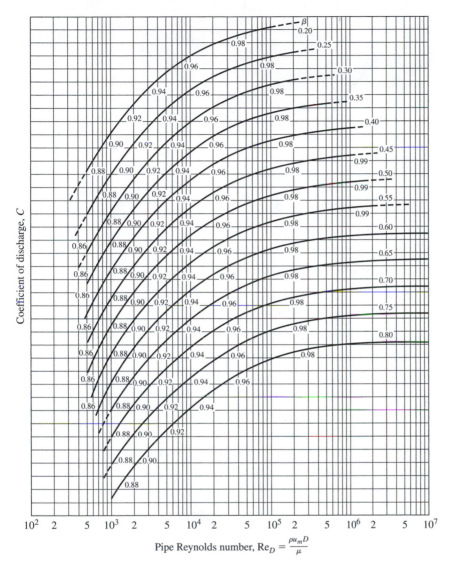

Figure 7.11 Discharge coefficients for ASME long-radius nozzles shown in Fig. 7.7, according to Ref. [1].

The recommended installations for concentric, thin-plate orifices are shown in Fig. 7.8. Note that three standard sets of pressure-tap locations are used:

1. Both pressure taps are installed in the flanges as shown.
2. The inlet pressure tap is located one pipe diameter upstream, and the outlet pressure tap is located one-half diameter downstream of the orifice as measured from the upstream face of the orifice.
3. The inlet pressure tap is located one pipe diameter upstream, and the outlet pressure tap is located at the vena contracta of the orifice, as given by Fig. 7.12.

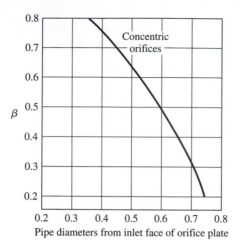

Figure 7.12   Location of outlet pressure connections for orifices with vena contracta taps, according to Ref. [1].

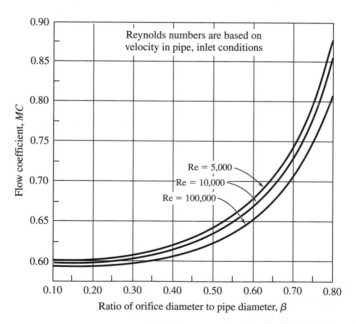

Figure 7.13   Flow coefficients for concentric orifices in pipes. Pressure taps one diameter upstream and one-half diameter downstream. Applicable for $1.25 < D < 3.00$ in. (*From Ref. [15]*.)

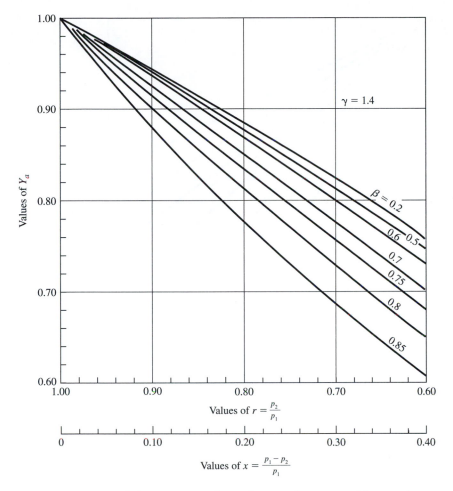

Figure 7.14 Adiabatic expansion factors for use with venturis and flow nozzles as calculated from Eq. (7.14). (*From Ref.* [2].)

Figure 7.13 gives the values of the orifice flow coefficient for pipe sizes $1\frac{1}{4}$ to 3 in with pressure taps located according to case 2 above. Flow coefficients for other cases are given in Ref. [1].

The various flow coefficients are plotted as a function of Reynolds number, defined by

$$\text{Re} = \frac{\rho u_m d}{\mu} \qquad\qquad \textbf{[7.21]}$$

where ρ = fluid density
μ = dynamic viscosity
u_m = mean flow velocity
d = diameter *at the particular section for which the Reynolds number is specified*

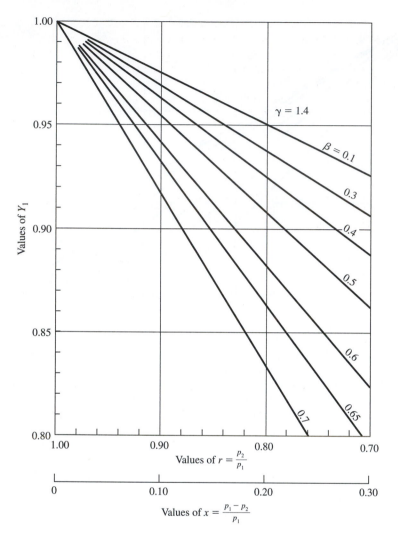

Figure 7.15   Expansion factors for square-edged orifices with pipe taps as calculated from Eq. (7.16). (*From Ref. [2].*)

Note that some charts, viz., Fig. 7.13, base the Reynolds number on upstream conditions, while others, viz., Fig. 7.10, base it on throat conditions. The product ρu_m may be calculated from the mass flow according to

$$\dot{m} = \rho u_m A_c \qquad\qquad \textbf{[7.22]}$$

where A_c is the cross-sectional area for the flow where u_m is measured. For a circular cross section $A_c = \pi d^2/4$. Further information on orifice and venturi meters is contained in Refs. [33] to [37].

DESIGN OF VENTURI METER.   A venturi tube is to be used to measure a maximum flow rate of water of 50 gpm (gallons per minute) at 70°F. The throat Reynolds number is to be at least 10^5 at these flow conditions. A differential pressure gage is selected which has an accuracy of 0.25 percent of full scale, and the upper scale limit is to be selected to correspond to the maximum flow rate. Determine the size of the venturi and the maximum range of the differential pressure gage and estimate the uncertainty in the mass flow measurement at nominal flow rates of 50 and 25 gpm. Use either Fig. 7.9 or Fig. 7.10 to determine the discharge coefficient.

Example 7.2

Solution

The properties of water are

$$\rho = 62.4 \ \text{lbm/ft}^3 = 8.33 \ \text{lbm/gal} \qquad \mu = 2.36 \ \text{lbm/h} \cdot \text{ft}$$

From the given maximum flow rate and throat Reynolds number we may calculate the maximum allowable throat diameter:

$$\text{Re}_d = \frac{\rho u_m d}{\mu} = \frac{\dot{m} d}{(\pi d^2/4)\mu} = \frac{4\dot{m}}{\pi d \mu} = 10^5$$

The maximum flow rate is

$$\dot{m} = (50)(8.33)(60) = 2.5 \times 10^4 \ \text{lbm/h} \ (3.027 \ \text{kg/s})$$

so that

$$d_{\text{max}} = \frac{(4)(2.5 \times 10^4)}{\pi(10^5)(2.36)} = 0.135 \ \text{ft} = 1.62 \ \text{in} \ (4.11 \ \text{cm})$$

We shall select a venturi with a 1.0-in throat diameter since we have a discharge coefficient curve for this size in Fig. 7.10. The upstream pipe diameter is taken as 2.0 in. From Fig. 7.10 we estimate the discharge coefficient for this size venturi as 0.976 for $8 \times 10^4 < \text{Re}_d < 3 \times 10^5$. The uncertainty in this coefficient will be taken as ±0.002 since Fig. 7.10 is a general set of curves. With this selection of venturi size, the maximum throat Reynolds number becomes

$$(\text{Re}_d)_{\text{max}} = (10^5)\left(\frac{1.62}{1.0}\right) = 1.62 \times 10^5$$

The minimum Reynolds number is thus one-half this value, or 8.1×10^4. The maximum pressure differential may be calculated with Eq. (7.17).

$$Q_{\text{actual}} = CMA_2 \sqrt{\frac{2g_c}{\rho}} \sqrt{\Delta p} \qquad\qquad \textbf{[7.17]}$$

or

$$\frac{(50)(231)}{(60)(1728)} = \frac{(0.976)\pi(1.0)^2}{(4)(144)\sqrt{1 - \left(\frac{1}{2}\right)^2}} \sqrt{\frac{(2)(32.2)}{62.4}} \sqrt{\Delta p}$$

This yields

$$\Delta p = 948 \ \text{psf} = 6.58 \ \text{psi} \ (45.4 \ \text{kPa})$$

Let us assume that a differential pressure gage with a maximum range of 1000 psf is at our disposal. In accordance with the problem statement the uncertainty in the pressure reading would be

$$w_{\Delta p} = \pm 2.5 \ \text{psf} \ (119.7 \ \text{Pa})$$

When the flow is reduced to 25 gpm, the pressure differential will be *one-fourth* of that at 50 gpm. To estimate the uncertainty in the flow measurement, we shall assume that the dimensions of the venturi are known exactly, as well as the density of the water. For the calculation we utilize Eq. (3.2). The quantities of interest are

$$\frac{\partial Q}{\partial C} = MA_2 \sqrt{\frac{2g_c}{\rho}} \sqrt{\Delta p}$$

$$\frac{\partial Q}{\partial \Delta p} = \frac{CMA_2}{2\sqrt{\Delta p}} \sqrt{\frac{2g_c}{\rho}}$$

$$w_c = \pm 0.002$$

Thus,

$$\frac{w_Q}{Q} = \left[\left(\frac{w_c}{C} \right)^2 + \frac{1}{4} \left(\frac{w_{\Delta p}}{\Delta p} \right)^2 \right]^{1/2}$$

For $Q = 50$ gpm

$$\frac{w_Q}{Q} = \left[\left(\frac{0.002}{0.976} \right)^2 + \frac{1}{4} \left(\frac{2.5}{948} \right)^2 \right]^{1/2}$$

$$= 0.002435 \qquad \text{or } 0.2435\%$$

For $Q = 25$ gpm

$$\frac{w_Q}{Q} = \left[\left(\frac{0.002}{0.976} \right)^2 + \frac{1}{4} \left(\frac{2.5}{984/4} \right)^2 \right]^{1/2}$$

$$= 0.00566 \qquad \text{or } 0.566\%$$

Example 7.3 | **UNCERTAINTY IN ORIFICE METER.**   An orifice with pressure taps one diameter upstream and one-half diameter downstream is installed in a 2.00-in-diameter pipe and used to measure the same flow of water as in Example 7.2. For this orifice, $\beta = 0.50$. The differential pressure gage has an accuracy 0.25 percent of full scale, and the upper scale limit is selected to correspond to the maximum flow rate. Determine the range of the pressure gage and the uncertainty in the flow-rate measurements at nominal flow rates of 50 and 25 gpm. Assume that the uncertainty in the flow coefficient is ± 0.002.

Solution

We first calculate the pipe Reynolds numbers. Using the properties from Example 7.2 we obtain

$$\text{Re}_d = \frac{(2.5 \times 10^4)(4)}{\pi(2.0/12)(2.36)} = 8.09 \times 10^4 \qquad \text{at 50 gpm (5.26 cm}^3\text{/s)}$$

$$\text{Re}_d = 4.05 \times 10^4 \qquad \text{at 25 gpm (2.63 cm}^3\text{/s)}$$

From Fig. 7.13 the flow coefficient is estimated as

$$K = 0.625 \qquad \text{at 50 gpm}$$

$$K = 0.630 \qquad \text{at 25 gpm}$$

The volumetric flow is

$$Q = \frac{(50)(231)}{(60)(1728)} = 0.1115 \text{ ft}^3/\text{s} \qquad \text{at 50 gpm}$$

$$Q = 0.0558 \text{ ft}^3/\text{s} \qquad\qquad \text{at 25 gpm}$$

The nominal values of the differential pressure are then calculated from Eq. (7.18) as

$$0.115 = (0.625)\frac{\pi(1)^2}{(4)(144)}\sqrt{\frac{(2)(32.2)}{(62.4)}}\sqrt{\Delta p} \qquad \text{at 50 gpm}$$

$$\Delta p = 1307 \text{ psf} = 7.21 \text{ psi (49.7 kPa)} \qquad \text{at 50 gpm}$$

$$\Delta p = 255 \text{ psf} = 1.77 \text{ psi (12.2 kPa)} \qquad \text{at 25 gpm}$$

A suitable differential pressure gage might be one with a maximum range of 1200 psf (57.5 kPa). The same equation for uncertainty applies in this problem as in Example 7.2, except that the flow coefficient K is used instead of the discharge coefficient. Thus,

$$\frac{w_Q}{Q} = \left[\left(\frac{w_K}{K}\right)^2 + \frac{1}{4}\left(\frac{w_{\Delta p}}{\Delta p}\right)^2\right]^{1/2}$$

with $w_K = 0.002$ and $w_{\Delta p} = (0.0025)(1200) = 3.0 \text{ psf (143.6 Pa)}$.
    For $Q = 50$ gpm

$$\frac{w_Q}{Q} = \left[\left(\frac{0.002}{0.625}\right)^2 + \frac{1}{4}\left(\frac{3.0}{1037}\right)^2\right]^{1/2}$$

$$= 0.00351 \qquad \text{or } 0.351\%$$

For $Q = 25$ gpm

$$\frac{w_Q}{Q} = \left[\left(\frac{0.002}{0.630}\right)^2 + \frac{1}{4}\left(\frac{3.0}{255}\right)^2\right]^{1/2} = 0.00669 \qquad \text{or } 0.669\%$$

7.5   THE SONIC NOZZLE

All the obstruction meters discussed above may be used with gases. When the flow rate is sufficiently high, the pressure differential becomes quite large, and eventually sonic flow conditions may be achieved at the minimum flow area. Under these conditions the flow is said to be "choked," and the flow rate takes on its maximum value for the given inlet conditions. For an ideal gas with constant specific heats it may be shown that the pressure ratio for this choked condition, assuming isentropic flow, is

$$\left(\frac{p_2}{p_1}\right)_{\text{critical}} = \left(\frac{2}{\gamma + 1}\right)^{\gamma/(\gamma-1)} \qquad\qquad \textbf{[7.23]}$$

This ratio is called the *critical pressure ratio.* Inserting this ratio in Eq. (7.8) gives for the mass flow rate

$$\dot{m} = A_2 p_1 \sqrt{\frac{2g_c}{RT_1} \left[\frac{\gamma}{\gamma + 1} \left(\frac{2}{\gamma + 1} \right)^{2/(\gamma-1)} \right]^{1/2}} \qquad \textbf{[7.24]}$$

Equation (7.24) is frequently applied to a nozzle when it is known that the pressure ratio p_2/p_1 is less than the critical value given by Eq. (7.23). Under these conditions the ideal flow is dependent only on the inlet stagnation conditions p_1 and T_1. These conditions are usually easy to measure so that the sonic nozzle offers a convenient method for measuring gas-flow rates. It may be noted, however, that a large pressure drop must be tolerated with the method. *Upstream stagnation conditions* must be used for p_1 and T_1 in the calculation.

The ideal sonic-nozzle flow rate given by Eq. (7.24) must be modified by an appropriate discharge coefficient which is a function of the geometry of the nozzle and other factors. There may be several complicating conditions, but discharge coefficients of about 0.97 are usually observed. A comprehensive survey of critical flow nozzles is presented by Arnberg [3], and the interested reader should consult this discussion for more information. More recent information is given in Refs. [32], [34], and [38].

The flow obstruction devices discussed above require the use of wall pressure taps. (Other devices also require the use of such taps.) Measurements with wall pressure taps can be subject to several influencing factors, which are discussed in detail by Rayle [19]. In general, the diameter of the pressure tap should be small enough in comparison with the diameter of the pipe.

Example 7.4 | **DESIGN OF SONIC NOZZLE.**   A sonic nozzle is to be used to measure a flow of air at 300 psia (2.07 MPa) and 100°F (37.8°C) in a 3-in-diameter pipe. The nominal flow rate is 1 lbm/s (0.454 kg/s). Calculate the throat diameter (nozzle size) such that just critical flow conditions are obtained.

Solution

We use Eq. (7.24) for this calculation with $\gamma = 1.4$ for air. The only unknown in this equation is A_2. Thus, we have

$$\dot{m} = C A_2 p_1 \sqrt{\frac{2g_c}{RT_1} \left[\frac{\gamma}{\gamma + 1} \left(\frac{2}{\gamma + 1} \right)^{2/(\gamma-1)} \right]^{1/2}}$$

$$1 = A_2 (300)(144) \sqrt{\frac{(2)(32.2)}{(53.35)(560)} \left[\left(\frac{1.4}{2.4} \right) \left(\frac{2}{2.4} \right)^{2/0.4} \right]^{1/2}}$$

and $A_2 = 0.001078 \ \text{ft}^2 = 0.1551 \ \text{in}^2$ (1.0 cm²).

The diameter at the throat is

$$d = \sqrt{\frac{4}{\pi}(0.1551)} = 0.444 \ \text{in} \ (1.13 \ \text{cm})$$

For the above calculations we have taken the given pressure and temperature as stagnation properties. The temperature would most likely be measured with a stagnation probe so that

100°F is probably the stagnation temperature. The pressure would probably be measured by a static tap in the side of the pipe upstream from the nozzle so that a static-pressure measurement is most likely the one which will be available. If the upstream pipe diameter is large enough, the static pressure will be very nearly equal to the stagnation pressure, and the error in the above calculation will be small. Let us examine the above situation, assuming that the 300 psia is a static-pressure measurement. The mass flow upstream is

$$\dot{m} = \frac{p_{1s}}{RT_{1s}} A_1 u_1 \qquad\qquad\text{[a]}$$

where the subscript s denotes static properties. The velocity upstream may be written in terms of the stagnation temperature as

$$u_1 = \sqrt{2g_c c_p \left(T_{1_0} - T_{1s}\right)} \qquad\qquad\text{[b]}$$

Combining Eqs. (*a*) and (*b*), we have

$$\dot{m} = \frac{p_{1s}}{RT_{1s}} A_1 \sqrt{2g_c c_p \left(T_{1_0} - T_{1s}\right)} \qquad\qquad\text{[c]}$$

Taking $p_{1s} = 300$ psia and $T_{1_0} = 100°F = 560°R$, we may solve Eq. (*c*) for T_{1s}. The result is

$$T_{1s} \approx 560°R \qquad\qquad\text{[d]}$$

or the upstream velocity is so small that the stagnation properties are very nearly equal to the static properties. This result may be checked by calculating the upstream velocity from Eq. (*a*) using the result from Eq. (*d*). We obtain

$$u_1 = 14.1 \text{ ft/s (4.3 m/s)}$$

The pressure difference $(p_{1_0} - p_{1s})$ corresponding to this velocity would be only 0.031 psia, while the temperature difference $(T_{1_0} - T_{1s})$ would be 0.017°F. Both of these values are negligible. It may be noted, however, that if the upstream pipe diameter were considerably smaller, say, 1.0-in diameter, it might be necessary to correct for a difference between the measured static pressure and the stagnation pressure that must be used in Eq. (7.24).

7.6   FLOW MEASUREMENT BY DRAG EFFECTS

ROTAMETER

The *rotameter* is a very commonly used flow-measurement device and is shown schematically in Fig. 7.16. The flow enters the bottom of the tapered vertical tube and causes the bob or "float" to move upward. The bob will rise to a point in the tube such that the drag forces are just balanced by the weight and buoyancy forces. The position of the bob in the tube is then taken as an indication of the flow rate. The device is sometimes called an *area meter* because the elevation of the bob is dependent on the annular area between it and the tapered glass tube; however, the meter operates on the physical principle of drag so that we choose to classify it in this category. A force balance on the bob gives

$$F_d + \rho_f V_b \frac{g}{g_c} = \rho_b V_b \frac{g}{g_c} \qquad\qquad\text{[7.25]}$$

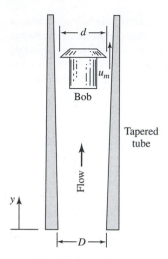

Figure 7.16   Schematic of a rotameter.

where ρ_f and ρ_b are the densities of the fluid and bob, V_b is the total volume of the bob, g is the acceleration of gravity, and F_d is the drag force, which is given by

$$F_d = C_d A_b \frac{\rho_f u_m^2}{2g_c}$$

[7.26]

C_d is a drag coefficient, A_b is the frontal area of the bob, and u_m is the mean flow velocity in the annular space between the bob and the tube.

Combining Eqs. (7.25) and (7.26) gives

$$u_m = \left[\frac{1}{C_d} \frac{2gV_b}{A_b} \left(\frac{\rho_b}{\rho_f} - 1 \right) \right]^{1/2}$$

[7.27]

or

$$Q = Au_m = A \left[\frac{1}{C_d} \frac{2gV_b}{A_b} \left(\frac{\rho_b}{\rho_f} - 1 \right) \right]^{1/2}$$

[7.28]

where A is the annular area and is given by

$$A = \frac{\pi}{4} [(D + ay)^2 - d^2]$$

[7.29]

D is the diameter of the tube at inlet, d is the maximum bob diameter, y is the vertical distance from the entrance, and a is a constant indicating the tube taper.

The drag coefficient is dependent on the Reynolds number and hence on the fluid viscosity; however, special bobs may be used that have an essentially constant drag coefficient, and thus offer the advantage that the meter reading will be essentially independent of viscosity. It may be noted that for many practical meters the quadratic area relation given by Eq. (7.29) becomes nearly linear for actual dimensions of the tube and bob that are used. Assuming such a linear relation, the equation for mass flow would become

$$\dot{m} = C_1 y \sqrt{(\rho_b - \rho_f)\rho_f}$$

[7.30]

where C_1 is now an appropriate meter constant.

For flow of a gas

$$\rho_f = \frac{p}{RT}$$

and for a bob density $\rho_b \gg \rho_f$,

$$\dot{m} \approx (\rho_f)^{-1/2} = \left(\frac{RT}{p}\right)^{1/2} \qquad \textbf{[7.30a]}$$

As noted before, it is common practice to rate gas flowmeters in terms of scfm or sccm. For rotameters their rating is usually in scfm of full scale. To determine the mass flow under inlet conditions other that 70°F and 1 atm, one must use a correction to both the relation in Eq. (7.30a) as well as a conversion to mass flow from volume flow, which is necessary to convert Eq. (7.28) to Eq. (7.30).

AIRFLOW IN A ROTAMETER   A rotameter is used for airflow measurement and has a | **Example 7.5**
rating of 8 scfm for full scale. The bob density has $\rho_b \gg \rho_f$. Calculate the mass rate of flow for inlet conditions of 80 psig and 100°F with a meter reading of 64 percent. The barometric pressure is 750 mmHg.

Solution

For the barometric pressure 750 mmHg = 14.5 psia. The inlet conditions are therefore

$$T = 100°F = 560°R$$
$$p = 80 \text{ psig} + 14.5 = 94.5 \text{ psia}$$

If the inlet were at standard conditions, the volume flow would be

$$Q = (8)(64\%) = 5.12 \text{ scfm} \qquad \textbf{[a]}$$

This value must be corrected because the measurement is made at other than standard conditions.

$$Q_{\text{corr}} = (5.12)\frac{(94.5)(530)}{(14.7)(560)} = 31.15 \text{ scfm} \qquad \textbf{[b]}$$

The corresponding mass flow at 70°F and 14.7 psia is

$$\dot{m} = \frac{pQ}{RT} = \frac{(14.7)(144)(31.15)}{(53.35)(530)} = 2.332 \text{ lbm/min} \qquad \textbf{[c]}$$

In Eq. (7.30a) the meter constant C_1 takes into account a conversion of volume flow to mass flow at standard conditions, but not an allowance for variation of ρ_f from standard conditions. So, we multiply the value in (c) by the pressure and temperature ratios to obtain

$$\dot{m}_{\text{corr}} = (2.232)\left[\frac{(560)(14.7)}{(530)(94.5)}\right]^{1/2} = 0.9454 \text{ lbm/min}$$

It is frequently advantageous to have a rotameter that gives an indication that is independent of fluid density; that is, we wish to have

$$\frac{\partial \dot{m}}{\partial \rho_f} = 0$$

Performing the indicated differentiation, we obtain

$$\rho_b = 2\rho_f \qquad \qquad \textbf{[7.31]}$$

and the mass flow is given by

$$\dot{m} = \frac{C_1 y \rho_b}{2} \qquad \qquad \textbf{[7.32]}$$

Thus, by special bob construction the meter may be used to compensate for density changes in the fluid. The error in Eq. (7.32) is less than 0.2 percent for a fluid-density deviation of 5 percent from that given in Eq. (7.31).

TURBINE METERS

A popular type of flow-measurement device is the turbine meter shown in Fig. 7.17. As the fluid moves through the meter, it causes a rotation of the small turbine wheel. In the turbine-wheel body a permanent magnet is enclosed so that it rotates with the wheel. A reluctance pickup attached to the top of the meter detects a pulse for each revolution of the turbine wheel. Since the volumetric flow is proportional to the

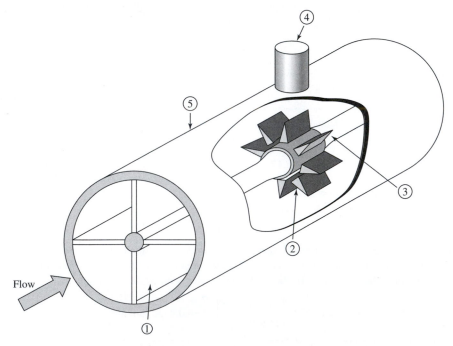

Figure 7.17   Schematic of turbine meter. (*1*) Inlet straightening vanes, (*2*) rotating turbine blades with embedded magnet, (*3*) smooth afterbody to reduce pressure drop, (*4*) reluctance pickup, (*5*) meter body for insert in pipe or flow channel.

number of wheel revolutions, the total pulse output may be taken as an indication of total flow. The pulse rate is proportional to flow rate, and the transient response of the meter is very good. A flow coefficient K for the turbine meter is defined so that

$$Q = \frac{f}{K} \qquad\qquad \textbf{[7.33]}$$

where f is the pulse frequency. The flow coefficient is dependent on flow rate and the kinematic viscosity of the fluid v. A calibration curve for a typical meter is given in Fig. 7.18. It may be seen that this particular meter will indicate the flow accurately within ±0.5 percent over a rather wide range of flow rates. The use of the turbine flowmeter in pulsating flow is discussed in Ref. [31].

FLOW COEFFICIENT FOR TURBINE METER.   Calculate the range of mass flow rates of liquid ammonia at 20°C for which the turbine meter of Fig. 7.18 would be within ±0.5 percent. Also, determine a flow coefficient for this fluid in terms of cycles per kilogram. | **Example 7.6**

Solution

From Fig. 7.18 the range for the ±0.5 percent calibration is approximately 55 to 700 cycles/ s · cSt. From the appendix the properties of ammonia at 20°C are

$$\rho = 612 \text{ kg/m}^3$$
$$v = 0.036 \times 10^{-5} \text{ m}^2/\text{s}$$

The centistoke is defined from

$$1 \text{ stoke (cSt)} = 10^{-4} \text{ square meter per second (m}^2/\text{s)}$$
$$1 \text{ centistoke (cSt)} = 10^{-6} \text{ square meter per second (m}^2/\text{s)}$$

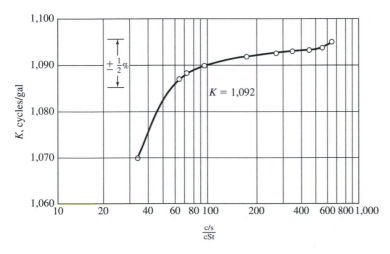

Figure 7.18   Calibration curve for 1-in-turbine flowmeter of the type shown in Fig. 7.17. Calibration was performed with water.

Therefore,
$$\nu = \frac{0.036 \times 10^{-5}}{10^{-6}} = 0.26 \text{ cSt}$$

The frequency range is then

$$f_{\text{low}} = (55)(0.36) = 19.8 \text{ cycles/s}$$
$$f_{\text{high}} = (700)(0.36) = 252 \text{ cycles/s}$$

Also, 1 gal $= 231$ in^3 so that for ammonia

$$1 \text{ gal} = \frac{(231 \text{ in}^3)(612 \text{ kg/m}^3)}{(39.36 \text{ in/m})^3} = 2.318 \text{ kg}$$

The flow coefficient would then be

$$K = 1092 \text{ cycles/gal} = 471.1 \text{ cycles/kg}$$

and the range of flow rates for the 0.50 percent calibration would be

$$\dot{m}_{\text{low}} = \frac{19.8 \text{ cycles/s}}{471.1 \text{ cycles/kg}} = 0.042 \text{ kg/s}$$
$$\dot{m}_{\text{high}} = \frac{252}{471.1} = 0.535 \text{ kg/s}$$

VORTEX-SHEDDING FLOWMETERS

Vortex flowmeters operate on the principle illustrated in Fig. 7.19. When a bluff body is placed in a flow stream, vortices are shed alternately from the back side.

The frequency of vortex shedding is directly proportional to the liquid velocity. A piezoelectric sensor mounted inside the vortex shedder detects the vortices, and subsequent amplification circuits can be used to indicate either the instantaneous flow rate or a totalized flow over a selected time interval. The meter is precalibrated by the manufacturer for a specific pipe size. It is generally unsuitable for use with highly viscous liquids. A number of special installation requirements must be met and are described in manufacturer's literature; see, for example, Ref. [49].

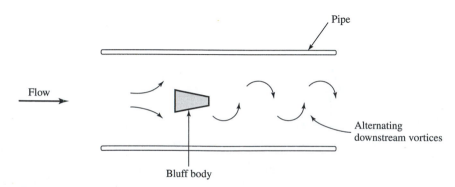

Figure 7.19   Vortex-shedding flowmeter.

The fluid parameter which governs the operation of the vortex-shedding meter is the Strouhal number S

$$S = \frac{f_s d}{u}$$

where f_s = shedding frequency

d = diameter or characteristic dimension of the bluff body

u = velocity

For the geometry shown in Fig. 7.19 the Strouhal number has an essentially constant value of 0.88, within 1 percent, for Reynolds numbers from 10^4 to 10^6.

The vortex-shedding principle can also be used to construct a velocity probe. The bluff body can be mounted along with the sensors in its own small section of a short tube and calibrated directly by the manufacturer. The assembly may then be inserted at various locations in the flow field to measure the local velocity.

ULTRASONIC FLOWMETERS

The Doppler effect is the basis for operation of the ultrasonic flowmeter illustrated in Fig. 7.20. A signal of known ultrasonic frequency is transmitted through the liquid. Solids, bubbles, or any discontinuity in the liquid will reflect the signal back to the receiving element. Because of the velocity of the liquid, there will be a frequency shift at the receiver which is proportional to velocity. Accuracies of about ±5 percent of full scale may be achieved with the device over a flow range of about 10 to 1. Most devices require that the liquid contain at least 25 parts per million (ppm) of particles or bubbles having diameters of 30 μm or more.

A microprocessor-based ultrasonic flowmeter has been developed which employs a Doppler signal reflected from turbulent eddies in the flow (Ref. [49]). As a result, it is suitable for operation with clean low-viscosity liquids. Accuracy of 2 percent of full scale may be achieved and the meter may be installed as few as three pipe diameters downstream from a 90° elbow. The price is quite high.

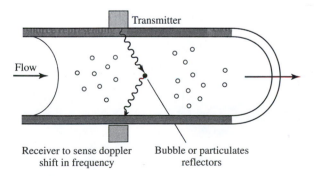

Figure 7.20   Ultrasonic Doppler flowmeter.

LAMINAR FLOWMETER

If the flow in a channel or tube is laminar, that is, the Reynolds number is less than about 2000, the volumetric flow rate is related to pressure drop by

$$Q = \frac{\pi d^4 (p_1 - p_2)}{128 \, \mu L}$$

[7.34]

for $\text{Re}_d = \rho u_m d / \mu < 2000$.

The nomenclature for such a laminar flow element is indicated in Fig. 7.21. The mass flow is, correspondingly,

$$\dot{m} = \rho Q$$

[7.35]

Thus, if laminar flow can be assured, the flow rate becomes a direct function of pressure difference. A laminar flow *meter* may be constructed of a collection of small tube elements as shown in Fig. 7.22. If the tubes are sufficiently small, laminar flow can be maintained. There are entrance and exit losses that occur with the tube assembly. These losses depend on the particular geometric configuration of the tube assembly, and its installation in the pipe. Manufacturers of commercial laminar flowmeters furnish calibration information that applies to their units. Uncertainties of $\pm 1/4$ percent for determination of flow rate are reported under careful operating conditions.

In contrast to the orifice, flow nozzle, and venturi, the laminar flowmeter has mass flow directly proportional to Δp instead of $(\Delta p)^{1/2}$. This fact allows for operation over a wider range of flow rates for a particular differential pressure-measuring device. Because of their relatively small size, the laminar tube elements are subject to clogging when used with dirty fluids. The overall pressure loss for the devices is high, of the order of 100 percent of the measured Δp.

We may note that

$$\text{Re} = \frac{\rho u_m d}{\mu} = \frac{\dot{m}}{A_c} \frac{d}{\mu}$$

$$= \frac{4\dot{m}}{\pi d\mu} = \frac{4Q\rho}{\pi d\mu}$$

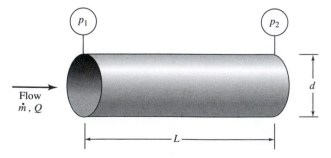

Figure 7.21   Laminar flow element.

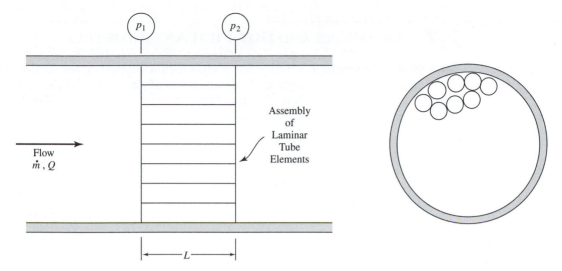

Figure 7.22   Laminar flowmeter.

Combining the latter relation with Eq. (7.34) gives

$$\Delta p = \frac{128 \text{Re} \, \mu^2 L}{4 \rho d^3}$$ **[7.34a]**

If acceptable pressure drop and Reynolds numbers are set, the proportions between L and d may be established for design purposes.

SIZING OF LAMINAR FLOWMETER   Size a laminar flowmeter used to measure the flow of air at 2 atm and 20°C with a maximum Reynolds number of 1000 and a pressure drop of 1000 Pa.   **Example 7.7**

Solution

We employ Eq. (7.34a) to determine the relationship between the length and diameter of the laminar flow elements. The properties of air are

$$\rho = (2)(1.18) = 2.36 \text{ kg/m}^3$$
$$\mu = 1.84 \times 10^{-5} \text{ kg/m·s}$$

Inserting these values in Eq. (7.34a) gives

$$L/d^3 = (1000)(2.36)/(32)(1000)(1.84 \times 10^{-5})^2 = 2.18 \times 10^8$$

If a tube diameter of 1.0 mm is selected, the required length of tube would be

$$L = (2.18 \times 10^8)(0.001)^3 = 0.218 \text{ m}$$

The mass flow through each flow element would be

$$\dot{m} = (Re)(\pi d \mu) = (1000)\pi(0.001)(1.84 \times 10^{-5}) = 5.78 \times 10^{-5} \text{ kg/s}$$

7.7 HOT-WIRE AND HOT-FILM ANEMOMETERS

The hot-wire anemometer is a device that is often used in research applications to study rapidly varying flow conditions. A fine wire is heated electrically and placed in the flow stream, and the early work of King [5] has shown that the heat-transfer rate from the wire can be expressed in the form

$$q = (a + bu^{0.5})(T_w - T_\infty) \qquad \textbf{[7.36]}$$

where T_w = wire temperature

T_∞ = free-stream temperature of fluid

u = fluid velocity

a, b = constants obtained from a calibration of the device

The heat-transfer rate must also be given by

$$q = i^2 R_w = i^2 R_0[1 + \alpha(T_w - T_0)] \qquad \textbf{[7.37]}$$

where i = electric current

R_0 = resistance of the wire at the reference temperature T_0

α = temperature coefficient of resistance

For measurement purposes the hot wire is connected to a bridge circuit, as shown in Fig. 7.23. The current is determined by measuring the voltage drop across the standard resistance R_s, and the wire resistance is determined from the bridge circuit. For steady-state measurements the null condition may be used, while an oscilloscope output may be used for transient measurements. With i and R_w determined, the flow velocity may be calculated with Eqs. (7.36) and (7.37). Hot-wire probes have been

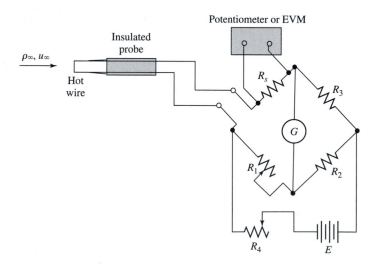

Figure 7.23   Schematic of hot-wire flow-measurement circuit.

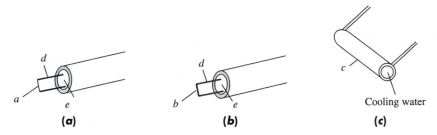

Figure 7.24 Schematic of hot-wire and hot-film probes: (a) hot-wire, (b) cylindrical hot-film, (c) cooled hot-film.

used extensively for measurement of transient flows, especially measurements of turbulent fluctuations. Time constants of the order of 1 ms may be obtained with 0.0001-in-diameter platinum or tungsten wires operating in air.

When the hot wire is to be employed for measurement of rapidly changing flow patterns, full account must be taken of the transient response of both the thermal and electrical resistance characteristics of the wire. Two types of electrical compensation are employed in practice: (1) a constant-current arrangement, where a large resistance is connected in series with the hot wire and a thermal compensating circuit is then applied to the output ac voltage, and (2) a constant-temperature arrangement, where a feedback control circuit is added to vary the current so that the wire temperature remains nearly constant. Response of the wire depends on the angle the flow velocity makes with the wire axis, and techniques have been developed to take this effect into account [6 and 44]. The length-to-diameter ratio of the wire (L/d) also has a significant effect on the measurement performance. L/d has a value of about 50 for typical hot wires.

A modification of the hot-wire method consists of a small insulating cylinder that is coated with a thin metallic film. This device is appropriately called a *hot-film probe,* and sketches of hot-wire and hot-film probes are shown in Fig. 7.24. Figure 7.24c shows a water-cooled probe which may be employed in high-temperature applications. The nomenclature is:

a. Tungsten hot-wire with platinum surface coating; $L \sim 1.2$ mm, $d \sim 4\ \mu$m

b. Cylindrical hot-film sensor with platinum film on glass rod; $L \sim 1.0$ mm, $d \sim 50\ \mu$m

c. Platinum film on hollow glass tube; $L \sim 1.5$ mm, $d \sim 0.15$ mm

d. Gold-plated sensor supports

e. Ceramic insulating core

Coatings for the hot-film probes are typically achieved with an electroplated gold layer about 5 μm thick.

Hot-film probes are extremely sensitive to fluctuations in the fluid velocity and have been used for measurements involving frequencies as high as 50,000 Hz. Sophisticated electronic instrumentation is required for such measurements, and the interested reader is referred to Refs. [6] and [44] for additional information, as well as to the literature of such manufacturers as Thermal Systems and DISA, Inc.

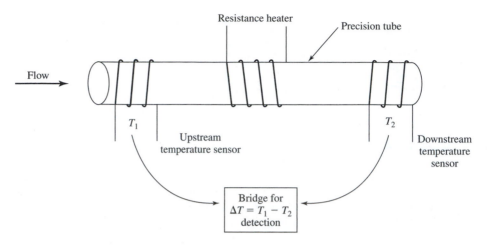

Figure 7.25   Mass flowmeter based on thermal energy transfer.

THERMAL MASS FLOWMETERS

A direct measurement of mass flow of gases may be accomplished using the principle illustrated in Fig. 7.25. A precision tube is constructed with upstream and downstream externally wound resistance temperature detectors. Between the sensors is an electric heater. The temperature difference $T_1 - T_2$ is directly proportional to the mass flow of the gas and may be detected with an appropriate bridge circuit. The device is restricted to use with very clean gases. Calibration is normally performed with nitrogen and a factor applied for use with other gases.

Another thermal mass flowmeter for gases utilizes two platinum resistance temperature detectors (RTDs; see Sec. 8.5). One sensor measures the temperature of the gas flow at the point of immersion. A second sensor is heated to a temperature of 60°F above the first sensor. As a result of the gas flow, the heating of the second sensor is transferred to the gas by convection. The heat-transfer rate is proportional to the mass velocity of the gas, defined as

$$\text{Mass velocity} = \rho u = \text{density} \times \text{velocity}$$

The two sensors are connected to a Wheatstone bridge and the output voltage or current is that required to maintain the 60°F temperature differential. This output is a nonlinear function of ρu, but special circuits may be used to produce a linearized output for readout purposes.

We must note, of course, that the immersion-type meter measures mass velocity *at the point of immersion*. For flow systems with varying velocities several measurements may be necessary to obtain an integrated mass flow across the channel. The probe containing both sensors is typically about 0.25 in in diameter. Velocities of gases at standard conditions between 0.08 and 160 ft/s can be measured with the device. Accuracies of about ±2 percent of full scale may be achieved.

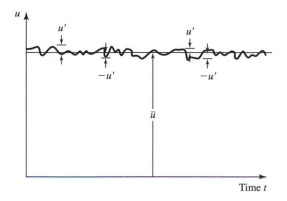

Figure 7.26 Turbulent fluctuations in direction of flow.

TURBULENCE MEASUREMENTS

In turbulent fluid flow, whether in a channel or in a boundary layer, the fluid-velocity components exhibit a random oscillatory character, which depends on the average fluid velocity, fluid density, viscosity, and other variables. The instantaneous-velocity components are expressed as

$$u = \bar{u} + u' \qquad (x \text{ component})$$
$$v = \bar{v} + v' \qquad (y \text{ component})$$
$$w = \bar{w} + w' \qquad (z \text{ component})$$

where the bar quantities are the integrated average-velocity components and the primed quantities represent the fluctuations from the average velocity. Figure 7.26 illustrates a typical plot of turbulent fluctuations. One measure of the *intensity of turbulence* is the root mean square of the fluctuations. For the x component

$$u'_{\text{rms}} = (\overline{u'^2})^{1/2} = \left[\frac{1}{T} \int_0^T u'^2 \, dt \right]^{1/2} \qquad \textbf{[7.38]}$$

where T is some time period that is large compared to the turbulence scale.

Turbulent fluctuations occur over an extremely broad range. Some typical values are:

Background turbulence in well-designed wind tunnels	0.05%
Turbulence created by grids	0.02–2%
Turbulent wakes	2–5%
Turbulent boundary layers and pipe flows	3–20%
Turbulent jets	20–100%

The average kinetic energy (KE) of turbulence per unit mass may be taken as a measure of the total turbulence intensity of the flow stream. Thus,

$$\frac{\text{Turbulence KE}}{\text{Mass}} = \frac{1}{2}(\overline{u'^2} + \overline{v'^2} + \overline{w'^2}) \qquad \textbf{[7.39]}$$

The hot-wire or hot-film anemometer is very useful for turbulence measurements because it can respond to very rapid changes in flow velocity. Two or more wires at one point in the flow can make simultaneous measurements of the fluctuating components. Various mean values and energy spectra can then be computed from the voltage outputs of the probes. In general, these operations can be performed electronically.

7.8   MAGNETIC FLOWMETERS

Consider the flow of a conducting fluid through a magnetic field, as shown in Fig. 7.27. Since the fluid represents a conductor moving in the field, there will be an induced voltage according to

$$E = BLu \times 10^{-8} \text{ V} \qquad \textbf{[7.40]}$$

where B = magnetic flux density, gauss
   u = velocity of the conductor, cm/s
   L = length of the conductor, cm

The length of the conductor is proportional to the tube diameter, and the velocity is proportional to the mean flow velocity. The two electrodes detect the induced voltage, which may be taken as a direct indication of flow velocity.

Two types of magnetic flowmeters are used commercially. One type has a nonconducting pipe liner and is used for fluids with low conductivities, like water. The electrodes are mounted so that they are flush with the nonconducting liner and make contact with the fluid. Alternating magnetic fields are normally used with these meters since the output is low and requires amplification. The second type of magnetic flowmeter is one which is used with high-conductivity fluids, principally liquid

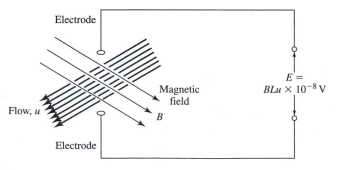

Figure 7.27   Flow of conducting fluid through a magnetic field.

metals. A stainless-steel pipe is employed in this case, with the electrodes attached directly to the outside of the pipe and diametrically opposed to each other. The output of this type of meter is sufficiently high that it may be used for direct readout purposes.

7.9   FLOW-VISUALIZATION METHODS

Fluid flow is a complicated subject with many areas that have not yet yielded to precise analytical techniques. Accordingly, flow-measurement problems are not always simple and precise because of the lack of analytical relations to use for calculation and reduction of experimental data. The interpretation of data involving turbulence or measurements of complicated boundary layer, viscous, and shock-wave effects is not easy. Frequently, the flow may be altered as a result of probes that are inserted to measure the pressure, velocity, and temperature profiles so that the experimentalist is uncertain about the effect that has been measured. Flow visualization by optical methods offers the advantage that when properly executed it does not disturb the fluid stream, and thus gives the experimentalist an extra tool to use in conjunction with other measurement devices. In some instances flow-visualization techniques may be employed for rather precise measurements of important flow parameters, while in other cases they may serve only to furnish qualitative information regarding the overall flow behavior. In the following paragraphs we shall discuss the principles of some of the basic flow-visualization schemes and indicate their applications. The interested reader should consult Refs. [8] and [17] for a rather complete survey of the subject. Additional advances are presented in Ref. [47].

Consider the gaseous flow field shown in Fig. 7.28. The flow is in a direction perpendicular to the figure, that is, the z direction. An incoming light ray is deflected

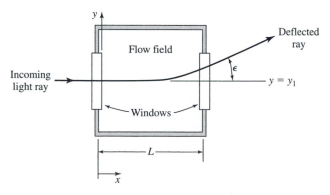

Figure 7.28   Basic optical effects used for flow visualization.

through an angle ϵ as a result of density gradients in the flow. It may be shown[2] that the deflection angle for small-density gradients is given as

$$\epsilon = \frac{\lambda}{n_1}\left(\frac{dn}{dy}\right)_{y=y_1} = \frac{L\beta}{\rho_s}\left(\frac{d\rho}{dy}\right)_{y=y_1}$$ **[7.41]**

where L is the width of the flow field, ρ is the local fluid density, ρ_s is a reference density which is usually taken at standard conditions, and n is the index of refraction, which for gases may be written as

$$n = \left(1 + \beta\frac{\rho}{\rho_s}\right)n_1$$ **[7.42]**

β is a dimensionless constant having a value of about 0.000292 for air; n_1 is the index of refraction outside the flow field and can be taken as very nearly unity in Eq. (7.41).

According to Eq. (7.41) the angular deflection of the light ray ϵ is proportional to the density gradient in the flow. This is the basic optical effect that is used for flow-visualization work. It may be noted that the deflection of the light ray is a measure of the average density gradient integrated over the x coordinate. Thus, the effect is primarily useful for indicating density variations in two dimensions (in this case the y and z dimensions) and will average the variations in the third dimension.

In the following sections we shall discuss several optical methods of flow visualization for use in gas systems. For liquid-flow visualization a typical experimental technique is to add a dye to the liquid in order to study the flow phenomena. Another technique which has received considerable attention for use with liquids is the so-called hydrogen-bubble method. The technique consists of using a fine wire, placed in water, as one end of a dc circuit to electrolyze the water. Very small hydrogen bubbles are generated in the liquid. The motion of the bubbles may be studied by illuminating the flow. The application of this method has been given a very complete description by Schraub et al. [16]. Figure 7.29 indicates the voltage-discharge

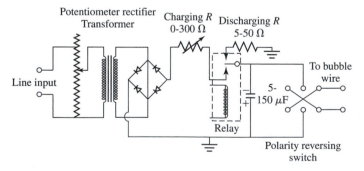

Figure 7.29   Voltage pulse circuits for hydrogen-bubble method, according to Ref. [16].

| 2See, e.g., Ref. [10].

circuits employed in the technique. A variable-input voltage from 10 to 250 V is provided and rectified in the diode circuit. A relay is used to charge and discharge the capacitor periodically, and a variable resistor is provided in the charging circuit to tune the system for optimum performance for each wire diameter and flow condition. In some cases it is necessary to add a small amount of sodium sulfate to the water in order to achieve a satisfactory electrolyte concentration.

7.10   THE SHADOWGRAPH

The shadow technique is a method for direct viewing of flow phenomena. Imagine the flow field as shown in Fig. 7.30 with a density gradient in the y direction. The parallel light rays enter the test section as shown. In the regions where there is no density gradient, the light rays will pass straight through the test section with no deflection. For the regions where a gradient exists the rays will be deflected. The net effect is that the rays will bunch together after leaving the test section to form bright spots and dark spots. The illumination will depend on the *relative deflection* of the light rays $d\epsilon/dy$, and hence on $d^2\rho/dy^2$. The illumination on a screen placed outside the test section is thus dependent on the second derivative of the density at the particular point.

The shadowgraph is a very simple optical tool, and its effect may be viewed in several everyday phenomena using only the naked eye and local room lighting. The free-convection boundary layer on a horizontal electric hot plate is clearly visible when viewed from the edge. This phenomenon is visible because of the density gradients that result from the heating of the air near the hot surface. It is almost fruitless to try to evaluate local densities using shadow photography; however, the shadowgraph is useful for viewing turbulent flow regions, and the method can be used

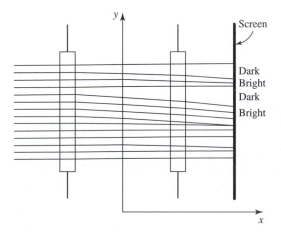

Figure 7.30   Shadowgraph flow-visualization device.

Figure 7.31   Shadowgraph of the free convection boundary layer on a 1.25-cm diameter horizontal cylinder. The cylinder size is indicated by the white circle placed on the photo. The dark area is the boundary layer and heated wake above the cylinder. The "halo" results from the refraction of the light pencils nearest the heated surface. The distance between the cylinder surface and the halo at the same angular position is proportional to the surface convection heat transfer coefficient. (See Sec. 9.7 for measurement of convection coefficients.)

to establish the location of shock waves with high precision. Figure 7.31 illustrates the free-convection boundary layer on a horizontal cylinder. Developments in digital imaging systems whereby the illumination of individual pixels may be accessed in the final image offer opportunities for further advances. (See Appendix B.)

7.11   THE SCHLIEREN

While the shadowgraph gives an indication of the second derivative of density in the flow field, the schlieren is a device which indicates the density gradient. Consider the schematic diagram shown in Fig. 7.32. Light from a slit source ab is collimated by the lens L_1 and focused at plane 1 in the test section. After the light passes through lens L_2, an inverted image of the source at the focal plane 2 is produced. Lens L_3 then focuses the image of the test section on the screen at plane 3. Now, let us consider the imaging process in more detail. The pencils of light originating at point a occupy a different portion of the various lenses from those originating from point b or any other point in the slit source. The regions in which these pencils overlap are shown in Fig. 7.32. Note that *all* pencils of light pass through the image plane cd in the test section and the source image plane $b'a'$. An image of the test section at $d'c'$ is then uniformly illuminated since the image at $b'a'$ is uniformly illuminated. This means that all points in the plane $b'a'$ are affected in the same manner by whatever fluid effects may take place in the test section.

If the test section is completely uniform in density, the pencils of light appear as shown in Fig. 7.32; a pencil originating at point c is deflected by the same amount as a pencil originating at point d. This is consistent with the observation that all pencils originating in plane cd completely fill the image plane $b'a'$. Now, consider the effect of the introduction of an obstruction at plane $b'a'$ under these circumstances. We immediately conclude that such an obstruction would uniformly decrease the illumination on the screen by a factor proportional to the amount of the area $b'a'$ intercepted.

Suppose now that a density gradient exists at the test section focal plane cd. This means that all pencils of light originating in this plane would no longer fill the image plane $b'a'$ completely. If, then, an obstruction is placed at plane $b'a'$, it will intercept more light from some points in the test section plane than from others, resulting in light and dark regions on the screen at plane 3. The obstruction is called a knife edge, and the resultant variation in illumination on the screen is called the *schlieren effect*.

Let us examine the variation in illumination in more detail. In Fig. 7.32b the total height of the source image is y and the portion not intercepted by the knife edge is

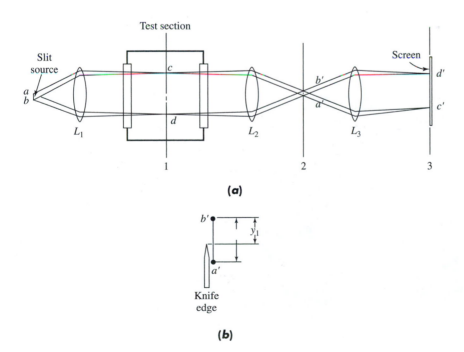

(a)

(b)

Figure 7.32　　(a) Schematic of schlieren flow visualization; (b) detail of knife edge.

y_1. Thus, the general illumination on the screen I is proportional to y_1. An angular displacement of a pencil of light in plane 1 is ϵ. This produces a vertical deflection at plane 2 of

$$\Delta y = f_2 \epsilon \qquad \qquad \textbf{[7.43]}$$

where f_2 is the focal length of lens L_2. As a result of this deflection, there is a fractional change in illumination on the screen. The contrast at any point on the screen may be defined as the ratio of the fractional change in illumination to the general illumination, or

$$C = \frac{\Delta I}{I} = \frac{\Delta y}{y_1} = \frac{f_2 \epsilon}{y_1} \qquad \qquad \textbf{[7.44]}$$

The angular deflection is given by Eq. (7.40), so that the expression for contrast may be written

$$C = \frac{f_2 L \beta}{y_1 \rho_s} \left(\frac{d\rho}{dy} \right)_{cd} \qquad \qquad \textbf{[7.45]}$$

Thus, the contrast on the screen is directly proportional to the density gradient in the flow. It may be observed that the contrast may be increased by reducing the distance y_1, that is, by intercepting more light at the source image plane. This also reduces the general illumination so that the contrast may not be increased indefinitely and a compromise must be accepted. Schlieren photographs are used extensively for location of shock waves and complicated boundary-layer phenomena in supersonic flow systems. A typical schlieren photograph is shown in Fig. 7.33. In actual practice most schlieren systems use mirrors instead of lenses for reasons of economy. Again, the development of digital imaging systems offers the possibility of pixel-by-pixel measurement of illumination and contrast in the final image.

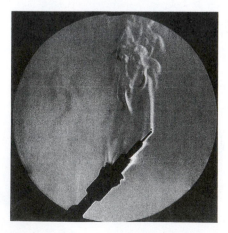

Figure 7.33   Schlieren photograph of the heated wake from an electric soldering iron. Tip of iron is approximately 0.125 inch in diameter. Note turbulence in wake.

7.12   THE INTERFEROMETER

The Mach–Zehnder interferometer is the most precise instrument for flow visualization. Consider the schematic representation in Fig. 7.34. The light source is collimated through lens L_1 onto the splitter plate S_1. This plate permits half of the light to be transmitted to mirror M_2 while reflecting the other half toward mirror M_1. Beam 1 passes through the test section, while beam 2 travels an alternative path of approximately equal length. The two beams are brought together again by means of the splitter plate S_2 and eventually focused on the screen. Now, if the two beams travel paths of different *optical lengths* because of either the geometry of the system or the refractive properties of any element of the optical paths, the two beams will be out of phase and will interfere when they are joined together at S_2. There will be alternate bright and dark regions called *fringes*. The number of fringes will be a function of the difference in optical path lengths for the two beams; for a difference in path lengths of one wavelength there will be one fringe, two fringes for a difference of two wavelengths, and so on.

The interferometer is used to obtain a direct measurement of density variations in the test section. If the density in the test section (i.e., beam 1) is different from that in beam 2, there will be a change in the refractive properties of the fluid medium. If the medium in the test section has the same optical properties as the medium in beam 2, there will be no fringe shifts except those resulting from the geometric arrangement of the apparatus. These fringe shifts may be neutralized by appropriate movements of the mirrors and splitter plates. Then, the appearance of fringes on the screen may be directly related to changes in density in the flow field within the test section by utilizing the following analysis.

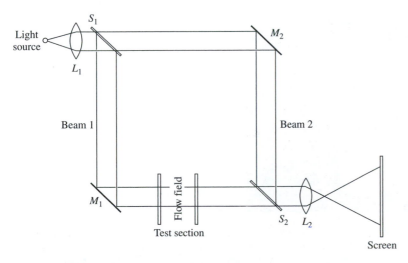

Figure 7.34   Schematic of Mach–Zehnder interferometer.

The change in optical path in the test section resulting from a change in refractive index is

$$\Delta L = L(n - n_0) \qquad \qquad \textbf{[7.46]}$$

where L is the thickness of the flow field in the test section. Using Eq. (7.42) the change in optical path may be related to change in density for gases by

$$\Delta L = \beta L \frac{\rho - \rho_0}{\rho_s} \qquad \qquad \textbf{[7.47]}$$

The number of fringe shifts N is then given by

$$N = \frac{\Delta L}{\lambda} = \frac{\beta L}{\lambda} \frac{\rho - \rho_0}{\rho_s} \qquad \qquad \textbf{[7.48]}$$

where λ is the wavelength of the light. In Eq. (7.48) it is to be noted that $\rho - \rho_0$ represents the change in the density from the zero-fringe condition. The subscript 0 refers to the zero-fringe condition, that is, conditions in the path followed by beam 2 in Fig. 7.34. ρ_s is the reference density at standard conditions.

The interferometer gives a direct quantitative indication of density changes in the test section, but these changes are represented as integrated values over the entire thickness of the flow field. It is applicable to a wide range of flow conditions ranging from the low-speed ($\sim$30 cm/s) flow in free-convection boundary layers to shock-wave phenomena in supersonic flow. Figure 7.35 shows a typical interferometer photograph. The caption explains the flow phenomena.

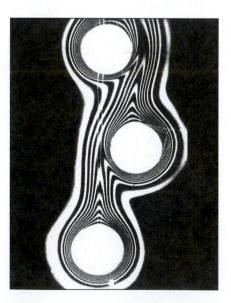

Figure 7.35   Interferometer photograph of the interaction of free-convection boundary layers on three horizontal heated cylinders. Fluid is air, and each fringe line represents a line of constant temperature. (*Photograph courtesy of E. Soehngen.*)

We have already given the frequency characteristics of some typical monochromatic light sources in Table 5.1. The mercury source is one commonly employed for interferometer work.

7.13   THE LASER DOPPLER ANEMOMETER (LDA)

We have seen how optical flow-visualization methods offer the advantage that they do not disturb the flow during the measurement process. The laser anemometer is a device that offers the nondisturbance advantages of optical methods while affording a very precise quantitative measurement of high-frequency turbulence fluctuations.

One possible schematic of an LDA is shown in Fig. 7.36a. The laser beam is focused on a small-volume element in the flow through lens L_1. In order for the device to function, the flow must contain some type of small particles to scatter the light, but the particle concentration required is very small. Ordinary tap water contains enough impurities to scatter the incident beam. Two additional lenses L_2 and L_3 are positioned to receive the laser beam that is transmitted through the fluid (lens L_3) and some portion of the beam that is scattered through the angle θ (lens L_2). The scattered light experiences a Doppler shift in frequency that is directly proportional to the flow velocity. The unscattered portion of the beam is reduced in intensity by the neutral density filter and recombined with the scattered beam through the beam splitter. The laser-anemometer device must be constructed so that the direct and scattered beams travel the same optical path in order that an interference will be observed at the photomultiplier tube that is proportional to frequency shift. This shift then gives an indication of the flow velocity. To retrieve the velocity data from the photomultiplier signal rather sophisticated electronic techniques must be employed for signal processing. A spectrum analyzer may be used to determine velocity in steady laminar flow as well as mean velocity and turbulence intensity in turbulent flow.

Alternative schemes for accomplishing the scattering and measurement process are shown in Fig. 7.36b and c. In (b) the laser beam is split outside the test section, and the two beams can be focused on the exact point to be studied in the flow field. The aperture acts as a shield for noncoherent scattered light and background light. The system in Fig. 7.36c is a further modification of the system and allows for easy adjustment of path length.

He–Ne gas lasers are most often employed for LDA work, although argon ion lasers provide a more intense beam output. The He–Ne laser operates at a wavelength of 632.8 nm ($\approx 5 \times 10^{14}$ Hz) with a bandwidth of about 10 Hz. Although the Doppler shift caused by the moving scattering centers is small compared to the laser source frequency, it is very large compared to the bandwidth and can be detected by heterodyne techniques. In this procedure the photocathode mixes the scattered beam with the reference beam to generate a current with a frequency equal to the difference in frequency of the two beams. The electronic processing requires a spectrum analysis

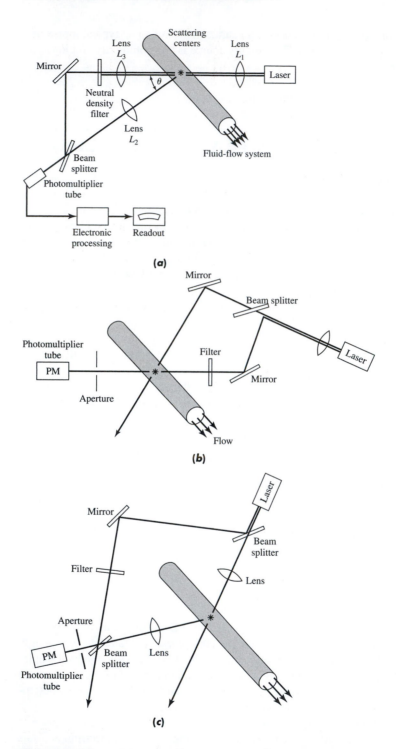

Figure 7.36   Schematic of laser-anemometer flow-measurement system.

of the photomultiplier current to determine the Doppler frequency and subsequently the flow velocity.

It is clear that the LDA measures the velocity of the scattering *particles*. If they are sufficiently small, the slip velocity between particles and fluid will be small, and thus an adequate indication of fluid velocity will be obtained. Laser anemometers that measure more than one velocity will be obtained. Laser anemometers that measure more than one velocity component simultaneously have been developed [41], but the optics and electronic-signal-processing techniques become quite complex and expensive. Even so, the technique offers unusual promise for detailed investigations of turbulence and other flow phenomena which may not be performed in any other way. Welch and Hines [27] have noted that the sample volume of a focused laser beam can be as small as 1.6×10^{-5} mm^3 with a spatial resolution of the order of a few tens of micrometers. The interested reader should consult Refs. [39] to [42] and [53] to [57] for additional information on the construction of laser anemometers and the measurements that have been obtained. Goldstein [46] gives a summary of LDA work. Progress in the development of LDA systems is very rapid, and those persons interested in the latest hardware and software will be well advised to consult current commercial suppliers[3] for information pertaining to their particular application.

INTERFEROMETER.   An interferometer is used for visualization of a free-convection boundary layer on a vertical flat plate in air. For this application the following data were collected:

Plate temperature $T_w = 50°C$

Free-stream air temperature $T_\infty = 20°C$

$\beta = 0.000293$

Depth of test section $L = 50$ cm

Wavelength of light source $\lambda = 5460$ Å

Reference density $= 20°C$

Pressure $= 1.0$ atm

Example 7.8

Calculate the number of fringes that will be viewed in the boundary layer.

Solution

We use Eq. (7.48) for this calculation. The reference density is the same as the zero-fringe density so that

$$N = \frac{\beta L}{\lambda}\left(\frac{\rho_w}{\rho_\infty} - 1\right)$$

---

[3] Dantec Measurement Technology, Inc., Mahwah, NJ; TSI Incorporated, St. Paul, MN; Aerometrics, Inc., Sunnyvale, CA; La Vision GMBH, Goettingen, Germany.

For air the density is calculated from the ideal-gas equation of state:

$$\rho = \frac{p}{RT}$$

Thus,

$$N = \frac{\beta L}{\lambda} \left(\frac{T_\infty}{T_w} - 1 \right)$$

$$= \frac{(0.000293)(50)}{(5460 \times 10^{-8})} \left(\frac{293}{323} - 1 \right)$$

$$= 24.92 \text{ fringes}$$

7.14 SMOKE METHODS

A very simple flow-visualization method utilizes the injection of smoke traces in a gas stream to follow streamlines. The method is primarily of qualitative utility in that direct measurements are difficult to obtain except for certain special phenomena. Figure 7.37 shows an example of a flow system where smoke visualization was used to verify an analytical calculation. In this case smoke is used to view the complicated secondary flow patterns in a channel through which a forced flow is coupled with a standing sound wave. The smoke patterns in Fig. 7.37*a* agree well with the analytical predictions in Fig. 7.37*b*.

In order for the smoke filaments to represent streamlines of the flow it is necessary that the individual smoke particles be of sufficiently small mass so that they are carried along freely at the flow velocity. Filtered smoke from burning rotten wood or cigars is generally suitable for smoke studies, as is smoke from the chemical titanium tetrachloride when it reacts with moisture in air to form hydrochloric acid and titanium oxide. This latter substance, however, is corrosive to many materials used for the construction of containers. Reference [14] discusses the use of titanium tetrachloride for low-speed flow measurements. One of the best fuels for producing nontoxic, noncorrosive, dense smoke is a product called Type-1964 Fog Juice. This fuel has a boiling temperature of approximately 530°F (276°C), contains petroleum hydrocarbon, and may be obtained from most theatrical supply houses. Some smoke-generation techniques are discussed in Ref. [28].

Particle image velocimetry (PIV) coupled with digital imaging systems offer the ability to track individual particle motions [56].

7.15 PRESSURE PROBES

A majority of fluid dynamic applications involve measuring the total flow rate by one or more of the methods discussed in the previous sections. These measurements ignore the local variations of velocity and pressure in the flow channel and permit

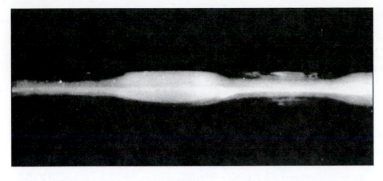

(a)

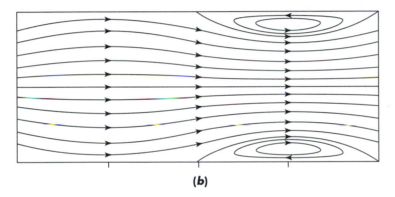

(b)

Figure 7.37   (a) Smoke photograph showing secondary flow effects resulting from a standing sound wave in a tube: (b) flow streamlines for the system in (a) as obtained from the theoretical analysis of Ref. [12]. (*Photograph courtesy of Dr. T. W. Jackson.*)

an indication of only the total flow through a particular cross section. In applications involving external flow situations, such as aircraft or wind-tunnel tests, an entirely different type of measurement is required. In these instances probes must be inserted in the flow to measure the local static and stagnation pressures. From these measurements the local flow velocity may be calculated. Several probes are available for such measurements, and summaries of characteristic behaviors are given in Refs. [4], [7], [13], and [18]. We shall discuss some of the basic probe types in this section.

The total pressure for isentropic stagnation of an ideal gas is given by

$$\frac{p_0}{p_\infty} = \left(1 + \frac{\gamma - 1}{2} M_\infty^2\right)^{\gamma/(\gamma-1)} \qquad \textbf{[7.49]}$$

where p_0 is the stagnation pressure p_∞ is the free-stream static pressure, and M_∞ is the free-stream Mach number given by

$$M_\infty = \frac{u_\infty}{a} \qquad \textbf{[7.50]}$$

a is the acoustic velocity and may be calculated with

$$a = \sqrt{\gamma g_c RT} \qquad \text{[7.51]}$$

for an ideal gas. It is convenient to express Eq. (7.49) in terms of dynamic pressure q defined by

$$q = \tfrac{1}{2}\rho u_\infty^2 = \tfrac{1}{2}\gamma p M_\infty^2 \qquad \text{[7.52]}$$

Equation (7.49) thus becomes

$$p_0 - p_\infty = \frac{2q}{\gamma M_\infty^2}\left[\left(1 + \frac{\gamma - 1}{2}M_\infty^2\right)^{\gamma/(\gamma-1)} - 1\right] \qquad \text{[7.53]}$$

This relation may be simplified to

$$p_0 - p_\infty = \frac{2q}{\gamma M_\infty^2}\left(1 + \frac{M_\infty^2}{4} + \frac{2 - \gamma}{24}M_\infty^4 + \cdots\right) \qquad \text{[7.54]}$$

when

$$M_\infty^2\left(\frac{\gamma - 1}{2}\right) < 1.$$

For very small Mach numbers Eq. (7.54) reduces to the familiar incompressible flow formula

$$p_0 - p_\infty = \tfrac{1}{2}\rho u_\infty^2 \qquad \text{[7.55]}$$

We thus observe that a measurement of static and stagnation pressures permits a determination of the flow velocity by either Eq. (7.55) or Eq. (7.53), depending on the fluid medium.

A basic total pressure probe may be constructed in several different ways, as shown in Fig. 7.38. In each instance the opening in the probe is oriented in a direction exactly parallel to the flow when a measurement of the total stream pressure is desired. If the probe is inclined at an angle θ to the free-stream velocity, a somewhat lower pressure will be observed. This reduction in pressure is indicated in Fig. 7.38 according to Ref. [7]. Configuration a represents an open-ended tube placed in the flow. Configuration b is called a shielded probe and consists of a venturi-shaped tube placed in the flow with an open-ended tube at the throat of the section to sense the stagnation pressure. It may be noted that this probe is rather insensitive to flow direction. Configuration c represents an open-ended tube with a chamfered opening. The chamfer is about 15°, and the ratio of OD to ID of the tube is about 5. Configuration d represents a tube having a small hole drilled in its side, which is placed normal to the flow direction. This type of probe, as might be expected, is the most sensitive to changes in a yaw angle. Also indicated in Fig. 7.38 is a portion of the curve for a Kiel probe, which is similar in construction to configuration b, except that a smoother venturi shape is used, as shown in Fig. 7.39. The Kiel probe is the least sensitive to the yaw angle.

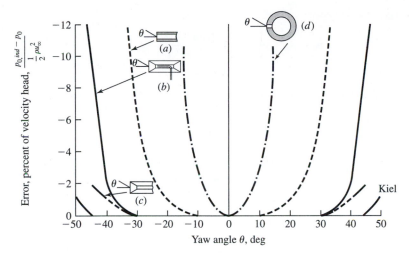

Figure 7.38 Stagnation pressure response of various probes to changes in yaw angle. (a) Open-ended tube; (b) channel tube; (c) chamfered tube; (d) tube with orifice in side. *(From Ref. [7].)*

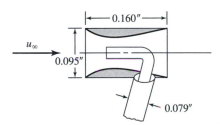

Figure 7.39 Kiel probe (Model 3696) for measurement of stagnation pressure. *(Courtesy of Airflow Instrument Co., Glastonbury, CT.)*

INFLUENCE OF YAW ANGLE. An open-ended tube probe is yawed at an angle of 30° from the flow direction in an airstream at 12 psia and 40°F having a free-stream velocity of 80 ft/s. Calculate the pressure indicated by the probe.

Example 7.9

Solution

We consult Fig. 7.40 for an open-ended probe at a yaw angle of 30° and obtain

$$\frac{p_{0,\text{ind}} - p_0}{\frac{1}{2}\rho u_\infty^2} = -0.06 \qquad \textbf{[a]}$$

We have $p_\infty = 12$ psia and $T_\infty = 40°\text{F} = 500°\text{R}$, so the density is calculated as

$$\rho = \frac{p}{RT} = \frac{(12)(144)}{(53.35)(500)} = 0.0648 \text{ lbm/ft}^3$$

From Eq. (7.55)

$$p_0 - p_\infty = \frac{1}{2}\rho u_\infty^2 = \frac{(0.0648)(80)^2}{(2)(32.17)}$$ **[b]**

$$= 6.446 \text{ psf}$$

Note that $g_c = 32.17$ lbm · ft/lbf · s^2 has been inserted to balance the units. Using Eq. (a), we obtain

$$p_{0,\text{ind}} - p_0 = (-0.06)(6.446) \text{ psf}$$ **[c]**

Adding (b) and (c), we have

$$p_{0,\text{ind}} - p_\infty = (0.94)(6.446) = 6.059 \text{ psf}$$

and

$$p_{0,\text{ind}} = 12 + \frac{6.059}{144} = 12.042 \text{ psia}$$

The measurement of static pressure in a flowstream is considerably more difficult than the measurement of stagnation pressure. A typical probe used for the measurement of both static and stagnation pressures is the *Pitot tube* shown in Fig. 7.40. The opening in the front of the probe senses the stagnation pressure, while the small holes around the outer periphery of the tube sense the static pressure. The static-pressure measurement with such a device is strongly dependent on the distance of the peripheral openings from the front opening as well as on the yaw angle. Figure 7.41 indicates the dependence of the static-pressure indication on the distance from the leading edge of the probe for both blunt subsonic and conical supersonic configurations. To alleviate this condition, the static-pressure holes are normally placed at least eight diameters downstream from the front of the probe. The dependence of the static and stagnation pressures on yaw angle for a conventional Pitot tube is indicated in Fig. 7.42. This device is quite sensitive to flow direction. The probe in Fig. 7.40 is sometimes called a *Pitot static tube* because it measures both static and stagnation pressure.

The static-pressure characteristics of three types of probes are shown in Figs. 7.43 and 7.44 as functions of Mach number and yaw angle. It may be noted that both the

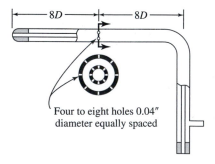

Figure 7.40   Schematic drawing of Pitot tube.

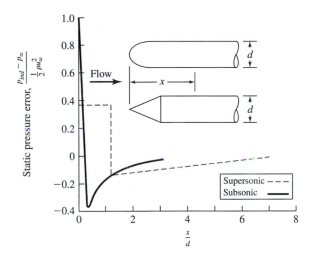

Figure 7.41   Variation of static pressure along standard subsonic and supersonic probe types. (*From Ref. [4].*)

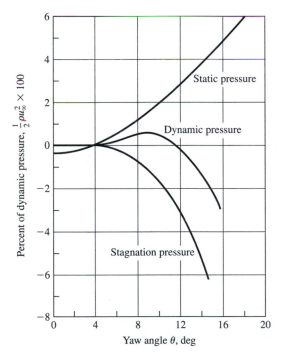

Figure 7.42   Variation of static, stagnation, and dynamic pressure with yaw angle for Pitot tube. (*Courtesy of Airflo Instrument Corp., Glastonbury, CT.*)

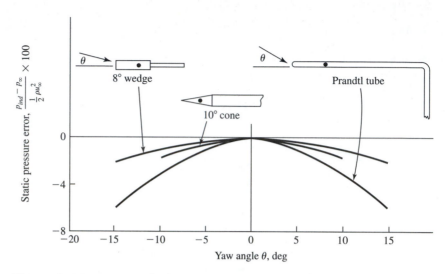

Figure 7.43 Yaw-angle characteristics of various static-pressure probes. (*From Ref. [7].*)

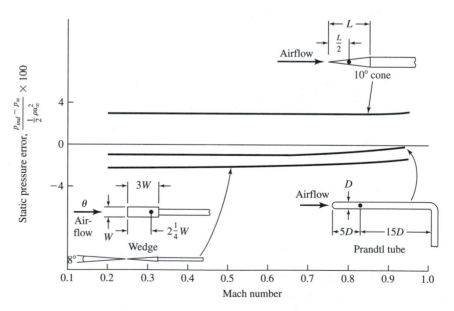

Figure 7.44 Mach-number characteristics of various static-pressure probes. (*From Ref. [7].*)

wedge and Prandtl tube indicate static-pressure values that are too low, while the cone indicates a value that is too high. The wedge is least sensitive to yaw angle. All three probes have two static-pressure holes located 180° apart.

VELOCITY MEASUREMENT WITH PITOT TUBE. A Pitot tube is inserted in a flow | **Example 7.10**
stream of air at 30°C and 1.0 atm. The dynamic pressure is measured as 1.12 in water when the tube is oriented parallel to the flow. Calculate the flow velocity at that point.

Solution

We use Eq. (7.55) for this calculation. The air density is calculated as

$$\rho = \frac{p_\infty}{RT_\infty} = \frac{1.0132 \times 10^5}{(287)(303)} = 1.165 \, \text{kg/m}^3$$

We also have

$$p_0 - p_\infty = 1.12 \, \text{in water} = 5.82 \, \text{psf} = 278.7 \, \text{Pa}$$

so that the velocity is

$$u_\infty = \sqrt{\frac{2(p_0 - p_\infty)}{\rho}} = \left[\frac{(2)(278.7)}{1.165}\right]^{1/2}$$

$$= 21.9 \, \text{m/s} \, (71.8 \, \text{ft/s})$$

DETERMINATION OF VOLUME FLOW RATE USING PITOT TUBE. A Pitot tube | **Example 7.11**
is used to measure the dynamic pressure $p_0 - p_\infty$ for air in a 4-in-diameter duct. The air conditions are 14.7 psia and 70°F (530°R). The readings are shown in the accompanying table for five radial positions measured from the center of the duct. No measurement is taken at the outside radius where the velocity is assumed zero. For the first reading at the center of the duct the velocity is calculated from

$$p_0 - p_\infty = \frac{\rho u^2}{2g_c} \qquad \textbf{[a]}$$

with the density of air calculated from

$$\rho = \frac{p}{RT} = (14.7)(144)/(53.35)(530) = 0.0748 \, \text{lbm/ft}^3 \qquad \textbf{[b]}$$

Then, with $p_0 - p_\infty = 2.87 \, \text{inH}_2\text{O} = 14.92 \, \text{lbf/ft}^2$,

$$u = \left[\frac{(2)(32.17)(14.92)}{0.748}\right]^{1/2} = 113 \, \text{ft/s} \qquad \textbf{[c]}$$

Velocities at the other radial positions are calculated in a similar manner and displayed in the table.

We want to calculate the volume rate of flow in the duct using this information.

Table Example 7.11

Index, i	r_i, in	$p_0 - p_\infty$, inH$_2$O	u_i, ft/s	r_m, in	ΔA_i, ft^2	$u_i \Delta A_i$, ft^3/s
0	0	2.87	113		0.00136	0.15
				0.25		
1	0.5	2.72	110		0.0109	1.20
				0.75		
2	1.0	2.41	104		0.0218	2.27
				1.25		
3	1.5	1.90	92		0.0235	2.16
				1.625		
4	1.75	1.60	84		0.014	1.18
				1.813		
5	1.875	1.30	76		0.0102	0.78
				1.9375		
6	2.0	0	0		0.0054	0

Solution

A numerical approximation for the volume flow may be written as

$$Q = \int 2\pi r u \, dr \approx \sum u_i \Delta A_i \qquad \text{[d]}$$

where the u_i are the calculated values at the respective r_i positions indicated in the table. An arithmetic mean radius r_m between each set of r_i's is also shown. We assume that the flow area ΔA_i for each r_i is

$$\Delta A_i = \pi \left(r_{m+1}^2 - r_{m-1}^2 \right)$$

or, for $r_i = 1.0$,

$$\Delta A_i = \frac{\pi(1.25^2 - 0.75^2)}{144} = 0.0218 \text{ ft}^2$$

At $r_i = 0$ the area is assumed to be $\pi(0.25)^2/144 = 0.00136$ ft^2 while that at $r_i = 2$ is $\pi(1.9375)^2/144 = 0.0054$ ft^2.

Performing the summation indicated in (d), we obtain

$$Q = 7.74 \text{ ft}^3/\text{s}$$

The area of the duct is

$$A_0 = \frac{\pi(2)^2}{144} = 0.0873 \text{ ft}^2$$

so the mean flow velocity is calculated as

$$u_{\text{mean}} = \frac{7.74}{0.0873} = 89 \text{ ft/s}$$

Comment

This calculation illustrates how a rather simple set of measurements may be used to estimate volumetric flow rate in a duct. The uncertainty in the final result will be a strong function of the uncertainties in the readings of $p_0 - p_\infty$. Since these readings will likely be taken in a time sequence, they are subject to flow variations over the total measurement time period. These fluctuations may be considerable.

7.16   IMPACT PRESSURE IN SUPERSONIC FLOW

Consider the impact probe shown in Fig. 7.45 which is exposed to a free stream with supersonic flow; that is, $M_1 > 1$. A shock wave will be formed in front of the probe as shown, and the total pressure measured by the probe will not be the free-stream total pressure before the shock wave. It is possible, however, to express the impact pressure at the probe in terms of the free-stream static pressure and the free-stream Mach number. The resulting expression as given in Refs. [10 and 51] is

$$\frac{p_\infty}{p_{0_2}} = \frac{\{[2\gamma/(\gamma+1)]M_1^2 - (\gamma-1)/(\gamma+1)\}^{1/(\gamma-1)}}{\{[(\gamma+1)/2]M_1^2\}^{\gamma/(\gamma-1)}} \qquad \textbf{[7.56]}$$

where p_∞ is the free-stream static pressure and P_{0_2} is the measured impact pressure behind the normal shock wave. This equation is valid for Reynolds numbers based on

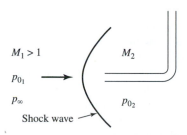

$M_1 > 1$

p_{0_1}

p_∞

Shock wave

M_2

p_{0_2}

Figure 7.45   Impact tube in supersonic flow.

probe diameter greater than 400. Equation (7.56) is called the *Rayleigh supersonic Pitot formula*. We see that in order to determine the value of the Mach number we must make a measurement of the free-stream static pressure. It is possible to make this measurement with special calibrated probes.

| **Example 7.12** | **IMPACT TUBE IN SUPERSONIC FLOW.**   A probe like that shown in Fig. 7.45 is placed in an airflow stream where $M_1 = 2.5$ and the free-stream static pressure is measured as 22 kPa. Determine the stagnation pressure before and after the shock wave. |

Solution

We employ Eq. (7.56) for this calculation of p_{0_2} with $M_1 = 2.5$ and $\gamma = 1.4$ for air. Thus,

$$\frac{p_\infty}{p_{0_2}} = \frac{\{[(2)(1.4)/(2.4)](2.5)^2 - (0.4)/(2.4)\}^{1/0.4}}{\{[(2.4)(2)](2.5)^2\}^{1.4/0.4}}$$

$$= 0.1173$$

and

$$p_{0_2} = \frac{22}{0.1173} = 187.6 \, \text{kPa}$$

The stagnation pressure upstream of the shock wave is calculated from Eq. (7.49):

$$\frac{p_{0_1}}{p_\infty} = \left(1 + \frac{\gamma - 1}{2} M_1^2\right)^{\gamma/(\gamma-1)}$$

$$= [1 + (0.2)(2.5)^2]^{1.4/0.4}$$

$$= 17.086$$

and

$$p_{0_1} = (22)(17.086) = 375.9 \, \text{kPa}$$

7.17   SUMMARY

Comparisons of the operating range, characteristics, and advantages of several flowmeters are presented in Table 7.2 and may be taken as a summary of our discussions in this chapter. The reader should bear in mind that the stated accuracies for the various flowmeters may be improved with suitable calibration. Of course, the overall accuracy of a flow-rate determination is also dependent on the accuracy of the readout equipment. As an example, a venturi might be carefully calibrated within 0.5 percent, but if it were used with a crude instrument for differential pressure measurements, a much poorer precision would result than that for which it was calibrated.

Table 7.2   Operating characteristics of several types of flowmeters

Meter Type		Maximum Flow Range $cm^3/s \times 10^{-3}$	Useful Range Ability	Maximum Pressure, Mpa	Temperature Range, °C	Maximum Viscosity, cSt $(10^{-6}\,m^2/s)$
Obstruction meters	Orifice	Liquid 0.012–220 Gas 23–43,000	5:1	42	−270–1100	4000
	Flow nozzle	Liquid 0.031–950 Gas 45–240,000	5:1	10	−50–800	4000
	Venturi	Liquid 0.03–950 Gas 45–240,000	5:1	10	−50–800	4000
Drag effects	Glass-tube rotameter	Liquid 0.1×10^{-3}–16 Gas 0.006–330	10:1	2	−45–200	100
	Metal-tube rotameter	Liquid 0.03–250 Gas 0.2–470	10:1	35	−180–870	100
	Turbine or propeller	Liquid 0.2×10^{-3}–3100 Gas 0.2–140,000	15:1 10:1	105	−270–540	30
		Clean liquid 0.02–63	10:1	1.0	−15–150	2000
Magnetic flowmeter		Conductive liquid only 0.1×10^{-3}–3100	20:1	4	−130–180	1000

7.18   REVIEW QUESTIONS

7.1. What basic methods are used for calibration of flow-measurement devices?

7.2. What is meant by a positive-displacement flowmeter?

7.3. What are the relative advantages of the venturi, orifice, and flow-nozzle meters?

7.4. What type of flow-measurement accuracy would you expect with an orifice; a venturi?

	Application Factors			Installation Factors				
Scale	**Indication**	**Standard Accuracy, %**	**Construction Materials**	**Pressure Loss, kPa**	**Line Size, cm**	**Special Piping Considerations**	**Typical Power Requirement**	**Relative Cost**
Square root rate	Remote differential pressure	1–2 full-scale differential pressure	Most metals	0.7–200	4–30	Straight pipe: 10 diameters upstream, 3 downstream	None	1.0
Square root rate	Remote differential pressure	1–2 full-scale differential pressure	Bronze, iron, steel, stainless steel	0.7–140	2–60	Straight pipe: 10 diameters upstream, 3 downstream	None	1.4
Square root rate	Remote differential pressure	1–2 full-scale differential pressure	Bronze, iron, steel, stainless steel, plastic	0.7–100	2–60	Straight pipe: 10 diameters upstream, 3 downstream	None	1.5
Linear rate	Local or remote electric or pneumatic	1–2 full-scale	Most metals, plastics, and ceramics	0.05–7	0.6–10	Vertical only	None	1.0
Linear rate	Local or remote electric or pneumatic	1–2 full-scale	Most metals, plastics, and ceramics	0.3–70	1.2–30	Vertical only	None	1.2
Linear total	Remote electric	0.5 reading	Aluminum, stainless steel	14–70	0.3–90	Any position	115 V, 60 Hz, 50 W	5.0
Linear total	Local or remote electric	1 reading	Bronze, iron, steel, stainless steel	7–140	1.2–15	Horizontal recommended	None	1.0
Linear rate	Remote electric	0.5–1 full-scale	Plastic, stainless steel with nonconductive liner	Very small	0.2–200	Any position	115 V, 60 Hz, 200 W	6.0

7.5. What is a sonic nozzle? How is it used? What are its advantages and disadvantages?

7.6. Why is a rotameter called a drag meter? Could it also be called an area meter?

7.7. What is the utility of the hot-wire anemometer?

7.8. Distinguish among the shadowgraph, schlieren, and interferometer flow-visualization techniques. What basic flow variable is measured in each technique?

7.9. Upon what does the sensitivity of the schlieren depend?

7.10. What is the primary advantage of the laser anemometer?

7.11. What is a Pitot tube?

7.12. Why must the tube be tapered in order for a rotameter to indicate flow rate?

7.13. What particular flow-measurement situations are adapted to the hot-wire anemometer?

7.14. How would one go about calibrating a venturi for liquid measurement? Could the calibration data be adapted to gas flow? If so, how?

7.19   PROBLEMS

7.1. Using Eq. (7.8) as a starting point, obtain Eqs. (7.9) and (7.10). Subsequently, calculate the error in these equations when $\Delta p = p_1/4$ for Eq. (7.9) and $\Delta p = p_1/10$ for Eq. (7.10).

7.2. In the compressible flow equation for a venturi the velocity of approach factor M cancels with a like term in the expansion factor Y_a. Construct a graphical plot which will indicate the error that would result if the approach factor given by Eq. (7.11) were used in conjunction with Eq. (7.8). Use the parameter $\Delta p/\gamma p_1$ as the abscissa for this plot.

7.3. A venturi is to be used for measuring the flow of air at 300 psia and 80°F. The maximum flow rate is 1.0 lbm/s. The minimum flow rate is 30 percent of this value. Determine the size of the venturi such that the throat Reynolds number is not less than 10^5. Calculate the differential pressure across the venturi for mass flows of 0.3, 0.5, 0.7, and 1.0 lbm/s. Assume $\beta = 0.5$ for the venturi.

7.4. Work Prob. 7.3 for an orifice with pressure taps one diameter upstream and one-half diameter downstream.

7.5. Work Prob. 7.3. for an ASME long-radius flow nozzle.

7.6. A sonic nozzle is to be used to measure a gas flow with an uncertainty of 1 percent. Assuming that the nozzle throat area and discharge coefficients are known exactly, derive an expression for the required relationship between the uncertainties in the stagnation pressure and temperature measurements. Which of the two measurements, pressure or temperature, is more likely to be the controlling factor?

7.7. Calculate the throat area for the sonic nozzle in Example 7.4 if the stated pressure is static pressure upstream, the temperature is total temperature upstream, and the pipe diameter is 1.0 in.

7.8. Show that the parameter that governs the use of a linear relation for a rotameter as in Eq. (7.30) is

$$\frac{ay}{D}$$

provided that $d \approx D$ to the extent that

$$1 - \left(\frac{d}{D}\right)^2 \ll \frac{ay}{D}$$

Under these restrictions, plot the error resulting from the linear approximation as a function of the parameter ay/D. Discuss the physical significance of this analysis and interpret its meaning in terms of specific design recommendations for rotameters.

7.9. Verify that the error in flow rate for a rotameter calculated from Eq. (7.32) is less than 0.2 percent for density variations of ±5 percent when the float density is designated according to Eq. (7.31).

7.10. The turbine flowmeter whose calibration is shown in Fig. 7.18 is to measure a nominal flow rate of water of 2.5 gpm at 60°F. A single value of the meter constant K is to be used in the data reduction. What deviations from the nominal flow rate are allowable in order that the nominal value of K is accurate within ±0.25 percent?

7.11. A rotameter is to be designed to measure a maximum flow of 10 gpm of water at 70°F. The bob has a 1-in diameter and a total volume of 1 in³. The bob is constructed so that the density is given in accordance with Eq. (7.31). The total length of the rotameter tube is 13 in and the diameter of the tube at inlet is 1.0 in. Determine the tube taper for drag coefficients of 0.4, 0.8, and 1.20. Plot the flow rate versus distance from the entrance of the tube for each of these drag coefficients. Determine the meter constant for use in Eq. (7.32) and estimate the error resulting from the use of this relation instead of the exact expression in Eq. (7.28).

7.12. A rotameter is to be used for measurement of the flow of air at 100 psia and 70°F. The maximum flow rate is 0.03 lbm/s, the inlet diameter of the meter is 1 in, and the length of the meter is 12 in. The bob is constructed so that its density is 5 times that of the air and its volume is 1 in³. Calculate the tube taper for drag coefficients of 0.4, 0.8, and 1.2 and determine the meter constant for use in Eq. (7.30). Plot the error resulting from the use of Eq. (7.30) as a function of flow rate.

7.13. Derive an expression for the product of density and velocity across a hot-wire anemometer in terms of the wire resistance, the current through the wire, and the empirical constants a and b. Subsequently, obtain an expression for the uncertainty in this product as a function of the measured quantities.

7.14. The sensitivity of a schlieren system is defined as the fractional deflection obtained at the knife edge per unit angular deflection of a light ray at the test section. Show that this sensitivity may be calculated with

$$S = \frac{f_2}{y_1}$$

Derive an expression for the contrast in terms of the sensitivity, the density gradient, and the test section width.

7.15. Show that the sensitivity of an interferometer, defined as the number of fringe shifts per unit change of density, may be written as

$$S = \frac{\beta L}{\lambda \rho_s}$$

Show that the maximum sensitivity of an interferometer, defined as the number of fringe shifts per unit change in Mach number, will be about 30 when $L = 15$ cm, $\lambda = 5400$ Å, $\beta = 0.000293$, and the stagnation density is that of air at standard conditions.

7.16. Calculate the temperatures corresponding to the four fringes nearest to the plate surface in Example 7.6.

7.17. The velocity in turbulent tube flows varies approximately as

$$\frac{u}{u_c} = \left(1 - \frac{r}{r_0}\right)^{1/7}$$

where u_c is the velocity at the center of the tube and r_0 is the tube radius. An experimental setup using air in a 30-cm-diameter tube is used to check this relation. The air temperature is 20°C and pressure is 1.0 atm. The maximum flow velocity is 15 m/s and a Pitot tube is used to traverse the flow and to obtain a measurement of the velocity distribution. Measurements are taken at radii of 0, 5, 10, and 12 cm, and the uncertainty in the dynamic-pressure measurement is ±5 Pa. Using the above relation and Eq. (7.55), calculate the nominal velocity and dynamic pressure at each radial location. Then, calculate the uncertainty in the velocity measurement at each location based on the uncertainty in the pressure measurement. Assuming that the above relation does represent the true velocity profile, calculate the uncertainty which could result in a mass flow determined from the experimental data. The mass flow would be obtained by performing the integration

$$\dot{m} = \int_0^{r_0} 2\pi r \rho u \, dr$$

7.18. Calculate the dynamic pressure measured by a Pitot tube in a flow stream of water at 20°C moving at a velocity of 3 m/s.

7.19. Using the supersonic Pitot-tube formula given by Eq. (7.56), obtain an expression for the uncertainty in the Mach number as a function of the percent uncertainty in the pressure ratio (p_∞/p_{0_2}) of 1 percent. Assume $\gamma = 1.4$.

7.20. A venturi with throat and upstream diameters of 8 and 16 in is used to measure the flow of water at 70°F. The flow rate is controlled by a motorized valve downstream from the venturi. The valve is operated so that a constant differential pressure of 12 inHg is maintained across the venturi. Suppose someone informs you that this type of control scheme is not very effective because it does not account for possible changes in temperature of the water. Reply to this criticism by plotting the error in the flow rate as a function of water temperature, taking the flow at 70°F as the reference value.

7.21. An obstruction meter is used for the measurement of the flow of moist air at low velocities. Suppose that the flow rate is calculated taking the density as that of dry air at 90°F. Plot the error in this flow rate as a function of relative humidity.

7.22. A small venturi is used for measuring the flow of water in a $\frac{1}{2}$-in-diameter line. The venturi has a throat diameter of $\frac{1}{4}$ in and is constructed according to ASME specifications. What minimum flow rate at 70°F should be used in order that the discharge coefficient remains in the flat portion of the curve? What pressure reading, in inches of mercury, would be experienced for this flow rate?

7.23. The venturi of Prob. 7.22 is used to measure the flow of air at 100 psia and 120°F (upstream flow conditions). The throat pressure is measured as 80 psia. Calculate the flow rate.

7.24. An orifice is to be used to indicate the flow rate of water in a 1.25-in-diameter line. The orifice diameter is 0.50 in. What pressure reading will be experienced on the orifice for a line-flow velocity of 10 ft/s? What would be the flow rate for a pressure reading of twice this value?

7.25. A small sonic nozzle having a throat diameter of 0.8 mm is used to measure and regulate the flow in a 7.5-cm-diameter line. The upstream pressure on the nozzle is varied in accordance with the demand requirements. The downstream pressure is always low enough to ensure sonic flow at the throat. What is the flow rate for upstream conditions at 20°C and 1.0 MPa?

7.26. An impact tube is used to measure the Mach number of a certain airflow in a wind tunnel. The static pressure is 3 psia, and the impact pressure at the probe is 116 kPa. What is the Mach number of the flow ($\gamma = 1.4$ for air)? If the free-stream air temperature is −40°C, what is the flow velocity?

7.27. A Pitot tube is used to measure the velocity of an airstream at 20°C and 1.0 atm. If the velocity is 2.5 m/s, what is the dynamic pressure in newtons per square meter? What is the uncertainty of the velocity measurement if the dynamic pressure is measured with a manometer having an uncertainty of 5 Pa?

7.28. A venturi is used to measure the flow of liquid Freon-12 at 20°C. The throat diameter is 1.2 cm and the inlet diameter is 2.4 cm. Calculate the pressure drop reading if the throat Reynolds number is 10^5.

7.29. Repeat Prob. 7.28 for a concentric orifice with the same diameter ratio.

7.30. A flow rate of 1 kg/s of air at 30 atm and 20°C is to be measured with a venturi and an orifice. Select appropriate size devices and specify suitable pressure instrumentation for each.

7.31. The flow rate in Prob. 7.30 is to be measured with a sonic nozzle. Calculate the exit diameter of the nozzle.

7.32. A rotameter is to be used to measure the flow rate of liquid Freon-12 at 20°C. You have at your disposal several specially constructed bobs of varying densities. What density would you select?

7.33. A test is to be conducted of the fuel economy of a certain automobile. The test is to be conducted under road conditions at varying speeds. The manufacturer states that the car will get 8.0 km/liter of fuel at a steady speed of 80 km/h. Select appropriate flow-measurement device(s) and specify the necessary pressure and temperature-measurement devices for a speed range of 30 to 90 km/h. Analyze your choice and estimate the uncertainties which may result in the final results of fuel economy. Give alternate suggestions for the measurements and show what result they would have on the uncertainty.

7.34. Water at 60°C flows in a 7.5-cm-diameter pipe at a mean flow velocity of 8 m/s. Calculate the flow rate in units of kg/s, lbm/min, gal/min, and liters/s.

7.35. Repeat Prob. 7.34 for liquid ammonia at 20°C.

7.36. Air at 400 kPa and 40°C flows in a circular tube having a diameter of 5.0 cm at a rate such that the Reynolds number is 50,000. Calculate the flow rate in units of kg/s, lbm/min, SCFM, cm^3/s, and SCCM.

7.37. A venturi is to be used to measure the airflow in Prob. 7.36. What size venturi would you use, and what would be its discharge coefficient and pressure differential?

7.38. A sonic flow nozzle is used to measure a nitrogen flow of 0.5 kg/s at conditions of 1.0 MPa and 100°C. What area should be used? What upstream diameter pipe would you recommend?

7.39. An open-ended stagnation pressure probe is inclined at a yaw angle of 20° with the flow velocity. For airflow at 20 m/s, 1 atm, and 20°C, what would be the indicated pressure?

7.40. A Pitot tube with a yaw angle of 10° is used for the airflow in Prob. 7.39. What would be the indicated stagnation and dynamic pressures? Express in units of Pa, lbf/in^2, and inH_2O.

7.41. Water flows in a 5-in-diameter pipe at 20°C with a mean flow velocity of 3 m/s. A sharp-edged orifice with a diameter of 2.5 in is to be used to measure the flow rate. What pressure differential would be indicated for a standard ASME installation? Express in units of Pa, inH_2O, and psia.

7.42. An impact tube is used with an airflow at 10°C, 40 kPa, and a velocity of 700 m/s. What pressure will be indicated by the probe?

7.43. A vortex-shedding flowmeter has a characteristic dimension of 3 mm and is used to measure velocity in a region where the Reynolds number is 10^5. If a velocity of 4 m/s is to be measured, what shedding frequency may be anticipated?

7.44. A venturi is constructed according to the specifications of Fig. 7.6 with throat and upstream diameters of 1.0 and 0.5 in. What flow rates of water will be measured when the performance of the meter lies in the flat portion of the curve for discharge coefficient in Fig. 7.10? What will be the differential pressure for a throat Reynolds number of 10^6?

7.45. An open-ended tube probe is oriented at an angle of 30° with the flow direction in an airflow at 200 kPa and 50°C. The air velocity is 20 m/s. Calculate the pressure indicated by the probe.

7.46. A Pitot tube is exposed to the same flow stream as in Prob. 7.45 but is yawed at an angle of 8° with the flow direction. What static, dynamic, and stagnation pressures will be indicated by the probe? What error would result in the velocity determination if these values were assumed to be at a zero yaw angle?

7.47. An impact probe is exposed to an airflow stream at 0°C, 20 kPa, and 600 m/s. What pressure will be indicated by the probe at a zero yaw angle?

7.48. The venturi of Prob. 7.44 is calibrated by direct weighing of a quantity of water over a 3-min interval. The time of weighing is accurate within +0.2 s and the weight is accurate within 0.1 kg. For a throat Reynolds number of 10^6 with water at 25°C, determine the allowable uncertainty of the differential pressure measurement such that the uncertainty in the discharge coefficient is +0.75 percent. Assume a nominal discharge coefficient in accordance with Fig. 7.10.

7.49. A positive-displacement meter is used to measure the flow of methane (CH_4) at 20°F and 0.95 atm. Temperature and pressure transducers are installed that feed an electronic circuit, which then indicates the flow in standard cubic feet per minute (ft³/min) for billing a customer. The meter itself senses actual volume flow. What factor must be multiplied by the actual volume flow to give SCFM?

7.50. If the uncertainties in temperature and pressure for Prob. 7.49 are ±1.5°F and ±2 kPa, respectively, what is the uncertainty of the billing if the actual volume flow measurement is exact?

7.51. A sharp-edge orifice is used to measure the flow of water at 25°C in a 1.5-in-diameter tube. The orifice diameter is 0.75 in. Pressure taps are 1 diameter upstream and $\frac{1}{2}$ diameter downstream. What pressure differential will be indicated for an upstream Reynolds number of 10,000?

7.52. A Pitot tube uses a manometer readout with a fluid having a specific gravity of 0.82. The tube is oriented parallel to an airstream at 1 atm, 25°C, having a velocity of 30 m/s. The accuracy of the dynamic pressure measurement is ±0.1 mm of the height of the manometer fluid. The static pressure is assumed exact, but the static temperature has an uncertainty of ±1.2°C. Assuming that the nominal value of the velocity is the value of 30 m/s, calculate the uncertainty in its determination.

7.53. A 10°-cone probe like that shown in Fig. 7.44 is used to measure the static pressure in a flow at −40°C and $M = 0.8$. The pressure indicated by the probe is 22 kPa. What is the true free-stream pressure?

7.54. What pressure would be indicated by an open-ended tube that is oriented parallel with the flow in Prob. 7.53?

7.55. A positive-displacement flowmeter is used to measure the flow of liquid Freon-12 at 20°C. A flow rate of 2.35 gal/min is indicated. What is the mass flow in kg/s?

7.56. A sonic nozzle is to be used to measure the mass flow of nitrogen at low-velocity upstream conditions of 50°C and 800 kPa. The discharge coefficient is 0.97 ± 0.5 percent. The upstream pressure is measured with a transducer having an upper range of 1 MPa and an uncertainty of ±1 percent of full scale. The upstream temperature is measured with a thermocouple having an uncertainty of ±1°C. The nozzle has a throat diameter of 2 cm and discharges into a large chamber maintained at a pressure of 100 kPa. Calculate the flow rate of nitrogen in kg/s for these conditions and its uncertainty.

7.57. Repeat Prob. 7.56 for upstream pressures of 220 and 400 kPa.

7.58. A Pitot tube is used to measure an airflow at 15 psia, 120°F, and 100 ft/s. If it is yawed at an angle of 10° with the flow direction, what static, dynamic, and stagnation pressures will be indicated? If these values are used for a velocity determination assuming a zero yaw angle, what percent error would result?

7.59. What pressure ratio would be indicated by Eq. (7.56) for $M_1 = 1.0$? Is this what would be expected?

7.60. An orifice is installed according to ASME specifications in a 5.0-cm-diameter pipe. The diameter of the orifice is 2.5 cm and the pressure taps are one diameter upstream and one-half diameter downstream. A manometer is used to measure the orifice pressure differential as 1932 mmH$_2$O and the upstream pressure is measured as 400 kPa gage pressure with the local barometer reading as 750 mmHg. The fluid is air at a temperature of 27°C. Calculate the airflow rate in kg/s.

7.61. The orifice in Prob. 7.60 is used to measure the same airstream and the differential pressure reading is observed as 190 mm with all other readings the same. What is the airflow under these conditions?

7.62. Suppose the flow system in Probs. 7.60 and 7.61 is fluctuating such that the differential pressure varies over a range of ±12 mmH$_2$O while the inlet pressure fluctuates by ±10 kPa. The temperature measurement is within ±2°C. Calculate the resultant uncertainties in the flow rates expressed as a percent of the calculated values.

7.63. The airflow system in Prob. 7.60 is measured with a venturi having like dimensions as the orifice, i.e., the same pipe diameter and throat diameter. Calculate the flow rate in kg/s for the same pressure and temperature measurements as in Prob. 7.60.

7.64. Repeat Prob. 7.63 for the conditions of Prob. 7.60.

7.65. Estimate the permanent pressure loss for the systems in Probs. 7.60, 7.61, 7.63, and 7.64.

7.66. Calculate the percent uncertainty in flow rate for the venturi measurements in Prob. 7.63 using the same uncertainties in the pressure and temperature measurements as in Prob. 7.62.

7.67. The uncertainty of the dynamic pressure measurement in Example 7.11 is estimated as ±0.03 inH$_2$O and the uncertainty of the radial position determination is estimated as ±0.03 in. Calculate the resultant percent uncertainty in the volumetric flow rate calculation for the flow element represented by index $i = 3$.

7.68. Repeat Prob. 7.67 for index $i = 4$. What do you conclude?

7.69. The calibration curve for the turbine flowmeter shown in Fig. 7.18 indicates linear response within ±0.5 percent over a rather wide range. A proposal is made to improve the calibration and eventual flow readout by fitting a polynomial relation to the calibration points and then inserting the resulting expression as a software adjustment for the computer readout. Using data correlation software available to you, obtain second-, third-, and fourth-degree polynomial fits to the data points shown and state recommendations for use.

7.70. Obtain a second-degree polynomial relationship between the orifice flow coefficient MC and Reynolds number for $\beta = 0.5$.

7.71. Fit a second-degree polynomial for the discharge coefficient C of Fig. 7.10 as a function of Reynolds number for $1/2 \times 1/4$ and $1 \times 1/2$ in venturis. Perform the fit for $5000 < \text{Re} < 100{,}000$. Use available computer software to perform a least-square calculation. Also calculate the value of r^2 for both relations.

7.72. A laminar flowmeter is to be designed for use with air and the Reynolds number in the flow channels is to be set at $\text{Re} = 100$. The permanent pressure drop is to be the same as for the orifice of Probs. 7.60 and 7.61 for the corresponding mass flow rates calculated in these problems. Obtain a relationship between L and d for the laminar flowmeter for each of the flow rates.

7.73. Is it possible to accommodate both flow rates in Prob. 7.72 with one L/d relation? What conditions would be necessary for such a design?

7.74. A rotameter is used for an airflow measurement at conditions of 10°C and 400 kPa gage pressure. The local barometer reads 750 mmHg and the rotameter is rated at 100 liters/min (full scale) at standard conditions of 1 atm and 20°C. Calculate the mass flow of air for a reading of 50 percent of full scale.

7.75. A turbine flowmeter having a calibration curve that matches Fig. 7.18 is placed in operation to measure a nominal flow rate of water of 3.0 gpm at 70°F. Determine the allowable deviations from the nominal flow rate such that the value of the flow constant K deviates ±0.4 percent.

7.76. Calculate the dynamic pressure measured by a Pitot tube in an airflow at 2 atm, 20°C with a flow velocity of 10 m/s.

7.77. The small venturi of Prob. 7.22 is used to measure the flow rate of air at 10 atm and 40°C. If the throat pressure is measured as 800 kPa, calculate the flow rate in kg/s.

7.78. A rotameter is available with several bobs of variable densities. If the meter will be used for measuring the flow of liquid ammonia at 20°C, what density of bob would you select?

7.79. An airflow of about 1.5 kg/s at 20 atm and 50°C (stagnation conditions) is to be measured using a sonic nozzle. What exit diameter should be designed for the nozzle?

7.80. Air at 500 kPa and 50°C flows in a circular tube having a diameter of 5.0 cm. The Reynolds number for the flow is 50,000. Suppose a venturi is to be used to measure the flow rate. Determine the size of the venturi and the differential pressure reading that may be expected.

7.81. The characteristic dimension for a vortex shedding flowmeter is 4.0 mm in a region of the flow where the Reynolds number is 1.1×10^5. What shedding frequency may be anticipated for measurement of a velocity of 3.9 m/s?

7.82. An open-ended tube-probe is installed in an airstream at M = 0.75 and −35°C. The probe is oriented parallel to the flow stream. The pressure of the airstream is 33 kPa. What pressure will be indicated by the probe?

7.20 REFERENCES

1. *Fluid Meters, Their Theory and Application,* 6th ed., ASME, New York, 1971.

2. *Flowmeter Computation Handbook,* ASME, New York, 1961.

3. Arnberg, B. T.: "Review of Critical Flowmeters for Gas Flow Measurements," *Trans. ASME,* vol. 84D, p. 447, 1962.

4. Gracey, W.: "Measurement of Static Pressure on Aircraft," *NACA Tech. Note* 4184, November 1957.

5. King, L. V.: "On the Convection of Heat from Small Cylinders in a Stream of Fluid, with Applications to Hot-Wire Anemometry," *Phil. Trans. Roy. Soc. London,* vol. 214, no. 14, p. 373, 1914.

6. Kovasznay, L. S. G.: "Hot-Wire Method," in *Physical Measurements in Gas Dynamics and Combustion,* p. 219, Princeton University Press, Princeton, NJ, 1954.

7. Krause, L. N., and C. C. Gettelman: "Considerations Entering into the Selection of Probes for Pressure Measurement in Jet Engines," *ISA Proc.,* vol. 7, p. 134, 1952.

8. Ladenburg, R. W. (ed.): *Physical Measurements in Gas Dynamics and Combustion,* Princeton University Press, Princeton, NJ, 1954.

9. Laurence, J. C., and L. G. Landes: "Application of the Constant Temperature Hot-Wire Anemometer to the Study of Transient Air Flow Phenomena," *J. Instr. Soc. Am.,* vol. 1, no. 12, p. 128, 1959.

10. Liepmann, H. W., and A. Roshko: *Elements of Gas Dynamics,* John Wiley & Sons, New York, 1957.

11. "Measurement of Fluid Flow in Pipes Using Orifices, Nozzle, and Venturi," ASME Standard MFC-3M-1985, American Society of Mechanical Engineers, New York, 1985.

12. Purdy, K. R., T. W. Jackson, and C. W. Gorton: "Viscous Fluid Flow under the Influence of a Resonant Acoustic Field," *J. Heat Transfer,* vol. 86C, p. 97, February 1964.

13. Schulze, W. M., G. C. Ashby, Jr., and J. R. Erwin: "Several Combination Probes for Surveying Static and Total Pressure and Flow Direction," *NACA Tech. Note* 2830, November 1952.

14. Smith, E., R. H. Reed, and H. D. Hodges: "The Measurement of Low Air Speeds by the Use of Titanium Tetrachloride," *Texas Eng. Exp. Sta. Res. Rpt.* 25, May 1951.

15. Tuve, G. L.: *Mechanical Engineering Experimentation,* McGraw-Hill, New York, 1961.

16. Schraub, F. A., and S. J. Kline, et al.: "Use of Hydrogen Bubbles for Quantitative Determination of Time-Dependent Velocity Fields in Low Speed Water Flows," ASME Paper No. 64-WA/FE-20, December 1964.

17. *Flow Visualization Symposium,* ASME, New York, 1960.

18. Dean, R. C. (ed.): *Aerodynamic Measurements,* Eagle Press, 1953.

19. Rayle, R. E.: *An Investigation of the Influence of Orifice Geometry on Static Pressure Measurements, S.M. thesis,* Dept. Mech. Eng., Massachusetts Institute of Technology, 1949. (See also Ref. [18].)

20. Yeh, Y., and H. Z. Cummins: "Localized Fluid Flow Measurements with an He–Ne Laser Spectrometer," *Appl. Phys. Lett.,* vol. 4, no. 10, pp. 176–178, May 1964.

21. Forman, J. W., E. W. George, and R. D. Lewis: "Measurement of Localized Flow Velocities in Gases with a Laser Doppler Flowmeter," *Appl. Phys. Lett.,* vol. 7, no. 4, pp. 77–78, August 1965.

22. Forman, J. W., R. D. Lewis, J. R. Thornton, and H. J. Watson: "Laser Doppler Velocimeter for Measurement of Localized Flow Velocities in Liquids," *Proc. IEEE,* vol. 54, pp. 424–425, March 1966.

23. Berman, N. S., and V. A. Santos: "Laminar Velocity Profiles in Developing Flows Using a Laser Doppler Technique," *AIChEJ,* vol. 15, no. 3, pp. 323–327, May 1969.

24. Goldstein, R. J., and D. K. Kried: "Measurements of Laminar Flow Development in a Square Duct Using a Laser Doppler Flowmeter," *J. Appl. Mech.,* vol. 34, no. 4, pp. 813–818, December 1967.

25. Goldstein, R. J., and W. F. Hagen: "Turbulent Flow Measurements Utilizing the Doppler Shift of Scattered Laser Radiation," *Phys. Fluids,* vol. 10, pp. 1349–1352, June 1967.

26. Welch, N. E., and W. J. Tomme: "The Analysis of Turbulence from Data Obtained with a Laser Velocimeter," *AIAA Paper* 67–179, January 1967.

27. Welch, N. E., and R. H. Hines: "The Practical Application of the Laser Anemometer for Fluid Flow Measurements," presented at the ISA Electro-Optical Systems Conf., New York, September 1969.

28. Hodges, A. E., and T. N. Pound: "Further Development of a Smoke Producer Using Vaporized Oil," *Aerodynamics Note* 233, Australian Defense Scientific Service Aeronautical Res. Labs., November 1964.

29. Benedict, R. P., J. S. Wyler, and G. B. Brandt: "The Effect of Edge Sharpness on the Discharge Coefficient of an Orifice," *J. Eng. Power,* p. 576, 1975.

30. Wyler, J. S., and R. P. Benedict: "Comparisons between Throat and Pipe Wall Tap Nozzles," *J. Eng. Power,* p. 569, 1975.

31. Lee, W. F., M. J. Kirik, and J. A. Bonner: "Gas Turbine Flowmeter Measurement of Pulsating Flow," *J. Eng. Power,* p. 531, 1975.

32. Szaniszlo, A. J.: "Experimental and Analytical Sonic Nozzle Discharge Coefficients for Reynolds Numbers up to 8×10^6," *J. Eng. Power,* p. 589, 1975.

33. Smith, R. V., and J. T. Leang: "Evaluations of Correlations for Two-Phase Flowmeters, Three Current—One New," *J. Eng. Power,* p. 589, 1975.

34. Benedict, R. P., and R. D. Schulte: "A Note on the Critical Pressure Ratio across a Fluid Meter," ASME Paper No. 72-WA/FM-1, December 1972.

35. Head, V. P.: "Improved Expansion Factors for Nozzles, Orifices, and Variable Area Meters," ASME Paper No. 73-WA/FM-1, December 1973.

36. Halmi, D.: "Metering Performance Investigation and Substantiation of the Universal Venturi Tube," Parts I and II, ASME Papers No. 73-WA/FM-3, 4, December 1973.

37. Miller, R. W., and O. Kneisel: "A Comparison between Orifice and Flow Nozzle Laboratory Data and Published Coefficients," ASME Paper, No. 73-WA/FM-5, December 1973.

38. Arnberg, B. T., C. L. Britton, and W. F. Seidl: "Discharge Coefficient Correlations for Circular Arc Venturi Flowmeters at Critical (Sonic) Flow," ASME Paper No. 73-WA/FM-8, December 1973.

39. Goldstein, R. J.: "Measurement of Fluid Velocity by Laser Doppler Techniques," *Appl. Mech. Rev.,* vol. 27, p. 753, 1974.

40. Orloff, K. L., G. R. Grant, and W. D. Gunter: "Laser Velocimeter for Simultaneous Two-Dimensional Velocity Measurements," NASA Tech. Brief. 73-10267, 1973.

41. Farmer, W. M.: "Determination of a Third Orthogonal Velocity Component Using Two Rotationally Displaced Laser Doppler Velocimeter Systems," *Appl. Opt.,* vol. 11, p. 770, 1972.

42. Durst, F., A. Melling, and J. H. Whitelaw: *Principles and Practice of Laser Doppler Anemometry,* Academic Press, New York, 1976.

43. Benedict, R. P.: *Fundamentals of Temperature, Pressure, and Flow Measurement,* 3d ed., John Wiley & Sons, New York, 1984.

44. Richards, B. E. (ed.): *Measurement of Unsteady Fluid Dynamic Phenomena,* chap. 6, "Hot-Wire and Hot-Film Anemometers," Hemisphere Publishing, Washington, DC, 1976.

45. James, D. F., and A. J. Acosta: "The Laminar Flow of Dilute Polymer Solutions around Circular Cylinders," *J. Fluid Mech.,* vol. 42, p. 269, 170.

46. Goldstein, R. J., and D. K. Kreid: "The Laser Doppler Anemometer," in *Measurements in Heat Transfer,* pp. 541–574, McGraw-Hill, New York, 1976.

47. Asanuma, T. (ed.): *Flow Visualization,* Hemisphere Publishing, Washington, DC, 1979.

48. Goldstein, R. J.: *Fluid Mechanics Measurements,* Hemisphere Publishing, Washington, DC, 1983.

49. Omega Engineering, Inc.: "Vortex Shedding Flowmeters," *Flow and Level Measurement Handbook,* p. G.3, 1991.

50. Drain, L. E.: *The Laser-Doppler Technique,* John Wiley & Sons, New York, 1980.

51. Anderson, J. D.: *Modern Compressible Flow: With Historical Perspective,* 2d ed., McGraw-Hill, New York, 1990.

52. White, F. M.: *Viscous Fluid Flow,* 2d ed., McGraw-Hill, New York, 1995.

53. Freymuth, P., and L. M. Fingerson: "Hot Wide Anemometry at Very High Frequencies: Effect of Electronic Noise," *Meas. Sci. Technol.,* vol. 8, p. 115, 1997.

54. Wang., T., and T. W. Simon: "Heat Transfer and Fluid Mechanics Measurements Transitional Boundary Layer Flows," ASME Paper No. 85-GT-113, 1985.

55. Lekakis, I. C., and R. J. Adrian: "Measurement of Velocity Vectors with Orthogonal and Non-Orthogonal Triple Sensor Probes," *Experiments in Fluids,* vol. 7, p. 228, 1989.

56. Bjorkquist, D. C., and L. M. Fingerson: "Particle Image Velocimetry," *Progress in Visualization,* vol. 1, Pergamon Press, New York, 1992.

57. Adrian. R. J.: "Multipoint Optical Measurements of Simultaneous Vectors in Unsteady Flow—A Review," *Int. J. Heat and Fluid Flow,* 1986.

8

THE MEASUREMENT OF TEMPERATURE

8.1 INTRODUCTION

To most people, temperature is an intuitive concept that tells whether a body is "hot" or "cold." In the exposition of the second law of thermodynamics temperature is related to heat, for it is known that heat flows only from a high temperature to a low temperature, in the absence of other effects. In the kinetic theory of gases and statistical thermodynamics it is shown that temperature is related to the average kinetic energy of the molecules of an ideal gas. Further extensions of statistical thermodynamics show the relationship between temperature and the energy levels in liquids and solids. We shall not be able to discuss the many theoretical aspects of the concept of temperature but may only note that it is important in every branch of physical science; hence, the experimental engineer should be familiar with the methods employed in temperature measurement. Detailed discussions of the thermodynamic meaning of temperature are given in Refs. [8], [9], [10], [12], and [18].

Since pressure, volume, electrical resistance, expansion coefficients, and so forth, are all related to temperature through the fundamental molecular structure, they change with temperature, and these changes can be used to measure temperature. Calibration may be achieved through comparison with established standards, as discussed in Chap. 2. The International Temperature Scale serves to *define* temperature in terms of observable characteristics of materials. A discussion of standards and calibration methods is given in Ref. [14].

8.2 TEMPERATURE SCALES

The two temperature scales are the Fahrenheit and Celsius scales. These scales are based on a specification of the number of increments between the freezing point and boiling point of water at standard atmospheric pressure. The Celsius scale has 100 units between these points, while the Fahrenheit scale has 180 units. The absolute Celsius scale is called the *Kelvin scale,* while the absolute Fahrenheit scale is termed the *Rankine scale.* Both absolute scales are so defined that they will correspond as

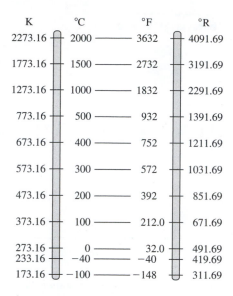

Figure 8.1   Relationship between Fahrenheit and Celsius temperature scales.

closely as possible with the absolute thermodynamic temperature scale. The zero points on both absolute scales represent the same physical state, and the ratio of two values is the same, regardless of the absolute scale used; i.e.,

$$\left(\frac{T_2}{T_1}\right)_{\text{Rankine}} = \left(\frac{T_2}{T_1}\right)_{\text{Kelvin}} \tag{8.1}$$

The boiling point of water at 1 atm is arbitrarily taken as 100° on the Celsius scale and 212° on the Fahrenheit scale. The relationship between the scales is indicated in Fig. 8.1. It is evident that the following relations apply:

$$°F = 32.0 + \tfrac{9}{5}°C \tag{8.2a}$$

$$°R = \tfrac{9}{5}K \tag{8.2b}$$

8.3   THE IDEAL-GAS THERMOMETER

The behavior of an ideal gas at low pressures furnishes the basis for a temperature-measurement device that may serve as a secondary experimental standard. The ideal-gas equation of state is

$$pV = mRT \tag{8.3}$$

where V is the volume occupied by the gas, m is the mass, and R is the gas constant for the particular gas, given by

$$R = \frac{\Re}{M}$$

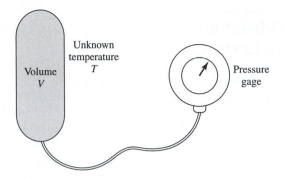

Figure 8.2   Schematic of ideal-gas thermometer.

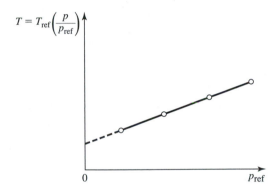

Figure 8.3   Results of measurements with ideal-gas thermometer.

where $\Re$ is the universal gas constant, having a value of 8314.5 J/kg · mol · K, and M is the molecular weight of the gas. For the gas thermometer a fixed volume is filled with gas and exposed to the temperature to be measured, as shown in Fig. 8.2. At the temperature T the gas-system pressure is measured. Next, the volume is exposed to a standard reference temperature (as discussed in Sec. 2.4), and the pressure is measured under these conditions. According to Eq. (8.3), at constant volume

$$T = T_{\text{ref}} \left(\frac{p}{p_{\text{ref}}} \right)_{\text{const vol}}$$ **[8.4]**

Now, suppose that some of the gas is removed from the volume and the pressure measurements are repeated. In general, there will be a slight difference in the pressure ratio in Eq. (8.4) as the quantity of gas is varied. However, regardless of the gas used, the series of measurements may be repeated and the results plotted as in Fig. 8.3. When the curve is extrapolated to zero pressure, the true temperature as defined by the ideal-gas equation of state will be obtained. A gas thermometer may be used to measure temperatures as low as 1 K by extrapolation.

8.4 TEMPERATURE MEASUREMENT BY MECHANICAL EFFECTS

Several temperature-measurement devices may be classified as mechanically opera-tive. In this sense we shall be concerned with those devices operating on the basis of a change in mechanical dimension with a change in temperature.

The liquid-in-glass thermometer is one of the most common types of temperature-measurement devices. The construction details of such an instrument are shown in Fig. 8.4. A relatively large bulb at the lower portion of the thermometer holds the major portion of the liquid, which expands when heated and rises in the capillary tube, upon which are etched appropriate scale markings. At the top of the capillary tube another bulb is placed to provide a safety feature in case the temperature range of the thermometer is inadvertently exceeded. Alcohol and mercury are the most commonly used liquids. Alcohol has the advantage that it has a higher coefficient of expansion than mercury, but it is limited to low-temperature measurements because it tends to boil away at high temperatures. Mercury cannot be used below its freezing point of $-38.78°F$ ($-37.8°C$). The size of the capillary depends on the size of the sensing bulb, the liquid, and the desired temperature range for the thermometer.

In operation, the bulb of the liquid-in-glass thermometer is exposed to the envi-ronment whose temperature is to be measured. A rise in temperature causes the liquid

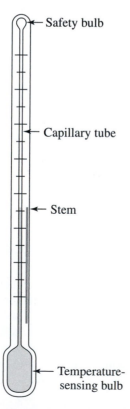

←— Safety bulb

←— Capillary tube

←— Stem

—— Temperature-sensing bulb

Figure 8.4   Schematic of a mercury-in-glass thermometer.

to expand in the bulb and rise in the capillary, thereby indicating the temperature. It is important to note that the expansion registered by the thermometer is the *difference* between the expansion of the liquid and the expansion of the glass. The difference is a function not only of the heat transfer to the bulb from the environment, but also of the heat conducted into the bulb from the stem; the more the stem conduction relative to the heat transfer from the environment, the larger the error. To account for such conduction effects the thermometer is usually calibrated for a certain specified depth of immersion. High-grade mercury-in-glass thermometers have the temperature scale markings engraved on the glass along with a mark which designates the proper depth of immersion. Very precise mercury-in-glass thermometers may be obtained from NIST with calibration information for each thermometer. With proper use these thermometers may achieve accuracies of $\pm 0.05°C$ and can serve as calibration standards for other temperature-measurement devices.

Mercury-in-glass thermometers are generally applicable up to about 600°F (315°C), but their range may be extended to 1000°F (538°C) by filling the space above the mercury with a gas like nitrogen. This increases the pressure on the mercury, raises its boiling point, and thereby permits the use of the thermometer at higher temperatures.

A very widely used method of temperature measurement is the bimetallic strip. Two pieces of metal with different coefficients of thermal expansion are bonded together to form the device shown in Fig. 8.5. When the strip is subjected to a temperature higher than the bonding temperature, it will bend in one direction; when it is subjected to a temperature lower than the bonding temperature, it will bend in the other direction. Eskin and Fritze [3] have given calculation methods for bimetallic strips. The radius of curvature r may be calculated as

$$r = \frac{t\{3(1+m)^2 + (1+mn)[m^2 + (1/mn)]\}}{6(\alpha_2 - \alpha_1)(T - T_0)(1+m)^2} \qquad \text{[8.5]}$$

where $t =$ combined thickness of the bonded strip, m or ft
         $m =$ ratio of thicknesses of low- to high-expansion materials
         $n =$ ratio of moduli of elasticity of low- to high-expansion materials
         $\alpha_1 =$ lower coefficient of expansion, per °C
         $\alpha_2 =$ higher coefficient of expansion, per °C
         $T =$ temperature, °C
         $T_0 =$ initial bonding temperature, °C

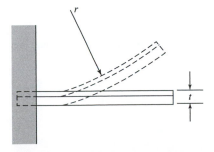

Figure 8.5   The bimetallic strip.

Table 8.1   Mechanical properties of some commonly used thermal materials

Material	Thermal Coefficient of Expansion Per °C	Modulus of Elasticity	
		psi	GN/m^2
Invar	1.7×10^{-6}	21.4×10^6	147
Yellow brass	2.02×10^{-5}	14.0×10^6	96.5
Monel 400	1.35×10^{-5}	26.0×10^6	179
Inconel 702	1.25×10^{-5}	31.5×10^6	217
Stainless-steel type 316	1.6×10^{-5}	28×10^6	193

The thermal-expansion coefficients for some commonly used materials are given in Table 8.1.

Bimetallic strips are frequently used in simple on-off temperature-control devices (thermostats). Movement of the strip has sufficient force to trip control switches for various devices. The bimetallic strip has the advantages of low-cost, negligible maintenance expense, and stable operation over extended periods of time. Alternate methods of construction can use a coiled strip to drive a dial indicator for temperatures.

Example 8.1 | **CURVATURE AND DEFLECTION OF BIMETALLIC STRIP.**   A bimetallic strip is constructed of strips of yellow brass and Invar bonded together at 30°C. Each has a thickness of 0.3 mm. Calculate the radius of curvature when a 6.0-cm strip is subjected to a temperature of 100°C.

Solution

We use Eq. (8.5) with properties from Table 8.1.

$$T - T_0 = 100 - 30 = 70°C$$

$$m = 1.0$$

$$n = \frac{147}{96.5} = 1.52$$

$$\alpha_1 = 1.7 \times 10^{-6}°C^{-1} \qquad \alpha_2 = 2.02 \times 10^{-5}°C^{-1}$$

$$t = (2)(0.3 \times 10^{-3}) = 0.6 \times 10^{-3} \text{ m}$$

Thus, $$r = \frac{(0.6 \times 10^{-3})[(3)(2)^2 + (1 + 1.52)(1 + 1/1.52)]}{6(2.02 - 0.17)(10^{-5})(70)(2)^2}$$

$$= 0.132 \text{ m}$$

From Fig. 8.5 we observe that the angle through which the strip is deflected is related to the strip length L and radius of curvature by

$$L = r\theta$$

where we assume the increase in length due to thermal expansion is small. Thus,

$$\theta = \frac{0.06}{0.132} = 0.454 \text{ rad} = 26.04°$$

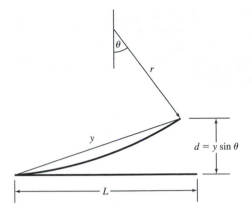

Figure Example 8.1

as indicated in the accompanying figure; the straight-line segment joining the ends of the strip has a length y of

$$y = 2r \sin \frac{\theta}{2} = (2)(0.132) \sin 13.02$$

$$= 0.0595 \text{ m}$$

The deflection d is related to y by

$$d = y \sin \theta = (0.0595) \sin 26.04 = 0.0261 \text{ m}$$

or a very substantial deflection.

Fluid-expansion thermometers represent one of the most economical, versatile, and widely used devices for industrial temperature-measurement applications. The principle of operation is indicated in Fig. 8.6. A bulb containing a liquid, gas, or vapor is immersed in the environment. The bulb is connected by means of a capillary tube

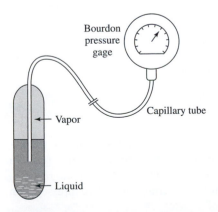

Figure 8.6   Fluid-expansion thermometer.

to some type of pressure-measuring device, such as the bourdon gage shown. An increase in temperature causes the liquid or gas to expand, thereby increasing the pressure on the gage; the pressure is thus taken as an indication of the temperature. The entire system consisting of the bulb, capillary, and gage may be calibrated directly. It is clear that the temperature of the capillary tube may influence the reading of the device because some of the volume of fluid is contained therein. If an equilibrium mixture of liquid and vapor is used in the bulb, however, this problem may be alleviated, provided that the bulb temperature is always higher than the capillary-tube temperature. In this circumstance the fluid in the capillary will always be in a sub-cooled liquid state, while the pressure will be uniquely specified for each temperature in the equilibrium mixture contained in the bulb.

Capillary tubes as long as 200 ft (60 m) may be used with fluid-expansion thermometers. The transient response is primarily dependent on the bulb size and the thermal properties of the enclosed fluid. The highest response may be achieved by using a small bulb connected to some type of electric-pressure transducer through a short capillary. Fluid-expansion thermometers are usually low in cost, stable in operation, and accurate within $\pm 1°C$.

8.5   TEMPERATURE MEASUREMENT BY ELECTRICAL EFFECTS

Electrical methods of temperature measurement are very convenient because they furnish a signal that is easily detected, amplified, or used for control purposes. In addition, they are usually quite accurate when properly calibrated and compensated.

ELECTRICAL-RESISTANCE THERMOMETER, OR RESISTANCE TEMPERATURE DETECTOR (RTD)

One quite accurate method of temperature measurement is the electrical-resistance thermometer. It consists of some type of resistive element, which is exposed to the temperature to be measured. The temperature is indicated through a measurement of the change in resistance of the element. Several types of materials may be used as resistive elements, and their characteristics are given in Table 8.2. The linear temperature coefficient of resistance α is defined by

$$\alpha = \frac{R_2 - R_1}{R_1(T_2 - T_1)} \qquad \textbf{[8.6]}$$

where R_2 and R_1 are the resistances of the material at temperatures T_2 and T_1, respectively. The relationship in Eq. (8.6) is usually applied over a narrow temperature range such that the variation of resistance with temperature approximates a linear relation. For wider temperature ranges the resistance of the material is usually expressed by a

Table 8.2   Resistance-temperature coefficients and resistivity at 20°C[1]

Substance	$\alpha\ (°C^{-1})$	$\rho\ (\mu\Omega \cdot cm)$
Nickel	0.0067	6.85
Iron (alloy)	0.002 to 0.006	10
Tungsten	0.0048	5.65
Aluminum	0.0045	2.65
Copper	0.0043	1.67
Lead	0.0042	20.6
Silver	0.0041	1.59
Gold	0.004	2.35
Platinum	0.00392	10.5
Mercury	0.00099	98.4
Manganin	±0.00002	44
Carbon	−0.0007	1400
Electrolytes	−0.02 to −0.09	Variable
Semiconductor (thermistors)	−0.068 to +0.14	10^9

| [1]According to Lion [6].

quadratic relation

$$R = R_0(1 + aT + bT^2) \qquad \textbf{[8.7]}$$

where R = resistance at temperature T
  R_0 = resistance at reference temperature T_0
  a, b = experimentally determined constants

   Various methods are employed for construction of resistance thermometers, depending on the application. In all cases care must be taken to ensure that the resistance
wire is free of mechanical stresses and so mounted that moisture cannot come in contact with the wire and influence the measurement.

   One construction technique involves winding the platinum on a glass or ceramic
bobbin followed by sealing with molten glass. This technique protects the platinum
RTD element but is subject to stress variations over wide temperature ranges.
Stress-relief techniques can alleviate the problem so that the element may be used for
temperature measurements within ±0.1°C.

   RTD sensors may also be constructed by depositing a platinum or metal-glass
slurry on a ceramic substrate. The film can then be etched and sealed to form the resistance element. This process is less expensive than the mechanical-winding ceramic
but is not as accurate. The thin-film sensor does offer the advantage of low mass and
therefore more rapid thermal response (see Sec. 8.8) and less chance of conduction
error (Sec. 8.7).

   The resistance measurement may be performed with some type of bridge circuit,
as described in Chap. 4. For steady-state measurements a null condition will suffice,
while transient measurements will usually require the use of a deflection bridge.
One of the primary sources of error in the electrical-resistance thermometer is the

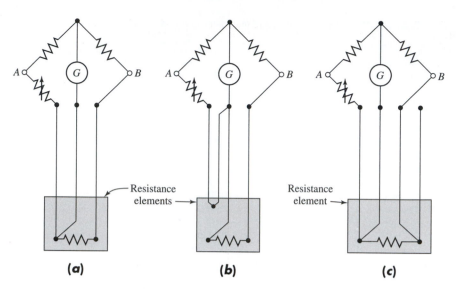

Figure 8.7   Methods of correcting for lead resistance with electrical-resistance thermometer. (*a*) Siemen's three-lead arrangement; (*b*) Callender four-lead arrangement; (*c*) floating-potential arrangement. Power connections made at *A* and *B*.

effect of the resistance of the leads which connect the element to the bridge circuit. Several arrangements may be used to correct for this effect, as shown in Fig. 8.7. The Siemen's three-lead arrangement is the simplest type of corrective circuit. At balance conditions the center lead carries no current, and the effect of the resistance of the other two leads is canceled out. The Callender four-lead arrangement solves the problem by inserting two additional lead wires in the adjustable leg of the bridge so that the effect of the lead wires on the resistance thermometer is canceled out. The floating-potential arrangement in Fig. 8.7*c* is the same as the Siemen's connection, but an extra lead is inserted. This extra lead may be used to check the equality of lead resistance. The thermometer reading may be taken in the position shown, followed by additional readings with the two right and left leads interchanged, respectively. Through this interchange procedure the best average reading may be obtained and the lead error minimized.

Practical problems which are encountered with RTDs involve lead error and relatively bulky size which sometimes give rise to poor transient response and conduction error discussed in Secs. 8.7 and 8.8. The RTD also has a relatively fragile construction. Because a current must be fed to the RTD for the bridge measurement, there is the possibility of self-heating ($i^2 R$ in the element) which may alter the temperature of the element. The importance one must assign to this effect depends on the thermal communication between the RTD and the medium whose temperature is to be measured. For measurement of the temperature of a block of metal the communication is good, while for an air temperature measurement the communication is poor. In still air the error due to self-heating is about $\frac{1}{2}^{\circ}$C per milliwatt.

Example 8.2

SENSITIVITY OF PLATINUM RESISTANCE THERMOMETER.   A platinum resistance thermometer is used at room temperature. Assuming a linear temperature variation with resistance, calculate the sensitivity of the thermometer in ohms per degrees Fahrenheit.

Solution

The meaning of a linear variation of resistance with temperature is

$$R = R_0[1 + \alpha(T - T_0)]$$

where R_0 is the resistance at the reference temperature T_0. The sensitivity is thus

$$S = \frac{dR}{dT} = \alpha R_0$$

R_0 depends on the length and size of the resistance wire. At room temperature $\alpha = 0.00392°\text{C}^{-1} = 0.00318°\text{F}^{-1}$ for platinum.

THERMISTORS

The *thermistor* is a semiconductor device that has a negative temperature coefficient of resistance, in contrast to the positive coefficient displayed by most metals. Furthermore, the resistance follows an exponential variation with temperature instead of a polynomial relation like Eq. (8.7). Thus, for a thermistor

$$R = R_0 \exp\left[\beta\left(\frac{1}{T} - \frac{1}{T_0}\right)\right] \qquad \textbf{[8.8]}$$

where R_0 is the resistance at the reference temperature T_0 and β is an experimentally determined constant. The numerical value of β varies between 3500 and 4600 K, depending on the thermistor material and temperature. The resistivities of three thermistor materials as compared with platinum are given in Fig. 8.8 according to Ref. [1]. A typical static voltage-current curve is shown in Fig. 8.9, while a typical set of transient voltage-current characteristics is illustrated in Fig. 8.10. The numbers on the curve in Fig. 8.9 designate the degrees Celsius rise in temperature above ambient temperature for the particular thermistor.

The thermistor is a very sensitive device, and consistent performance within 0.01°C may be anticipated with proper calibration. A rather nice feature of the thermistor is that it may be used for temperature compensation of electric circuits. This is possible because of the negative temperature characteristic that it exhibits so that it can be used to counteract the increase in resistance of a circuit with a temperature increase.

We have noted that the thermistor is an extremely sensitive device because its resistance changes so rapidly with temperature; however, it has the disadvantage of highly nonlinear behavior. This is not a particularly severe problem because data acquisition systems can employ firmware computing programs to provide a direct temperature readout from the resistance measurement.

Because the resistance of the thermistor is so high, the error due to lead resistance is small compared to that for the RTD, and four-wire leads are usually not required.

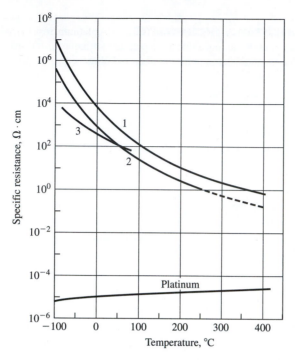

Figure 8.8   Resistivity of three thermistor materials compared with platinum, according to Ref. [1].

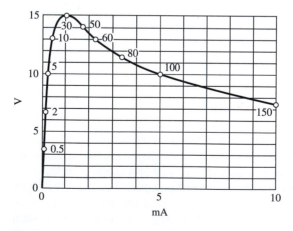

Figure 8.9   Static voltage-current curve for a typical thermistor, according to Ref. [1].

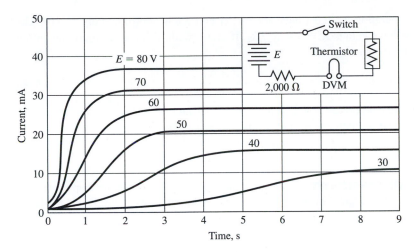

Figure 8.10   Typical set of transient voltage-current curves for a thermistor, according to Ref. [1]. Circuit for measurement is shown in insert.

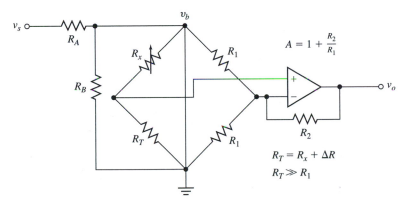

Figure 8.11   Bridge amplifier circuit, according to Ref. [24].

In addition, the high resistance of the thermistor means that smaller currents are required for the measurement, and thus errors due to self-heating are very small.

The thermistor is a semiconductor device and therefore is subject to deterioration at high temperatures; for this reason they are limited to temperature measurements below about 300°C.

A bridge amplifier circuit which may be employed with resistance-thermometer or thermistor devices is shown in Fig. 8.11. In this arrangement R_T is the resistance of the thermometer and is expressed as

$$R_T = R_x = \Delta R$$

For linear operation we must have $R_T \gg R_1$ and R_B is approximately $R_1/10$. The amplifier voltage gain is $A = 1 + R_2/R_1$. If these conditions are satisfied, the output voltage can be expressed as

$$v_o = \frac{A v_b \alpha \, \Delta T}{4}$$ [8.9]

where α is the temperature coefficient of resistance and ΔT is the temperature difference from balanced conditions, i.e., from $\Delta R = 0$.

Example 8.3

THERMISTOR SENSITIVITY.   Calculate the temperature sensitivity for thermistor No. 1 in Fig. 8.8 at 100°C. Express the result in ohm-centimeters per degree Celsius. Take $\beta = 4120\,\mathrm{K}$ at 100°C.

Solution

The sensitivity is obtained by differentiating Eq. (8.8).

$$S = \frac{dR}{dT} = R_0 \exp\left[\beta\left(\frac{1}{T} - \frac{1}{T_0}\right)\right]\left(\frac{-\beta}{T^2}\right)$$

We wish to express the result in resistivity units; thus, the resistivity at 100°C is inserted for R_0. Also,

$$T = T_0 = 100°\mathrm{C} = 373\ \mathrm{K}$$

so that

$$S = -\rho_{100°\mathrm{C}} \frac{4120}{(373)^2}$$

$$= -\frac{(110)(4120)}{(373)^2} = -3.26\ \Omega \cdot \mathrm{cm}/°\mathrm{C}$$

THERMOELECTRIC EFFECTS (THERMOCOUPLES)

The most common electrical method of temperature measurement uses the thermocouple. When two dissimilar metals are joined together as in Fig. 8.12, an emf will exist between the two points A and B, which is primarily a function of the junction

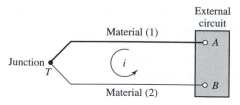

Figure 8.12   Junction of two dissimilar metals indicating thermoelectric effect.

temperature. This phenomenon is called the *Seebeck effect*. If the two materials are connected to an external circuit in such a way that a current is drawn, the emf may be altered slightly owing to a phenomenon called the *Peltier effect*. Further, if a temperature gradient exists along either or both of the materials, the junction emf may undergo an additional slight alteration. This is called the *Thomson effect*. There are, then, three emfs present in a thermoelectric circuit: the Seebeck emf, caused by the junction of dissimilar metals; the Peltier emf, caused by a current flow in the circuit; and the Thomson emf, which results from a temperature gradient in the materials. The Seebeck emf is of prime concern since it is dependent on junction temperature. If the emf generated at the junction of two dissimilar metals is carefully measured as a function of temperature, then such a junction may be utilized for the measurement of temperature. The main problem arises when one attempts to measure the potential. When the two dissimilar materials are connected to a measuring device, there will be another thermal emf generated at the junction of the materials and the connecting wires to the voltage-measuring instrument. This emf will be dependent on the temperature of the connection, and provision must be made to take account of this additional potential.

Two rules are available for analysis of thermoelectric circuits:

1. If a third metal is connected in the circuit as shown in Fig. 8.13, the net emf of the circuit is not affected as long as the new connections are at the same temperature. This statement may be proved with the aid of the second law of thermodynamics and is known as the *law of intermediate metals*.

2. Consider the arrangements shown in Fig. 8.14. The simple thermocouple circuits are constructed of the same materials but operate between different temperature

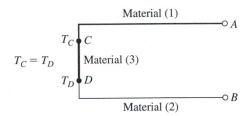

Figure 8.13   Influence of a third metal in a thermoelectric circuit; law of intermediate metals.

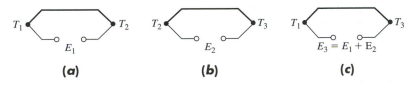

Figure 8.14   Circuits illustrating the law of intermediate temperatures.

limits. The circuit in Fig. 8.14*a* develops an emf of E_1 between temperatures T_1 and T_2; the circuit in Fig. 8.14*b* develops an emf of E_2 between temperatures T_2 and T_3. The *law of intermediate temperatures* states that this same circuit will develop an emf of $E_3 = E_1 + E_2$ when operating between temperatures T_1 and T_3, as shown in Fig. 8.14*c*.

It may be observed that all thermocouple circuits must involve at least two junctions. If the temperature of one junction is known, then the temperature of the other junction may be easily calculated using the thermoelectric properties of the materials. The known temperature is called the *reference temperature*. A common arrangement for establishing the reference temperature is the ice bath shown in Fig. 8.15. An equilibrium mixture of ice and air-saturated distilled water at standard atmospheric pressure produces a known temperature of 32°F. When the mixture is contained in a Dewar flask, it may be maintained for extended periods of time. Note that the arrangement in Fig. 8.15*a* maintains both thermocouple wires at a reference temperature of 32°F, whereas the arrangement in Fig. 8.15*b* maintains only one at the reference

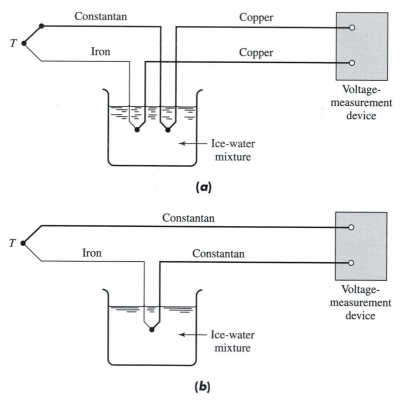

Figure 8.15　　Conventional methods for establishing reference temperature in thermocouple circuit. Iron-constantan thermocouple illustrated.

temperature. The system in Fig. 8.15a would be necessary if the binding posts at the voltage-measuring instrument were at different temperatures, while the connection in Fig. 8.15b would be satisfactory if the binding posts were at the same temperature. To be effective the measurement junctions in Fig. 8.15a must be of the same material.

It is common to express the thermoelectric emf in terms of the potential generated with a reference junction at 32°F (0°C). Standard thermocouple tables have been prepared on this basis, and a summary of the output characteristics of the most common thermocouple combinations is given in Table 8.3. These data are shown graphically in Fig. 8.16. The output voltage E of a simple thermocouple circuit is usually written in the form

$$E = AT + \tfrac{1}{2}BT^2 + \tfrac{1}{3}CT^3 \qquad \textbf{[8.10]}$$

where T is the temperature in degrees Celsius and E is based on a reference junction temperature of 0°C. The constants A, B, and C are dependent on the thermocouple material. Powell [19] gives an extensive discussion of the manufacture of materials

Table 8.3a   Thermal emf in absolute millivolts for commonly used thermocouple combinations, according to ITS(90) (Reference junction of 0°C)[1]

Temperature, °C	Copper vs. Constantan (T)	Chromel vs. Constantan (E)	Iron vs. Constantan (J)	Chromel vs. Alumel (K)	Platinum vs. Platinum–10% Rhodium (S)	Nicosil vs. Nisil (N)
−150	−4.648	−7.279	−6.500	−4.913		−1.530
−100	−3.379	−5.237	−4.633	−3.554		−1.222
−50	−1.819	−2.787	−2.431	−1.889	−0.236	−0.698
−25	−0.940	−1.432	−1.239	−0.968	−0.127	−0.368
0	0	0	0	0	0	0
25	0.992	1.495	1.277	1.000	0.143	0.402
50	2.036	3.048	2.585	2.023	0.299	0.836
75	3.132	4.657	3.918	3.059	0.467	1.297
100	4.279	6.319	5.269	4.096	0.646	1.785
150	6.704	9.789	8.010	6.138	1.029	2.826
200	9.288	13.421	10.779	8.139	1.441	3.943
300	14.862	21.036	16.327	12.209	2.323	6.348
400	20.872	28.946	21.848	16.397	3.259	8.919
500		37.005	27.393	20.644	4.233	11.603
600		45.093	33.102	24.906	5.239	14.370
800		61.017	45.494	33.275	7.345	20.094
1000		76.373	57.953	41.276	9.587	26.046
1200			69.553	48.838	11.951	32.144
1500					15.582	
1750					18.503	

[1]Composition of Thermocouple Alloys:
Alumel: 94% nickel, 3% manganese, 2% aluminum, 1% silicon
Chromel: 90% nickel, 10% chromium
Constantan: 55% copper, 45% nickel
Nicosil: 84% nickel, 14% chromium, 1.5% silicon
Nisil: 95% nickel, 4.5% silicon, 0.1% Mg

Table 8.3b   Error limits for commercial thermocouples

Type Thermocouple	Error, Standard Grade*	Error, Special Grade*
E	1.7°C or 0.5% above 0°C	1.0°C or 0.4%
J	2.2°C or 0.75% above 0°C	1.1°C or 0.4%
K	2.2°C or 0.75% above 0°C	1.1°C or 0.4%
N	2.2°C or 0.75% above 0°C	1.1°C or 0.4%
R	1.5°C or 0.25	0.6°C or 0.1%
S	1.5°C or 0.25%	0.6°C or 0.1%
T	1.0°C or 0.75% above 0°C	0.5°C or 0.4%

| *Whichever is greater.

Table 8.3c   Properties of common thermocouple materials

Material	Thermal Conductivity, W/M · °C	Specific Heat, kJ/kg · °C	Density, kg/m³	Electric Resistivity, $\mu\Omega \cdot cm$	Temperature Coefficient of Expansion, $°C^{-1} \times 10^6$	Melting Point, °C
Alumel	29.8	0.52	8600	29	12	1400
Constantan	21.7	0.39	8900	49	−0.1	1220
Chromel	19.2	0.45	8700	70	13	1450
Copper	377	0.385	8900	1.56	17	1080
Iron	68	0.45	7900	8.6	12	1490

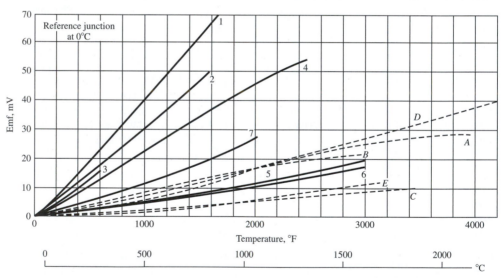

Legend:
1 Chromel-constantan (type E)
2 Iron-constantan (type J)
3 Copper-constantan (type T)
4 Chromel-alumel (type K)
5 Platinum-platinum rhodium (type R)
6 Platinum-platinum rhodium (type S)
7 Nicosil-Nisil (type N)

A Rhenium-molybdenum
B Rhenium-tungsten
C Iridium-iridium rhodium
D Tungsten-tungsten rhenium
E Plat. rhodium-plat. 10% rhodium

Figure 8.16   Emf temperature relations for thermocouple materials, positive electrode listed first.

Table 8.4   Thermoelectric sensitivity $S = dE/dT$ of thermoelement made of materials listed against platinum, $\mu V^{\circ}C^{-1}*$
(Reference junction kept at a temperature of $0^{\circ}C$)

Bismuth	−72	Silver	6.5
Constantan	−35	Copper	6.5
Nickel	−15	Gold	6.5
Potassium	−9	Tungsten	7.5
Sodium	−2	Cadmium	7.5
Platinum	0	Iron	18.5
Mercury	0.6	Nichrome	25
Carbon	3	Antimony	47
Aluminum	3.5	Germanium	300
Lead	4	Silicon	440
Tantalum	4.5	Tellurium	550
Rhodium	6	Selenium	900

| *According to Lion [6].

for thermocouple use, inhomogeneity ranges, and power series relationships for thermoelectric voltages of various standard thermocouples. The NIST publication which gives these data is Ref. [23].

The sensitivity, or thermoelectric power, of a thermocouple is given by

$$S = \frac{dE}{dT} = A + BT + CT^2 \qquad \textbf{[8.11]}$$

Table 8.4 gives the approximate values of the sensitivity of various materials relative to platinum at $0^{\circ}C$. A summary of the operating range and characteristics of the most common thermocouple materials is given in Figs. 8.16 and 8.17.

If computer processing of thermocouple data is to be performed, the power-series relations like Eq. (8.10) or the extensive collection in Refs. [14] and [23] will certainly be of value. But a firm word of caution must be given here. When one buys a roll of commercial thermocouple wire, there are different grades available (at different prices, of course). Precision-grade wire will usually follow the NIST tables by $\pm 0.5^{\circ}C$. A "commercial" grade might not be better than $\pm 2^{\circ}F$. Therefore, the experimentalist should not be lulled into security by a 10-digit computer printout based on the NIST equations. If better precision is required, samples of the thermocouple wire should be calibrated directly against known temperature standards as discussed in Chap. 2. The individual can then use this calibration data to determine the constants in Eq. (8.10). For precise data analysis it is the specific calibration which must be employed for calculations. The typical error limits for commercial grades of thermocouple wires are given in Table 8.3b [25]. Table 8.3c gives some of the physical properties of the more common thermocouple materials. These properties are useful in calculation of heat-transfer effects as described in Sec. 8.7.

The need for precise calibration is particularly acute when two thermocouples are used for a small-differential temperature measurement. For example, suppose two "precision" thermocouples having an uncertainty of $\pm 1^{\circ}F$ each are used for

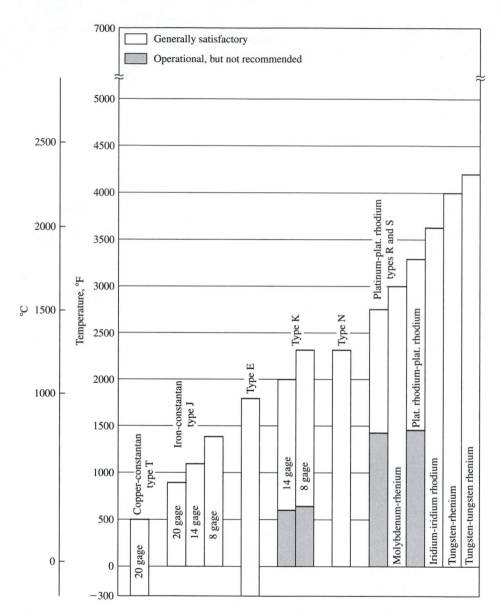

Figure 8.17   Summary of operating range of thermocouples. See also Fig. 8.16.

a temperature difference measurement of 10°F. Using the uncertainty analysis of Chap. 3 the uncertainty in the temperature difference would be

$$w(\Delta T) = [(1)^2 + (1)^2]^{1/2} = 1.414°F$$

or *14* percent! We must note that if both thermocouples come from the same roll,

Table 8.5   Polynomial coefficients for Eq. (8.12) for several standard thermocouple combinations

	Type E	Type J	Type K	Type R	Type S	Type T
	Chromel(+) vs. Constantan(−)	Iron(+) vs. Constantan(−)	Chromel(+) vs. Nickel-5%(−) (Aluminum Silicon)	Platinum-13% Rhodium(+) vs. Platinum(−)	Platinum-10% Rhodium(+) vs. Platinum(−)	Copper(+) vs. Constantan(−)
	−100°C to 1000°C* ±0.5°C 9th Order	0°C to 760°C* ±0.1°C 5th Order	0°C to 1370°C* ±0.7°C 8th Order	0°C to 1000°C* ±0.5°C 8th Order	0°C to 1750°C* ±1°C 9th Order	−160°C to 400°C* ±0.5°C 7th Order
a_0	0.104967248	−0.048868252	0.226584602	0.263632917	0.927763167	0.100860910
a_1	17189.45282	19873.14503	24152.10900	179075.491	169526.5150	25727.94369
a_2	−282639.0850	−218614.5353	67233.4248	−48840341.37	−31568363.94	−767345.8295
a_3	12695339.5	11569199.78	2210340.682	$1.90002E+10$	8990730663	78025595.81
a_4	−448703084.6	−264917531.4	−860963914.9	$-4.82704E+12$	$-1.63565E+12$	−9247486589
a_5	$1.10866E+10$	2018441314	$4.83506E+10$	$7.62091E+14$	$1.88027E+14$	$6.97688E+11$
a_6	$-1.76807E+11$		$-1.18452E+12$	$-7.20026E+16$	$-1.37241E+1$	$-2.66192E+13$
a_7	$1.71842E+12$		$1.38690E+13$	$3.71496E+18$	$6.17501E+17$	$3.94078E+14$
a_8	$-9.19278E+12$		$-6.33708E+13$	$-8.03104E+19$	$-1.56105E+19$	
a_9	$2.06132E+13$				$1.69535E+20$	

they should have essentially the same characteristics, and this result would be highly unlikely. Also, the accuracy of the thermocouples for a direct *temperature-difference* measurement will be much better than for an *absolute* temperature measurement.

For those persons wishing to design software to calculate temperatures from thermocouple voltages, a ninth-order polynomial can be used in the form

$$T = a_0 + a_1 x + a_2 x^2 + \cdots + a_9 x^9 \qquad \textbf{[8.12]}$$

where T = temperature, °C

x = thermocouple voltage, volts, reference junction at 0°C

a = polynomial coefficients given in Table 8.5 for various thermocouple combinations

The accuracy with which each polynomial fits the NIST tables is indicated in Table 8.5. From the standpoint of computer calculations the equations are usually written in a nested polynomial form to minimize execution time. For example, the fifth-order polynomial for iron-constantan would then be written

$$T = a_0 + x(a_1 + x(a_2 + x(a_3 + x(a_4 + a_5 x)))) \qquad \textbf{[8.13]}$$

A number of commercial instruments are available which measure thermocouple voltages and employ an internal microprocessor to calculate temperatures for a digital readout, either visually or on a strip-chart printer.

The output of thermocouples is in the millivolt range and may be measured by a digital millivoltmeter. The voltmeter is basically a current-sensitive device, and hence the meter reading will be dependent on both the emf generated by the thermocouple and the total circuit resistance, including the resistance of the connecting wires.

The complete system, consisting of the thermocouple lead wires and millivoltmeter, may be calibrated directly to furnish a reasonably accurate temperature determination. For precision measurements, the thermocouple output may be determined with a potentiometer circuit similar to the one described in Sec. 4.4. For very precise laboratory work a microvolt potentiometer is used, which can read potentials within $1 \, \mu V$. It may be noted that the resistance of the lead wires is of no consequence when a potentiometer is used at balance conditions since the current flow is zero in the thermocouple circuit.

Obviously, a large number of electronic voltmeters are suitable for thermocouple measurements. Many provide a digital output which can be used for direct computer processing of the data. These instruments typically have very high input impedance and therefore do not draw an appreciable current in the thermocouple circuit.

The problem of the reference junction can be alleviated with the circuit shown in Fig. 8.18. A thermistor is placed in thermal contact with the terminal strip to which the thermocouple wires are attached. The voltage v_b and temperature coefficient of the thermistor must be adjusted so that V_c will match the thermocouple temperature coefficient in millivolts/degree. The value of R_x is adjusted so that the voltage output ΔV is zero at 0°C. A convenient value for R_A is about 1 kΩ.

The compensator shown in Fig. 8.18 is called a *hardware*-compensation device. Others which may be employed can be based on RTD or other types of solid-state temperature sensors. They are limited to a single type of thermocouple because, as shown in Table 8.5, each thermocouple has its own emf-temperature characteristic. An alternate technique is to provide *firmware* compensation in the data acquisition system. In this technique the temperature of the measurement junction(s) is (are) measured with a RTD or thermistor and compensation provided with a built-in microprocessor. Direct digital readout of temperature is then provided. With appropriate switching one can use the same junction(s) for several types of thermocouples. We must caution the reader that such compensation assumes a polynomial relation like Eq. (8.12), and if the thermocouple wires and/or junctions do not conform to the NIST standard, then an error in measurement will result. To alleviate this difficulty, the wires and

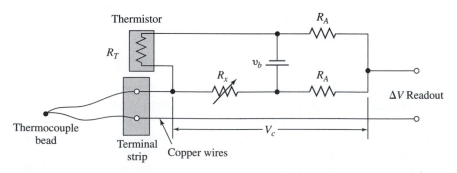

Figure 8.18   Reference junction compensation using thermistor.

junctions may be calibrated against known standards and a correction factor CF determined as

$$CF = \text{true temperature} - \text{indicated temperature}$$
$$= \text{fcn (indicated temperature)} \qquad \textbf{[8.14]}$$

For some acquisition systems a simple polynomial expression for CF (perhaps a quadratic) may be programmed to furnish a correct readout. In other cases the output may be fed to a computer where the correction is performed.

THERMOCOUPLE MEASUREMENT.   An iron-constantan thermocouple is connected to a potentiometer whose terminals are at 25°C. The potentiometer reading is 3.59 mV. What is the temperature of the thermocouple junction?

Example 8.4

Solution

The thermoelectric potential corresponding to 25°C is obtained from Table 8.3a as

$$E_{75} = 1.277 \text{ mV}$$

The emf of the thermocouple based on a 0°C reference temperature is thus

$$E_T = 1.277 + 3.59 = 4.867 \text{ mV}$$

From Table 8.3a, the corresponding temperature is 92.5°C.

   In order to provide a more sensitive circuit, thermocouples are occasionally connected in a series arrangement as shown in Fig. 8.19. Such an arrangement is called a *thermopile,* and for a three-junction situation the output would be three times that of a single thermocouple arrangement *provided* the temperatures of the hot and cold junctions are uniform. The thermopile arrangement is useful for obtaining a substantial emf for measurement of a small temperature difference between the two junctions. In this way a relatively insensitive instrument may be used for voltage measurement, whereas a microvolt potentiometer might otherwise be required. When a thermopile is installed, it is important to ensure that the junctions are electrically insulated from one another. We have seen that the typical thermocouple measures the difference in temperature between a certain unknown point and another point designated as the reference temperature. The circuit could just as well be employed for the measurement of a differential temperature. For small differentials the thermopile circuit is frequently used to advantage. Now, consider the series thermocouple arrangement shown in Fig. 8.20. The four junctions are all maintained at different temperatures and connected in series. Since there are an even number of junctions, it is not necessary to install a reference junction because the same type of metal is connected to both terminals of the potentiometer. If we note that the current will flow from plus to minus and assume that junction A produces a potential drop in this direction, then

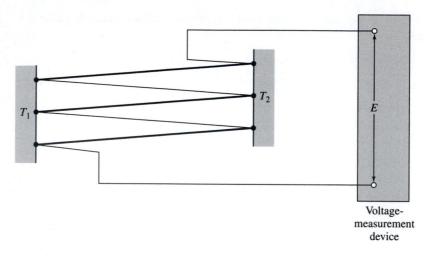

Figure 8.19   Thermopile connection.

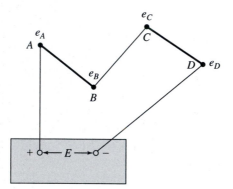

Figure 8.20   Series connection of thermocouples.

junctions B and D will produce a potential drop in the opposite direction and junction C will generate a potential drop in the same direction as junction A. Thus, the total emf measured at the potentiometer terminals is

$$E = e_A - e_B + e_C - e_D \qquad \textbf{[8.15]}$$

The reading will be zero when all the junctions are at the same temperature and will take on some other value at other conditions. Note, however, that the emf of this series connection is *not* indicative of any particular temperature. It is *not* representative of an average of the junction temperatures because the emf vs. T relationship is nonlinear.

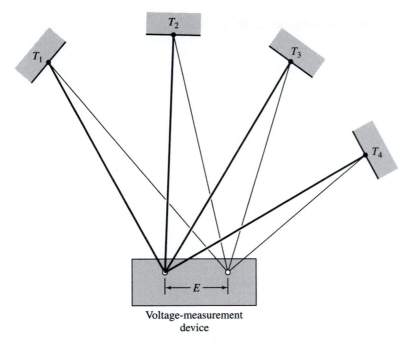

Figure 8.21   Parallel connection of thermocouples.

The parallel connection in Fig. 8.21 may be used for obtaining the average tem-perature of a number of points. Each of the four junctions may be at a different temperature and hence will generate a different emf. The bucking potential furnished by the potentiometer will be the average of the four junction potentials. There can be a small error in this reading, however, because there is a small current flow in the lead wires as a result of the difference in potential between the junctions. Thus, the resistance of the lead wires will influence the reading to some extent.

A more suitable way of obtaining an average temperature is to use the thermopile circuit in Fig. 8.19. Each of the "hot" junctions may be at a different temperature, while all the "cold" junctions may be maintained at a fixed reference value. The average emf is then given by

$$E_{\mathrm{av}} = \frac{E}{n}$$
[8.16]

where n is the number of junction pairs and E is the total reading of the thermopile. The average temperature corresponding to the average emf given in Eq. (8.13) may then be determined.

Example 8.5

EFFECT OF COLD JUNCTION LEVEL OF THERMOPILE.   A thermopile consisting of five junction pairs of Chromel-constantan is used to measure a temperature difference of 50°C with the cold junctions at 25°C. Determine the voltage output of the thermopile. Suppose the cold-junction temperature is incorrectly stated as 75°C (in reality, the hot-junction temperature). What error in temperature-difference measurement would result from this incorrect statement?

Solution

From Table 8.3*a* the data needed are:

$$E_{25} = 1.495 \text{ mV}$$
$$E_{75} = 4.657 \text{ mV}$$
$$E_{125} = 8.054 \text{ mV}$$

all referenced to 0°C. For the correct statement of the cold junction at 25°C, with a hot-junction temperature of 75°C, the voltage output reading of the thermopile would be

$$E_{\Delta T} = (5)(4.657 - 1.495) = 15.810 \text{ mV}$$

By incorrectly stating the cold-junction temperature as 75°C, we have not changed the *actual* reading value of 15.810 mV; however, the reading now indicates a different value of ΔT. To obtain the new high temperature we would calculate a new high-temperature emf as

$$E_T = 4.657 + \frac{15.810}{5} = 7.819 \text{ mV}$$

Interpolating, using the above data

$$T_{\text{hot}} = 75 + \frac{(125 - 75)(7.819 - 4.657)}{8.054 - 4.657}$$
$$= 121.54°C$$
$$\Delta T = 121.54 - 75 = 46.54°C$$

Not only is the temperature *level* of the hot junction grossly in error (121.54°C instead of 75°C) but the temperature difference is also off by a substantial amount—46.54°C instead of 50°C.

Comment

When using thermopiles for temperature-difference determinations, it is important that the temperature *level* of either the hot or cold junction be expressed correctly.

Example 8.6

RESULT OF INSTALLATION MISTAKE.   A heat-exchanger facility is designed to use type J thermocouples to sense an outlet gas temperature. A safety device is installed to shut down the flow heating system when the gas temperature reaches 800°C. During a periodic maintenance inspection, the thermocouple is judged to need replacement because of oxidation. By mistake, a type K thermocouple is installed as the replacement. What may be the results of such an installation?

Solution

The voltage output of a type J thermocouple at 800°C is

$$E_{800}(J) = 45.494 \text{ mV}$$

Presumably, the measurement system is calibrated for this type of thermocouple and will indicate a temperature of 800°C. For this same voltage output from a type K thermocouple the corresponding temperature would be

$$T_{45.494}(K) = 1110°C$$

Thus, the safety device would not be activated until a temperature 310°C higher than the design value is reached. This could easily result in material failure of parts of the equipment. Despite such consequences, faulty installations do occur in practice, and sometimes result in equipment breakdowns.

From the above discussion it is clear that the thermocouple measures the temperature at the last point of electric contact of the two dissimilar materials. Consider the situations shown in Fig. 8.22. The thermocouple installation in Fig. 8.22*a* is made so that only the junction bead makes contact with the metal plate whose temperature is to be measured. The installation in Fig. 8.22*b* allows contact at two points. If there is a temperature gradient in the metal plate, then the emf of the thermocouple will be indicative of the average of the temperatures of these two points. In Fig. 8.22*c* only the junction bead contacts the metal plate, but the two thermocouple wires are in electric contact a short distance away from the plate. The temperature indicated by the thermocouple will be that temperature at the shorted electric contact.

The transient response of thermocouples depends on junction size; the smaller the size, the faster the response. A number of commercial junctions are available which provide rapid transient response. Preformed butt-bonded junctions of about 0.1 mm thickness are encased in a polymer-glass laminate that can respond to temperature changes within 5 ms [25].

Thermocouples are used in applications ranging from measurement of room-air temperature to that of a liquid metal bath. Despite the simplicity, low cost, and ready availability of these thermocouples, there are problems which may be encountered such as:

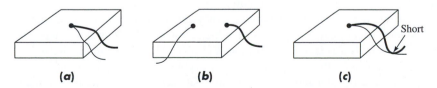

| (a) | (b) | (c) |

Figure 8.22   Installation of thermocouples on a metal plate. (*a*) Only junction bead contacts plate; (*b*) contact at two points: (*c*) contact at bead and along wires.

1. Junctions formed by users may involve excessive temperatures or faulty soldering techniques such that the thermocouple does not conform to the standard emf-temperature tables.

2. Thermocouples may be used outside their applicable range and become decalibrated over a period of time.

3. Faulty reference junction compensation may be employed.

4. Installation faults, as shown in Fig. 8.22, may occur.

5. The user may install the *wrong-type* thermocouple for the readout equipment. Gross errors can result.

QUARTZ-CRYSTAL THERMOMETER

A novel and highly accurate method of temperature measurement is based on the sensitivity of the resonant frequency of a quartz crystal to temperature change. When the proper angle of cut is used with the crystal, there is a very linear correspondence between the resonant frequency and temperature. Commercial models for the device utilize electronic counters and digital readout for the frequency measurement. For absolute temperature measurements usable sensitivities of $0.001°C$ are claimed for the device. Since the measurement process relies on a frequency measurement, the device is particularly insensitive to noise pickup in connecting cables. The device has a normal operating range of -40 to $230°C$ (-40 to $440°F$) and a frequency-temperature relationship of about 100 Hz/°C.

LIQUID-CRYSTAL THERMOGRAPHY

Cholesteric liquid crystals, formed from esters of cholesterol, exhibit an interesting response to temperature. Over a reproducible temperature range the liquid crystal will exhibit all colors of the visible spectrum. The phenomenon is reversible and repeatable. By varying the particular formulation, one can make cholesteric liquid crystals operate from just below $0°C$ to several hundred degrees Celsius. The *event temperature range,* or temperature at which the color changes are displayed, can vary from 1 to $50°C$. Thus, the liquid crystals afford the opportunity for rather precise temperature indication through observation of the color changes.

   To prevent deterioration of the crystals, they can be coated with polyvinyl alcohol, producing encapsulated liquid crystals which are available either as a water-based slurry or precoated on a blackened substrate of paper or Mylar. Resolution of the order of $0.1°C$ is claimed for the technique. Digital imaging, whereby pixel-by-pixel readings may be taken of color and illumination intensity across the image, are applicable to the technique.

   A description of the application of liquid crystals to heat-transfer measurements is given in Refs. [21] and [22].

THERMOELECTRIC POWER. The thermoelectric effect has been used for modest electric-power generation. Calculate the thermoelectric sensitivity of a device using bismuth and tellurium as the dissimilar materials and estimate the maximum voltage output for a 100°F temperature difference at approximately room temperature using one junction.

Example 8.7

Solution

The sensitivity is calculated from the data of Table 8.4 as

$$S = S_{\text{tellurium}} - S_{\text{bismuth}}$$
$$= 500 - (-72) = 572 \ \mu V/°C$$

The voltage output for a 100°F temperature difference is calculated as

$$E = S \ \Delta T = (572 \times 10^{-6})\left(\tfrac{5}{9}\right)(100)$$
$$= 3.18 \times 10^{-2} \text{ V}$$

8.6 TEMPERATURE MEASUREMENT BY RADIATION

In addition to the methods described in the preceding sections, it is possible to determine the temperature of a body through a measurement of the thermal radiation emitted by the body. Two methods are commonly employed for measurement: (1) optical pyrometry and (2) emittance determination. Before discussing these methods, we need to describe the nature of thermal radiation.

Thermal radiation is electromagnetic radiation emitted by a body as a result of its temperature. This radiation is distinguished from other types of electromagnetic radiation such as radio waves and X-rays, which are not propagated as a result of temperature. Thermal radiation lies in the wavelength region from about 0.1 to 100 μm (1 μm = 10^{-6} m). The total thermal radiation emitted by a *blackbody* (ideal radiation) is given as

$$E_b = \sigma T^4 \qquad \textbf{[8.17]}$$

where σ = Stefan-Boltzmann constant
$$= 0.1714 \times 10^{-8} \text{ Btu/h} \cdot \text{ft}^2 \cdot °R^4$$
$$= 5.669 \times 10^{-8} \text{ W/m}^2 \cdot K^4$$
E_b = emissive power, Btu/h · ft^2 or W/m^2
T = absolute temperature, °R or K

The emissive power of the blackbody varies with wavelength according to the Planck distribution equation

$$E_{b\lambda} = \frac{C_1 \lambda^{-5}}{e^{C_2/\lambda T} - 1} \qquad \textbf{[8.18]}$$

where $E_{b\lambda}$ = monochromatic blackbody emissive power

   = Btu/h · ft² · μm

   = W/m² · μm

λ = wavelength, μm

T = temperature, °R or K

$C_1 = 1.187 \times 10^8$ Btu · μm⁴/h · ft²

   = 3.743×10^8 W · μm⁴/m²

$C_2 = 2.5896 \times 10^4$ μm · °R

   = 1.4387×10^4 μm · K

A plot of Eq. (8.18) is given in Fig. 8.23 for two temperatures.

When thermal radiation strikes a material surface, the following relation applies:

$$\alpha + \rho + \tau = 1 \qquad\qquad \textbf{[8.19]}$$

where α = absorptivity or the fraction of the incident radiation absorbed

   ρ = reflectivity or the fraction reflected

   τ = transmissivity or the fraction transmitted

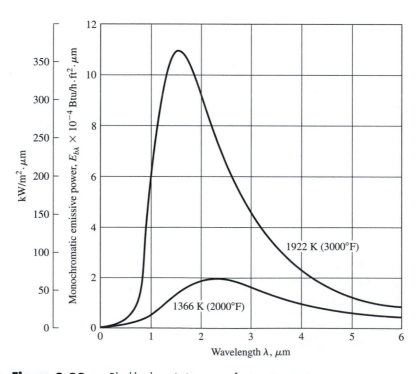

Figure 8.23   Blackbody emissive power for two temperatures.

For most solid materials $\tau = 0$ so that

$$\alpha + \rho = 1 \qquad \qquad \textbf{[8.20]}$$

The emissivity ϵ is defined as

$$\epsilon = \frac{E}{E_b} \qquad \qquad \textbf{[8.21]}$$

where E is the emissive power of an actual surface and E_b is the emissive power of a blackbody at the same temperature. Kirchhoff's identity furnishes the additional relationship

$$\epsilon = \alpha \qquad \qquad \textbf{[8.22]}$$

under conditions of thermal equilibrium. A gray body is one for which the emissivity is constant for all wavelengths; i.e.,

$$\epsilon_\lambda = \frac{E_\lambda}{E_{b\lambda}} = \epsilon \qquad \qquad \textbf{[8.23]}$$

Actual surfaces frequently exhibit highly variable emissivities over the wavelength spectrum. Figure 8.24 illustrates the distinctive features of blackbody and gray-body radiation. For purposes of analysis the real surface is frequently approximated by a

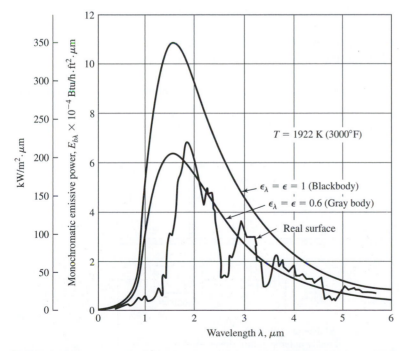

Figure 8.24 Comparison of emissive powers of blackbody, ideal gray body, and actual surfaces.

gray body having an emissivity equal to the average total emissivity of the real surface as defined by Eq. (8.21).

Let us now consider the measurement of temperature through the use of optical pyrometry. This methods refers to the identification of the temperature of a surface with the color of the radiation emitted. As a surface is heated, it becomes dark red, orange, and finally white in color. The maximum points in the blackbody radiation curves shift to shorter wavelengths with increase in temperature according to Wien's law.

$$\lambda_{max}T = 5215.6 \ \mu m \cdot {}^{\circ}R(2897.6 \ \mu m \cdot K) \qquad \textbf{[8.24]}$$

where λ_{max} is the wavelength at which the maximum points in the curves in Fig. 8.23 occur. The shift in these maximum points explains the change in color as a body is heated; that is, higher temperatures result in a concentration of the radiation in the shorter-wavelength portion of the spectrum. The temperature-measurement problem consists of a determination of the variation of temperature with color of the object. For this purpose an instrument is constructed as shown schematically in Fig. 8.25. The radiation from the source is viewed through the lens and filter arrangement. An absorption filter at the front of the device reduces the intensity of the incoming radiation so that the standard lamp may be operated at a lower level. The standard lamp is placed in the optical path of the incoming radiation. By an adjustment of the lamp current, the color of the filament may be made to match the color of the incoming radiation. The red filter is installed in the eyepiece to ensure that comparisons are made for essentially monochromatic radiation, thus eliminating some of the uncertainties resulting from variation of radiation properties with wavelength.

Figure 8.26 illustrates the appearance of the lamp filament as viewed from the eyepiece. When balance conditions are achieved, the filament will seem to disappear in the total incoming radiation field. Temperature calibration is made in terms of the lamp heating current.

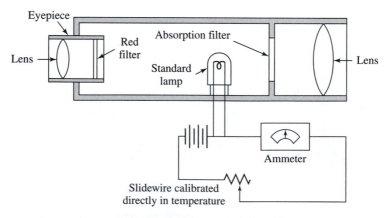

Figure 8.25   Schematic of optical pyrometer.

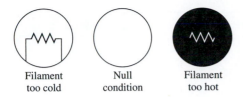

Figure 8.26 Appearance of lamp filament in eyepiece of optical pyrometer.

The temperature of a body may also be measured by determining the total emitted energy from the body and then calculating the temperature from

$$E = \epsilon\sigma T^4 \qquad \textbf{[8.25]}$$

In order to determine the temperature, the emissivity of the material must be known so that

$$T = \left(\frac{E}{\epsilon\sigma}\right)^{1/4} \qquad \textbf{[8.26]}$$

The *apparent* blackbody temperature is the value as calculated from Eq. (8.26) with $\epsilon = 1$, or

$$T_a = \left(\frac{E}{\sigma}\right)^{1/4} \qquad \textbf{[8.27]}$$

If the apparent temperature is taken as the measured value, the *error* in temperature due to nonblackbody conditions is thus

$$\text{Error} = \frac{T - T_a}{T} = 1 - \frac{T_a}{T} = 1 - \epsilon^{1/4} \qquad \textbf{[8.28]}$$

Figure 8.27 gives this error as a function of emissivity.

Regrettably, the emissivities of surfaces are subject to a great amount of uncertainty because they depend on surface finish, color, oxidation, aging, and a number of other factors. The uncertainty in the temperature of Eq. (8.26) resulting *only* from the uncertainty in emissivity is thus

$$\frac{w_T}{T} = \frac{1}{4}\frac{w_\epsilon}{\epsilon}$$

and the effect of absolute uncertainty in emissivity is more pronounced at low values of ϵ. For example,

$$\text{At } \epsilon = 0.2 \pm 0.05 \qquad \frac{w_T}{T} = \frac{0.05}{(4)(0.2)} = 0.0625$$

$$\text{At } \epsilon = 0.9 \pm 0.05 \qquad \frac{w_T}{T} = \frac{0.05}{(4)(0.9)} = 0.0139$$

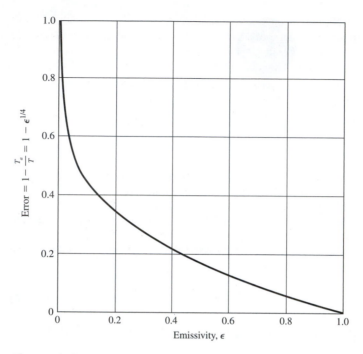

The vertical axis is labeled $\text{Error} = 1 - \dfrac{T_a}{T} = 1 - \epsilon^{1/4}$ and the horizontal axis is labeled Emissivity, ϵ.

Figure 8.27   Temperature error due to nonblackbody surface conditions.

Several methods are available to measure the emitted thermal energy from a body, and some of these methods will be discussed in Chap. 12. For now, it is important to realize the temperature may be calculated from the above equations once this measurement is made.

One way to employ the radiant flux measurement is to measure the actual surface temperature at some modest temperature condition using a noninvasive thermocouple and compare with the blackbody temperature indicated by the radiometer. The emissivity may then be calculated from Eq. (8.28). Unless surface conditions change appreciably with temperature (such as oxidation of a polished copper surface), the calculated value of emissivity may be used over a modest range of temperatures. Some radiometers (see Ref. [25]) provide internal electronic circuits for entering assumed values of emissivity during the measurement process so that a direct readout of temperature is given.

In practice, the optical pyrometer is the more widely used of the two radiation temperature methods for high temperatures, since it is relatively inexpensive and portable and the determination does not depend strongly on the surface properties of the material. The measurement of radiant energy from a surface can be quite accurate, however, when suitable instruments are employed. If the surface emissive properties are accurately known, this measurement can result in a very accurate determination of temperature.

Close proximity devices are available (see Ref. [25]) to operate within 2 mm of a surface and sense surface temperature by emitted radiation. Internal electronic circuits convert the signals to a voltage output which corresponds to standard emf-temperature relationships for thermocouples as given in Table 8.5. In this way standard thermocouple data loggers may be employed for recording and manipulation of the data.

The interested reader should consult Refs. [2], [4], and [7] for more information on thermal radiation and temperature measurements by the above methods.

EFFECT OF EMISSIVITY ON TEMPERATURE MEASUREMENT.   The energy emit- | **Example 8.8**
ted from a piece of metal is measured, and the temperature is determined to be 1050°C, assuming a surface emissivity of 0.82. It is later found that the true emissivity is 0.75. Calculate the error in the temperature determination.

Solution

The emitted energy is given by

$$\frac{q}{A} = \epsilon \sigma T^4$$

when $T = 1050°C = 1323$ K, $\epsilon = 0.82$. We wish to calculate the value of the true temperature T', such that

$$\frac{q}{A} = \epsilon' \sigma (T')^4$$

where $\epsilon' = 0.75$. Thus,

$$(0.82)(1323)^4 = (0.75)(T')^4$$

and

$$T' = (1323)\left(\frac{0.82}{0.75}\right)^{1/4} = 1352 \text{ K}$$

so that the temperature error is

$$\Delta T = 1352 - 1323 = 29°C$$

Comment

From this calculation we see that a relatively large error in estimation of emissivity (0.82 is 9.3 percent too high) causes a smaller error in temperature; i.e., 1323 K is 2.1 percent below 1352 K. Considering the *level* of the temperature measurement, this is quite good. For a measurement near room temperature of 300 K the error would be in the order of 6°C.

UNCERTAINTY IN RADIOMETER TEMPERATURE MEASUREMENT.   A radio- | **Example 8.9**
meter is used for a temperature measurement at 400 K and 800 K. The emissivity of the surface being measured is estimated as 0.02 ± 0.05 and the absolute uncertainty in the measurement of the emitted energy is estimated as 1 percent of the value of E at 800 K. Determine the uncertainty in the determination of the two temperatures.

Solution

The temperature is obtained from Eq. (8.26) as

$$T = \left(\frac{E}{\epsilon\sigma}\right)^{1/4}$$

Using the uncertainty relation for a product grouping from Eq. (3.2a) we obtain

$$\frac{W_T}{T} = \left[\left(\frac{w_\epsilon}{4\epsilon}\right)^2 + \left(\frac{w_E}{E}\right)^2\right]^{1/2} \qquad\qquad [a]$$

We have

$$\frac{w_\epsilon}{\epsilon} = \frac{0.05}{0.2}$$

$$w_E = 0.01 E_{800}$$

which is stated to be an *absolute value*.

Assuming that the emitted energy varies with T^4,

$$E_{400} = E_{800}\left(\frac{400}{800}\right)^4$$

and

$$\frac{w_E}{E_{400}} = (0.01)\left(\frac{800}{400}\right)^4 = 0.16$$

At 800 K Eq. (*a*) yields

$$\frac{w_T}{T} = \left\{\left[\frac{0.05}{4(0.2)}\right]^2 + 0.01^2\right\}^{1/2} = 0.0633$$

and $w_T = (800)(0.0633) = 50.6°C$.

In contrast, at 400 K the fractional uncertainty becomes

$$\frac{w_T}{T} = \left\{\left[\frac{0.05}{4(0.2)}\right]^2 + 0.16^2\right\}^{1/2} = 0.1718$$

and

$$w_T = (400)(0.1718) = 68.7°C$$

The uncertainty that results from w_ϵ alone is 50°C at 800 K and 25°C at 400 K.

8.7   EFFECT OF HEAT TRANSFER ON TEMPERATURE MEASUREMENT

A heat-transfer process is associated with all temperature measurements. When a thermometer is exposed to an environment, the temperature indicated by the thermometer is determined in accordance with the total heat-energy exchange with the temperature-sensing element. In some instances the temperature of the thermometer can be substantially different from the temperature which is to be measured. In this section we discuss some of the methods that may be used to correct the temperature readings. It may be noted that the errors involved are classified as fixed errors.

Heat transfer may take place as a result of one or more of three modes: conduction, convection, or radiation. In general, all three modes must be taken into account in analyzing a temperature-measurement problem. Conduction is described by Fourier's law:

$$q = -kA\frac{\partial T}{\partial x} \qquad \textbf{[8.29]}$$

where k = thermal conductivity, W/m · °C or Btu/h · ft · °F

         A = area through which the heat transfer takes place, m² or ft²

         q = heat-transfer rate in the direction of the decreasing temperature gradient, W or Btu/h

If a temperature gradient exists along a thermometer, heat may be conducted into or out of the sensing element in accordance with this relation.

Convection heat transfer is described in accordance with Newton's law of cooling:

$$q = hA(T_s - T_\infty) \qquad \textbf{[8.30]}$$

where h = convention heat-transfer coefficient, W/m² · °C or Btu/h · ft² · °F

         A = surface area exchanging heat with the fluid, m² or ft²

         T_s = surface temperature, °C or °F

         T_∞ = fluid temperature, °C or °F

The radiation heat transfer between two surfaces is proportional to the difference in absolute temperatures to the fourth power according to the Stefan-Boltzmann law of thermal radiation:

$$q_{1-2} = \sigma F_G F_\epsilon \left(T_1^4 - T_2^4\right) \qquad \textbf{[8.31]}$$

where F_G is a geometric factor and F_ϵ is a factor that describes the radiation properties of the surfaces.

Consider the temperature-measurement problem illustrated in Fig. 8.28. A thermocouple junction is installed in the flat plate whose temperature is to be measured. The plate has a thickness δ, is exposed to a convection environment on the top side, and insulated on the back side. The thermocouple is exposed to the same environment. The thermocouple wires are covered with insulating material as shown. If the plate temperature is higher than the convection environment, heat will be conducted out along the thermocouple wire and the temperature of the junction will be lower than the true plate temperature.

A solution to a similar problem, neglecting radiation effects, has been given by Schneider [11]. The effects of the thermocouple installation are shown in Fig. 8.29.

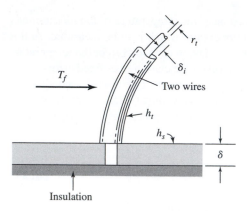

Figure 8.28   Schematic of thermocouple installation in a finite-thickness flat plate.

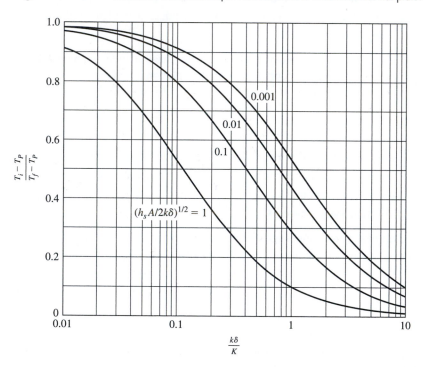

Figure 8.29   Temperature-compensation curves for installation in Fig. 8.28.

The nomenclature for the parameters in Fig. 8.29 is as follows:

h_s = convection heat-transfer coefficient on the top side of the plate as indicated, Btu/h · ft^2 · °F or W/m^2 · °C

h_t = convection heat-transfer coefficient from the thermocouple wires, Btu/h · ft^2 · °F or W/m^2 · °C

k = thermal conductivity of the plate material, Btu/h · ft · °F or W/m · °C

δ = plate thickness, ft or m

T_f = fluid temperature surrounding the thermocouple wire, °F or °C

T_i = temperature indicated by the thermocouple, °F or °C

T_p = true plate temperature (temperature a large distance away from the thermocouple junction), °F or °C

A = total cross-sectional area of the wires = $2\pi r_t^2$

$$K = A\left(k_A^{1/2} + k_B^{1/2}\right)\left(\frac{1}{h_t} + \frac{\delta_i}{k_i}\right) \Big/ (2r_t)^{1/2} \qquad \textbf{[8.32]}$$

where r_t is the radius of each thermocouple wire, k_A and k_B are the thermal conductivities of the two thermocouple materials, δ_i is the thickness of the thermocouple insulation, and k_i is the thermal conductivity of the insulation.

If the solid is a relatively massive one, the temperature correction can be made in accordance with calculations developed by Hennecke and Sparrow [16] and presented in graphical form in Fig. 8.30. In this figure k is the thermal conductivity of the solid, r is the radius of the wire or $\sqrt{2}$ times the radius for two wires, L is the length of the thermocouple leads, h_s is the convection heat-transfer coefficient between the solid and the fluid, $\overline{kA}$ is the equivalent conductivity area product for axial conduction in the wire, and R is the radial thermal resistance of the wire, insulation, and convection to the fluid. For a wire of radius r_w, covered by a layer of insulation having an outside radius r_i and thermal conductivity k_i, the value of R is calculated from

$$R = \frac{1}{h2\pi r_i} + \frac{\ln(r_i/r_w)}{2\pi k_i} \qquad \textbf{[8.33]}$$

where h is the convection heat-transfer coefficient from the outside of the insulation to the fluid. We should note that the length L is sufficiently long in many applications to make the hyperbolic tangent term unity in the equation for X in Fig. 8.30.

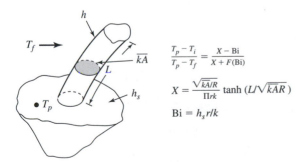

$$\frac{T_p - T_i}{T_p - T_f} = \frac{X - \text{Bi}}{X + F(\text{Bi})}$$

$$X = \frac{\sqrt{\overline{kA}/R}}{\Pi rk} \tanh (L/\sqrt{\overline{kA}R})$$

$$\text{Bi} = h_s r/k$$

Figure 8.30   Temperature correction for installation on a relatively massive plate.

The term $\text{Bi} = h_s r/k$ is called the Biot number or Biot modulus and is important in heat-transfer applications which involve combined conduction-convection systems like the thermocouple installations illustrated here. This parameter indicates the relative magnitude of the resistance to heat transfer resulting from convection to the resistance to heat transfer from conduction. Large values of Bi indicate a small convection resistance in comparison to conduction while small values of Bi indicate just the opposite. The function $F(\text{Bi})$ is given by

$$F(\text{Bi}) = 1.27 + 1.08\text{Bi} - 0.5\text{Bi}^2 \qquad \text{for Bi} < 1.0 \qquad \textbf{[8.34]}$$

This is an approximation by the present writer to more complicated expressions in Ref. [16]. It can be extended somewhat beyond the range of Bi < 1.0; however, larger values of Bi would indicate a larger effect of convection and thus a larger temperature error. In these cases, the installation should be modified.

The conduction lead error may be reduced by laying the wire along the solid to reduce the temperature gradient. Obviously, the wire must be electrically insulated from the solid if the solid is a metal.

Example 8.10 | **ERROR IN LOW-CONDUCTIVITY SOLID.**   A thermocouple wire having an effective diameter of 1.5 mm is attached to a ceramic solid having the properties $\rho = 2500$ kg/m^3, $c = 0.7$ kJ/kg · °C, and $k = 0.9$ W/m · °C. The thermocouple has an effective conductivity of 80 W/m · °C. The wire is very long and essentially bare, with a convection coefficient of 250 W/m^2 · °C. The convection coefficient h_s is 20 W/m^2 · °C. Calculate the true plate temperature when the thermocouple indicates 200°C and the fluid temperature is 90°C.

Solution

We make use of Fig. 8.30 and Eq. 8.33 and calculate the quantities

$$R = \frac{1}{h2\pi r_i} = \frac{1}{(250)(2\pi)(0.75 \times 10^{-3})} = 0.849$$

$$\overline{kA} = (80)\pi(0.75 \times 10^{-3})^2 = 1.414 \times 10^{-4}$$

Because $L \to \infty$, $\tanh(\infty) \to 1.0$, and

$$X = \frac{\sqrt{kA/R}}{\pi r k} = \frac{[(1.414 \times 10^{-4})/0.849]^{1/2}}{\pi(0.75 \times 10^{-3})(0.9)} = 6.085$$

$$\text{Bi} = \frac{h_s r}{k} = \frac{(20)(0.75 \times 10^{-3})}{0.9} = 0.0167 \qquad F(\text{Bi}) = 1.288$$

We then calculate

$$\frac{T_p - T_i}{T_p - T_f} = \frac{6.085 - 0.0167}{6.085 + 1.288} = 0.823$$

and, with $T_i = 200$°C, $T_f = 90$°C, the true plate temperature is

$$T_p = 711°C$$

This installation results in a very large error.

ERROR IN HIGH-CONDUCTIVITY SOLID.   Repeat Example 8.10 for installation on | **Example 8.11**
a block of aluminium with $k = 200$ W/m $\cdot$ °C.

Solution

We recalculate the appropriate parameters as

$$X = \frac{\sqrt{kA/R}}{\pi r k} = \frac{[(1.414 \times 10^{-4})/0.849]^{1/2}}{\pi (0.75 \times 10^{-3})(200)} = 0.0274$$

$$\text{Bi} = \frac{h_s r}{k} = \frac{(20)(0.75 \times 10^{-3})}{200} = 7.49 \times 10^{-5} \qquad F(\text{Bi}) = 1.271$$

Then,

$$\frac{T_p - T_i}{T_p - T_f} = \frac{0.0274 - 9.36 \times 10^{-4}}{0.0274 + 1.271} = 0.021$$

and

$$T_p = 202.3°\text{C}$$

Of course, this indicates the extreme importance of the conductivity of the solid to the measurement error.

Now, let us consider the general problem of the temperature measurement of a gas stream and the influence of radiation on this measurement. The situation is illustrated in Fig. 8.31. The temperature of the thermometer is designated as T_t, the true temperature of the gas is T_g, and the effective radiation temperature of the surroundings is T_s. If it is assumed that the surroundings are very large, then the following energy balance may be made:

$$hA(T_g - T_t) = \sigma A \epsilon \left(T_t^4 - T_s^4 \right) \qquad \text{[8.35]}$$

where $h =$ convection heat-transfer coefficient from the gas to the thermometer

$A =$ surface area of the thermometer

$\epsilon =$ surface emissivity of the thermometer

Equation (8.35) may be used to determine the true gas temperature.

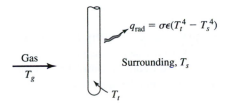

$q_{rad} = \sigma \epsilon (T_t^4 - T_s^4)$

Gas T_g

Surrounding, T_s

T_t

Figure 8.31   Schematic illustrating influence of radiation on temperature of thermometer.

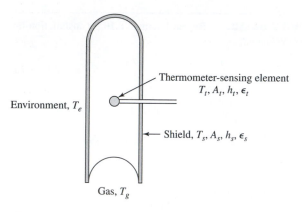

Figure 8.32   A simple radiation-shield arrangement for a thermometer.

In practice, the radiation error in a temperature measurement is reduced by placing a radiation shield around the thermometer, which reflects most of the radiant energy back to the thermometer. A simple radiation-shield arrangement is shown in Fig. 8.32. The environment is assumed to be very large, and we wish to find the true gas temperature knowing the indicated temperature T_t and the other heat-transfer parameters, such as the heat-transfer coefficients and emissivities. The thermometer size may vary considerably, from the rather substantial dimensions of a mercury-in-glass thermometer to a tiny thermistor bead embedded in the tip of a hypodermic needle. An energy balance that neglects conduction effects may be made for both the thermometer and the shield using a radiation-network analysis to obtain expressions for the radiation heat transfer.[1] The results of the analysis are

THERMOMETER

$$h_t(T_g - T_t) = \frac{\epsilon_t}{1 - \epsilon_t}(E_{bt} - J_t) \qquad [8.36]$$

SHIELD

$$2h_s(T_g - T_s) = \frac{E_{bs} - J_t}{(1/F_{ts})(A_s/A_t) + 1/\epsilon_s - 1} + \frac{E_{bs} - E_{be}}{1/\epsilon_s - 1 + 1/F_{se}} \qquad [8.37]$$

where

$$J_t = \frac{\{E_{bt}[\epsilon_t/(1-\epsilon_t)] + F_{te}E_{be}\}A_t/A_s + E_{bs}/[(1/F_{ts})(A_s/A_t) + 1/\epsilon_s - 1]}{(A_t/A_s)[F_{te} + \epsilon_t/(1-\epsilon_t)] + 1/[(1/F_{ts})(A_s/A_t) + 1/\epsilon_s - 1]} \qquad [8.38]$$

ϵ_t and ϵ_s are the emissivities of the thermometer and shield, respectively, h_t and h_s are the convection heat-transfer coefficients from the gas to the thermometer and shield, respectively, A_t is the surface area of the thermometer for convection and radiation, A_s is the surface area of the radiation shield *on each side,* and the blackbody emissive

[1] An explanation of the radiation-network method is given in Ref. [5].

powers E_b are given by

$$E_{bt} = \sigma T_t^4 \qquad \textbf{[8.39]}$$

$$E_{bs} = \sigma T_s^4 \qquad \textbf{[8.40]}$$

$$E_{be} = \sigma T_e^4 \qquad \textbf{[8.41]}$$

The radiation-shape factors F are defined as

F_{ts} = fraction of radiation that leaves the thermometer and arrives at the shield

F_{se} = fraction of radiation that leaves the outside surface of shield and gets to the environment (this fraction is 1.0)

F_{te} = fraction of radiation that leaves the thermometer and gets to the environment

The analysis that arrives at Eqs. (8.36) and (8.37) has been simplified somewhat by assuming that all the radiation exchange between the shield and environment may be taken into account by adjusting the value of F_{se} to include radiation from the inside surface of the shield. Some type of convection heat-transfer analysis must be used to determine the values of h_t and h_s, depending on whether natural or forced convection is involved, etc. An iterative solution must usually be performed on Eqs. (8.36) and (8.37) except in special cases. Some useful charts are given in Ref. [15] for correcting temperature measurements for the effects of conduction and radiation.

In all cases the radiation error will be reduced by using a radiation shield, which is as reflective as possible (ϵ very small).

When convection from the shield can be neglected, it can be shown that if the shield placed around the thermometer essentially surrounds the sensing element, the radiation term on the right-hand side of Eq. (8.35) should be multiplied by a factor F_s.

$$F_s = \frac{1}{1 + (A\epsilon/A_s)(2/\epsilon_s - 1)} \qquad \textbf{[8.35a]}$$

where A = area of the thermometer element

A_s = surface area of the shield

ϵ_s = emissivity of the shield

ϵ = emissivity of the sensor

It should be noted again that the installation of *any* shield will reduce the radiation heat transfer and thereby improve the temperature measurement.

In general, the determination of convection heat-transfer coefficients requires very detailed consideration of the fluid dynamics of the problem, and empirical relations must often be used for the calculation. For more information the reader should consult Refs. [2] and [5]. Some frequently used formulas for calculating convection heat transfer are summarized in Table 8.6.

Uncertainties in the values of h of ± 25 percent are not uncommon. This should not alarm the reader, however, because even an approximate value can provide a very useful calculation for correcting the temperature measurement.

Table 8.6 Frequently used convection heat-transfer formulas, adapted from Holman [5]

Physical Situation	Type of Fluid	Range of Validity	Heat-Transfer Relation	Fluid Properties Evaluated At
Forced convection over flat plate, plate heated over entire length	Gas or liquid	Laminar: $\mathrm{Re}_x < 5 \times 10^5$ Turbulent: $\mathrm{Re}_x > 5 \times 10^5$	$\mathrm{Nu}_x = 0.332\,\mathrm{Re}_x^{1/2}\mathrm{Pr}^{1/3}$ $\mathrm{St}_x\,\mathrm{Pr}^{2/3} = 0.0296\,\mathrm{Re}_x^{1/5}$	Film temperature T_f
		Laminar: $\mathrm{Re}_L < 5 \times 10^5$ Turbulent: $\mathrm{Re}_L > 5 \times 10^5$	$\overline{\mathrm{Nu}}_L = 0.664\,\mathrm{Re}_L^{1/2}\mathrm{Pr}^{1/3}$ $\overline{\mathrm{Nu}}_L = -.037\,\mathrm{Re}_L^{0.8} \qquad 871/\mathrm{Pr}^{1/3}$	
Forced convection in smooth circular tube	Gas or liquid	$\mathrm{Re}_d > 3000$	$\overline{\mathrm{Nu}}_d = 0.023\,\mathrm{Re}_d^{0.8}\mathrm{Pr}^n$ $n = 0.4$ for heating $n = 0.3$ for cooling	Average bulk temperature of fluid
		$\mathrm{Re}_d < 2100$ and $\left.\begin{array}{l}\\ \mathrm{Re}_d\mathrm{Pr}\left(\dfrac{d}{L}\right) > 10 \end{array}\right\}$	$\overline{\mathrm{Nu}}_d = 1.86\,\mathrm{Re}_d\,\mathrm{Pr}^{1/3}\left(\dfrac{d}{L}\right)^{1/3}\left(\dfrac{\pi}{\pi_w}\right)^{0.14}$	
Forced-convection crossflow over cylinder	Gas or liquid	$40 < \mathrm{Re}_d < 4000$ $4000 < \mathrm{Re}_d < 40{,}000$ $40{,}000 < \mathrm{Re}_d < 400{,}000$	$\overline{\mathrm{Nu}}_d = 0.683\,\mathrm{Re}_d^{0.466}\mathrm{Pr}^{1/3}$ $\overline{\mathrm{Nu}}_d = 0.193\,\mathrm{Re}_d^{0.618}\mathrm{Pr}^{1/3}$ $\overline{\mathrm{Nu}}_d = 0.0266\,\mathrm{Re}_d^{0.805}\mathrm{Pr}^{1/3}$	Film temperature T_f Film temperature T_f
Forced convection over spheres	Gas	$17 < \mathrm{Re}_d < 70{,}000$	$\overline{\mathrm{Nu}}_d = 0.37\,\mathrm{Re}_d^{0.6}$	Film temperature T_f
	Liquid	$1 < \mathrm{Re}_d < 200{,}000$	$\overline{\mathrm{Nu}}_d\,\mathrm{Pr}^{-0.3}\left(\dfrac{\pi}{\pi_w}\right)^{0.25} = 1.2 + 0.5\,\mathrm{Re}_d^{0.54}$	Free-stream temperature T_∞
Free convection from vertical flat plate	Gas or liquid	$10^4 < \mathrm{Gr}_L\mathrm{Pr} < 10^9$ $10^9 < \mathrm{Gr}_L\mathrm{Pr} < 10^{12}$	$\overline{\mathrm{Nu}}_L = 0.59\,\mathrm{Gr}_L\mathrm{Pr}^{1/4}$ $\overline{\mathrm{Nu}}_L = 0.10\,\mathrm{Gr}_L\mathrm{Pr}^{1/3}$	Film temperature T_f Film temperature T_f
Free convection from horizontal cylinders	Gas or liquid	$10^4 < \mathrm{Gr}_d\mathrm{Pr} < 10^9$ $10^9 < \mathrm{Gr}_d\mathrm{Pr} < 10^{12}$	$\overline{\mathrm{Nu}}_d = 0.53\,\mathrm{Gr}_d\mathrm{Pr}^{1/4}$ $\overline{\mathrm{Nu}}_d = 0.13\,\mathrm{Gr}_d\mathrm{Pr}^{1/3}$	Film temperature T_f Film temperature T_f

Definition of symbols: All quantities in consistent set of units so that Nu, Re, Pr, Gr, and St are dimensionless.

$$\overline{\mathrm{Nu}}_d = \frac{hd}{k} \qquad \overline{\mathrm{Nu}}_L = \frac{hL}{k} \qquad \mathrm{Nu}_x = \frac{h_x x}{k} \qquad \mathrm{Re}_d = \frac{\rho u_\infty d}{\pi} \text{ for flow over cylinders or spheres} \qquad \mathrm{Re}_d = \frac{\rho u_m d}{\pi} \text{ for flow in tubes} \qquad \mathrm{Re}_x = \frac{\rho u_\infty x}{\pi}$$

$$\mathrm{Re}_l = \frac{\rho u_\infty L}{\pi} \qquad \mathrm{Pr} = \frac{c_p \mu}{k} \qquad \mathrm{Gr}_l = \frac{\rho^2 g \cdot T_w \quad T_x/l^3}{\mu^2} \qquad \mathrm{Gr}_d = \frac{\rho^2 g \cdot T_w \quad T_\infty/d^3}{\mu^2} \qquad \mathrm{St}_x = \frac{h_x}{\rho u_\infty c_p}$$

where

d = diameter of tube
g = acceleration of gravity
h = average heat-transfer coefficient over entire surface
h_x = local heat-transfer coefficient on flat plate
k = thermal conductivity of fluid
L = total length or height of flat plate or length of tube

T_f = film temperature $\cdot T_w + T_\infty//2$
T_w = wall or surface temperature
T_∞ = free-stream temperature
u_m = mean fluid velocity in tube
u_∞ = free-stream velocity past flat plate, cylinder, or sphere

x = distance from leading edge of flat plate
β = volume coefficient of expansion of fluid
μ = dynamic viscosity of fluid
μ_w = viscosity evaluated at wall temperature
ρ = fluid density

CALCULATION OF RADIATION ERROR.   A mercury-in-glass thermometer is placed | **Example 8.12**
inside a cold room in a frozen-food warehouse to measure the change in air temperature when
the door is left open for extended periods of time. In one instance the thermometer reads 1°C,
while the automatic temperature-control system for the room indicates that the *wall temperature*
of the room is −10°C. The convection heat-transfer coefficient for the thermometer is estimated
at 10 W/m^2 · °C, and $\epsilon = 0.9$ for glass. Estimate the true air temperature.

Solution

We use Eq. (8.35) to calculate the true air temperature T_g:

$$h(T_g - T_t) = \epsilon\sigma\left(T_t^4 - T_s^4\right)$$
$$(10)(T_g - 274) = (5.669 \times 10^{-8})(0.9)(274^4 - 263^4)$$
$$T_g = 278.3 \text{ K} = 5.3°\text{C}$$

We may note, of course, the rather substantial difference between the indicated air temperature
of 1°C and the actual temperature of 5.3°C.

IMPROVEMENT WITH RADIATION SHIELD.   A shield having an emissivity of 0.1 | **Example 8.13**
is placed around the thermometer in Example 8.12 such that the area ratio A/A_s is 0.3. The
temperature indicated by the thermometer under these conditions is 5°C. The convection heat-
transfer coefficient is assumed to be the same, and the shield may be assumed essentially to
surround the thermometer. Calculate the true air temperature under these circumstances.

Solution

This is a condition where the factor F_s may be applied to the right-hand side of Eq. (8.35). We
have

$$\epsilon_s = 0.1, \qquad \epsilon = 0.9, \qquad\qquad\qquad h = 10 \text{ W/m}^2 - °\text{C}$$
$$\frac{A}{A_s} = 0.3, \qquad T_t = 273 + 5 = 278 \text{ K}, \qquad T_s = 263 \text{ K}$$

and
$$E_s = \frac{1}{1 + (0.3)(0.9)[(2/0.1) - 1]} = 0.1631$$

Inserting the values in Eq. (8.35) modified with F_s gives

$$(10)(T_g - 278) = (5.669 \times 10^{-8})(0.9)(278^4 - 263^4)(0.1631)$$

and
$$T_g = 279 \text{ K} = 6°\text{C}$$

So, by use of the shield, the difference in the indicated temperature (5°C) and the true temperature (6°C) has been reduced to −1°C.

8.8   TRANSIENT RESPONSE OF THERMAL SYSTEMS

When a temperature measurement is to be made under non-steady-state conditions,
it is important that the transient response characteristics of the thermal system be

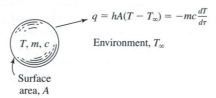

$$q = hA(T - T_\infty) = -mc\frac{dT}{d\tau}$$

Environment, T_∞

Figure 8.33   Simple thermal system exposed to a sudden change in ambient temperature.

taken into account. Consider the simple system shown in Fig. 8.33. The thermometer is represented by the mass m and specific heat c and is suddenly exposed to a convection environment of temperature T_∞. The convection heat-transfer coefficient between the thermometer and fluid is h, and radiation heat transfer is assumed to be negligible. It is also assumed that the thermometer is substantially uniform in temperature at any instant of time; that is, the thermal conductivity of the thermometer material is sufficiently large in comparison with the surface conductance because of the convection heat-transfer coefficient. The energy balance for the transient process may be written as

$$hA(T_\infty - T) = mc\frac{dT}{d\tau} \qquad [8.42]$$

For this equation appropriate units are:

h in W/m^2 · °C or Btu/h · ft^2 · °F

A in m^2 or ft^2

T in °C or °F

m in kg or lbm

c in J/kg · °C or Btu/lbm · °F

τ in s or h

The solution of Eq. (8.39) gives the temperature of the thermometer as a function of time:

$$\frac{T - T_\infty}{T_0 - T_\infty} = e^{(-hA/mc)\tau} \qquad [8.43]$$

where T_0 is the thermometer temperature at time zero. Equation (8.43) represents the familiar exponential behavior of first-order systems subjected to a step input as discussed in Chap. 2. The time constant for the system in Fig. 8.33 is

$$\tau = \frac{mc}{hA} \qquad [8.44]$$

When the heat-transfer coefficient is sufficiently large, there may be substantial temperature gradients within the thermometer itself and a different type of analysis must be used to determine the temperature variation with time. In many cases the heat-transfer coefficient h may vary with both surface position and surface temperature, and an appropriate average value must be used to evaluate the time constant from Eq. (8.44). The interested reader should consult Ref. [5] for more information on this subject.

8.9 THERMOCOUPLE COMPENSATION

Suppose a thermocouple is used to measure a transient temperature variation. The response of the thermocouple will be dependent on several factors, as outlined above, and will follow a variation like that of Eq. (8.43) when subjected to a step change in environment temperature. If a compensating electric network is applied to the system, it is possible to increase the frequency response of the thermocouple. The disadvantage of the compensating network is that it reduces the thermocouple output; however, if the measuring instrument is sufficiently sensitive, this problem is not too critical.

A typical thermocouple-compensation network is shown in Fig. 8.34. The thermocouple *input voltage* is represented by E_i, and the *output voltage* is represented by E_o. This particular network will attenuate low frequencies more than high frequencies and has a frequency response to a step input that is approximately opposite to that of a thermocouple. The network and thermocouple may thus be used in combination to produce a more nearly flat response over a wider frequency range. The output response of the network is given by

$$\frac{E_o}{E_i} = \alpha \frac{1 + j\omega\tau_c}{1 + \alpha j\omega\tau_c} \qquad \textbf{[8.45]}$$

where $\alpha = \dfrac{R}{R + R_c}$

$\tau_c = R_c C$

$\omega = $ frequency of the input signal

$j = \sqrt{-1}$

The amplitude response is given by the absolute value of the function in Eq. (8.45). Thus,

$$\left(\frac{E_o}{E_i}\right)_{\text{amplitude}} = \alpha \sqrt{\frac{1 + \omega^2\tau_c^2}{1 + \alpha^2\omega^2\tau_c^2}} \qquad \textbf{[8.46]}$$

The high-frequency compensation of the network is improved as the value of α is decreased, but this brings about a decreased output since the steady-state output is

$$\left(\frac{E_o}{E_i}\right)_{\text{steady state}} = \frac{R}{R + R_c} = \alpha \qquad \textbf{[8.47]}$$

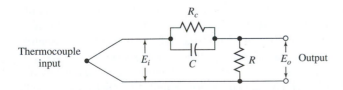

Figure 8.34   Typical thermocouple-compensation network.

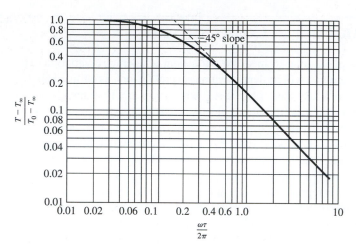

Figure 8.35   Thermocouple response to step change in environmental temperature.

In practice, the compensating circuit is usually designed so that τ_c is equal to the time constant for the thermocouple. Variable resistors may be used to change the compensation with a change in thermal environment conditions.

The overall problem we have described above is one of compensating a thermocouple output so that higher-frequency response may be obtained. The step response of the thermocouple given by Eq. (8.43) will hold, of course, only for the conditions described. Therefore, the compensating network of Fig. 8.34 will achieve the desired objectives in varying degrees, depending on the exact thermal environment to which the thermocouple is exposed. Let us examine the thermocouple and compensating network response in more detail. Figure 8.35 shows the thermocouple response to a step change in temperature of the environment T_∞. We assume that the voltage output of the thermocouple is directly proportional to temperature over this range so that the voltage output would also follow a similar curve. In Fig. 8.36 we have plotted the output voltage of the compensating network for a step-voltage input and for $\alpha = 0.1$. The thermocouple response is obtained by solving the electric-network equivalent of the thermocouple. In that network the thermal capacitance is analogous to electric capacitance, and the convection heat-transfer coefficient is analogous to thermal or electric resistance. If Figs. 8.35 and 8.36 are added together, we obtain the plot shown in Fig. 8.37, where the response is shown for different values of the ratio τ_c/τ. For $\tau_c/\tau > 1$, the response curve "overshoots" the steady-state voltage ratio of 0.1. It is clear that the curve $\tau = \tau_c$ offers a broader frequency response than the uncompensated curve. Careful inspection will also show that some overshoot can be desirable. The curve for $\tau_c/\tau = 1.1$ indicates a substantial extension of the frequency response within a plus or minus range of about 5 percent. Excessive overshoot, of course, gives a decided nonlinear response. Optimal dynamic response of the network will be obtained when $\tau_c/\tau \approx 1.1$, provided the thermocouple input behaves nicely. In practice, variable resistors will be provided in the circuits to adjust the compensation in accordance with the actual variations in the thermal environment.

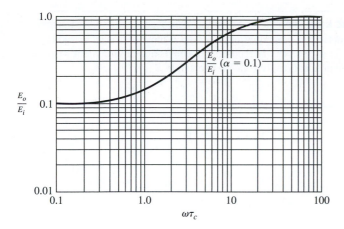

Figure 8.36   Response of the network in Fig. 8.34 for a step voltage input
with $\alpha = 0{:}1$.

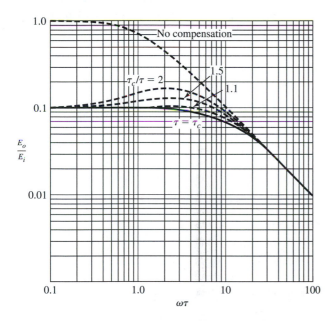

Figure 8.37   Combined response of thermocouple and compensating
network to step change in environment temperature.

We note again that the foregoing discussion pertains to a step-function input, and
the actual input may be markedly different. In practice, one should make the thermocouple as small as possible and calibrate against some known transient temperature.
Adjustments to the network can be made to obtain the best calibration. Further information on transient response of thermocouples is given in Ref. [17].

Example 8.14 | **THERMOCOUPLE TIME CONSTANT.**   A thermocouple bead has the approximate shape of a sphere 1.5 mm in diameter. The properties may be taken as those of iron ($\rho = 7900$ kg/m^3, $c = 450$ J/kg · °C). Suppose such a bead is exposed to a convection environment where $h = 75$ W/m^2 · °C. Estimate the time constant for the thermocouple.

Solution

We use Eq. (8.44) to calculate the time constant with

$$m = \rho V = \rho \left(\frac{4}{3}\pi r^3 \right)$$

$$A = 4\pi r^2$$

so that $$\tau = \frac{(7900)(450)(0.75 \times 10^{-3})}{(3)(75)} = \frac{\rho \left(\frac{4}{3}\pi r^3 \right) c}{h4\pi r^2}$$

$$= 11.85 \text{ s}$$

Example 8.15 | **THERMOCOUPLE COMPENSATION.**   Design a thermocouple-compensation network like that shown in Fig. 8.34 having a steady-state attenuation of a factor of 10. Let the network have the same time constant as the thermocouple bead in Example 8.14.

Solution

We take

$$\alpha = \frac{R}{R + R_c} = 0.1 \qquad \text{and} \qquad \tau_c = R_c C = 11.85$$

We must arbitrarily choose a value for one of the resistances. Let us choose a value of R_c which will make C a reasonable number. Take

$$R_c = 1 \text{ M}\Omega = 10^6 \text{ }\Omega$$

Then $C = 11.85/10^6 = 1.185 \times 10^{-5} = 11.85$ μF. Using the above value of R_c, we obtain

$$R = 0.111 \text{ M}\Omega$$

A more practical set of values would be

$$C = 10 \text{ }\mu\text{F}$$
$$R_c = 1.185 \text{ M}\Omega$$
$$R = 0.132 \text{ M}\Omega$$

8.10   TEMPERATURE MEASUREMENTS IN HIGH-SPEED FLOW

The measurements of temperatures in high-speed gas-flow streams are sometimes difficult because of the fact that a stationary probe must measure the temperature of the gas that is brought to rest at the probe surface. As the gas velocity is reduced to zero, the kinetic energy is converted to thermal energy and evidenced as an increase in temperature. If the gas is brought to rest adiabatically, the resulting stagnation temperature is given by

$$T_0 = T_\infty + \frac{u_\infty^2}{2c_p g_c} \qquad [8.48]$$

where T_0 = stagnation temperature

T_∞ = free-stream or *static* temperature

u_∞ = flow velocity

c_p = constant-pressure specific heat of the gas

The stagnation temperature of the gas may also be expressed in terms of the Mach number as

$$\frac{T_0}{T_\infty} = 1 + \frac{\gamma - 1}{2} M^2 \qquad [8.49]$$

where $\gamma = c_p/c_v$ and has a value of 1.4 for air. An inspection of Eqs. (8.48) and (8.49) shows that the stagnation temperature is substantially the same as the static temperature for low-speed flow. The difference in these temperatures is only 4°F (2.2°C) for a velocity of 226 ft/s (68.9 m/s) in air.

For the actual case of a probe inserted in a high-speed flow stream the temperature indicated by the probe will *not,* in general, be equal to the stagnation temperature given by Eq. (8.48). The measured temperature is called the *recovery temperature* and is strongly dependent on the probe configuration. The *recovery factor* is defined by

$$r = \frac{T_r - T_\infty}{T_0 - T_\infty} \qquad [8.50]$$

where T_r is the recovery temperature. In practice, the recovery factor must be determined from a calibration of the probe at different flow conditions. It is usually in the range $0.75 < r < 0.99$. One of the important objectives of a probe design is to achieve a configuration which produces a near constant recovery factor over a wide range of flow velocities.

A typical high-speed temperature probe is shown in Fig. 8.38 according to Winkler [13]. The gas stream enters the front of the probe and is diffused to a lower velocity in the enclosing shield arrangement. The flow subsequently leaves the probe through the side ventholes. The shield also serves to reduce radiation losses from the thermocouple-sensing element. The recovery factor for the probe is essentially 1.0.

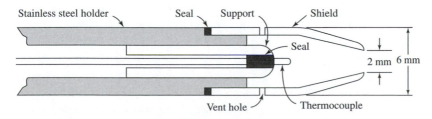

Figure 8.38   High-speed temperature probe, according to Winkler [13].

Table 8.7 Summary of characteristics of temperature-measurement devices

Device	Applicable Temperature Range		Approximate Accuracy		Transient Response	Cost	Remarks
	°F	°C	°F	°C			
Liquid-in-glass thermometer:							
a. Alcohol	−90–150	−70–65	±1	±0.5	Poor	Low	Used as inexpensive low-temperature thermometers
b. Mercury	−35–600	−40–300	±0.5	±0.25	Poor	Variable	Accuracy of ±0.1° F .0.05°C) may be obtained with special calibrated thermometers
c. Gas-filled mercury	−35–1000	−40–550	±0.5	±0.25	Poor	Variable	
Fluid-expansion thermometer:							
a. Liquid or gas	−150–1000	−100–550	±2	±1	Poor	Low	Widely used for industrial temperature measurements
b. Vapor-pressure	−20–400	−6–200	±2	±1	Poor	Low	
Bimetallic strip	−100–1000	−70–550	±0.5	±0.25	Poor	Low	Widely used in simple temperature-control devices
Electrical-resistance thermometer	−300–1800	−180–1000	±0.005[1]	±0.0025	Fair to good, depending on size of element	Readout equipment can be rather expensive for high-precision work	Most accurate of all methods
Thermistor	−100–500	−70–250	±0.02[1]	±0.01	Very good	Low, but readout equipment may be expensive for high-precision work	Useful for temperature-compensation circuits; thermistor beads may be obtained in very small sizes

Instrument	Range	Range	Accuracy ±	Accuracy ±	Response	Cost	Remarks
Copper-constantan thermocouple	−300–650	−180–350	±0.5	±0.25	Good, depends on wire size	Low	
Iron-constantan thermocouple	−300–1200	−180–650	±0.5	±0.25	Good, depends on wire size	Low	Superior in reducing atmospheres
Chromel-alumel thermocouple	−300–2200	−180–1200	±0.5	±0.25	Good, depends on wire size	Low	Resistant to oxidation at high temperatures
Platinum-platinum 10% rhodium thermocouple	−0–3000	−15–1650	±0.5	±0.25	Good, depends on wire size	High	Low output; most resistant to oxidation at high temperatures; accuracy of $0.15°F$ ($0.07°C$) may be obtained in carefully controlled conditions
Optical pyrometer	1200 up	650 up	±20	±10	Poor	Medium	Widely used for measurement of industrial furnace temperatures
Radiation pyrometer	0 up	−15 up	±0.5°C at low ranges, 2.5 to 10°C at high temperatures; depends on blackbody conditions and type of pyrometer; see Chap. 12		Good, depending on type of pyrometer	Medium to high	Increased applications resulting from new high-precision devices being developed

[1] Accuracy is that which *may* be achieved. Inexpensive versions of the devices may not give the optimum accuracy. Figures do not take into account inaccuracies which may result from thermal conduction or convection at the point of measurement.

Example 8.16 | **STAGNATION AND RECOVERY TEMPERATURES.**   A thermocouple is enclosed in a 3.0-mm-diameter sheath and exposed to a high-speed airstream at $M = 2.5$ and $20°C$. The recovery factor for the sheath is 0.85 and the emissivity is 0.9. The effective radiation temperature of the surroundings is $20°C$. Estimate the temperature indicated by the sensor. The convection heat-transfer coefficient is approximately $150 \ W/m^2 - °C$.

Solution

We first calculate the stagnation temperature from Eq. (8.49). Taking $T_\infty = 20°C = 293 \ K$ and $\gamma = 1.4$ for air, we have

$$T_0 = (293)\left[1 + \frac{1.4 - 1}{2}(2.5)^2\right] = 659 \ K$$

The recovery temperature for *adiabatic* stagnation is then obtained from Eq. (8.50):

$$r = 0.85 = \frac{T_r - 293}{659 - 293}$$

and $T_r = 604 \ K$

This is the temperature the sensor would indicate if there were no radiation heat transfer. To take radiation into account, we write Eq. (8.35) as

$$\sigma \epsilon A\left(T_t^4 - T_\infty^4\right) = hA(T_r - T_t) \qquad \textbf{[a]}$$

where the convection heat transfer is based on the difference between the adiabatic recovery temperature and the actual thermocouple temperature T_t. We have $\epsilon = 0.9$, $T_\infty = 293 \ K$, and $T_r = 604 \ K$ so that a solution of Eq. (*a*) gives

$$T_t = 571 \ K = 298°C$$

8.11 SUMMARY

A summary of the range and application of the various temperature-measurement devices is given in Table 8.7. It will be noted that many of the devices overlap in their range of application. Thus, the selection of a device for use in a particular application is often a matter of preference. In many cases the selection is based on a consideration of the type of readout equipment that is already available at the installation where the device is to be used.

8.12 REVIEW QUESTIONS

8.1 Distinguish between the Fahrenheit and Celsius temperature scales.

8.2 Upon what principle does the mercury-in-glass thermometer operate?

8.3 Describe the resistance characteristics of thermistors.

8.4 State the law of intermediate metals for thermocouples.

8.5 What is the Seebeck effect?

8.6 State the law of intermediate temperatures for thermocouples.

8.7 Why is a reference temperature necessary when using thermocouples?

8.8 Why does the resistance of thermocouple wires not influence a temperature reading when a potentiometer is used?

8.9 In general, why does heat transfer influence the accuracy of a temperature measurement?

8.10 In general, would one classify the response of thermal systems as fast, slow, or intermediate?

8.11 Distinguish between hardware and software compensation for thermocouple reference junctions.

8.12 What kinds of problems can be encountered in thermocouple use?

8.13 Why does self-heating cause a problem with RTDs? Why is it of lesser importance with thermistors?

8.14 Why is emissivity important in radiation temperature measurements?

8.15 Where might one use a bimetallic strip thermometer?

8.16 What would be the major cause of error in the mercury-in-glass thermometer?

8.17 What are the major disadvantages of RTDs?

8.13   PROBLEMS

8.1. Calculate the temperature at which the Fahrenheit and Celsius scales coincide.

8.2. A certain mercury-in-glass thermometer has been calibrated for a prescribed immersion depth. The thermometer is immersed too much, such that the extra depth is equal to a distance of 10° on the scale. The true temperature reading may be calculated with

$$T_{\text{true}} = T_{\text{ind}} - 0.000088(T_{\text{ind}} - T_{\text{amb}})D$$

where T_{ind} = indicated temperature

T_{amb} = ambient temperature of the exposed stem

D = extra immersion depth of the thermometer past the correct mark

Calculate the thermometer error for an indicated temperature of 210°F and an ambient temperature of 70°F.

8.3. The bimetallic strip material of Example 8.1 is to be used in an on-off-temperature-control device, which will operate at a nominal temperature of 200°F. Calculate the deflection at the end of a 4-in strip for deviations of ±1 and 2°F from the nominal temperature.

8.4. Suppose the bimetallic strip in Prob. 8.3 is used to indicate temperatures of 150, 200, and 300°F. The tip of the strip is connected to an appropriate mechanism, which amplifies the deflection so that it may be read accurately. The uncertainty in the length of the strip is ±0.01 in, and the uncertainty in the thickness of each material is ±0.0002 in. The perpendicular deflection from the 100°F position is taken as an indication of the temperature. Calculate the uncertainty in each of the above temperatures, assuming zero uncertainty in all the material properties; that is, calculate the uncertainty in the deflection at each of these temperatures.

8.5. A bimetallic strip of yellow brass and Monel 400 is bonded at 120°F. The thickness of the yellow brass is 0.014 ± 0.0002 in, and the thickness of the Monel is 0.010 ± 0.0001 in. The length of the strip is 5 in. Calculate the deflection sensitivity defined as the deflection per degree Fahrenheit temperature difference. Estimate the uncertainty in this deflection sensitivity.

8.6. The specific volume of mercury is given by the relation

$$v = v_0(1 + aT + bT^2)$$

where T is in degrees Celsius and

$$a = 0.1818 \times 10^{-3}$$
$$b = 0.0078 \times 10^{-6}$$

A high-temperature thermometer is constructed of a Monel-400 tube having an inside diameter of 0.8 mm ± 5 nm. After the inside is evacuated, mercury is placed on the inside such that a column height of 10 cm ± 0.25 mm is achieved when the thermometer temperature is 260°C. If the tube is to be used for a temperature measurement, calculate the uncertainty at 260°C if the uncertainty in the height measurement is ±0.25 mm.

8.7. A fluid-expansion thermometer uses liquid Freon ($C = 0.22$ Btu/lbm · °F, $\rho = 85.2$ lbm/ft^3) enclosed in a 0.25-in-ID copper ($C = 0.092$, $\rho = 560$) cylinder with a wall thickness of 0.032 in. The thermometer is exposed to a crossflow of air at 400°F, 15 psia, and 20 ft/s. Estimate the time constant for such a thermometer. Neglect all effects of the capillary tube. Repeat for a crossflow of liquid water at 180°F and 20 ft/s. Recalculate the time constant for both of these situations for a stainless-steel bulb ($C = 0.11$, $\rho = 490$).

8.8. It is desired to measure a temperature differential of 5°F using copper-constantan thermocouples at a temperature level of 200°F. A millivolt recorder is available for the emf measurement which has an uncertainty of 0.004 mV. Precision thermocouple wire is available, which may be assumed to match the characteristics in Table 8.3 exactly. How many junction pairs in a thermopile must be used in order that the uncertainty in the temperature differential measurement does not exceed 0.05°F? Neglect all errors due to heat transfer.

8.9. A thermopile of chromel-alumel with four junction pairs is used to measure a differential temperature of 4.0°F at a temperature level of 400°F. The sensitivity of the thermocouple wire is found to match that in Table 8.3 within ±0.5 percent, and the millivolt potentiometer has an uncertainty of 0.002 mV. Calculate the uncertainty in the differential temperature measurement.

8.10. For a certain thermistor $\beta = 3420$ K, and the resistance at 200°F is known to be 1010 ± 3 Ω. The thermistor is used for a temperature measurement, and the resistance is measured as 2315 ± 4 Ω. Calculate the temperature and the uncertainty.

8.11. Calculate the thermoelectric sensitivity for iron-constantan and copper-constantan at 0°C. Plot the error which would result if these values were assumed constant over the temperature range from -50 to $+600$°F.

8.12. A chromel-alumel thermocouple is exposed to a temperature of 1560°F. The potentiometer is used as the cold junction, and its temperature is estimated to be 83°F. Calculate the emf indicated by the potentiometer.

8.13. Four iron-constantan thermocouple junctions are connected in series. The temperatures of the four junctions are 200, 300, 100, and 32°F. Calculate the emf indicated by a potentiometer.

8.14. A millivolt recorder is available with a total range of 0 to 10 mV and an accuracy of ± 0.25 percent of full scale. Calculate the corresponding temperature ranges for use with iron-constantan, copper-constantan, and chromel-alumel thermocouples, and indicate the temperature accuracies for each of these thermocouples.

8.15. When a material with spectral emissivity less than unity is viewed with an optical pyrometer, the apparent temperature will be somewhat less than the true temperature and will depend on the wavelength at which the measurements are made. The error is given by

$$\frac{1}{T} - \frac{1}{T_a} = \frac{\lambda}{C_2} \ln \epsilon_\lambda$$

where $T =$ true temperature, K
$T_a =$ apparent temperature, K
$\epsilon_\lambda =$ spectral emissivity at the particular value of λ
$C_2 =$ constant from Eq. (8.15)

For a measurement at 0.655 μm and an apparent temperature of 1300°C, plot the error as a function of ϵ_λ.

8.16. A radiant energy measurement is made to determine the temperature of a hot block of metal. The emitted energy from the surface of the metal is measured as 28 ± 0.4 kW/m² and the surface emissivity is estimated as $\epsilon = 0.90 \pm 0.05$. Calculate the surface temperature of the metal and estimate the uncertainty.

8.17. A radiometer is used to measure the radiant energy flux from a material having a temperature of 280 ± 0.5°C. The emissivity of the material is 0.95 ± 0.03. The radiometer is then used to view a second material having exactly the same geometric shape and orientation as the first material and an emissivity of 0.72 ± 0.05. The uncertainty in the radiant energy flux measurement is ± 225 W/m² for both measurements, and the second measurement is 2.23 times as large as the first measurement. Calculate the temperature of the second material and the uncertainty.

8.18. Consider the thermometer-shield arrangement shown in Fig. 8.32. Assuming that the shield area is very large compared with the thermometer area and that all radiation leaving the thermometer is intercepted by the shield, calculate the true gas temperature using the following data:

$$h_t = 2.0 \text{ Btu/h} \cdot \text{ft}^2 \cdot {}^\circ\text{F}$$
$$h_s = 1.5 \text{ Btu/h} \cdot \text{ft}^2 \cdot {}^\circ\text{F}$$
$$\epsilon_t = 0.8$$
$$\epsilon_s = 0.2$$
$$T_t = 600{}^\circ\text{F}$$
$$T_s = 100{}^\circ\text{F}$$

(It is assumed from the statement of the problem that $F_{ts} = 1.0$, $F_{se} = 1.0$, and $F_{te} = 0$.) What temperature would the thermometer indicate if the shield were removed? Calculate the error reduction owing to the use of the radiation shield.

8.19. Rework Prob. 8.18 assuming that $A_s/A_t = 100$ and that the radiation shape factors are given as $F_{ts} = 0.9$, $F_{te} = 0.1$, and $F_{se} = 1.1$ (recall the assumption pertaining to F_{se}).

8.20. A platinum resistance thermometer is placed in a duct to measure the temperature of an airflow stream. The thermometer is placed inside a cylindrical shell 6 mm in diameter which has a polished outside surface with $\epsilon = 0.08 \pm 0.02$. The airstream velocity is known to be 3 m/s and the pressure is 1.0 atm. The thermometer indicates a temperature of 115°C. The duct wall temperature is measured at 193°C. Calculate the true air temperature and estimate the uncertainty in this temperature. Assume that the uncertainty in the convection heat-transfer coefficient calculated with the appropriate formula from Table 8.6 is ±15 percent. Assume that the uncertainty in the resistance-thermometer measurement is ±0.03°C.

8.21. Repeat Prob. 8.20, but assume that the cylinder is a chrome-plated copper rod with a thermocouple embedded in the surface. The uncertainty in the thermocouple measurement is assumed to be ±0.3°C.

8.22. A small copper sphere is constructed with a thermocouple embedded in the center and is used to measure the air temperature in an oven. The walls of the oven are at 1200 ± 20°F. The emissivity of the copper surface is 0.57 ± 0.04, and the temperature indicated by the thermocouple is 1015 ± 1.0°F. The convection heat-transfer coefficient is 5.0 Btu/h · ft² · °F ± 15 percent. Calculate the true air temperature and the uncertainty. Calculate the true air temperature if the surface of the sphere had been chromeplated with $\epsilon = 0.06 \pm 0.02$. Estimate the uncertainty in this circumstance.

8.23. Calculate the time constant for the copper-rod thermometer of Prob. 8.21. Consider only the convection heat transfer from the gas in calculating this value.

8.24. A thermocouple is placed inside a $\frac{1}{8}$-in-OD, $\frac{1}{16}$-in-ID copper tube. The tube is then placed inside a furnace whose walls are at 1500°F. The air temperature in the furnace is 1200°F. Calculate the temperature indicated by the thermocouple and estimate the time τ to obtain

$$\frac{T - T_i}{T_t - T_i} = 0.5$$

where $T_i =$ initial temperature of the thermocouple before it is placed in the furnace

$T_t =$ indicated temperature after a long time

$T =$ temperature at time τ

Assume $\epsilon = 0.78$ for copper.

8.25. A certain thermocouple has a time constant of 1.2 s. Design a compensation network like that shown in Fig. 8.34 having a steady-state attenuation factor of 8. Select the value of the resistance so that the capacitor has an even multiple value.

8.26. A certain high-speed temperature probe having a recovery factor of 0.98 ± 0.01 is used to measure the temperature of air at Mach 3.00. The thermocouple installed in the probe is accurate within $\pm 1.0°C$ and indicates a temperature of 380°C. Calculate the free-stream temperature and the uncertainty.

8.27. A 4-in-square copper plate is suspended vertically in a room, and a thermocouple is attached to measure the plate temperature. The walls of the room are at 110°F, and the thermocouple indicates a temperature at 95°F. Assuming the copper has an emissivity of 0.7, calculate the air temperature in the room. Determine the convection heat-transfer coefficient from Table 8.6. If the uncertainty in the surface emissivity is ± 0.05, what is the resulting uncertainty in the calculated air temperature?

8.28. Repeat Prob. 8.27 for polished copper having an emissivity of $\epsilon = 0.05 \pm 0.01$.

8.29. A thermometer having an emissivity of 0.9 is placed in a large room and indicates a temperature of 25°C. The walls of the room are at 35°C and the convection heat-transfer coefficient is 5.0 W/m^2 · °C. Calculate the true temperature of the air.

8.30. A stainless-steel sphere having a diameter of 3.0 mm is embedded with a copper-constantan thermocouple and is suddenly exposed to air at 50°C with a convection coefficient of 20 W/m^2 · °C. The initial temperature of the sphere is 20°C. Plot the output of the thermocouple as a function of time. What is the time constant?

8.31. A thermopile is to be constructed for measuring small temperature differences. Devise arrangements to measure temperature differences of 1, 3, and 5°C using copper-constantan and chromel-constantan thermocouples. Discuss the uncertainties which may be encountered.

8.32. Two platinum resistance thermometers are used to measure a temperature difference of 8°C. Discuss the uncertainties which may be involved. Would a thermopile arrangement be better and, if so, why?

8.33. Using the polynomial coefficients of Table 8.5, determine the voltage outputs for chromel-constantan at 10, 50, and 120°C. Then fit a parabolic relation to these points. What error would result from using the parabolic relation at 20, 60, and 100°C?

8.34. A heavily oxidized aluminum plate is exposed to a flow stream having a heat-transfer coefficient of 50 W/m^2°C. The plate temperature is measured as 100°C, ±0.5°C, and the room temperature surrounding the plate is at 20°C, ±0.5°C. Using emissivity data from the appendix, calculate the flow-stream temperature and estimate its uncertainty.

8.35. Repeat Prob. 8.34 for a plate covered with highly polished aluminum.

8.36. A bimetallic strip is to be constructed of yellow brass and Invar for the purpose of measuring temperature over the range from −10°C to 120°C. The length of the strip is 2.5 cm and each sheet has a thickness of 0.3 mm. Assuming that the strip is bonded at 30°C, calculate the deflection at the extremes of the design temperature range.

8.37. Assuming that R_0 is the resistance of a thermistor at 0°C, obtain the value of β for thermistor 1 in Fig. 8.8.

8.38. The following junctions are connected in series: copper-constantan at 20°C, constantan-iron at 100°C, iron-constantan at 60°C, constantan-chromel at 15°C, and chromel-alumel at 20°C. Calculate the voltage output of the series arrangement.

8.39. Consider a thin thermocouple that behaves like iron having dimensions of 0.5 × 0.1 × 0.005 in. Calculate the time constant for such a thermocouple when exposed to an airstream with a convection heat-transfer coefficient of 100 W/m^2 · °C.

8.40. For the thermocouple of Prob. 8.39, calculate the temperature difference across the thermocouple when exposed to a heat flux of 100 W/cm^2.

8.41. A mercury-in-glass thermometer having a diameter of 6 mm is placed in a vertical position to measure the air temperature in a metal warehouse in the summer. The walls of the warehouse are at 50°C. The convection coefficient is 7 W/m^2 − °C, and the emissivity of the thermometer may be taken as $\epsilon = 0.9$. Calculate the temperature indicated by the thermometer if the air temperature is 32°C.

8.42. For the situation in Prob. 8.41, calculate the temperature which will be indicated if the thermometer is enclosed by a long radiation shield having $\epsilon_s = 0.1$ and a diameter of 12 mm. For a long shield the area ratio A/A_s, is very nearly d_t/d_s.

8.43. A thermocouple sensor enclosed in a metal sheath has a diameter of 3 mm and is placed in an oven the walls of which are maintained at 550°C. The convection heat-transfer coefficient between the thermocouple and the surrounding air is 30 W/m^2 − °C. If the air temperature is 400°C, calculate the temperature which will be indicated by the thermocouple. Take the emissivity of the thermocouple as 0.8.

8.44. A radiation shield having a diameter of 7 mm and emissivity of 0.15 is placed around the thermocouple of Prob 8.43, such that it essentially covers the sensor from the standpoint of intercepting radiation. Calculate the temperature indicated by the thermocouple under these conditions.

8.45. A thermopile is constructed with chromel-alumel materials. Five junctions are maintained at 400°F and five junctions are maintained at 100°F. The output of the thermopile is connected to a potentiometer. What voltage will be indicated if both terminals of the potentiometer are at the same temperature?

8.46. A thermocouple data logger has a built-in circuit to compensate for the temperature of the connecting board which is used for mounting the thermocouples. A test is to be performed on the circuit by measuring the temperature of the board and the compensating voltage generated internally. If the temperature of the board is measured as 50°F, what should be the compensating voltage, and to what tolerance should it be maintained so that the compensation is accurate within ±0.2°F? Assume iron-constantan thermocouple.

8.47. The radiation from a metal part is measured with a pyrometer which indicates a temperature of 300°C. The actual part temperature is measured with a thermocouple which indicates 315°C. Both measurements have an uncertainty of ±1°C. From these measurements, determine the emissivity of the metal part and its uncertainty.

8.48. Using the results of Prob. 8.47, determine the true metal temperature when the pyrometer indicates 450°C and the uncertainty in this temperature.

8.49. A long, bare, thermocouple wire having a diameter of 1.0 mm is attached to a ceramic solid having a density of 2400 kg/m^3, specific heat of 650 J/kg $-$ °C, and a thermal conductivity of 0.85 W/m $-$ °C. The conductivity of the thermocouple wire is 75 W/m $-$ °C and it is exposed to an environment with $h = 275$ W/m^2 $-$ °C and a temperature of 50°C. The convection coefficient to the ceramic is 25 W/m^2 $-$ °C. The thermocouple indicates a temperature of 220°C. What is the true temperature of the solid?

8.50. Rework Prob. 8.49 for installation of the thermocouple on (a) an aluminum and (b) an 18-8 stainless-steel plate.

8.51. A thermocouple is enclosed in a 2.0-mm-diameter sheath having an emissivity of 0.9 and is exposed to a high-speed airstream at 20°C and $M = 2.0$. If the recovery factor for the sensor is 0.9, calculate the temperature which will be indicated. Take the convection heat-transfer coefficient as 180 W/m^2 $-$ °C.

8.52. Repeat Prob. 8.51 for a Mach number of 3.0.

8.53. Suppose the sheath in Probs. 8.51 and 8.52 is coated with a highly reflective substance having an emissivity of 0.07. What temperatures would be indicated under these conditions?

8.54. Highly polished copper has an emissivity of about 0.05. After accumulation of a heavy oxide layer the emissivity rises to about 0.8. A radiometer with an internal setting for emissivity is used to measure the polished copper temperature as 210°F. After oxidation the radiometer is used for the same measurement without changing the internal emissivity setting. What temperature will be indicated under the oxidized conditions?

8.55. A copper sphere having a diameter of 12.0 mm is initially at a uniform temperature of 100°C. It is then suddenly exposed to an air-forced-convection environment at 20° which produces a convection heat-transfer coefficient of

97 W/m^2 – °C. Embedded in the sphere is a chromel-alumel thermocouple to indicate the temperature as a function of time. How long will it take for the thermocouple to indicate a temperature of 25°C? What is the time constant for this system?

8.56. Assume that a chromel-alumel thermocouple has approximately the same thermal properties as 18-8 stainless steel. By some suitable means the junction is formed between two 0.05-mm bare wires and the thermocouple is suddenly exposed to an air environment at 0°C. The thermocouple wires are initially at 25°C. Assuming that the junction and bare wires behave as a cylinder, determine the time constant for a convection heat-transfer coefficient of 3800 W/m^2 – °C. How long will it take the thermocouple to indicate a temperature of 1°C?

8.57. Four chromel-alumel thermocouples are connected in series. The four junctions are at 100, 150, 200, and 250°C. The series arrangement is hooked to a potentiometer, the copper-binding posts of which are at 30°C. What voltage will be indicated by the potentiometer?

8.58. What could be the thermoelectric sensitivity of a nichrome-aluminum combination? Of a nichrome-constantan combination?

8.59. A bimetallic strip is to be designed to trip an electric contact. It is to be constructed of Invar and yellow brass and must produce a deflection of 1 mm at its end for a temperature variation of ±1.0°C from the set point of the strip. Consider several alternate lengths and strip thicknesses which will accomplish the design objective. State criteria which might be used to select from the alternative choices.

8.60. A thermopile is constructed of type N thermocouple wire having 10 junction pairs. One set of junctions is assembled and exposed to a temperature of 50°C while the other set of junctions is immersed in boiling water at 100°C. What is the voltage output of the thermopile?

8.61. What would be the voltage output of the thermopile in Prob. 8.60 if the cooler junction was maintained at 0°C?

8.62. The thermopile of Prob. 8.61 is exposed to a temperature differential with the junction sets at 500 and 600°C. What is the voltage output of the thermopile? How much does it differ from the output calculated in Prob. 8.61 for the same temperature differential?

8.63. A thermocouple is installed in the exhaust of an incinerator used to deodorize fumes from a corn chip fryer. The electronic instrumentation is designed to employ type K thermocouples and the temperature is in the range of 600°C. The instrumentation is at 25°C. What error would result if an iron-constantan thermocouple is installed by mistake? Speculate on what mistake(s) in process control might result from this error.

8.64. A temperature-sensing element for the air in a conditioned space is located close to an exterior wall. The wall is poorly insulated such that the inside surface becomes quite cold in the winter. In the northern city locale it may be as low as 0°C. The walls of the room behave nearly like a black surface and the thermometer has $\epsilon = 0.9$ and a convection heat-transfer coefficient of 14 W/m$^2 \cdot$ °C. When the thermometer reads 28°C, what will be the true air temperature?

8.65. A person sitting in proximity to the wall in Prob. 8.64 will experience considerable "chill" even when the air temperature is "comfortable." Explain why. Suppose the wall were covered with reflective aluminum foil having $\epsilon = 0.07$. What improvement in comfort might result? Why? (Show calculation.)

8.66. A thermocouple wire has an effective thermal conductivity of 70 W/m $\cdot$ °C and is installed vertically in a solid having a thermal conductivity of 5 W/m $\cdot$ °C. The thermocouple effective diameter is 2 mm and it is exposed to a convection environment at 20° with $h = 125$ W/m$^2 \cdot$ °C. The convection coefficient to the solid is 15 W/m$^2 \cdot$ °C. The thermocouple indicates a solid temperature of 125°C. Calculate the true temperature of the solid.

8.67. Repeat Prob. 8.66 for a solid with $k = 70$ W/m $\cdot$ °C.

8.68. Use available computer software to obtain second-, third-, and fourth-degree polynomial fits to the data entries of Table 8.3a for type N thermocouples between 0 and 1000°C. Compute the corresponding values of r^2 for each least-squares fit.

8.69. The temperature of a surface is to be measured with a radiometer at temperature levels of 350 and 600 K. The emissivity of the surface is estimated as $\epsilon = 0.3 \pm 0.03$ and the absolute uncertainty of the measurement of emitted energy is estimated or specified by the instrument manufacturer as ± 1.0 percent of the emitted energy the instrument would measure from a black surface at 500 K. Determine the uncertainty in the determination of the two specified temperatures. State the relative influence of instrument uncertainty and surface emissivity uncertainty.

8.70. Repeat Prob. 8.69 for $\epsilon = 0.8 \pm 0.04$.

8.71. Alternate thermocouple measurement techniques are proposed for measuring the temperature difference between inlet and exit for a heated fluid in a pipe. (a) Each temperature is measured separately with a type E thermocouple with "Special Grade" error specifications. (b) The temperature difference is measured directly with a thermopile constructed of five junction pairs of type E thermocouple wire, all wire from the same spool. Comment on the relative merits of the two techniques for actual inlet and exit fluid temperatures of 0°C and 25°C. Repeat for 0°C and 5°C. Assume the electronic measurements of thermocouple and thermopile emf's are exact.

8.72. Comment on the influence which uncertainties in the electronic measurements of emf's may have on the conclusions of Prob. 8.71.

8.73. A small block of aluminum 4 mm on a side is embedded with an iron-constantan thermocouple and is suddenly exposed to air at 65°C with a convection coefficient of 24 W/m^2–°C. The initial temperature of the block is 20°C. Calculate the time constant and determine the output of the thermocouple at $\tau = 0.5 \times$ (time constant).

8.74. A plate of aluminum which has become heavily oxidized over a period of time is placed in a large enclosure and exposed to a convection environment with

$h = 40\,\mathrm{W/m^2 \cdot {}^\circ C}$ and $T_\infty = 30°C \pm 0.5°C$. The plate temperature is measured as $86°C \pm 0.7°C$. Calculate the effective temperature of the large enclosure and estimate its uncertainty.

8.75. Repeat Prob. 8.74 for a highly polished aluminum surface.

8.76. Assuming that R_0 is the resistance of a thermistor at $0°C$, obtain the value of β for thermistor 2 in Fig. 8.8.

8.77. A thermocouple sensor is enclosed in a metal sheath having a surface emissivity of 0.7, and placed in a large oven, the walls of which are maintained at $600°C$. The air temperature in the oven is $350°C$, and the convection coefficient between the air and the sensor is $40\,\mathrm{W/m^2 \cdot {}^\circ C}$. What temperature will be indicated by the thermocouple?

8.78. Using the results from Prob. 8.47 determine the true metal temperature and its uncertainty when the pyrometer indicates $400°C$.

8.79. A polished surface has an emissivity of 0.08. After exposure to an oxidizing environment the emissivity rises to 0.7. A radiometer is used to measure the surface temperature when its internal setting for emissivity is set for the polished value. Later, after the surface has oxidized, the measurement is repeated, but the internal setting for sensitivity is maintained at the polished value. What error in temperature will result from neglecting to change the internal setting for emissivity?

8.80. Calculate the thermoelectric sensitivity of (a) a platinum-constantan, (b) a platinum-nickel combination.

8.81. A thermopile is constructed of chromel-constantan wire with 15 junction pairs. What will be the voltage output from the thermopile if one set of junction is exposed to an ice bath at $0°C$ while the other set of junctions is exposed to boiling water at $100°C$?

8.82. A thermometer having a surface emissivity of 0.85 is placed in an air space in a metal-walled building, the walls of which may attain a temperature of $-10°C$ during winter climate conditions. The convection coefficient between the thermometer and the surrounding air is estimated as $h = 10\,\mathrm{W/m^2 \cdot {}^\circ C}$. If the thermometer indicates a temperature of $17°C$, estimate the true air temperature in the building.

8.14   REFERENCES

1. Becker, J. A., C. B. Green, and G. L. Pearson: "Properties and Uses of Thermistors," *Trans. AIEE,* vol. 65, pp. 711–725, November 1946.

2. Eckert, E. R. G., and R. M. Drake: *Heat and Mass Transfer,* 2d ed., McGraw-Hill, New York, 1959.

3. Eskin, S. G., and J. R. Fritze: "Thermostatic Bimetals," *Trans. ASME,* vol. 62, pp. 433–442, July 1940.

4. Hackforth, H. L.: *Infrared Radiation,* McGraw-Hill, New York, 1960.

5. Holman, J. P.: *Heat Transfer,* 10th ed., McGraw-Hill, New York, 2010.

6. Lion, K. S.: *Instrumentation in Scientific Research,* McGraw-Hill, New York, 1959.

7. Siegel, R., and J. R. Howell: *Thermal Radiation Heat Transfer,* 2d ed., McGraw-Hill, New York, 1981.

8. Obert, E. F.: *Concepts of Thermodynamics,* McGraw-Hill, New York, 1960.

9. Rossini, E. F. (ed.): *Thermodynamics and Physics of Matter,* Princeton University Press, Princeton, NJ, 1955.

10. Sears, F. W.: *Thermodynamics,* Addison-Wesley, Reading, MA, 1953.

11. Schneider, P. J.: *Conduction Heat Transfer,* Addison-Wesley, Reading, MA, 1955.

12. Tribus, M.: *Thermostatics and Thermodynamics,* D. Van Nostrand, Princeton, NJ, 1961.

13. Winkler, E. M.: "Design and Calibration of Stagnation Temperature Probes for Use at High Supersonic Speeds and Elevated Temperatures," *J. Appl. Phys.,* vol. 25, p. 231, 1954.

14. McGee, T. D.: *Principles and Methods of Temperature Measurements,* Wiley, New York, 1988.

15. West, W. E., and J. W. Westwater: "Radiation-Conduction Correction for Temperature Measurements in Hot Gases," *Ind. Eng. Chem.,* vol. 45, pp. 2, 152, 1953.

16. Hennecke, D. K., and E. M. Sparrow: "Local Heat Sink on a Convectively Cooled Surface, Application to Temperature Measurement Error," *Int. J. Heat Mass Trans.,* vol. 13, p. 287, February 1970.

17. Shepard, C. E., and I. Warshawsky: "Electrical Techniques for Compensation of Thermal Time Lag of Thermocouples and Resistance Thermometer Elements," *NACA Tech. Note* 2703, May 1952.

18. Holman, J. P.: *Thermodynamics,* 4th ed., McGraw-Hill, New York, 1988.

19. Powell, R. L.: "Thermocouple Thermometry," in E. R. G. Eckert and R. Goldstein (eds.), *Measurements in Heat Transfer,* 2d ed., McGraw-Hill, New York, 1976.

20. Schultz, D. L., and T. V. Jones: "Heat Transfer Measurements in Short Duration Hypersonic Facilities," *AGARDograph* 165, 1973.

21. Cooper, T. E., J. F. Meyer, and R. J. Field: "Liquid Crystal Thermography and Its Application to the Study of Convective Heat Transfer," *ASME Paper* 75-HT-15, August 1975.

22. Cooper, T. E., and J. P. Groff: "Thermal Mapping via Liquid Crystals of the Temperature Field Near a Heated Surgical Probe," *J. Heat Trans.,* vol. 95, p. 250, 1973.

23. Burns, G. W., M. G. Scroger, and G. F. Strouse: "Temperature Electromotive Force Reference Functions and Tables for the Letter Designated Thermocouple Types Based on ITS-90," *NIST Monograph* 175, April 1993 (supersedes *NBS Monograph* 125).

24. Wobschall, D.: *Circuit Design for Electronic Instrumentation,* McGraw-Hill, New York, 1979.

25. Omega Engineering: *Temperature Measurement Handbook,* Omega Engineering, Stamford, CT, 1991.

26. Moffat, R. J.: "The Gradient Approach to Thermocouple Circuitry," in *Temperature—Its Measure and Control in Sciences and Industry,* Reinhold, New York, 1962.

27. Benedict, R. P.: *Fundamentals of Temperature, Pressure and Flow Measurement,* 3d ed., Wiley, New York, 1984.

28. Bedford, R. E.: "Calculation of Effective Emissivities of Cavity Sources of Thermal Radiation," in D. P. Dewitt, and G. D. Nutter (eds.), *Theory and Practice of Thermal Radiation Thermometry,* chap. 12, Wiley, New York, 1988.

29. Gu, S., G. Fu, and Q. Zhang: "3500 K High Frequency Induction Heated Blackbody Source," *J. Thermophysics Heat Trans.,* vol. 3, no. i, pp. 83–85, 1989.

9

THERMAL- AND TRANSPORT-PROPERTY MEASUREMENTS

9.1   INTRODUCTION

Several types of thermal properties are essential for energy-balance calculations in heat-transfer applications. Values of these properties for a variety of substances and materials are already available in tabular form in various handbooks; however, for new materials, which appear regularly, it is important that the engineer be familiar with some basic methods of measuring these properties.

Most thermal-property measurements involve a determination of heat flow and temperature. We have already discussed several types of temperature-measuring devices in Chap. 8 and shall have occasion to refer to them from time to time in discussing thermal-property measurements. Heat flow is usually measured by making an energy balance on the device under consideration. For example, a metal plate might be heated with an electric heater and the plate immersed in a tank of water during this heating process. The convection heat loss from the plate could thus be determined by making a measurement of the electric power dissipated in the heater. As another example, consider the heating of water by passing it through a heated pipe. The convection heat transfer from the pipe wall to the water may be determined by measuring the mass flow rate of water and the inlet and exit water temperatures to the heated section of pipe. The energy gained by the water is therefore the heat transfer from the pipe, provided that the outside surface of the pipe is insulated so that no losses are incurred. The techniques for measurement of heat transfer by thermal radiation are discussed in Chap. 12.

Thermal conductivity may be classified as a transport property since it is indicative of the energy transport in a fluid or solid. In gases and liquids the transport of energy takes place by molecular motion, while in solids transport of energy by free electrons and lattice vibration is important. Fluid viscosity is also classified as a transport property because it is dependent on the momentum transport that results from molecular motion in the fluid. Mass diffusion is similarly classified as a transport process because it results from molecular movement. The diffusion coefficient is the transport property in this case. In this chapter we shall consider some simple methods for measurement of these transport properties.

The measurement of heat flow falls under the general subject of calorimetry. In this chapter we shall discuss some simple calorimetric determinations that may be performed. The broad subject of thermodynamic-property measurement is beyond the scope of our discussion.

9.2   THERMAL-CONDUCTIVITY MEASUREMENTS

Thermal conductivity is defined by the Fourier equation

$$q_x = -kA\frac{\partial T}{\partial x}$$ **[9.1]**

where q_x = heat-transfer rate, Btu/h or W
    A = area through which the heat is transferred, ft^2 or m^2
    $\partial T/\partial x$ = temperature gradient in the direction of the heat
        transfer, °F/ft or °C/m
    k = thermal conductivity, Btu/h · ft · °F or W/m · °C

Experimental determinations of thermal conductivity are based on this relationship. Consider the thin slab of material shown in Fig. 9.1. If the heat-transfer rate through the material, the material thickness, and the difference in temperature are measured, then the thermal conductivity may be calculated from

$$k = \frac{q\Delta x}{A(T_1 - T_2)}$$ **[9.2]**

In an experimental setup the heat may be supplied to one side of the slab by an electric heater and removed from the other side by a cooled plate. The temperatures on each side of the slab may be measured with thermocouples or thermistors, whichever is more appropriate.

We must note that the standard English and SI units for thermal conductivity given above are not always the ones employed in practice. Frequently, the thickness of material x is expressed in inches, while the area is expressed in square feet. The units for

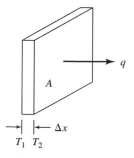

Figure 9.1   Simple thermal-conductivity measurement.

thermal conductivity appropriate to such dimensions then become Btu · in/h · ft^2 · °F. Not infrequently, a mixture of units is employed such as W · in · /ft^2 · °C.

The main problem of the above method for determining thermal conductivity is that heat may escape from the edges of the slab, or if the edges are covered with insulation, a two-dimensional temperature profile may result, which can cause an error in the determination. This problem may be alleviated by the installation of guard heaters, as shown in Fig. 9.2. In this arrangement the heater is placed in the center and a slab of the specimen is placed on each side of the heater plate. A coolant is circulated through the device to remove the energy, and thermocouples are installed at appropriate places to measure the temperatures. A guard heater surrounds the main heater, and its temperature is maintained at that of the main heater. This prevents heat transfer out from the edges of the specimen, and thus maintains a one-dimensional heat transfer through the material whose thermal conductivity is to be determined. The guarded hot plate, as it is called, is widely used for determining the thermal conductivity of nonmetals, that is, solids of rather low thermal conductivity. For materials of high thermal conductivity a small temperature difference would be encountered which would require much more precise temperature measurement.

A very simple method for the measurement of thermal conductivities of metals is depicted in Fig. 9.3. A metal rod A of known thermal conductivity is connected to a rod of the metal B whose thermal conductivity is to be measured. A heat source and heat sink are connected to the ends of the composite rod, and the assembly is surrounded by insulating material to minimize heat loss to the surroundings and to ensure one-dimensional heat flow through the rod. Thermocouples are attached, or embedded, in both the known and unknown materials as shown. If the temperature gradient through the known material is measured, the heat flow may be determined. This heat flow is subsequently used to calculate the thermal conductivity of the unknown material. Thus,

$$q = -k_A A \left(\frac{dT}{dx} \right)_A = -k_B A \left(\frac{dT}{dx} \right)_B \qquad \textbf{[9.3]}$$

The temperatures may be measured at various locations along the unknown and the variation of thermal conductivity with temperature determined from these measurements. Van Dusen and Shelton [6] have used this method for the determination of thermal conductivities of metals up to 600°C.

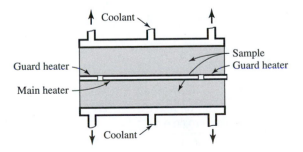

Figure 9.2   Schematic of guarded hot plate apparatus for measurement of thermal conductivity.

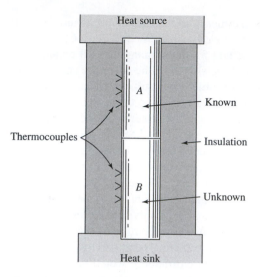

Heat source

A

Known

Thermocouples

Insulation

B

Unknown

Heat sink

Figure 9.3   Apparatus for measurement of thermal conductivity of metals.

Example 9.1 | **GUARDED HOT PLATE FOR THERMAL CONDUCTIVITY OF A METAL.**   A guarded hot-plate apparatus is used to measure the thermal conductivity of a metal having $k = 50$ Btu/h · ft · °F. The thickness of the specimen is 0.125 ± 0.002 in, and the heat flux may be measured within 1 percent. Calculate the accuracy necessary on the ΔT measurement in order to ensure an overall uncertainty in the measurement of k of 5 percent. If one of the plate temperatures is nominally 300°F, calculate the nominal value of the other plate temperature and the tolerable uncertainty in each temperature measurement, assuming a nominal heat flux of 20,000 Btu/h · ft^2.

Solution

We use Eq. (9.2) to estimate the uncertainty in conjunction with Eq. (3.2)

$$k = \frac{(q/A)\Delta x}{\Delta T} \qquad\qquad \text{[a]}$$

Equation (3.2) may be written as

$$\frac{w_R}{R} = \left[\left(\frac{\partial R}{\partial x_1} \right)^2 \left(\frac{w_1}{R} \right)^2 + \left(\frac{\partial R}{\partial x_2} \right)^2 \left(\frac{w_2}{R} \right)^2 + \left(\frac{\partial R}{\partial x_3} \right)^2 \left(\frac{w_3}{R} \right)^2 \right]^{1/2} \qquad \text{[b]}$$

We have

$$\frac{w_k}{k} = 0.05$$

$$\frac{\partial k}{\partial (q/A)} = \frac{\Delta x}{\Delta T} \qquad\qquad \frac{w_{q/A}}{k} = \frac{0.01 q/A(\Delta T)}{(q/A)\Delta x} = \frac{0.01 \Delta T}{\Delta x}$$

$$\frac{\partial k}{\partial (\Delta x)} = \frac{q/A}{\Delta T} \qquad\qquad \frac{w_{\Delta x}}{k} = \frac{(0.002)\Delta T}{(q/A)(0.125)}$$

$$\frac{\partial k}{\partial (\Delta T)} = -\frac{(q/A)\Delta x}{(\Delta T)^2} \qquad\qquad \frac{w_{\Delta T}}{k} = \frac{w_{\Delta T}\Delta T}{(q/A)\Delta x}$$

Inserting these expressions, Eq. (b) gives

$$0.05 = \left[(0.01)^2 + (0.002/0.125)^2 + \left(\frac{w_{\Delta T}}{\Delta T} \right)^2 \right]^{1/2}$$

or $w_{\Delta T}/\Delta T = 0.0146 = 1.46$ percent. We now calculate the nominal value of ΔT as

$$\Delta T = \frac{(q/A)\Delta x}{k} = \frac{(2 \times 10^4)(0.125)}{(12)(50)} = 4.1667°F$$

Since $\Delta T = T_1 - T_2$, the nominal value of T_1 is

$$T_1 = 300 + 4.1667 = 304.1667°F$$

The tolerable uncertainty in each temperature measurement is calculated from

$$\frac{w_{\Delta T}}{\Delta T} = \left\{ \left[\frac{\partial(\Delta T)}{\partial T_1} \frac{w_{T_1}}{\Delta T} \right]^2 + \left[\frac{\partial(\Delta T)}{\partial T_2} \frac{w_{T_2}}{\Delta T} \right]^2 \right\}^{1/2}$$

$$w_{\Delta T} = \left(w_{T_1}^2 + w_{T_2}^2 \right)^{1/2}$$

and we assume the uncertainties are the same for both temperatures so that

$$w_T = \frac{1}{\sqrt{2}} w_{\Delta T} = 0.707 w_{\Delta T}$$

$$w_T = (0.707)(0.0146)(4.1667) = \pm 0.043°F$$

We thus see that a very accurate measurement of temperature is necessary in order to give a 5 percent accuracy in the determination of k when this measurement technique is employed. Such high accuracy is very difficult to achieve.

A BETTER MEASUREMENT TECHNIQUE.   An apparatus like that in Fig. 9.3 is used to measure the thermal conductivity of the metal of Example 9.1. The same heat flux is used with a bar 3.0 ± 0.005 in long. Determine the accuracy for the determination of ΔT and the tolerable uncertainty in the temperature measurements. Assume the same conditions as in Example 9.1.

Example 9.2

Solution

For a 5 percent uncertainty in k the uncertainty in ΔT is given

$$0.05 = \left[(0.01)^2 + \left(\frac{0.005}{3.0} \right)^2 + \left(\frac{w_{\Delta T}}{\Delta T} \right)^2 \right]^{1/2}$$

$$\frac{w_{\Delta T}}{\Delta T} = 0.049 = 4.9\%$$

The nominal value of ΔT is given as

$$\Delta T = \frac{(2 \times 10^4)(3)}{(12)(50)} = 100°F$$

The allowable uncertainty in the determination of the temperatures is

$$w_T = \frac{1}{\sqrt{2}} w_{\Delta T} = (0.707)(0.049)(100) = \pm 3.46°F$$

This value is very easy to obtain. Even if the heat flux were reduced to 2×10^3 Btu/h · ft^2, the nominal temperature difference would be $10°$F and the allowable uncertainty would be $\pm 0.346°$F. This is still a value that may be attained with careful experimental techniques.

9.3 THERMAL CONDUCTIVITY OF LIQUIDS AND GASES

Kaye and Higgins (3) have used a guarded hot-plate method for determining thermal conductivity of liquids. Their apparatus is shown in Fig. 9.4. The diameter of the plates is 5 cm, and the thickness of the liquid layer is approximately 0.05 cm. This layer must be sufficiently thin so that convection currents are minimized. An annular arrangement, as shown in Fig. 9.5, may also be used for the determination of liquid thermal conductivities. Again, the thickness of the liquid layer must be thin enough to minimize thermal-convection currents.

A concentric-cylinder arrangement may also be used for the measurement of the thermal conductivity of gases. Keyes and Sandell [4] have used such a device for the measurement of thermal conductivity of water vapor, oxygen, nitrogen, and other gases. The inner and outer cylinders were both constructed of silver with a length of 5 in and an outside diameter of $1\frac{1}{2}$ in. The gap space for the gas was 0.025 in. Vines [7] has utilized such a device for the measurement of high-temperature-gas thermal conductivities. A schematic of his apparatus is shown in Fig. 9.6. The emitter serves

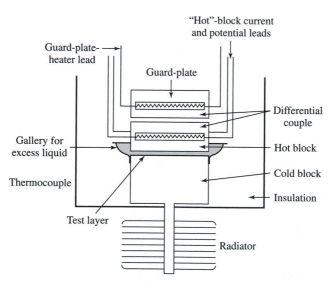

Figure 9.4 Guarded hot-plate apparatus for measurement of thermal conductivity of liquids, according to Ref. [3].

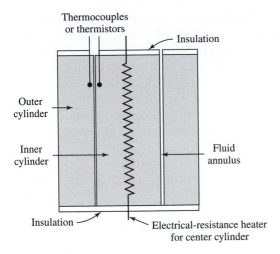

Figure 9.5   Concentric-cylinder method for measurements of thermal conductivity of liquids.

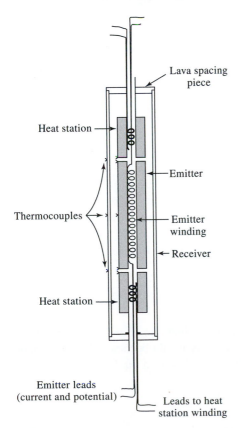

Figure 9.6   Apparatus for determining thermal conductivity of gases at high temperatures, according to Vines [7].

as the heat source, while the heat stations on either end act as guard heaters. The emitter has an outside diameter of 6 mm and a length of 50 mm, while the receiver has an inside diameter of 10 mm and a length of 125 mm with a wall thickness of 1 mm. For most of the tests it was possible to maintain a temperature difference of 5 to 10°C between the emitter and the receiver. The heat-transfer rate is measured by determining the electric-power input to the emitter, while thermocouples installed on the surface of the emitter and receiver determine the temperature difference.

For the concentric-cylinder device the relation that is used to calculate the thermal conductivity is

$$k = \frac{q \ln(r_2/r_1)}{2\pi L(T_1 - T_2)} \qquad \textbf{[9.4]}$$

where q = heat-transfer rate
r_2, r_1 = outside and inside radii of the annular space containing the fluid, respectively
T_2, T_1 = temperatures measured at these radii

An investigation by Leidenfrost [8] proposes an apparatus for measuring thermal conductivities of gases and liquids from −180 to +500°C. Complete experimental data are not available, but accuracies of 0.1 percent are claimed for the device.

Thermal conductivities of several materials are given in the appendix, and additional information is given in Ref. [20].

9.4   MEASUREMENT OF VISCOSITY

The defining equation for *dynamic* or *absolute viscosity* is

$$\tau = \mu \frac{du}{dy} \qquad \textbf{[9.5]}$$

where τ = shear stress between fluid layers in laminar flow, N/m^2 or lbf/ft^2
μ = dynamic viscosity, N · s/m^2 or lbf · s/ft^2
du/dy = normal velocity gradient as indicated in Fig. 9.7

The *kinematic* viscosity is defined by

$$\nu = \frac{\mu}{\rho}$$

where ρ is the fluid density. If we insert the above units for μ and either lbm/ft^3 or kg/m^3 for density, the units for kinematic viscosity would become lbf · s · ft/lbm or N · s · m/kg. The normal practice is to define lbf and N from Eq. (2.2) as

$$1 \text{ lbf} = \frac{1}{g_c} \text{ lbm} \cdot \text{ft/s}^2$$

$$1 \text{ N} = \frac{1}{g_c} \text{ kg} \cdot \text{m/s}^2$$

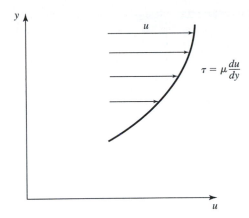

Figure 9.7   Diagram indicating relation of viscosity gradient and fluid shear.

Then the units of kinematic viscosity become ft²/s and m²/s. In reality, a very large number of units are employed for viscosity in both the English and metric systems. Some of the more common units are summarized below along with appropriate conversion factors:

Dynamic viscosity:

$$1 \text{ lbf} \cdot \text{s/ft}^2 = 47.8803 \text{ N} \cdot \text{s/m}^2$$
$$= 478.803 \text{ poise (P)}$$
$$= 1.158 \times 10^5 \text{ lbm/h} \cdot \text{ft}$$
$$= 47.8803 \text{ kg/m} \cdot \text{s}$$
$$1 \text{ N} \cdot \text{s/m}^2 = 2.08854 \times 10^{-2} \text{ lbf} \cdot \text{s/ft}^2$$
$$= 10 \text{ P} = 1000 \text{ centipoise (cP)}$$
$$= 2419 \text{ lbm/h} \cdot \text{ft}$$
$$= 1 \text{ kg/m} \cdot \text{s}$$
$$1 \text{ P} = 100 \text{ cP} = 1 \text{ dyn} \cdot \text{s/cm}^2$$
$$= 0.1 \text{ N} \cdot \text{s/m}^2$$
$$= 2.08854 \times 10^{-3} \text{ lbf} \cdot \text{s/ft}^2$$
$$= 241.0 \text{ lbm/h} \cdot \text{ft}$$
$$= 0.1 \text{ kg/m} \cdot \text{s}$$

Kinematic viscosity:

$$1 \text{ ft}^2\text{/s} = 9.2903 \times 10^{-2} \text{ m}^2\text{/s}$$
$$= 929.03 \text{ stokes (St)}$$
$$= 929.03 \text{ cm}^2\text{/s}$$
$$1 \text{ m}^2\text{/s} = 10.7639 \text{ ft}^2\text{/s}$$
$$= 10^4 \text{ St}$$
$$1 \text{ St} = 1 \text{ cm}^2\text{/s}$$
$$= 100 \text{ centistokes (cSt)}$$
$$= 10^{-4} \text{ m}^2\text{/s}$$
$$= 10.7639 \times 10^{-4} \text{ ft}^2\text{/s}$$

Various methods are employed for measurement of the viscosity. The two most common methods are the rotating-concentric-cylinder method and the capillary-flow method. We shall discuss the rotating-cylinder method first.

Consider the parallel plates as shown in Fig. 9.8. One plate is stationary, and the other moves with constant velocity u. The velocity profile for the fluid between the two plates is a straight line, and the velocity gradient is

$$\frac{du}{dy} = \frac{u}{b} \qquad\qquad \textbf{[9.6]}$$

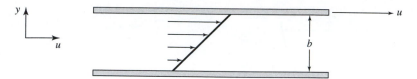

Figure 9.8   Velocity distribution between large parallel plates.

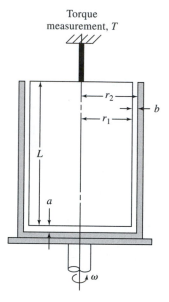

Figure 9.9   Rotating concentric-cylinder apparatus for measurement of viscosity.

The system could be used to measure the viscosity by measuring the force required to maintain the moving plate at the constant velocity u. The system is impractical from a construction standpoint, however, and the conventional approach is to approximate the parallel flat-plate situation with the rotating concentric cylinders shown in Fig. 9.9. The inner cylinder is stationary and attached to an appropriate torque-measuring device, while the outer cylinder is driven at a constant angular velocity ω. If the annular space b is sufficiently small in comparison with the radius of the inner cylinder, then the rotating cylinder arrangement approximates the parallel-plate situation, and the velocity profile in the gap space may be assumed to be linear. Then,

$$\frac{du}{dy} = \frac{r_2\omega}{b}$$ **[9.7]**

where, now, the y distance is taken in the radial direction and it is assumed that $b \ll r_1$.

Now, if the torque T is measured, the fluid shear stress is expressed by

$$\tau = \frac{T}{2\pi r_1^2 L} \qquad \textbf{[9.8]}$$

where L is the length of the cylinder. The viscosity is determined by combining Eqs. (9.5), (9.7), and (9.8) to give

$$\mu = \frac{Tb}{2\pi r_1^2 r_2 L\omega} \qquad \textbf{[9.9]}$$

If the concentric-cylinder arrangement is constructed such that the gap space a is small, then the bottom disk will also contribute to the torque and influence the calculation of the viscosity. The torque on the bottom disk is

$$T_d = \frac{\mu\pi\omega}{2a} r_1^4 \qquad \textbf{[9.10]}$$

where a is the gap spacing, as shown in Fig. 9.9. Combining the torques due to the bottom disk and the annular space gives

$$T = \mu\pi\omega r_1^2 \left(\frac{r_1^2}{2a} + \frac{2Lr_2}{b} \right) \qquad \textbf{[9.11]}$$

Once the torque, angular velocity, and dimensions are measured, the viscosity may be calculated from Eq. (9.11).

Perhaps the most common method of viscosity measurement consists of a measurement of the pressure drop in laminar flow through a capillary tube. Consider the tube cross section shown in Fig. 9.10. If the Reynolds number defined by

$$\mathrm{Re}_d = \frac{\rho u_m d}{\mu} \qquad \text{(dimensionless)} \qquad \textbf{[9.12]}$$

is less than 1000, laminar flow will exist in the tube and the familiar parabolic-velocity profile will be experienced as shown. If the fluid is incompressible and the flow is

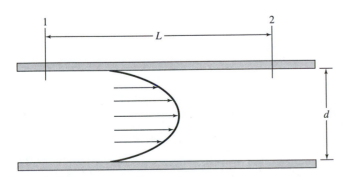

Figure 9.10   Laminar flow through a capillary tube.

steady, it can be shown that the volume rate of flow Q can be written as

$$Q = \frac{\pi r^4 (p_1 - p_2)}{8 \mu L} \qquad (\text{m}^3/\text{s or ft}^3/\text{s}) \qquad \textbf{[9.13]}$$

A viscosity determination may be made by measuring the volume rate of flow and pressure drop for flow in such a tube. To ensure that laminar flow will exist, a small-diameter capillary tube is used; the small diameter reduces the Reynolds number as calculated from Eq. (9.12). In Eq. (9.12) the product ρu_m may be calculated from

$$\rho u_m = \frac{\dot{m}}{\pi r^2} \qquad \textbf{[9.14]}$$

where $\dot{m}$ is the mass rate of flow.

When a viscosity measurement is made on a gas, the compressibility of the gas must be taken into account. The resulting expression for the mass flow of the gas under laminar flow conditions in the capillary is

$$\dot{m} = \frac{\pi r^4}{16 \mu R T} \left(p_1^2 - p_2^2 \right) \qquad \textbf{[9.15]}$$

where R is the gas constant for the particular gas.

Care must be taken to ensure that the flow in the capillary is *fully developed*, i.e., the parabolic-velocity profile has been established. This means that the pressure measurements should be taken far enough downstream from the entrance of the tube to ensure that developed flow conditions persist. It may be expected that the flow will be fully developed when

$$\frac{L}{d} > \frac{\text{Re}_d}{8} \qquad \textbf{[9.16]}$$

where L is the distance from the entrance of the tube.

Example 9.3 | **CAPILLARY MEASUREMENT OF VISCOSITY.**   The viscosity of liquid water at 20°C is to be measured using a capillary tube. Design a suitable apparatus and specify the ranges of instruments required.

Solution

From the appendix the viscosity of liquid water at 20°C is $\mu = 1.01 \times 10^{-3}$ kg/m · s and the density is 998 kg/m³. We will select a tube such that $\text{Re}_d = 500$ with a tube diameter of 1.0 mm. Then, from Eq. (9.14) we have

$$\rho u_m = \frac{4\dot{m}}{\pi d^2} = 500 \frac{\mu}{d}$$

$$= \frac{(500)(1.01 \times 10^{-3})}{0.001} = 505 \text{ kg/m}^2 \cdot \text{s}$$

and the mass flow is

$$\dot{m} = \frac{(505)\pi(0.001)^2}{4} = 3.966 \times 10^{-4} \text{ kg/s}$$

The volume flow is

$$Q = \frac{\dot{m}}{\rho} = \frac{3.966 \times 10^{-4}}{998} = 3.974 \times 10^{-7} \text{ m}^3/\text{s}$$

Fully developed flow will result when

$$\frac{L}{d} > \frac{\text{Re}_d}{8}$$

or $L > (0.001)(500/8) = 0.0625$ m, or in a rather short length. The volume or mass flow of water may be measured with a weighing technique or a standard flowmeter. Suppose we wish to have a pressure differential of at least 50 kPa in order to operate with a modest pressure gage. From Eq. (9.15) the required length over which the pressure differential is measured would be

$$L = \frac{\pi r^4 (p_1 - p_2)}{8\mu Q}$$

$$= \frac{\pi (0.0005)^4 (50000)}{(8)(1.01 \times 10^{-3})(3.974 \times 10^{-7})}$$

$$= 3.057 \text{ m}$$

Of course, this is a length that is easily constructed in a laboratory.

The Saybolt viscosimeter is an industrial device that uses the capillary-tube principle for measurement of viscosities of liquids. A schematic of the device is shown in Fig. 9.11. A cylinder is filled to the top with the liquid and enclosed in a

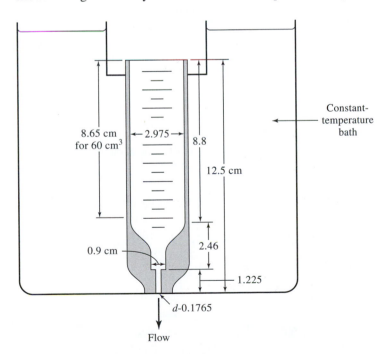

Figure 9.11   Schematic of the Saybolt viscosimeter.

constant-temperature bath to ensure uniformity of temperature during the measurements. The liquid is then allowed to drain from the bottom through the short capillary tube. The time required for 60 mL to drain is recorded, and this time is taken as indicative of the viscosity of the liquid. Since the capillary tube is short, a fully developed laminar-viscosity profile is not established, and it is necessary to apply a correction to account for the actual profile. If the velocity profile were fully developed, the kinematic viscosity would vary directly with the time for drainage; that is,

$$\nu = \frac{\mu}{\rho} = c_1 t$$

To correct for the nonuniform profile, another term is added to give

$$\nu = c_1 t + \frac{c_2}{t}$$

With the constants inserted, the relation is

$$\nu = \left(0.00237t - \frac{1.93}{t}\right) \times 10^{-3} \ \text{ft}^2/\text{s} \qquad \textbf{[9.17]}$$

The symbol t designates the drainage time in seconds for 60 mL of liquid, and it is common to express the viscosity in units of *Saybolt seconds* when this method of measurement is used. Equation (9.17) provides a method of converting to more useful units for kinematic viscosity. Viscosities for several fluids are given in the appendix.

Another technique for viscosity measurements is to observe the decrement in the period of oscillation when spheres or disks are suspended in the fluid with an elastic wire and allowed to vibrate from some initially deflected position. Such systems are particularly applicable to measurements of viscosities at high pressures. Kestin and Leidenfrost [18] describe an oscillating-disk system which has met with good success.

Example 9.4 | **SAYBOLT VISCOSITY MEASUREMENT.**   A Saybolt viscosimeter is used to measure the viscosity of a certain motor oil. The time recorded for drainage of 60 mL is estimated at 183 ± 0.5 s. Calculate the percentage uncertainty in the viscosity.

Solution

Equation (3.2) is used in conjunction with Eq. (9.17) to estimate the uncertainty. We have

$$\frac{\partial \nu}{\partial t} = c_1 - \frac{c_2}{t^2} = \left(0.00237 + \frac{1.93}{t^2}\right) \times 10^{-3}$$

so that the uncertainty in the viscosity is given as

$$w_\nu = \frac{\partial \nu}{\partial t} w_t$$

$$= \left(0.00237 + \frac{1.93}{t^2}\right)(10^{-3})(0.5) \ \text{ft}^2/\text{s}$$

The nominal value of the viscosity is

$$v = \left(0.00237t - \frac{1.93}{t}\right)10^{-3} \text{ ft}^2/\text{s}$$

At $t = 183$ s

$$w_v = 1.21 \times 10^{-6} \text{ ft}^2/\text{s} \ (1.12 \times 10^{-7} \text{ m}^2/\text{s})$$

$$v = 4.22 \times 10^{-4} \text{ ft}^2/\text{s} \ (3.92 \times 10^{-5} \text{ m}^2/\text{s})$$

so that $w_v/v = 2.87 \times 10^{-3} = 0.287$ percent.

9.5 GAS DIFFUSION

Consider a container of a certain gas (2) as shown in Fig. 9.12. At one end of the container another gas (1) is introduced and allowed to diffuse into gas (2). The diffusion rate of gas (1) at any instant of time is given by Fick's law of diffusion as

$$n_1 = -D_{12}A\frac{\partial \overline{N}}{\partial x} \qquad \textbf{[9.18]}$$

where
n_1 = molal rate of diffusion
A = cross-sectional area for diffusion
$\overline{N}_1$ = molal concentration of component 1 at any point
D_{12} = diffusion coefficient

When n_1 is in moles per second, $\overline{N}_1$ is in moles per cubic meter, and x is in meters, the diffusion coefficient will have the units of square meters per second, the same units as kinematic viscosity. It may be noted that Eq. (9.18) is similar to the Fourier law for heat conduction. When a mass balance is made on an element, the one-dimensional diffusion equation results:

$$\frac{\partial \overline{N}_1}{\delta t} = D_{12}\frac{\partial^2 \overline{N}_1}{\partial x^2} \qquad \textbf{[9.19]}$$

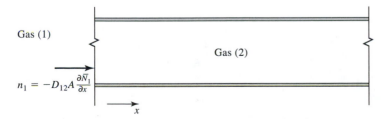

Figure 9.12   Schematic of gas-diffusion process.

A solution to this equation may be used as the basis for an experimental determination of the diffusion coefficient D_{12}.

A simple experimental setup for the measurement of gas-diffusion coefficients is shown in Fig. 9.13. Two sections of glass tubing are connected by a flexible Tygon tubing connection, which may be clamped as shown. The sections of tubing are charged with the two gases whose diffusion characteristics are to be investigated. After the initial charging process the flexible connection is opened and the two gases

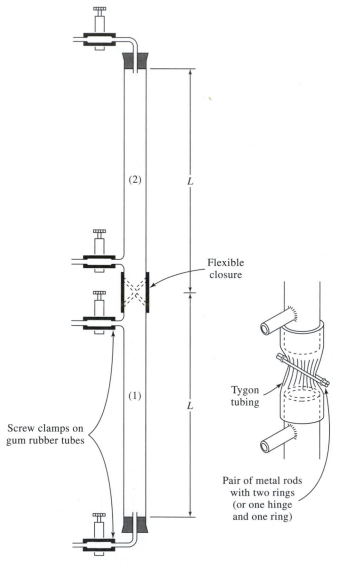

Figure 9.13   Loschmidt apparatus for measurement of diffusion coefficients in gases. (a) Apparatus; (b) detail of flexible connection.

are allowed to diffuse into one another. After a period of time the Tygon connection is closed again, the exact time is recorded, and the two sections are allowed to reach equilibrium concentrations. These equilibrium concentrations are then measured with appropriate analytical equipment. It is only necessary to measure the concentration of one of the gases in both sections since the concentration of the other gas will be obtained from a mole balance. However, measuring both gives a check and information about the accuracy of the experiment. Some check of this kind is always very desirable, although not always feasible.

A solution of Eq. (9.19) gives these concentrations in terms of the diffusion coefficients and dimensions of the tubes. Assuming that the lower portion is filled with gas (1) and the upper portion with gas (2) initially, the solution is

$$F = \frac{N_{1A} - N_{1B}}{N_{1A} + N_{1B}} = \frac{8}{\pi^2} \sum_{k=0}^{\infty} \frac{1}{(2k+1)^2} \exp\left[\frac{-\pi^2 D_{12}t(2k-1)^2}{4L^2}\right] \qquad \textbf{[9.20]}$$

where the subscripts A and B refer to the lower and upper sections, respectively, t is the time the two gases are allowed to diffuse, and L is the length of *each* tube. N_{1A} and N_{1B} represent the number of moles of component 1 in the lower and upper parts of the apparatus, respectively, after time t. For an experimental determination of D_{12} the optimum time to be allowed for the diffusion process is

$$t_{\text{opt}} = \frac{4L^2}{\pi^2 D_{12}} \qquad \textbf{[9.21]}$$

assuming that the only experimental errors are those involved in the determination of F. If the diffusion process is allowed to run for t_{opt} or longer, it is found that the first term in the series of Eq. (9.20) is the primary term and that the higher-order terms are negligible. Then,

$$F = 0.2982 \qquad \text{at } t = t_{\text{opt}} \qquad \textbf{[9.22]}$$

If the second- and higher-order terms are neglected, it is easy to calculate the value of D_{12} once a determination of F and t has been made. Neglecting these higher-order terms gives

$$F = \frac{8}{\pi^2} \exp\frac{-\pi^2 D_{12}t}{4L^2} \qquad \textbf{[9.23]}$$

or

$$D_{12} = \frac{-4L^2}{\pi^2 t} \ln\frac{\pi^2 F}{8} \qquad \textbf{[9.24]}$$

It is necessary to know only an approximate value for D_{12} in order to calculate the value of t_{opt} from Eq. (9.21). Then, the experiment is conducted using a diffusion time greater than this value. The exact value of D_{12} is then calculated from the experimental data using Eq. (9.24).

The simple experimental setup shown in Fig. 9.13 is called a *Loschmidt appara-tus*. In the apparatus the heavier gas is usually contained in the lower section, while the lighter gas is placed in the upper section. Extreme care must be exerted to ensure that the system is free of leaks and that each section contains a pure component at the start of the experiment. Care must also be exerted to ensure that both gases are initially at the same temperature and pressure. The apparatus should be constructed so that the volume of the inlet tubing is negligible in comparison with the volume of the main apparatus. While the diffusion process is in progress, the temperature of the apparatus should be maintained constant. Diffusion coefficients for several gases are given in the appendix.

Example 9.5

DIFFUSION COEFFICIENT FOR CO_2–AIR.   An apparatus like that shown in Fig. 9.13 is constructed so that $L = 50$ cm. CO_2 is placed in the lower tube and air in the upper tube. Both gases are at standard atmospheric pressure and 77°F. The diffusion coefficient for CO_2 in air is 0.164 cm^2/s. Calculate the mole fraction of CO_2 in each tube for (1) 5 min and (2) 2 h after the connection between the tubes is opened.

Solution

The mole fractions are given by

$$x_{1A} = \frac{N_{1A}}{N_{1_0}} \qquad x_{1B} = \frac{N_{1B}}{N_{1_0}}$$

where N_{1_0} is the initial number of moles in each tube and equal to $N_{1A} + N_{1B}$. Gas (1) is the CO_2 in this case. From Eq. (9.20)

$$F = x_{1A} - x_{1B} \qquad\qquad \textbf{[a]}$$

We calculate the optimum time as

$$t_{\text{opt}} = \frac{(4)(50)^2}{\pi^2 (0.164)} = 6180 \text{ s}$$

The 5-min condition is far short of this so that the series solution of Eq. (9.20) must be used to find F. Thus, with $t = 300$ s,

$$F = \frac{8}{\pi^2}\left[e^{-300/6180} + \frac{1}{9}e^{-(9)(300)/6180} + \frac{1}{25}e^{-(25)(300)/6180} + \cdots \right]$$

$$= \frac{8}{\pi^2}(0.9526 + 0.0718 + 0.0119 + \cdots)$$

$$= 0.8400$$

Using Eq. (a) and

$$x_{1A} + x_{1B} = 1.0 \qquad\qquad \textbf{[b]}$$

gives $x_{1A} = 0.92$ and $x_{1B} = 0.08$.

     For the 2-h time $t = (2)(3600) = 7200$ s, and only the first term of the series needs to be used. Thus, from Eq. (9.23)

$$F = \frac{8}{\pi^2}e^{-7200/6180} = 0.2525$$

Again, using Eqs. (a) and (b) gives

$$x_{1A} = 0.6262 \qquad \text{and} \qquad x_{1B} = 0.3738$$

9.6 CALORIMETRY

The subject of calorimetry is concerned with a determination of energy quantities. These quantities may be classified in terms of the thermodynamic properties of a system, such as enthalpy, internal energy, specific heat, and heating value, or in terms of energy flow that results from a transport of mass across the boundaries of the thermodynamic system. Calorimetry is a very broad subject covering a majority of the experimental determinations that are used for measurement of thermodynamic properties. We shall give an example of calorimetry which shows the overall features of energy and mass balances and their importance; this is the flow calorimeter, which is used for the measurement of heating values of gaseous or liquid fuels.

The actual device usually used for the experiment is called the Junkers calorimeter. A schematic of the calorimeter is given in Fig. 9.14. The gaseous fuel is burned inside the calorimeter, where it gives up heat to the cooling water. The water flow rate is determined by weighing, and the inlet and outlet water temperatures are measured with precision mercury-in-glass thermometers as shown. The products of combustion are cooled to a sufficiently low temperature so that water vapor is condensed. This condensate is collected in the graduated flask as shown. The gas-flow rate is usually measured with a positive-displacement flowmeter. We shall discuss the system in a general way and indicate an analysis which may be performed to determine the heating value of the fuel.

The flow schematic for the flow calorimeter is shown in Fig. 9.15. For convenience, all streams entering the device are designated with the subscript 1, while all streams leaving the device are designated with the subscript 2. The fuel and air are burned inside the calorimeter, and the major portion of the heat of combustion is removed by the cooling water.

The following experimental measurements are made: the inlet and exit cooling water temperatures T_{w_1} and T_{w_2}, the mass flow rate of fuel $\dot{m}_f$, the mass flow rate of cooling water $\dot{m}_w$, the condensate temperature T_{c_2}, the entering fuel and air temperatures T_{f_1} and T_{a_1}, and the relative humidity of the inlet air ϕ_1. In addition, an analysis of the products of combustion is made to determine the oxygen, carbon dioxide, and carbon monoxide content. From all these data mass and energy balances may be made to determine the heating value of the fuel. Let us consider the simple case of the combustion of methane, CH_4. We write the chemical balance equation as

$$aCH_4 + xO_2 + 3.76xN_2 + bH_2O \rightarrow dCO_2 + eCO + fO_2 + gH_2O + 3.76xN_2$$

[9.25]

An analysis of the products gives the values of d, e, and f for 100 mol of dry products. The mass balance then becomes

CARBON

$$a = d + e$$

[9.26]

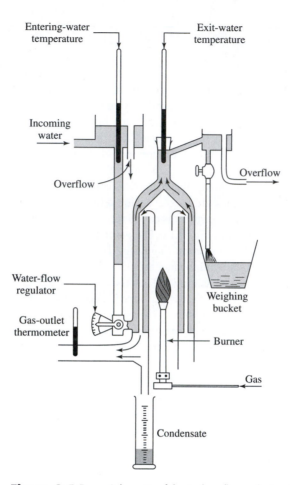

Figure 9.14   Schematic of the Junkers flow calorimeter.

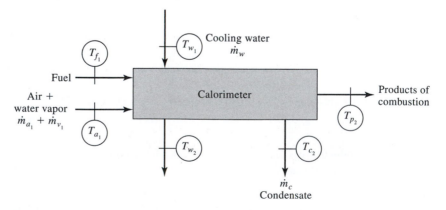

Figure 9.15   Flow schematic for calorimeter of Fig. 9.14.

HYDROGEN

$$4a + 2b = 2g \qquad \textbf{[9.27]}$$

OXYGEN

$$2x + b = 2d + e + 2f + g \qquad \textbf{[9.28]}$$

Equation (9.26) permits an immediate determination of a. The relative humidity of the incoming air ϕ_1 gives a relationship for b as

$$\frac{b}{4.76x} = \frac{p_{g_1}\phi_1}{p_1 - \phi_1 p_{g_1}} \qquad \textbf{[9.29]}$$

where p_{g_1} is the saturation pressure of water vapor at the inlet air temperature and p_1 is the total pressure of the incoming air–water-vapor mixture. We thus have three unknowns (b, g, and x) and three equations [(9.27) to (9.29)] which may be solved for the unknowns. With the mass balance now complete, an energy balance may be made to determine the chemical energy of the incoming fuel. We observe that

$$\dot{m}_{f_1} = 16a \qquad \dot{m}_{a_1} = 32x + (28)(3.76x)$$
$$\dot{m}_{v_1} = 18b$$

The water vapor in the products of combustion is calculated from

$$\dot{m}_{v_2} = 18g - \dot{m}_c \qquad \textbf{[9.30]}$$

The energy balance becomes

$$\dot{m}_{f_1}\left(h_{f_1} + E_{f_1}\right) + \dot{m}_{a_1}h_{a_1} + \dot{m}_{v_1}h_{v_1} - \dot{m}_{p_2}h_{p_2} - \dot{m}_{c_2}h_{c_2} = \dot{m}_w\left(h_{w_2} - h_{w_1}\right) \qquad \textbf{[9.31]}$$

where E_{f_1} is designated as chemical energy of the fuel. The energy term for the products $\dot{m}_p h_{p_2}$ is calculated from the energies of the individual constituents. Thus,

$$\dot{m}_p h_{p_2} = \dot{m}_{CO_2}h_{CO_2} + \dot{m}_{CO}h_{CO} + \dot{m}_{O_2}h_{O_2} + \dot{m}_{N_2}h_{N_2} + \dot{m}_{v_2}h_{v_2} \qquad \textbf{[9.32]}$$

The mass flow rate of these constituents is determined from the balance conditions in Eqs. (9.27) to (9.29).

The purpose of the above discussion is to illustrate the number of experimental measurements that must be made in even a simple calorimetric determination like the one shown and the fact that all these measurements influence the accuracy of the final result. Of course, the calculated value of heating value (or chemical energy content) will be more sensitive to some measurements than to others. For details on the measurement techniques the reader should consult Ref. [5]. It may be noted that there are standardized procedures available for this particular test which make

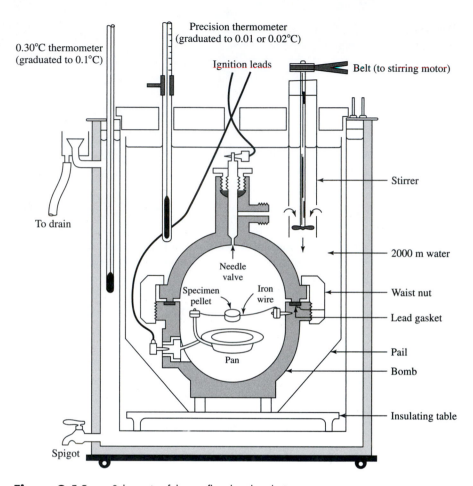

Figure 9.16   Schematic of the nonflow bomb calorimeter.

it unnecessary to perform all the analysis indicated above; nevertheless, this analysis indicates the theory behind the measurement process.

The *bomb calorimeter* is another device frequently used for heating-value determinations in solid and liquid fuels. In contrast to the flow calorimeter described above, the measurements are made under constant-volume, nonflow conditions as shown in Fig. 9.16. A measured sample of the fuel is enclosed in a metal container which is subsequently charged with oxygen at high pressure. The bomb is then placed in a container of water and the fuel is ignited through external electric connections. The temperature of the water is measured as a function of time after the ignition process; and from a knowledge of the mass of water in the system, the mass and specific heat of the container, and transient heating and cooling curves the energy release during combustion may be determined. A motor-driven stirrer ensures uniformity in temperature of the water surrounding the bomb. In some circumstances external heating may be supplied to the jacket water to maintain a uniform temperature, while in other instances the jacket may be left empty in order to maintain nearly adiabatic conditions on the inner water container. A compensation for the heat lost to the environment may be made through an analysis of the transient heating and cooling curve.

If the bomb is constructed of a Dewar flask enabling the assumption of an adiabatic combustion process, and the work input of the stirring motor may be neglected, we have

$$Q + W = 0 = \Delta U$$

$$\Delta U = \Delta U_{bomb} + H_{ignition \ wire} + H_{fuel} + \Delta U_{water} \qquad \textbf{[9.33]}$$

where $H_{ignition \ wire}$ and H_{fuel} are the heats of combustion of the ignition wire and fuel, repectively. For a nickel wire $H = 4113$ kJ/kg. We also have

$$\Delta U_{bomb} = m_{bomb} c_{v,bomb} \Delta T$$

$$\Delta U_{water} = m_{water} c_{v,water} \Delta T \text{ with } c_{v,water} = 4.19 \text{ kJ/kg} \cdot^{\circ} \text{C}$$

The mass of water, ignition wire consumed, and fuel sample may be determined by weighing, and the value of $\Delta T = T_{final} - T_{initial}$ measured during the combustion process. The initial temperature is the steady value measured before ignition and the final temperature is the value when the temperature-time curve levels out after ignition. The value of $m_{bomb} c_{v,bomb} = c_{v,bomb}$, which is dependent on the specific construction materials for the bomb may be determined through a calibration process where a substance with known H_{fuel} is burned and the valued of bomb-specific heat determined from Eq. (9.33). Benzoic acid crystals [C_6H_5COOH, $\rho = 1.08$ g/cm^3, $H_{fuel} = 26980$ kJ/kg] are typically used for such calibration.

HEATING VALUE OF FUELS

The energy release when a fuel is burned with oxygen, or air, is called the heating value of the fuel and may be measured with a flow or bomb calorimeter as described above. In industrial applications the quantity that is of economic interest is the higher heating value, which is defined as the energy released per unit mass of fuel when the fuel is burned with the theoretical amount of air and the water in the products due to combustion is in the liquid state.

In a large pipeline carrying a gaseous or liquid fuel, it is not convenient to extract fuel samples on a frequent basis and conduct the time-consuming flow calorimeter test to determine the heating value of the fuel. Yet the heating value must be determined to account for any changes if fuel composition, or possible leakage of air into the pipeline in the case of a gaseous fuel. The solution to the problem is to measure the chemical composition of the fuel on a continuous basis using spectroscopic methods. The heating value of the fuel is then computed from known values of the enthalpies of formation of the individual constituents. Basic data are stored in a computer and an immediate calculation is performed to give the heating value of the fuel. If any particularly unusual anomalies occur, a physical sample of the fuel may be tested to confirm the calculation.

Table 9.1   Approximate Higher Heating Values of Fuels

Gaseous fuels at 1 atm (101.32 kPa) and 20°C	
Methane, CH_4	37.3 kJ/liter
Propane, C_3H_8	93.1
Butane, C_4H_{10}	119.2
Natural gas (largely methane)	38.4
Liquid fuels at 20°C	
Propane, C_3H_8	25800 kJ/liter
Butane, C_4H_{10}	36200
Methanol, CH_3OH	15900
Ethanol, C_2H_5OH	21200
Gasoline	35000
Vegetable oil	34200
Motor oil	35000
Kerosene	38500
#2 Fuel oil	39500
#4 Fuel oil	41300
Solid fuels	
Coal	
Anthracite	30200 kJ/kg
Bituminous	28000
Lignite	11600–14000
Wood	
Softwood	11200
Hardwood	10000
Corn (avg, shelled and raw)	18600

Some typical heating values for hydrocarbon fuels are given in Table 9.1. We should note that the higher heating values do not represent the actual energy release in a typical industrial application of burning of a hydrocarbon fuel. The products of combustion will almost always be at a temperature high enough to cause the water in the products to be in a vapor state, and probably at a much higher temperature. This higher temperature consumes part of the higher heating value, leaving less energy available for operating the specific industrial process. Nevertheless, the price of fuels is based on higher heating values, and economic evaluations of industrial processes must take this fact into account.

9.7   CONVECTION HEAT-TRANSFER MEASUREMENTS

Determination of convection heat-transfer coefficients covers a very broad range of experimental activities. In this section we shall illustrate only two simple experimental setups; one for a forced-convection system and one for a free-convection system. The interested reader should consult Ref. [2] for more information.

Consider the experimental setup shown in Fig. 9.17. It is desired to obtain convection heat-transfer coefficients for the flow of water in a smooth tube. Heat is supplied to the tube by electric heating as shown. The tube is usually made of a high-resistance material like stainless steel to reduce the electric current necessary for heating. Thermocouples are spot-welded or cemented to the outside surface of the tube to measure the wall temperature. Either thermocouples or thermometers are inserted in the flow to measure the water temperature at inlet and outlet to the heated section. A voltmeter and ammeter measure the power input to the tube, while some type of flowmeter measures the water flow rate. The electrically heated tube delivers a constant heat flux to the water (constant Btu/h · ft² or W/m² of tube surface) so that it is reasonable to assume a straight-line variation of water bulk temperature from inlet to outlet. Thus, the wall and bulk temperatures are known along the length of the tube, and the heat-transfer coefficient may be calculated at any axial location from

$$q = hA(T_w - T_B) \qquad \textbf{[9.34]}$$

where A is the total *inside-heated* surface area of the tube, T_w and T_B are the wall and bulk temperatures at the perpendicular location, and q is the total heat-transfer rate, given by

$$q = EI \qquad \textbf{[9.35]}$$

E and I are the voltage and current impressed on the test section. The heated surface area is

$$A = \pi d_i L \qquad \textbf{[9.36]}$$

The outside surface of the tube must be covered with insulation to ensure that all the electric heating is dissipated in the water.

In Example 16.1 of Chap. 16 we apply a design protocol to arrive at very specific test section and equipment specifications for this type of convection experiment.

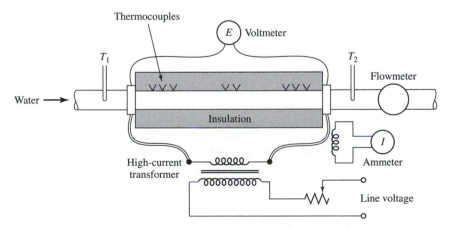

Figure 9.17   Schematic of apparatus for determination of forced convection heat-transfer coefficients in smooth tubes.

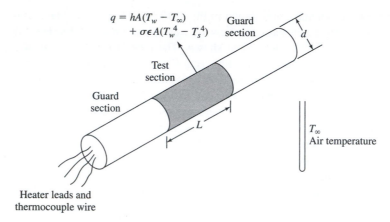

Figure 9.18   Schematic of heated cylinder arrangement for determination of free-convection heat-transfer coefficients.

The interested reader may wish to skip to that section to see how such an experiment is conducted in some detail, including applications of uncertainty analysis in the design protocol, and preliminary estimates to determine ranges for experimental data.

A simple experimental setup for determination of natural-convection heat-transfer coefficients on a horizontal cylinder is shown in Fig. 9.18. The heated horizontal cylinder consists of three sections. The center section is the test element, while the two end sections are guard heaters that eliminate end losses from the center section. All three sections are electrically heated and instrumented with thermocouples which measure their respective surface temperatures. The electric current to the guard heaters is adjusted so that their temperature is just equal to the temperature of the center section. At this balance condition all the electric heating of the center section is dissipated to the surrounding air or enclosure. The total heat loss from the test element is given by

$$q = hA(T_w - T_\infty) + \sigma\epsilon A\left(T_w^4 - T_s^4\right) \qquad [9.37]$$

where h is the free-convection heat-transfer coefficient, T_w is the test section temperature, ϵ is the surface emissivity, T_∞ is the surrounding air temperature, T_s is the temperature of the surrounding room or enclosure, and A is the surface area, given by

$$A = \pi dL \qquad [9.38]$$

The heat-transfer rate q is determined from measurement of the electric-power input to the test section. To determine the value of h from Eq. (9.37), the surface emissivity must be known. In practice, the test-section surface is usually nickel- or chrome-plated so that the radiation loss is small compared with the convection loss.

Example 9.6   **FREE-CONVECTION MEASUREMENT.**   A 30 × 30-cm square metal plate is used for a determination of free-convection heat-transfer coefficients. The plate is placed in a vertical position and exposed to room air at 20°C. The plate is electrically heated to a uniform surface temperature of 50°C, and the heating rate is measured as 15.0 ± 0.2 W. The emissivity of the surface is estimated as 0.07 ± 0.02. Determine the nominal value and uncertainty of the heat-transfer coefficient, assuming that the temperature measurements are exact. Assume that the effective radiation temperature of the surroundings is 20°C.

Solution

The nominal value of the heat-transfer coefficient is obtained from Eq. (9.37):

$$\frac{q}{A} = h(T_w - T_\infty) + \sigma\epsilon\left(T_w^4 - T_s^4\right) \qquad \text{[a]}$$

where

$$T_w = 50°C \qquad T_\infty = T_s = 20°C$$

$$\epsilon = 0.07 \qquad \frac{q}{A} = \frac{15}{(0.3)^2} = 166.7 \text{ W/m}^2$$

The resulting value of h is

$$h = 5.091 \text{ W/m}^2 \cdot °C$$

From Eq. (a) the explicit relation for h is

$$h = \frac{1}{T_w - T_\infty}\left[\frac{q}{A} - \sigma\epsilon\left(T_w^4 - T_s^4\right)\right] \qquad \text{[b]}$$

The uncertainty may be obtained from this expression using Eq. (3.2). We have

$$\frac{\partial h}{\partial(q/A)} = \frac{1}{T_w - T_\infty} \qquad w_{q/A} = \frac{0.2}{(0.3)^2} = 2.22 \text{ W/m}^2$$

$$\frac{\partial h}{\partial\epsilon} = \frac{-\sigma\left(T_w^4 - T_s^4\right)}{T_w - T_\infty} \qquad w_\epsilon = 0.02$$

The uncertainty in h is thus

$$W_h = \left\{\left(\frac{1}{30}\right)^2 (2.22)^2 + \left[\frac{(5.669 \times 10^{-8})(323^4 - 293^4)}{30}\right]^2 (0.02)^2\right\}^{1/2}$$

$$= (5.476 \times 10^{-3} + 0.01764)^{1/2}$$

$$= 0.152 \text{ W/m}^2 \cdot °C$$

or

$$\frac{W_h}{h} = 0.0299 = 2.99\%$$

9.8 HUMIDITY MEASUREMENTS

The water-vapor content of air is an important parameter in many processes. Proper control of critical operations with fabrics, papers, and chemicals frequently hinges upon a suitable control of the humidity of the surrounding environment. We shall discuss four basic techniques for measurement of the water-vapor content of air. First, let us consider some definitions. The *specific humidity*, or *humidity ratio*, is the mass of water vapor per unit mass of dry air. The *dry-bulb temperature* is the temperature of the air–water-vapor mixture as measured by a thermometer exposed to the mixture. The *wet-bulb temperature* is the temperature indicated by a thermometer covered with a wicklike material saturated with liquid after the arrangement has been allowed to reach evaporation equilibrium with the mixture, as indicated in Fig. 9.19. The *dew point* of the mixture is the temperature at which vapor starts to condense when the mixture is cooled at constant pressure.

The *relative humidity* ϕ is defined as the ratio of the actual mass of vapor to the mass of vapor required to produce a saturated mixture at the same temperature. If the vapor behaves like an ideal gas,

$$\phi = \frac{m_v}{m_{\text{sat}}} = \frac{p_v V / R_v T}{p_g V / R_v T} = \frac{p_v}{p_g} \qquad \text{[9.39]}$$

where p_v is the actual partial pressure of the vapor and p_g is the saturation pressure of the vapor at the temperature of the mixture. The *specific humidity* is

$$\omega = \frac{m_v}{m_a} \qquad \text{[9.40]}$$

which, for ideal-gas behavior, becomes

$$\omega = 0.622 \frac{p_v}{p_a} \qquad \text{[9.41]}$$

where p_a is the partial pressure of the air.

The most fundamental method for measuring humidity is a gravimetric procedure, which is used by NIST for calibration purposes. A sample of the air–water-vapor mixture is exposed to suitable chemicals until the water is absorbed. The chemicals

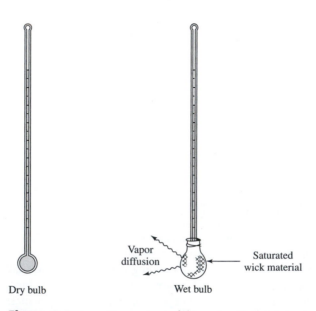

Dry bulb   Wet bulb

Figure 9.19   Measurement of dry- and wet-bulb temperatures.

are then weighed to determine the amount of vapor absorbed. Uncertainties as low as 0.1 percent can be obtained with this method.

There is a definite analytical relationship between the dry-bulb, wet-bulb, and dew-point temperatures of a mixture and its humidity. A determination of any two of these temperatures may be used to calculate the humidity. The classic method used for humidity determination in large open spaces is a measurement of the dry- and wet-bulb temperatures with a sling psychrometer. Both thermometers are rotated with a speed of about 5 m/s and the temperatures are recorded. The vapor pressure of the mixture may then be calculated with Carrier's equation:

$$p_v = p_{gw} - \frac{(p - p_{gw})(T_{DB} - T_{WB})}{K_w - T_{WB}} \qquad \textbf{[9.42]}$$

where
$\quad p_v$ = actual vapor pressure
$\quad p_{gw}$ = saturation pressure corresponding to wet-bulb temperature
$\quad p$ = total pressure of mixture
$\quad T_{DB}$ = dry-bulb temperature, °F or °C
$\quad T_{WB}$ = wet-bulb temperature, °F or °C
$\quad K_w$ = 2800 when T is in °F
$\quad\quad$ = 1537.8 when T is in °C

The specific or relative humidity is then calculated from Eq. (9.39) or (9.41). For calculation purposes a plot of the saturation pressure for water vapor as a function of temperature is given in Fig. 9.20.

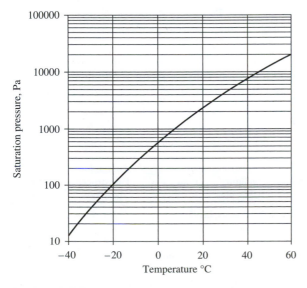

Figure 9.20   Saturation pressure for water vapor.

An alternate to the sling psychrometer uses two precision-matched temperature sensors, one for dry bulb and one for wet bulb. The sensors are enclosed in a portable chamber which also contains a fan to blow the ambient air across the temperature elements. Based on Eq. (9.42), equations for the thermodynamic properties of water vapor, and the measured temperatures, a microprocessor circuit may be used to calculate the relative humidity and to furnish a visual display or electrical readout. This instrument requires periodic changes of wick material for the wet-bulb sensor as well as frequent recharging of the wick with distilled water.

The sling psychrometer is not suitable to automation or continuous-recording requirements. For such applications a Dunmore-type electrical humidity transducer is used. This transducer consists of a resistance element which is constructed by winding dual noble-metal elements on a plastic form with a carefully controlled spacing between them. The windings are then coated with a lithium chloride solution, which forms a conducting path between the wires. The resistance of this coating is a very strong function of relative humidity and thus may be used to sense the humidity. An ac bridge circuit, as discussed in Chap. 4, is normally used for sensing the resistance of the transducer. Because of wide nonlinear variation of the resistance of the coating with humidity, a single transducer is normally suitable for use only over a narrow range of humidities. If measurements over a wide range of humidities are desired, multiple sensors can be employed, each with a coating suitable for some segment of the humidity spectrum. In general, a single transducer is not used over much more than a 10 percent relative-humidity range. Accuracies for the device are about 1.5 percent relative humidity and may be used over an operating range of −40 to +150°F (−40 to 65°C).

Calibration of these humidity sensors may be accomplished by exposing the sensor to standard samples wherein a closed air space is maintained in equilibrium with a saturated aqueous salt solution. Some example solutions are shown in Table 9.2. A zero relative-humidity level may be maintained in a closed air sample with a molecular sieve desiccant.

The humidity of a mixture may also be determined with a measurement of dew-point temperature. For this purpose the conventional technique is to cool a highly polished metal surface (mirror) until the first traces of fogging or condensation appear

Table 9.2   Relative humidities over saturated salt solutions

Temperature	Lithium Chloride Monohydrate, %	Magnesium Chloride Mexahydrate, %	Sodium Dichromate Dihydrate, %	Sodium Chloride, %	Potassium Chromate, %	Ammonium Dihydrogen Phosphate, %	Potassium Sulfate, %
15°C (59°F)	—	34	56	76	—	—	97
20°C (68°F)	11	34	55	76	87	93	97
25°C (77°F)	11	33	54	75	86	93	97
30°C (86°F)	11	33	52	75	86	92	97
35°C (95°F)	11	33	51	75	86	92	96

when the surface is in contact with the mixture. The temperature of the surface for the onset of fogging is then the dew point. The sensitivity of the measurement process may be improved by using a photocell to detect a beam of light that is reflected from the surface. The intensity of the reflected beam will decrease at the onset of fogging, and this point may be used to mark the temperature signal from the mirror as the dew point. For continuous monitoring of dew point the mirror-photocell technique may be modified by connecting the mirror to a low-temperature source like a dry-ice-acetone bath and simultaneously to an electric-heating element. An appropriate electronic circuit senses the output of the photocell and causes the electric heater to be actuated when the output drops as a result of fogging. The heater then remains on until the mirror surface is clear and the photocell output rises, thereby reducing the heater current. Basically, the system just described is a control system to maintain the mirror at the dew-point temperature regardless of changes in environment. A thermocouple attached to the mirror may then be used to monitor the dew point continuously. A summary of humidity-measurement techniques is given in Refs. [9] to [11]. The thermodynamics of air–water-vapor mixtures is discussed in Ref. [12].

Example 9.7

A sling psychrometer is employed to measure humidity in an air space maintained at a total pressure of 0.9 atm. The device indicates dry-bulb and wet-bulb temperatures of 35°C and 20°C, respectively. Calculate the dew point and relative humidity.

Solution

The actual vapor pressure for the water in the air may be calculated with Eq. (9.42), with the value of p_{gw} obtained from Fig. 9.20 as

$$p_{gw} = 2300 \text{ Pa}$$

We also have

$$p = 0.9 \text{ atm} = 91200 \text{ Pa}$$

Inserting the values in Eq. (9.42) gives

$$p_v = 2300 - (91200 - 2300)(35 - 20)/(1537.8 - 20)$$
$$= 1421 \text{ Pa}$$

This is the saturation pressure corresponding to the dew-point temperature. Consulting Fig. 9.20 we find

$$T_{DP} = 12°C$$

At the dry-bulb temperature we find

$$p_g = 5800 \text{ Pa}$$

so that the relative humidity may be calculated from Eq. (9.39) as

$$\phi = 1421/5800 = 24.5 \text{ percent}$$

9.9   HEAT-FLUX METERS

There are many applications where a direct measurement of heat flux is desired. In this section we shall discuss a few devices for such measurements. The first type of heat-flux meter we shall consider is the slug sensor shown in Fig. 9.21. A slug of high thermal conductivity is installed in the wall as shown with a surrounding layer of insulation. When the slug is subjected to a heat flux at the surface, its temperature will rise and give an indication of the magnitude of the heat flux. Assuming that the temperature of the slug remains uniform at any instant of time, the transient-energy balance becomes

$$\frac{q}{A} = \frac{mc}{A}\frac{dT_s}{d\tau} - U(T_s - T_w)$$

[9.43]

where q/A = imposed heat flux

      m = mass of the slug

      c = specific heat of the slug

      U = coefficient expressing the conduction loss to the surrounding wall

      T_s = temperature of the slug

      T_w = surrounding wall temperature

It is clear that a solution of Eq. (9.43) will give the slug temperature as a function of the heat flux and the properties of the slug. Thus, a measurement of T_s may be taken as an indication of the heat flux.

Some problems are encountered in using slug sensors. First, for very high heat fluxes the temperature of the slug may not remain uniform and a more cumbersome analysis must be employed to calculate the heat flux from the temperature measurements. Second, the presence of the sensor alters the temperature profile in the wall; in particular, the temperature at the sensor location is higher than it might have been otherwise (for slug conductivity greater than wall conductivity), thereby influencing the net heat flux at the point of measurement, whether it occurs by convection, radiation, or a combination of both. Third, the slug sensor is clearly adapted only to

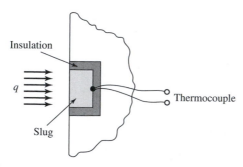

Figure 9.21   Slug-type heat-flux meter.

transient measurements. As soon as the wall reaches equilibrium, the temperature of the slug will no longer be indicative of the heat flux. Additional information on the response of slug-type sensors is available in Refs. [13], [14], and [19].

The Gardon [15] heat-flux meter is depicted in Fig. 9.22. A thin constantan disk is mounted to a copper heat sink, as shown, with the heat sink installed in the wall material. A small copper wire is then attached to the center of the disk and, along with a wire connected to the copper sink, forms the two wires of a copper-constantan thermocouple to measure the temperature difference between the center of the disk and the sides. A heat-flux incident on the disk is absorbed and conducted radially outward, thereby creating the temperature difference that is sensed by the thermocouple. The only losses in the gage are the radiation losses from the back side of the disk to the copper heat sink, but by careful calibration this may be taken into account with little difficulty. Heat fluxes over a range of 5×10^4 to 10^6 Btu/h $\cdot$ ft^2 (0.15 to MW/m^2) may be measured with the device. Time constants of the order of 0.1 s are experienced. Increased sensitivity may be obtained by using a copper disk, along with a positively doped bismuth-telluride center connection [17].

A very versatile heat-flux meter has been described by Hager [16] and is marketed by the RdF Corp., Hudson, NH. The meter operates on the principle of measuring the temperature drop across a thin insulating material of about a 0.0003-in (7.6-μm) thickness. The temperature drop is directly proportional to the heat conducted through the material in accordance with Fourier's law [Eq. (9.1)]. The unique feature of these meters is the butt-bonding process, which allows a distortion-free joining of copper-constantan on both sides of the material. The detailed construction is indicated in Fig. 9.23, with a number of junctions connected in series to produce a thermopile. As many as 40 junctions may be produced in this manner within a space of about 1 in^2. The junctions are covered with a protective coating, but the total thickness of the heat-flux meter is never more than 0.012 in (0.3 mm). The maximum heat flux that may be accommodated with the meter is about 2×10^5 Btu/h $\cdot$ ft^2 (0.63 MW/m^2) at a maximum operating temperature of 500°F (260°C). Sensitivity ranges from 0.07 to 27 mV $\cdot$ ft^2 $\cdot$ s/Btu (0.062 to 23.8 mV $\cdot$ cm^2/W), depending on the number of

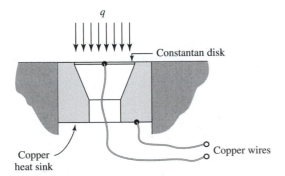

Figure 9.22   Gardon foil-type heat-flux meter, according to Ref. [15].

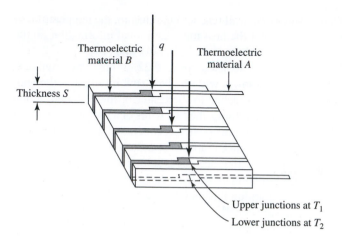

Figure 9.23   Detail of heat-flux meter operating on conduction principle.
(*Courtesy of RdF Corp., Hudson, NH.*)

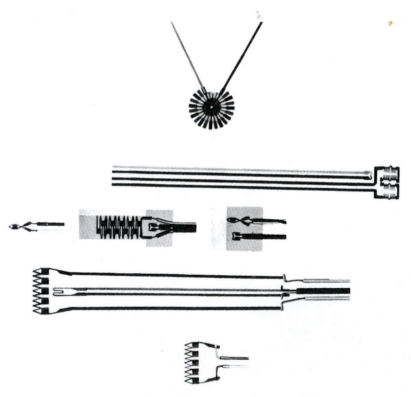

Figure 9.24   Photographs of assembled microfoil heat-flux meters. (*Courtesy of RdF
Corp., Hudson, NH.*)

junctions employed. Time constants range between 0.02 and 0.4 s. Figure 9.24 shows some photographs of typical meter configurations.

An excellent summary of heat-flux measurement techniques is given in Ref. [23], including comparative discussions of the methods previously described and general comparisons of differential temperature and calorimetric methods. Tabular comparisons of the response times, suitable operating temperature ranges, and other installation and manufacturing considerations are also given.

9.10 pH Measurement

The pH of a solution is a measure of the hydrogen ion concentration C_H and is defined by

$$pH = \log C_H \qquad \textbf{[9.44]}$$

A typical pH meter is shown in Fig. 9.25. A special glass electrode is used which is permeable only to hydrogen ions. The very thin bottom therefore acts as a membrane. A reference electrode of calomel is also placed in the solution. Reversible electrodes are attached to both the glass and reference electrode, and hydrochloric acid (1 N) is contained in the sealed glass electrode.

The net potential across the membrane is given by

$$\Delta v_i = -2.30 \frac{\Re T}{\mathcal{F}} \log \frac{C_H}{C_R} \qquad \textbf{[9.45]}$$

where
$\Re$ = universal gas constant, 8314 J/kg · mol · K
T = absolute temperature, K
$\mathcal{F}$ = Faraday's constant, 0.647×10^7 C/kg · mol
C_H = hydrogen ion concentration in solution
C_R = concentration in glass electrode
$(C_R = 1.0$ for 1 N HCl)

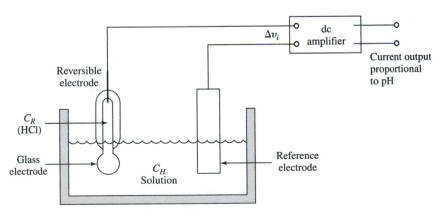

Figure 9.25 Schematic of pH meter.

Table 9.3   Approximate pH values of common substances

Substance	pH
Battery acid	0.3
Lemon juice	2.3
Vinegar	2.9
Orange juice	4.3
Boric acid	5.0
Corn	6.2
Milk	6.7
Distilled water	7.0
Blood	7.5
Sea water	8.0
Baking soda	8.4
Milk of magnesia	10.3
Ammonia	11.4
Bleach	12.6

Inserting the constants and making use of the definition of pH gives

$$\Delta v_i = -1.98 \times 10^{-4} T \qquad \text{(pH units)} \qquad \textbf{[9.46]}$$

Because the resistance of the glass membrane is very high (on the order of 100 MΩ) a dc amplifier with high-input impedance must be used. Figure 4.32 j is an example of an operational amplifier configuration which might be employed. Calibration is achieved by inserting the electrodes in a solution of known pH and adjusting the output meter for a proper reading.

For reference purposes pH values of some common substances are given in Table 9.3. A value of 7.0 is neutral (distilled water), while values less than 7 are acidic and values greater than 7 are basic (alkaline).

9.11   REVIEW QUESTIONS

9.1 Write the defining equation for thermal conductivity.

9.2 What is a guarded hot-plate apparatus? Under what conditions is it applied?

9.3 How are thermal conductivities of metals measured?

9.4 What are some of the considerations that must be applied for gas and liquid thermal conductivity that are not important in thermal-conductivity measurements for solids?

9.5 Write the defining equation for viscosity.

9.6 What is the operating principle of the Saybolt viscosimeter?

9.7 Write the defining equation for diffusion coefficient.

9.8 How does a Loschmidt apparatus function?

9.9 Describe the flow calorimeter.

9.10 Describe the bomb calorimeter.

9.11 Write the defining equation for the heat-transfer coefficient.

9.12 How does a sling psychrometer measure relative humidity?

9.13 Describe a technique for measuring dew point.

9.14 What is the operating principle of the slug-type heat-flux meter?

9.15 What is a Gardon heat-flux meter?

9.12   PROBLEMS

9.1. The following data are taken from Ref. [6] on an apparatus like that shown in Fig. 9.3. The measurements were made on a lead sample whose thermal conductivity is taken to be 0.352 W/cm · °C at 0°C.

Thermocouple Positions	Temperature, °C	Distance between Adjacent Thermocouples, cm
1	277.3	3.14
2	231.6	3.14
3	186.5	3.14
4	143.0	3.15
5	100.3	
6	78.6	3.16
7	37.3	

Using these data, construct a graph of the thermal conductivity of lead versus temperature.

9.2. The apparatus of Fig. 9.3 is used for measurement of the thermal conductivity of an unknown metal. The lengths of the unknown and known bars are the same, and five thermocouples are equally spaced on each bar. From the temperature and heat-flux measurement the thermal conductivity of the unknown is to be established as a function of temperature. Design the apparatus to measure the thermal conductivity of a material, with $k \sim 100$ Btu/h · ft · °F. Consider the influence of thermocouple spacing, dimensions of the bars, heat-flux measurement, and temperature measurement on accuracy. Take the thermal conductivity of the known material as exactly 20 Btu/h · ft · °F.

9.3. A concentric-cylinder device like that in Fig. 9.5 is used for a measurement of the thermal conductivity of water at 35°C. Using the dimensions given in the text, discuss the influence of the heat-flux and temperature measurements on the accuracy of the thermal-conductivity determination.

9.4. The viscosity of water is 1.65 lbm/h · ft at 100°F. A small capillary tube 50 ± 0.01 ft long is to be used to check this value. Calculate the maximum allowable flow rate for an inside tube diameter of 0.100 ± 0.001 in. Calculate the pressure drop for this flow rate and estimate the allowable uncertainty in the pressure drop in order that the uncertainty in the viscosity does not exceed 5 percent. Assume that the flow rate is measured within ±0.01 lbm/h.

9.5. The rotating concentric-cylinder apparatus of Fig. 9.9 is used to measure the viscosity of water at 100°F. The inner cylinder radius is 1.50 ± 0.001 in, and its length is 4.00 ± 0.002 in. The outer cylinder is rotated at a speed of 1800 ± 2 rpm. The outer cylinder radius is 1.60 ± 0.001 in and the gap spacing at the bottom of the cylinder is 0.500 ± 0.005 in. Calculate the nominal torque which will be measured and estimate the allowable uncertainty in this measurement in order that the overall uncertainty in the viscosity does not exceed 5 percent. Repeat the calculation for glycerin at 100°F.

9.6. The tolerance limits on the radii of the inner and outer cylinders of Fig. 9.9 exert a strong influence on the accuracy of the viscosity measurement. Establish a relation expressing the percent uncertainty in the viscosity measurement as a function of the uncertainty in these radius measurements. Assume that the torque, angular velocity, and bottom-spacing measurements are exact for this calculation. Assume that the uncertainty (tolerance) is the same for both cylinders.

9.7. The viscosity of a gas is to be determined by measuring the flow rate and pressure drop through a capillary tube. The experimental variables which are measured are:

> Inlet pressure p_1
> Pressure differential $\Delta p = p_1 - p_2$
> Temperature T
> Mass flow rate $\dot{m}$
> Tube diameter $d = 2r$

Derive an expression for the percent uncertainty in the viscosity measurement in terms of the uncertainties in these five primary measurements. In view of the information in Chaps. 5 to 8, discuss the relative influence of each of the measurements, i.e., which primary measurement will probably have the most effect on the final estimated uncertainty in the viscosity.

9.8. The viscosity of an oil having a density of 51.9 lbm/ft^3 is measured as 200 Saybolt seconds. Determine the dynamic viscosity in units of centipoise and pounds-mass per hour-foot.

9.9. A Loschmidt apparatus is used to measure the diffusion coefficient for benzene in air. At 25°C the diffusion coefficient is 0.088 cm^2/s for a pressure of 1 standard atmosphere. Both tubes have a length of 60 cm. Calculate the mole fraction of benzene in each tube 10 min and 3 h after the connection between the tubes is opened.

9.10. A Loschmidt apparatus with $L = 50$ cm is used to measure the diffusion coefficient for CO_2 in air. After a time of 6.52 min the mole fraction of CO_2 in the upper tube is 0.0912. Calculate the value of the diffusion coefficient.

9.11. For $t = t_{\text{opt}}$, calculate the relative magnitudes of the first four terms in the series expansion of Eq. (9.20). Repeat for $t = \frac{1}{2}t_{\text{opt}}$ and $t = 2t_{\text{opt}}$.

9.12. Obtain a simplified expression for the heating value of methane as measured with a flow calorimeter with the following conditions:

$$T_{a_1} = T_{f_1} = T_{p_2} = T_{c_2} \qquad \phi_1 = 0$$

The incoming gaseous fuel is saturated with water vapor.

9.13. Design an apparatus like that shown in Fig. 9.17 to measure forced-convection heat-transfer coefficients for water at about 200°F in a range of Reynolds numbers from 50,000 to 100,000 based on tube diameter; i.e.,

$$\text{Re}_d = \frac{\rho u_m d}{\mu}$$

where u_m is the mean flow velocity in the tube. Be sure to specify the accuracy and range of all instruments required and estimate the uncertainty in the calculated values of the heat-transfer coefficients.

9.14. Design an apparatus like that shown in Fig. 9.18 to measure free-convection heat-transfer coefficients for air at atmospheric pressure and ambient temperature of 20°C. The apparatus is to produce data over a range of Grashof numbers from 10^5 to 10^7, where

$$\text{Grashof number} = \text{Gr} = \frac{\rho^2 g \beta (T_w - T_\infty) d^3}{\mu^2}$$

where β = volume coefficient of expansion for air
 d = tube diameter
 T_w = tube surface temperature
 T_∞ = ambient air temperature

The properties of the air are evaluated at the film temperature defined by (see Table 8.5)

$$T_f = \frac{T_w + T_\infty}{2}$$

The Grashof number is dimensionless when a consistent set of units is used. Be sure to specify the accuracy and range of all instruments used and estimate the uncertainty in the calculated values of the heat-transfer coefficient.

9.15. A copper sphere 2.5 cm in diameter is used as a slug-type heat-flux meter. A thermocouple is attached to the interior and the transient temperature recorded when the sphere is suddenly exposed to a convection environment. Plot the time response of the system for sudden exposure to an environment with $h = 570$ W/m$^2 \cdot$ °C. Assume the sphere is initially at 95°C and the environment temperature remains constant at 35°C. What is the time constant of this system?

9.16. A heat-flux meter is to be constructed of a thin layer of glasslike substance having a thermal conductivity of 0.4 Btu/h · ft · °F. The temperature drop across the layer will be measured with a thermopile consisting of 10 copper-constantan junctions. The glass thickness is 0.008 in. Calculate the voltage output of the thermopile when a flux of 10^5 Btu/h · ft^2 is imposed on the layer.

9.17. The dry- and wet-bulb temperatures for a quantity of air at 14.7 psia are measured as 95 and 75°F, respectively. Calculate the relative humidity and the dew point of the air–water-vapor mixture.

9.18. An air–water-vapor mixture is contained at a total pressure of 206 kPa. The dry-bulb temperature is 32°C. What will be the wet-bulb temperature if the relative humidity is 50 percent?

9.19. The thermal conductivity of an insulating material is to be determined using a guarded hot-plate apparatus. A 30×30-cm sample is used, and the approximate thermal conductivity is about 1.7 W/m · °C. A differential-thermocouple arrangement is used to measure the temperature drop across the sample with an uncertainty of ±0.3°C. A heat input of 5 kW ± 1 percent is supplied, with a resulting temperature drop of 55°C. The sample thickness is 2.0 mm. What is the thermal conductivity and its uncertainty, assuming no heat losses and that the sample dimensions are known exactly?

9.20. Wet-bulb and dry-bulb thermometers of air at 1 atm read 20°C and 35°C, respectively, within ±0.5°C. Calculate the relative humidity and dew point and their uncertainties.

9.21. Devise an apparatus for measuring the thermal conductivity of window glass. Specify heating requirements and temperature measurement techniques. Estimate uncertainty in the measurement.

9.22. Devise a capillary tube arrangement for measuring the viscosity of air at 20°C. Specify dimensions, flow rates, and temperature and pressure measurement requirements and techniques.

9.23. The dew point of air at 40°C is measured as 25°C ± 0.5°C. Calculate the relative humidity and wet-bulb temperatures and their uncertainties.

9.24. Show that 1 lbf · s/ft^2 = 1.158×10^5 lbf/h · ft.

9.25. The convection heat-transfer coefficient for free convection from a sphere is given by

$$\frac{\bar{h}d}{k_f} = 2 + 0.43(\mathrm{Gr}_f \, \mathrm{Pr}_f)^{1/4}$$

where the nomenclature is as defined in Table 8.6. Suppose a copper sphere having a diameter of 10 mm is initially heated to a uniform temperature of 200°C and then allowed to cool by a free convection in room air at 20°C. The sphere is coated with a highly reflective material such that radiation can be neglected.

Assuming the sphere behaves as a "lump" or "slug," plot its temperature as a function of time during the cooling process.

9.26. Suppose the sphere in Prob. 9.25 is to be used to measure values of the convection coefficient h at temperatures of 150, 100, and 50°C, with an uncertainty of ±5 percent. Specify temperature and time measurements and their acceptable uncertainties to accomplish these objectives.

9.27. Suppose the sphere in Prob. 9.25 is blackened such that its emissivity is 1.0 and the surrounding radiation temperature is 20°C. Plot the sphere temperature as a function of time under this condition. Note that

$$q_{\text{rad}} = \sigma A \epsilon \left(T_{\text{sphere}}^4 - T_{\text{surr}}^4 \right)$$

where $\epsilon = 1.0$ and A is the surface area.

9.28. Repeat Prob. 9.27 for the radiation condition of Prob. 9.26.

9.29. Construct a matrix table for converting units of dynamic viscosity with the following units: lbf-s/ft², lbf/h-ft, centipoise, kg/m-s, N-s/m².

9.30. A composite layer is formed of 3 mm of copper, 2 mm of constantan, and 3 mm of copper. Constantan has the thermal properties of 18-8 stainless steel. If a heat flux of 100 W/cm² is imposed on one of the copper surfaces, what emf would be generated by the arrangement?

9.31. An apparatus like that shown in Fig. 9.17 is to be designed to measure the heat-transfer coefficients for air flowing in a 12-mm-diameter tube having a wall thickness of 0.8 mm. The length of the tube is 2.0 m. Using the forced convection relations of Chap. 8 with an approximate Reynolds number of 100,000, design the experiment; that is, specify the voltages, currents, flow rates, temperatures, and heat fluxes which may be experienced. Specify the devices to be used for the various measurements, estimate the uncertainties in the measurements, and subsequently estimate the uncertainties which will result in the final determination of the convection heat-transfer coefficients.

9.32. A Saybolt viscosimeter is used to measure the viscosity of an oil having $v = 50$ centistokes. To what uncertainty must the drainage time be determined so that v is measured within ±1 percent?

9.33. A capillary tube apparatus is to be used for measuring viscosities of oils with $v \sim 0.001$ m²/s and $\rho \sim 890$ kg/m³. The design Reynolds number is 500. Calculate the length of a 1.0-mm-diameter tube which will produce a pressure drop of 40 kPa.

9.34. The apparatus of Fig. 9.6 is used to measure the thermal conductivity of air at 400 K. The temperature difference between the emitter and receiver is measured as 10°C. For the dimensions given in the text, determine the required power input.

9.35. An apparatus like that shown in Fig. 9.3 is to be used to measure the thermal conductivity of a solid having an approximate value of $k = 15$ W/m · °C. The "standard" metal used for comparison has $k = 69$ W/m · °C. If sample and standard both have nominal lengths of 10 cm, calculate the heat flux for an overall temperature difference of 25°C.

9.36. An insulating material has a thermal conductivity of approximately 0.05 W/m · °C that is to be measured with a guarded hot-plate apparatus. The material is rather fragile so that the thickness is difficult to determine exactly. A sample having a nominal thickness of 10 cm is to be exposed to a heat flux in the apparatus that will produce a temperature difference across the insulation sample of 15 ± 0.2°C. Assuming the heat-flux determination is within ± 1 percent, what is the allowable uncertainty in the thickness measurement that will produce an overall uncertainty of ± 5 percent in the determination of k? Comment on the result.

9.37. A Saybolt viscosimeter is used to measure the viscosity of a certain oil. The time for drainage of the standard 60-mL sample is 140 ± 1 s. Calculate the dynamic viscosity of the oil in units of kg/m · s if the density is 880 kg/m³.

9.38. The dry-bulb and dew-point temperatures of an air–water-vapor sample at 1 atm total pressure are measured as 30°C and 10°C, respectively. The uncertainty in each measurement is ± 1°C. Using an appropriate set of steam tables, determine the relative humidity of the mixture and percent uncertainty. *Note:* p_v is the saturation pressure of water vapor corresponding to the dew-point temperature.

9.39. A slug-type heat-flux sensor operating on the principle described by Eq. (9.43) has $U = 0.1$ W/m² · °C. The slug is a copper cube 2 mm on a side. What heat flux would be necessary to produce a rise in temperature of the sensor of 50°C in a time of 20 s? State assumptions.

9.40. Using properties given in the appendix, comment on the merits of using a capillary tube for measurement of the dynamic viscosity of motor oil or air.

9.41. A horizontal copper cylinder having a diameter of 2 cm is heated internally and dissipates the energy by free convection to surrounding air at 20°C and 1 atm. Using the appropriate formula of Table 8.8, calculate the free convection heat loss for a cylinder length of 1 m and surface temperature of 70°C. What temperature difference along a 3-m length of solid copper bar would be required to produce the same heat-transfer rate? How does this result relate to the use of guard sections shown in Fig. 9.18?

9.42. Suppose the slug sensor of Prob. 9.39 has a U value of 100 W/m² · °C. What heat flux would be needed to produce the 50°C temperature rise in this circumstance? Should any problems be anticipated? State assumptions.

9.43. Calculate the dew point and relative humidity for an air–water vapor mixture at 1 atm, with dry-bulb and wet-bulb temperatures of 40°C and 20°C, respectively.

9.44. Calculate the drainage time for an oil having a kinematic viscosity of 60 centistokes when placed in a Saybolt viscosimeter.

9.45. A capillary tube is to be designed to operate as a measuring device for oils having ν in the range of 0.0015 m²/s and density of about 750 kg/m³. Assuming a design Reynolds number of 600, what length of a 1.0 mm diameter tube would be required to produce a pressure drop of 45 kPa?

9.46. Consider the combustion of methane and gaseous propane with the theoretical air. Calculate the ratio (kg of CO_2 generated) (kJ of higher heating value).

9.47. The series-bars apparatus shown in Fig. 9.3 is used to measure the thermal conductivity of solid having $k = 12$ W/m-°C and the standard metal bar has $k = 69$ W/m-°C. An overall temperature difference of 30°C is imposed on the assembly. Both bars have a length of 9 cm. Calculate the heat flux that will be experienced.

9.48. The drainage time for a standard sample of a certain oil is 150 s from a Saybolt viscosimeter. The specific gravity of the oil is 0.8. Calculate the dynamic viscosity of the oil.

9.49. A small copper cube 2.5 mm on a side operates as a slug-type heat-flux sensor. It is installed in a conduction sink as in Fig. 9.21 having $U = 0.15$ W/m²-°C. What heat flux would be indicated for rise in sensor temperature of 40°C in a time of 15 s?

9.50. The slug-sensor of Prob. 9.49 is installed in a different conduction sink having $U = 50$ W/m²-°C. Calculate the heat flux which would be required to produce the 40°C temperature rise in 15 s.

9.13 REFERENCES

1. Doolittle, J. S.: *Mechanical Engineering Laboratory,* McGraw-Hill, New York, 1957.

2. Holman, J. P.: *Heat Transfer,* 10th ed., McGraw-Hill, New York, 2010.

3. Kaye, G. W. C., and W. F. Higgins: "The Thermal Conductivities of Certain Liquids," *Proc. Roy. Soc. London,* vol. 117, no. 459, 1928.

4. Keyes, F. G., and D. J. Sandell: "New Measurements of Heat Conductivity of Steam and Nitrogen," *Trans. ASME,* vol. 72, p. 768, 1950.

5. Shoop, C. F., and G. L. Tuve: *Mechanical Engineering Practice,* 5th ed., McGraw-Hill, New York, 1956.

6. Van Dusen, M. S., and S. M. Shelton: "Apparatus for Measuring Thermal Conductivity of Metals up to 600°C," *J. Res. Natl. Bur. Std.,* vol. 12, no. 429, 1934.

7. Vines, R. G.: "Measurements of Thermal Conductivities of Gases at High Temperatures," *Trans. ASME,* vol. 82C, p. 48, 1960.

8. Leidenfrost, W.: "An Attempt to Measure the Thermal Conductivity of Liquids, Gases and Vapors with a High Degree of Accuracy over Wide Ranges of Temperature and Pressure," *Int. J. Heat and Mass Transfer,* vol. 7, no. 4, pp. 447–476, 1964.

9. Fraade, D. J.: "Measuring Moisture in Gases," *Instr. Cont. Syst.,* p. 100, April 1963.

10. "Moisture and Humidity," *Instr. Cont. Syst.,* p. 121, October 1964.

11. Amdur, E. J.: "Humidity Sensors," *Instr. Cont. Syst.,* p. 93, June 1963.

12. Holman, J. P.: *Thermodynamics,* 4th ed., McGraw-Hill, New York, 1988.

13. Kirchhoff, R. H.: "Calorimetric Heating-Rate Probe for Maximum Response Time Interval," *AIAA J.,* p. 966, May 1964.

14. Hornbaker, D. R., and D. L. Rall: "Thermal Perturbations Caused by Heat Flux Transducers and Their Effect on the Accuracy of Heating Rate Measurements," *ISA Trans.,* p. 100, April 1963.

15. Gardon, R.: "An Instrument for the Direct Measurement of Intense Thermal Radiation," *Rev. Sci. Instr.,* p. 366, May 1953.

16. Hager, N. E., Jr.: "Thin Foil Heat Meter," *Rev. Sci. Instr.,* vol. 36, pp. 1564–70, November 1965.

17. Carey, R. M.: *Experimental Test of a Heat Flow Meter,* M.S. thesis, Southern Methodist University, Dallas, TX, July 1963.

18. Kestin, J., and W. Leidenfrost: "An Absolute Determination of the Viscosity of Eleven Gases over a Range of Pressure," *Physica,* vol. 25, p. 1033, 1959.

19. Schultz, D. L., and T. V. Jones: "Heat Transfer Measurements in Short-Duration Hypersonic Facilities," *AGARDograph* 165, 1973.

20. Leidenfrost, W.: "Measurement of Thermophysical Properties," in *Measurements in Heat Transfer,* E. R. G. Eckert and R. J. Goldstein, eds., pp. 457–517. McGraw-Hill, New York, 1976.

21. Wexler, A.: "A Study of the National Humidity and Moisture Measurement System," NBS IR75-933, National Bureau of Standards (U.S.), 1975.

22. Borman, G. L., and K. W. Ragland: *Combustion Engineering,* McGraw-Hill, New York, 1998.

23. Childs, P. R. N., J. R. Greenwood, and C. A. Long: "Heat Flux Measurement Techniques," *Proc. Instn. Mech. Engrs.,* vol. 213, Part C, pp. 655–677, 1999.

10

FORCE, TORQUE, AND STRAIN MEASUREMENTS

10.1 INTRODUCTION

Some types of force measurements applicable to pressure-sensing devices have been discussed in Chap. 6. We now wish to consider other methods for measuring forces and torques and to relate these methods to basic strain measurements and experimental stress analysis.

Force is represented mathematically as a vector with a point of application. Physically, it is a directed push or pull. According to Newton's second law of motion, as written for a particle of constant mass, force is proportional to the product of mass and acceleration. Thus,

$$F = \frac{1}{g_c} ma \qquad \textbf{[10.1]}$$

where the $1/g_c$ term is the proportionality constant. When force is expressed in lbf, mass in lbm, and acceleration in ft/s², g_c has the value of 32.1739 lbm · ft/lbf · s² and is numerically equal to the acceleration of gravity at sea level. In the SI system $g_c = 1.0$ kg · m/N · S². The weight of a body is the force exerted on the body by the acceleration of gravity at sea level so that

$$F = W = \frac{mg}{g_c} \qquad \textbf{[10.2]}$$

At sea level the weight in lbf units is numerically equal to the mass in lbm units. In Sec. 2.5 we discussed the relationships between different unit systems. In the SI system the newton is the unit of force and the kilogram is the standard unit of mass. Thus, the weight of a body should be expressed in newtons in the SI system. There will be users of metric units who will express force in kilograms-*force* (kilopond), and the reader should be aware of this possible practice.

Torque is represented as a moment vector formed by the cross product of a force and radius vector. In Fig. 10.1 the torque that the force F exerts about the point 0 is

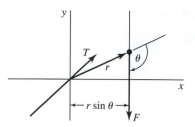

Figure 10.1   Vector representation for torque.

given mathematically as

$$\mathbf{T} = \mathbf{r} \times \mathbf{F} = |r||F| \sin \theta \qquad \text{[10.3]}$$

In the SI system torque is expressed in units of newton-meters, while in the English system the common unit is foot-pounds-force.

Mass represents a quantity of matter and is an inertial property related to force through Eq. (10.1). As such, mass is not a physical property that is measured directly; it is determined through a force measurement or by comparison with mass standards. Even with a comparison of two masses, as with a simple balance, a force equivalency is still involved.

In this chapter we shall be concerned with force measurements as related to mechanical systems. Forces arising from electrical and magnetic effects are not considered.

10.2   MASS BALANCE MEASUREMENTS

Consider the schematic of an analytical balance as shown in Fig. 10.2. The balance arm rotates about the fulcrum at point 0 (usually a knife edge) and is shown in an exaggerated unbalanced position as indicated by the angle ϕ. Point G represents the center of gravity of the arm, and d_G is the distance from 0 to this point. W_B is the weight of the balance arm and pointer. When $W_1 = W_2$, ϕ will be zero and the weight of the balance arms will not influence the measurements. The sensitivity of the balance is a measure of the angular displacement ϕ per unit unbalance in the two weights W_1 and W_2. Expressed analytically this relation is

$$S = \frac{\phi}{W_1 - W_2} = \frac{\phi}{\Delta W} \qquad \text{[10.4]}$$

We now wish to determine the functional dependence of this sensitivity on the physical size and mass of the balance. According to Fig. 10.2, the moment equation for equilibrium is

$$W_1(L \cos \phi - d_B \sin \phi) = W_2(L \cos \phi + d_B \sin \phi) + W_B d_G \sin \phi \qquad \text{[10.5]}$$

For small deflection angles $\sin \phi \approx \phi$ and $\cos \phi \sim 1.0$ so that Eq. (10.5) becomes

$$W_1(L - d_B \phi) = W_2(L + d_B \phi) + W_B d_G \phi$$

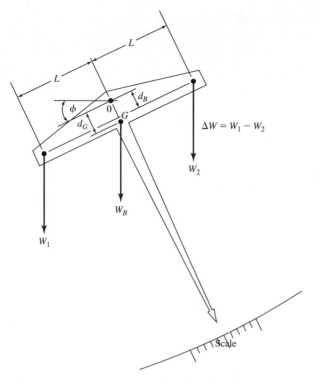

Figure 10.2   Schematic of analytical balance.

or
$$\frac{\phi}{\Delta W} = \frac{L}{(W_1 + W_2)d_B + W_B d_G} \qquad \textbf{[10.6]}$$

Near equilibrium $W_1 \approx W_2$, and S becomes

$$S = \frac{\phi}{\Delta W} = \frac{L}{W_B d_G + 2W d_B} \qquad \textbf{[10.7]}$$

where, now, we have used the single symbol W to designate the loading on the balance. If the balance arm is constructed so that $d_B = 0$, we obtain

$$S = \frac{L}{W_B d_G} \qquad \textbf{[10.8]}$$

an expression that is independent of the loading of the balance. Precision balances are available that have an accuracy of about 1 part in 10^8. For precision instruments optical methods are usually employed to sense the deflection angle and to determine the equilibrium position.

UNCERTAINTY DUE TO ANGULAR SENSITIVITY OF BALANCE.   A balance is | **Example 10.1**
constructed with the following dimensions:

$W_B = 50$ g
$d_G = 0.3$ cm
$L = 21.2$ cm
$d_B = 0.01$ cm

The pointer scale is graduated so that readings may be taken within one-quarter of a degree. Estimate the uncertainty due to sensitivity in determining the weight of a mass of 1000 g. It may be noted that other uncertainties exist in the measurement, such as frictional effects, errors in reading, and so forth.

Solution

From Eq. (10.7) we have

$$S = \frac{\phi}{\Delta W} = \frac{L}{2Wd_B + W_B d_G}$$

$$= \frac{21.2}{(2)(1000)(0.01) + (50)(0.3)} = 0.606 \text{ rad/g}$$

For an uncertainty in ϕ of 0.25° the uncertainty in W is then given by

$$\Delta W = \frac{\phi}{S} = \frac{0.25\,\pi}{(180)(0.606)} = 0.00072 \text{ g}$$

or about 1 part in 100,000.

When a balance is used for mass determinations, errors may result if corrections are not made for buoyancy forces of the air surrounding the samples. Typically, an unknown mass is balanced with a group of standard brass weights. The forces that the instrument senses are not the weight forces of the unknown mass and brass weights but the weight forces *less* the buoyancy force on each mass. If the measurement is conducted in a vacuum or the unknown brass and mass weights have equal volume, the buoyancy forces will cancel out and there will be no error. Otherwise the error may be corrected with the following analysis. The two forces exerted on the balance arms will be

$$W_1 = (\rho_u - \rho_a)V_u \qquad\qquad \textbf{[10.9]}$$
$$W_2 = (\rho_s - \rho_a)V_s \qquad\qquad \textbf{[10.10]}$$

where
ρ_u = density of the unknown
ρ_s = density of the standard weights
ρ_a = density of the surrounding air
V_u, V_s = volumes of the unknown and standard weights, respectively

The true weights of the unknown and standard weights are

$$W_u = \rho_u V_u \qquad\qquad \textbf{[10.11]}$$
$$W_s = \rho_s V_s \qquad\qquad \textbf{[10.12]}$$

At balance $W_1 = W_2$ and there results

$$W_u = W_s[1 + (\rho_a/\rho_s)(\rho_s - \rho_u)/(\rho_u - \rho_a)] \qquad\qquad \textbf{[10.13]}$$

For unknown sample densities that are substantially lower than the density of the standard weights, the buoyancy error in the weight measurement can be significant.

BUOYANCY ERROR IN A BALANCE.   A quantity of a plastic material having a density | **Example 10.2**
of about 1280 kg/m³ is weighed on a standard equal-arm balance. Balance conditions are
achieved with brass weights totaling 152 g in a room where the ambient air is at 20°C and
1 atm. The density of the brass weights may be taken as 8490 kg/m³. Calculate the true weight
of the plastic material and the percent error that would result if the balance reading were taken
without correction.

Solution

The true weight of the brass weights is 152 g. The density of the air may be calculated as

$$\rho_a = \frac{p}{RT} = \frac{1.0132 \times 10^5}{(287)(293)} = 1.2049 \text{ kg/m}^2$$

The true weight of the plastic is then calculated from Eq. (10.13) as

$$W_u = 152\left[1 + \frac{(1.2049)(8490 - 1280)}{(8490)(1280 - 1.2049)}\right] = 152.122 \text{ g}$$

The error is $(0.122/152) \times 100$ or 0.08 percent.

Comment

We should note that the buoyancy error becomes larger for lower-density samples. For balsa
wood having a density of 140 kg/m³ the error would be 0.85 percent. For a lightweight insulating material having a density of 24 kg/m³ (fiberglass) the error would be 5.27 percent. The
effect on the error of measurement is shown in Figure Example 10.2 as a function of ρ_u/ρ_a.

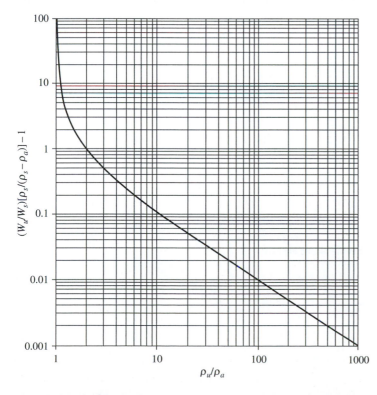

Figure Example 10.2

10.3   ELASTIC ELEMENTS FOR FORCE MEASUREMENTS

Elastic elements are frequently employed to furnish an indication of the magnitude of an applied force through a displacement measurement. The simple spring is an example of this type of force-displacement transducer. In this case the force is given by

$$F = ky \qquad \textbf{[10.14]}$$

where k is the spring constant and y is the displacement from the equilibrium position. For the simple bar shown in Fig. 10.3 the force is given by

$$F = \frac{AE}{L} y \qquad \textbf{[10.15]}$$

where A = cross-sectional area
        L = length
        E = Young's modulus for the bar material

The deflection of the cantilever beam shown in Fig. 10.4 is related to the loading force by

$$F = \frac{3EL}{L^3} y \qquad \textbf{[10.16]}$$

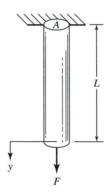

Figure 10.3   Simple elastic element.

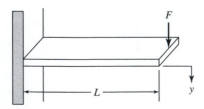

Figure 10.4   Cantilever elastic element.

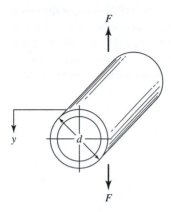

Figure 10.5 Thin-ring elastic element.

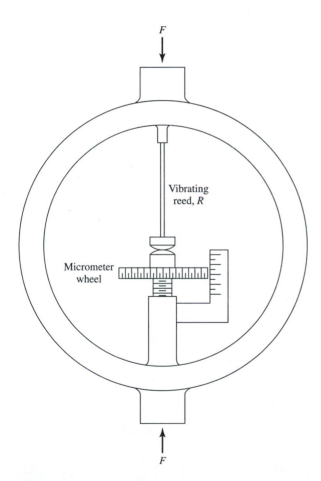

Figure 10.6 Proving ring.

where I is the moment of inertia of the beam about the centroidal axis in the direction of the deflection. Any one of the three devices mentioned above is suitable for use as a force transducer provided that accurate means are available for indicating the displacements. The differential transformer (Sec. 4.21), for example, may be useful for measurement of these displacements, as well as capacitance and piezoelectric transducers (Secs. 4.22 and 4.23).

Another elastic device frequently employed for force measurements is the thin ring shown in Fig. 10.5. The force-deflection relation for this type of elastic element is

$$F = \frac{16}{\pi/2 - 4/\pi} \frac{EI}{d^3} y \qquad \textbf{[10.17]}$$

where d is the outside ring diameter and I is the moment of inertia about the centroidal axis of the ring section. The *proving ring* is a ring transducer that employs a sensitive micrometer for the deflection measurement, as shown in Fig. 10.6. To obtain a precise measurement, one edge of the micrometer is mounted on a vibrating-reed device R, which is plucked to obtain a vibratory motion. The micrometer contact is then moved forward until a noticeable damping of the vibration is observed. Deflection measurements may be made within ± 0.00002 in (0.5 μm) with this method. The proving ring is used as a calibration standard for large tensile-testing machines.

Example 10.3 | **CANTILEVER DEFLECTION MEASUREMENT.**   A small cantilever beam is constructed of spring steel having $E = 28.3 \times 10^6$ psi. The beam is 0.186 in wide and 0.035 in thick, with a length of 1.00 ± 0.001 in. An LVDT is used for the displacement-sensing device, and it is estimated that the uncertainty in the displacement measurement is ± 0.001 in. The uncertainties in the bar dimensions are ± 0.0003 in. Calculate the indicated force and uncertainty due to dimension tolerances when $y = 0.100$ in.

Solution

The moment of inertia is calculated with

$$I = \frac{bh^3}{12} \qquad \textbf{[a]}$$

where b is the width and h is the thickness. Thus,

$$I = \frac{(0.186)(0.035)^3}{12} = 6.64 \times 10^{-7} \text{ in}^4$$

Equation (10.16) may be rewritten as

$$F = \frac{3Ebh^3}{12L^3} y \qquad \textbf{[b]}$$

Using Eq. (3.2) the uncertainty in the force measurement can be written

$$\frac{w_F}{F} = \left[\left(\frac{w_b}{b}\right)^2 + 9\left(\frac{w_h}{h}\right)^2 + 9\left(\frac{w_L}{L}\right)^2 + \left(\frac{w_y}{y}\right)^2 \right]^{1/2} \qquad \textbf{[c]}$$

The nominal value of the force becomes

$$F = \frac{(3)(28.3 \times 10^6)(6.64 \times 10^{-7})}{(1.00)^3} 0.100$$

$$= 5.64 \text{ lbf } (25.09 \text{ N})$$

From Eq. (c) the uncertainty is

$$\frac{w_F}{F} = \left[\left(\frac{0.0003}{0.186}\right)^2 + 9\left(\frac{0.0003}{0.035}\right)^2 + 9\left(\frac{0.001}{1.00}\right)^2 + \left(\frac{0.001}{0.100}\right)^2 \right]^{1/2}$$

$$= 0.0278 \text{ or } 2.78\%$$

In this instance 1.0 percent uncertainty would be present even if the dimensions of the beam were known exactly.

The surface strain (deformation) in elastic elements like those discussed above is, of course, a measure of the deflection from the no-load condition. This surface strain may be measured very readily by the electrical-resistance strain gage to be discussed in subsequent paragraphs. The output of the strain gage may thus be taken as an indication of the impressed force. The main problem with the use of these gages for the force-measurement applications is that a moment may be impressed on the elastic element because of eccentric loading. This would result in an alteration of the basic strain distribution as measured by the strain gage. There are means for compensating for this effect through installation of multiple gages, properly interconnected to cancel out the deformation resulting from the impressed moment. The interested reader should consult Refs. [4] and [5] for a summary of these methods.

Complete assemblies incorporating the elastic element, gages, and signal conditioning are available commercially for different ranges of force measurements. The electric outputs may then be used to drive recording equipment. These devices operate on the principle of the displacement of a linear elastic element with impressed force. The displacements are typically measured by bonded strain gages[1] (Sec. 10.7), LVDT transducers[2] (Sec. 4.21) [7], or change in capacitance due to the displacement of a diaphragm.[3] Force ranges for the load cells may be as low as 10 g full scale to hundreds of thousands of kg. The frequency response for most load cells is usually low but may be as high as 2 kHz for sensitive LVDT devices. Linearity of 0.2 percent of full scale is typical for some devices.

10.4 TORQUE MEASUREMENTS

Torque, or moment, may be measured by observing the angular deformation of a bar or hollow cylinder, as shown in Fig. 10.7. The moment is given by

$$M = \frac{\pi G \left(r_o^4 - r_i^4\right)}{2L} \phi \qquad\qquad \textbf{[10.18]}$$

---

[1] Interface, Inc., Scottsdale, AZ.
[2] Schaevitz Engineering, Pennsauken, NJ.
[3] Kavlico Corp., Chatsworth, CA.

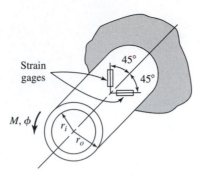

Figure 10.7   Hollow cylinder as an elastic element for torque measurement.

where M = moment, in-lbf or N · m
G = shear modulus of elasticity, psi or N/m^2
r_i = inside radius, in or m
r_o = outside radius, in or m
L = length of cylinder, in or m
ϕ = angular deflection, rad

The shear modulus of elasticity may be calculated from

$$G = \frac{E}{2(1 + \mu)}$$

where μ is Poisson's ratio, which has a value of about 0.3. Strain gages attached at 45° angles as shown will indicate strains of

$$\epsilon_{45°} = \pm \frac{Mr_0}{\pi G \left(r_o^4 - r_i^4 \right)} \qquad \textbf{[10.19]}$$

Either the deflection or the strain measurement may be taken as indication of the applied moment. Multiple strain gages may be installed and connected so that any deformation due to axial or transverse load is canceled out in the final readout circuit.

Again, complete assemblies are available commercially for different torque ranges.

A very old device for the measurement of torque and dissipation of power from machines is the Prony brake.[4] A schematic diagram is shown in Fig. 10.8. Wooden or synthetic friction blocks are mounted on a flexible band or rope, which is connected to the arm. Some arrangement is provided to tighten the rope to increase the frictional resistance between the blocks and the rotating flywheel of the machine. The torque exerted on the Prony brake is given by

$$T = FL \qquad \textbf{[10.20]}$$

The force F may be measured by conventional platform scales or other methods discussed in the previous paragraphs.

| [4]Named for G. C. F. M. Riche, Baron de Prony (1755–1839), French hydraulic engineer.

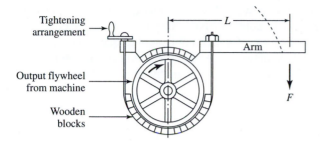

Tightening
arrangement

L

Arm

Output flywheel
from machine

F

Wooden
blocks

Figure 10.8 Schematic of a Prony brake.

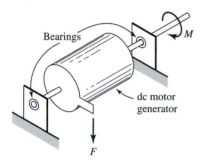

Bearings

M

dc motor
generator

F

Figure 10.9 Schematic of a cradled dynamometer.

The power dissipated in the brake is calculated from

$$P = \frac{2\pi TN}{33,000} \text{ hp} \qquad [10.21]$$

where the torque is in foot-pounds-force (ft-lb$_f$) and N is the rotational speed in revolutions per minute (rpm).

Various other types of brakes are employed for power measurements on mechanical equipment. The *water brake,* for example, dissipates the output energy through fluid friction between a paddle wheel mounted inside a stationary chamber filled with water. The chamber is freely mounted on bearings so that the torque transmitted to it can be measured through an appropriate moment arm similar to that used with the Prony brake.

The dc cradled dynamometer is perhaps the most widely used device for power and torque measurements on internal-combustion engines, pumps, small steam turbines, and other mechanical equipment. The basic arrangement of this device is shown in Fig. 10.9. A dc motor generator is mounted on bearings as shown, with a moment arm extending from the body of the motor to a force-measurement device, which is usually a pendulum scale. When the device is connected to a power-producing machine, it acts as a dc generator whose output may be varied by dissipating the power in resistance racks. The torque impressed on the dynamometer is measured with the moment arm and the output power calculated with Eq. (10.21). The dynamo-meter may also be used as an electric motor to drive some power-absorbing device like a pump. In this case it furnishes a means for measurement of torque and power input to the machine. Commercial dynamometers are equipped with controls for precise variation of the load and speed of the machine and are available with power ratings as high as 3700 kW (5000 hp).

MEASUREMENT OF ANGULAR VELOCITY

One form of tachometer, or angular velocity measurement device, employs a small dc permanent-magnet generator to generate a voltage proportional to the rotational speed. A measurement of the output voltage then serves to indicate angular velocity of the shaft. The device may be attached permanently to the shaft for continuous monitoring of speed, or connected by friction to measure speed at some desired time.

Optical methods may also be used to measure angular velocity. A strobe light is flashed on the rotating shaft or object. A marking on the shaft, or the object itself, for example, a propeller, will appear stationary when the strobe frequency coincides with the rotational speed (frequency). A reading of strobe frequency then determines angular speed. A very complete discussion of the measurement of angular velocity is given in Ref. [14].

Example 10.4 | **RATINGS OF AN ENGINE.**   An automobile engine is rated at a 290-ft-lb torque at 2400 rpm and 180 hp at 4200 rpm. Are these ratings consistent and, if not, how might they be explained?

Solution

The relation governing horsepower, torque, and rpm is given by Eq. (10.21):

$$P(\text{hp}) = \frac{2\pi TN}{33{,}000}$$

For the torque specification we obtain

$$P = \frac{2\pi(290)(2400)}{33{,}000} = 132.5 \text{ hp}$$

while for the horsepower specification we obtain the torque as

$$T = \frac{(33{,}000)(180)}{2\pi(4200)} = 225 \text{ ft-lb}$$

Of course, the two calculations are not consistent. The explanation is that both the torque (T) and horsepower (hp) vary with speed for most engines. Although not normally stated, the ratings here give the *maximum* values of T and hp and the values of rpm at which they occur.

Example 10.5 | **TORQUE AND POWER MEASUREMENT.**   A solid cylinder like that shown in Fig. 10.7 is to be designed to measure the torque of the engine in Example 10.4 by measuring the angular deflection at the 290 ft-lb condition. Spring steel with $E = 28.3 \times 10^6$ psi and $\mu = 0.3$ is employed for construction. To obtain sufficient resolution, an angular deflection of at least 10° is desired. Determine a suitable length and cylinder diameter combination for this measurement.

Solution

For a solid cylinder we have $r_i = 0$ and Eq. (10.18) becomes

$$M = \frac{\pi G r_o^4}{2L} \qquad\qquad [a]$$

The value of G is calculated as

$$G = \frac{E}{2(1 + \mu)} = \frac{28.3 \times 10^6}{(2)(1.3)} = 1.088 \times 10^6 \text{ psi}$$

We also have

$$M = 290 \text{ ft-lb} = 3480 \text{ in-lb}$$
$$\phi = 10° = 0.1745 \text{ rad}$$

Inserting the values in Eq. (a) yields

$$\frac{r_o^4}{L} = \frac{(2)(3480)}{(1.088 \times 10^6)(0.1745)}$$
$$= 0.01167$$

There are an infinite number of combinations. For $L = 1$ in we would have

$$r_o = 0.3287 \text{ in}$$

or a diameter of 0.657 in. A length of 4 in would require a diameter of 0.93 in.

10.5 STRESS AND STRAIN

Stress analysis involves a determination of the stress distribution in materials of various shapes and under different loading conditions. Experimental stress analysis is performed by measuring the deformation of the piece under load and inferring from this measurement the local stress which prevails. The measurement of deformation is only one facet of the overall problem, and the analytical work that must be applied to the experimental data in order to determine the local stresses is of equal importance. Our concern in the following sections is with the methods that may be employed for deformation measurements. Some simple analysis of these measurements will be given to illustrate the reasoning necessary to obtain local stress values. For more detailed considerations the reader should consult Refs. [2] and [5], the periodical publication [9], and Refs. [10–13].

Consider the bar shown in Fig. 10.10 subjected to the axial load T. Under no-load conditions the length of the bar is L and the diameter is D. The cross-sectional area of the bar is designated by A. If the load is applied such that the stress does not exceed

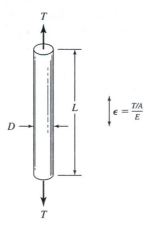

$$\epsilon = \frac{T/A}{E}$$

Figure 10.10   Simple bar in axial strain.

the elastic limit of the material, the axial strain is given by

$$\epsilon = \frac{T/A}{E} = \frac{\sigma_a}{E} \qquad\qquad \textbf{[10.22]}$$

where σ_a is the axial stress and E is Young's modulus for the material. The unit axial strain ϵ_a is defined by the relation

$$\epsilon_a = \frac{dL}{L} \qquad\qquad \textbf{[10.23]}$$

i.e., it is the axial deformation per unit length. Resulting from the deformation in the axial direction is a corresponding deformation in the cross-sectional area of the bar. The change in area is evidenced by a change in the diameter or, more specifically, by a change in the transverse dimension. The ratio of the unit strain in the transverse direction to the unit strain in the axial direction is defined as *Poisson's ratio* and must be determined experimentally for various materials.

$$\mu = -\frac{\epsilon_t}{\epsilon_a} = -\frac{dD/D}{dL/L} \qquad\qquad \textbf{[10.24]}$$

A typical value for Poisson's ratio for many materials is 0.3. If the material is in the plastic state, the volume remains constant with the change in strain so that

$$dV = L\,dA + A\,dL = 0$$

or

$$\frac{dA}{A} = -\frac{dL}{L}$$

Expressed in terms of the diameter, this relation is

$$2\frac{dD}{D} = -\frac{dL}{L}$$

so that $\mu = 0.5$ under these conditions.

10.6   STRAIN MEASUREMENTS

Let us first consider some basic definitions. Any strain measurements must be made over a finite length of the workpiece. The smaller this length, the more nearly the measurement will approximate the unit strain at a point. The length over which the average strain measurement is taken is called the *base length*. The *deformation sensitivity* is defined as the minimum deformation that can be indicated by the appropriate gage. *Strain sensitivity* is the minimum deformation that can be indicated by the gage per unit base length.

A simple method of strain measurement is to place some type of grid marking on the surface of the workpiece under zero-load conditions and then to measure the deformation of this grid when the specimen is subjected to a load. The grid may be scribed on the surface, drawn with a fine ink pen, or photoetched. Rubber threads have also been used to mark the grid. The sensitivity of the grid method depends on the accuracy with which the displacement of the grid lines may be measured. A micrometer microscope is frequently employed for such measurements. An alternative method is to photograph the grid before and after the deformation and to make the measurements on the developed photograph. Photographic paper can have appreciable shrinkage so that glass photographic plates are preferred for such measurements. The grid may also be drawn on a rubber model of the specimen and the local strain for the model related to that which would be present in the actual workpiece. Grid methods are usually applicable to materials and processes having appreciable deformation under load. These methods might be applicable to a study of the strains encountered in sheet-metal-forming processes. The grid could be installed on a flat sheet of metal before it is formed. The deformation of the grid after forming gives the designer an indication of the local stresses induced in the material during the forming process.

The use of digital photography whereby deformations may be detected on the basis of individual pixels offers advantages over conventional film photographic measures. See Appendix B for further information.

Brittle coatings offer a convenient means for measuring the local stress in a material. The specimen or workpiece is coated with a special substance having very brittle properties. When the specimen is subjected to a load, small cracks appear in the coating. These cracks appear when the state of tensile stress in the workpiece reaches a certain value and thus may be taken as a direct indication of this local stress. The brittle coatings are valuable for obtaining an overall picture of the stress distribution over the surface of the specimen. They are particularly useful for determination of stresses at stress concentration points that are too small or inconveniently located for installation of electrical-resistance or other types of strain gages.

10.7   ELECTRICAL-RESISTANCE STRAIN GAGES

The electrical-resistance strain gage is the most widely used device for strain measurement. Its operation is based on the principle that the electrical resistance of a conductor changes when it is subjected to mechanical deformation. Typically, an electric conductor is bonded to the specimen with an insulating cement under no-load conditions. A load is then applied, which produces a deformation in both the specimen and the resistance element. This deformation is indicated through a measurement of the change in resistance of the element and a calculation procedure which is described below.

Let us now develop the basic relations for the resistance strain gage. The resistance of the conductor is

$$R = \rho \frac{L}{A} \qquad \text{[10.25]}$$

where L = length
        A = cross-sectional area
        ρ = resistivity of the material

Differentiating Eq. (10.25), we have

$$\frac{dR}{R} = \frac{d\rho}{\rho} + \frac{dL}{L} - \frac{dA}{A} \qquad \text{[10.26]}$$

The area may be related to the square of some transverse dimension, such as the diameter of the resistance wire. Designating this dimension by D, we have

$$\frac{dA}{A} = 2\frac{dD}{D} \qquad \text{[10.27]}$$

Introducing the definition of the axial strain and Poisson's ratio from Eqs. (10.23) and (10.24), we have

$$\frac{dR}{R} = \epsilon_a(1 + 2\mu) + \frac{d\rho}{\rho} \qquad \text{[10.28]}$$

The gage factor F is defined by

$$F = \frac{dR/R}{\epsilon_a} \qquad \text{[10.29]}$$

so that
$$F = 1 + 2\mu + \frac{1}{\epsilon_a}\frac{d\rho}{\rho} \qquad \text{[10.30]}$$

We may thus express the local strain in terms of the gage factor, the resistance of the gage, and the change in resistance with the strain:

$$\epsilon = \frac{1}{F}\frac{\Delta R}{R} \qquad \text{[10.31]}$$

The value of the gage factor and the resistance are usually specified by the manufacturer so that the user only needs to measure the value of ΔR in order to determine the local strain. For most gages the value of F is constant over a rather wide range of strains. It is worthwhile, however, to examine the influence of various physical properties of the resistance material on the value of F. If the resistivity of the material does not vary with the strain, we have from Eq. (10.30)

$$F = 1 + 2\mu \qquad \text{[10.32]}$$

Taking a typical value of μ as 0.3, we would obtain $F = 1.6$. In this case the change in resistance of the material results solely from the change in physical dimensions. If the resistivity decreases with strain, the value of F will be less than 1.6. When the resistivity increases with strain, F will be greater than 1.6. Gage factors for various materials have been observed from -140 to $+175$. If the resistance material is strained to the point that it is operating in the plastic region, $\mu = 0.5$ and the resistivity remains essentially constant. Under these conditions the gage factor approaches a value of 2. For most commercial strain gages the gage factor is the same for both compressive and tensile strains. A high gage factor is desirable in practice because a larger change in resistance ΔR is produced for a given strain input, thereby necessitating less sensitive readout circuitry.

Three common types of resistance strain gages are shown in Fig. 10.11. The bonded-wire gage employs wire sizes varying between 0.0005 and 0.001 in (12 and 25 μm). The foil gage usually employs a foil less than 0.001 in thick and is available in a wide variety of configurations which may be adapted to different stress-measurement situations. Because of this flexibility, it is the most commonly used gage.

The semiconductor gage employs a silicon base material that is strain-sensitive and has the advantage that very large values of F may be obtained ($F \sim 100$). The material is usually produced in brittle wafers having a thickness of about 0.01 in (0.25 mm). Semiconductor gages also have very high temperature coefficients of resistance. Table 10.1 presents a summary of characteristics of several strain-gage materials. Most commercial gages employ constantan (or other alloys) or isoelastic for materials of construction. Reference [8] furnishes additional information.

Wire and foil gages may be manufactured in various ways, but the important point is that the resistance element must be securely bonded to its mounting. It is essential that the bond between the resistance element and the cement joining it to the workpiece be stronger than the resistance wire itself. In this way the strength of the resistance element is the smaller, and hence the deformation of the entire gage is governed by the deformation of the resistance element. Most wire strain gages employ either a nitrocellulose cement or a phenolic resin for the bonding agent with

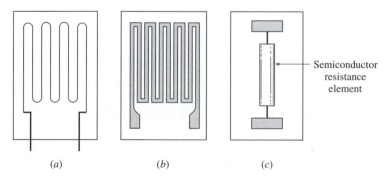

Semiconductor resistance element

(a)   (b)   (c)

Figure 10.11   Three types of resistance strain gages. (a) Wire gage; (b) foil gage; (c) semiconductor gage.

Table 10.1   Characteristics of some resistance strain-gage materials

Material	Trade Name	Approx. Gage Factor, F	Approx. Resistivity at 20°C, $\mu\Omega \cdot cm$	Temp. Coeff. of Resistance, $°C^{-1} \times 10^6$	Remarks
55% Cu, 45% Ni	Advance, Constantan, Copel	2.0	49	11	F constant over wide range of strain; low-temperature use below 360°C
4% Ni, 12% Mn, 84% Cu	Manganin	0.47	44	20	Same
80% Ni, 20% Cu	Nichrome V	2.0	108	400	Suitable for high-temperature use to 800°C
36% Ni, 8% Cr, 0.5% Mo, 55.5% Fe	Isoelastic	3.5	110	450	Used for low temperatures to 300°C
67% Ni, 33% Cu	Monel	1.9	40	1900	Useful to 750°C
74% Ni, 20% Cr, 3% Al, 3% Fe	Karma	2.4	125	20	Useful to 750°C
95% Pt, 5% Ir	—	5.0	24	1250	Useful to 1000°C
Silicon semiconductor	—	−100 to +150	10^9	90,000	Brittle but has high-gage factor; not suitable for large strain measurements; can be used for miniature pressure and force transducers

a thin paper backing to maintain the wire configuration. Such gages may be used up to 150°C (300°F). A Bakelite mounting is usually employed for temperatures up to 260°C (500°F). Foil gages are manufactured by an etching process similar to that used with printed circuit boards and use base materials of paper, Bakelite, and epoxy film. Epoxy cements are also employed for both wire and foil gages.

When strain gages are mounted on a specimen, two notes of caution should be observed: (1) The surface must be absolutely clean. Cleaning with an emery cloth followed by acetone is usually satisfactory. (2) Sufficient time must be allowed for the cement to dry and harden completely. Even though the cement is dry around the edge of the gage, it may still be wet under the gage. If possible, 24 h should be allowed for drying at room temperature. Drying time may be reduced for higher temperatures.

For low-temperature applications (−100 to +100°C) Duco cement (nitrocellulose) is normally employed with paper-covered gages and Eastman 910 (cyanoacrylate) with foil gages mounted on epoxy. These cements are discussed in Ref. [2], along with rather detailed instructions for mounting the various types of gages. The interested reader should consult this reference and the literature of various manufacturers of strain gages for more information.

Problems associated with strain-gage installations generally fall into three categories: (1) temperature effects, (2) moisture effects, and (3) wiring problems. It is assumed that the gage is properly mounted. Temperature problems arise because of differential thermal expansion between the resistance element and the material to which it is bonded. Semiconductor gages offer the advantage that they have a lower expansion coefficient than either wire or foil gages. In addition to the expansion problem, there is a change in resistance of the gage with temperature, which must be adequately compensated for. We shall see how this compensation is performed in a subsequent section. Moisture absorption by the paper and cement can change the electrical resistance between the gage and the ground potential and thus affect the output-resistance readings. Methods of moistureproofing are discussed in Refs. [1] and [2]. Wiring problems are those situations that arise because of faulty connections between the gage-resistance element and the external readout circuit. These problems may develop from poorly soldered connections or from inflexible wiring, which may pull the gage loose from the test specimen or break the gage altogether. Proper wiring practices are discussed in Refs. [2] and [6].

Electrical-resistance strain gages cannot be easily calibrated because once they are attached to a calibration workpiece, removal cannot be made without destroying the gage. In practice, then, the gage factor is taken as the value specified by the manufacturer and a semicalibration effected by checking the bridge measurement and readout system.

In addition to the general references on strain gages given at the end of this chapter, excellent technical manuals are available from manufacturers. Balwin-Lima Hamilton, Micro-Measurements Group, and Omega Engineering are three such companies.

10.8 MEASUREMENT OF RESISTANCE STRAIN-GAGE OUTPUTS

Consider the bridge circuit of Fig. 4.25. The voltage present at the detector is given by Eq. (4.26) as

$$E_g = E_D = E\left(\frac{R_1}{R_1 + R_4} - \frac{R_2}{R_2 + R_3}\right) \qquad \textbf{[10.33]}$$

If the bridge is in balance, $E_D = 0$. Let us suppose that the strain gage represents R_1 in this circuit and a high-impedance readout device, like a digital voltmeter, is employed so that the bridge operates as a voltage-sensitive deflection circuit. We assume that the bridge is balanced at zero-strain conditions and that a strain on the gage of a ϵ results in a change in resistance ΔR_1. R_1 will be used to represent the resistance of the gage at zero-strain conditions. We immediately obtain a voltage due to the strain as

$$\frac{\Delta E_D}{E} = \frac{R_1 + \Delta R_1}{R_1 + \Delta R_1 + R_4} - \frac{R_2}{R_2 + R_3} \qquad \textbf{[10.34]}$$

Solving for the resistance change gives

$$\frac{\Delta R_1}{R_1} = \frac{(R_4/R_1)[\Delta E_D/E + R_2/(R_2 + R_3)]}{1 - \Delta E_D/E - R_2/(R_2 + R_3)} - 1 \qquad \textbf{[10.35]}$$

Equation (10.35) expresses the resistance changes as a function of the voltage unbalance at the detector ΔE_D.

The bridge circuit may also be used as a current-sensitive device. Combining Eqs. (4.24) to (4.26) gives

$$i_g = \frac{E(R_1 R_3 - R_2 R_4)}{R_1 R_2 R_4 + R_1 R_3 R_4 + R_1 R_2 R_3 + R_2 R_3 R_4 + R_g(R_1 + R_4)(R_2 + R_3)}$$

$$[\textbf{10.36}]$$

Again, we assume the bridge to be balanced under zero-strain conditions and take R_1 as the resistance of the gage under these conditions. Thus,

$$R_1 R_3 = R_2 R_4 \qquad [\textbf{10.37}]$$

The galvanometer current ΔI_g is that value which results from a change in resistance ΔR_1 from the balanced condition. It may be shown that the denominator of Eq. (10.36) is not very sensitive to small changes in R_1 and hence is very nearly a constant which we shall designate as C. Thus,

$$\Delta I_g = \frac{E}{C}[(R_1 + \Delta R_1)R_3 - R_2 R_4] \qquad [\textbf{10.38}]$$

Applying the balance conditions from Eq. (10.37) gives

$$\Delta I_g = \frac{E}{C} R_3 \, \Delta R_1 \qquad [\textbf{10.39}]$$

Introducing the gage factor from Eq. (10.31), we have

$$\Delta I_g = \frac{E}{C} R_3 R_1 F \epsilon = \text{const} \times \epsilon \qquad [\textbf{10.40}]$$

Thus, the deflection current may be taken as a direct indication of the strain imposed on the gage.

We have already mentioned a bridge amplifier circuit in connection with electrical-resistance thermometers (Fig. 8.11). This same arrangement can be employed for use with strain-gage bridge circuits. Numerous readout devices for strain gages are available commercially.

10.9   TEMPERATURE COMPENSATION

It is generally not possible to calculate corrections for temperature effects in strain gages. Consequently, compensation is made directly by means of the experimental setup. Such a compensation arrangement is shown in Fig. 10.12. Gage 1 is installed on the test specimen, while gage 2 is installed on a like piece of material that remains unstrained throughout the tests but at the same temperature as the test piece. Any changes in the resistance of gage 1 due to temperature are thus canceled out by similar changes in the resistance of gage 2, and the bridge circuit detects an unbalanced condition resulting only from the strain imposed on gage 1. Of course, care must be exerted to ensure that both gages are installed in exactly the same manner on their respective workpieces.

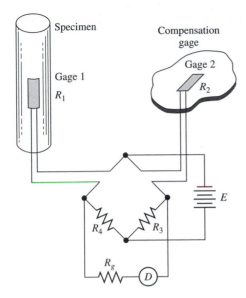

Figure 10.12   Temperature-compensation arrangement for electrical-resistance strain gages.

DETECTOR CURRENT FOR STRAIN GAGE.   A resistance strain gage with $R = 120\ \Omega$ and $F = 2.0$ is placed in an equal-arm bridge in which all resistances are equal to $120\ \Omega$. The power voltage is 4.0 V. Calculate the detector current in microamperes per microinch of strain. The galvanometer resistance is $100\ \Omega$.

| | **Example 10.6** |

Solution

The denominator of Eq. (10.36) is calculated as

$$C = (4)(120)^3 + (100)(240)^2 = 1.267 \times 10^7$$

We calculate the current sensitivity from Eq. (10.40) as

$$\frac{\Delta I_g}{\epsilon} = \frac{E}{C} R_3 R_1 F = \frac{(4.0)(120)^2(2.0)}{1.267 \times 10^7} = 9.08 \times 10^{-3}\ \text{A/in}$$

$$= 9.08 \times 10^{-3}\ \mu\text{A}/\mu\text{in}\ (0.357\ \text{mA/mm})$$

DETECTOR VOLTAGE FOR STRAIN GAGE.   For the gage and bridge in Example 10.6, calculate the voltage indication for a strain of 1.0 μin/in when a high-impedance detector is used.

| | **Example 10.7** |

Solution

We have $R_1 = R_2 = R_3 = R_4 = 120\ \Omega$. Under these conditions Eq. (10.34) becomes

$$\frac{\Delta E_D}{E} = \frac{1 + (\Delta R_1/R_1)}{2 + (\Delta R_1/R_1)} - \frac{1}{2} = \frac{\Delta R_1/R_1}{4 + 2(\Delta R_1/R_1)}$$

The resistance change is calculated from Eq. (10.31):

$$\frac{\Delta R_1}{R_1} = F\epsilon = 2.0 \times 10^{-6}$$

so that

$$\frac{\Delta E_D}{E} = \frac{2.0 \times 10^{-4}}{4 + 4 \times 10^{-6}} = 0.5 \times 10^{-6}$$

and

$$\Delta E_D = (4.0)(0.5 \times 10^{-6}) = 2.0 \; \mu V$$

10.10   STRAIN-GAGE ROSETTES

The installation of a strain gage on a bar specimen like that shown in Fig. 10.10 is a useful application of the gage, but it is quite restricted. The strain that is measured in such a situation is a principal strain since we assumed that the bar is operating under only a tensile load. Obviously, a more general measurement problem will involve strains in more than one direction, and the orientation of the principal stress axes will not be known. It would, of course, be fortunate if the strain gages were installed on the specimen so that they were oriented exactly with the principal stress axes. We now consider the methods that may be used to calculate the principal stresses and strains in a material from three strain-gage measurements. The arrangements for strain gages in such application are called *rosettes*. We shall give only the final relations that are used for calculation purposes. The interested reader should consult Refs. [2], [5], and [17] for the derivations of these equations.

Consider the rectangular rosette shown in Fig. 10.13. The three strain gages are oriented as shown, and the three strains measured by these gages are ϵ_1, ϵ_2, and ϵ_3. The principal strains for this situation are

$$\epsilon_{max}, \epsilon_{min} = \frac{\epsilon_1 + \epsilon_3}{2} \pm \frac{1}{\sqrt{2}}[(\epsilon_1 - \epsilon_2)^2 + (\epsilon_2 - \epsilon_3)^2]^{1/2} \qquad \textbf{[10.41]}$$

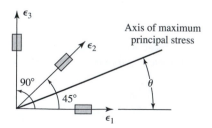

Figure 10.13   Rectangular strain-gage rosette.

The principal stresses are

$$\sigma_{max}, \sigma_{min} = \frac{E(\epsilon_1 + \epsilon_3)}{2(1 - \mu)} \pm \frac{E}{\sqrt{2}(1 + \mu)}[(\epsilon_1 - \epsilon_2)^2 + (\epsilon_2 - \epsilon_3)^2]^{1/2} \quad \textbf{[10.42]}$$

The maximum shear stress is designated by τ_{max} and calculated from

$$\tau_{max} = \frac{E}{\sqrt{2}(1 + \mu)}[(\epsilon_1 - \epsilon_2)^2 + (\epsilon_2 - \epsilon_3)^2]^{1/2} \qquad \textbf{[10.43]}$$

The principal stress axis is located with the angle θ according to

$$\tan 2\theta = \frac{2\epsilon_2 - \epsilon_1 - \epsilon_3}{\epsilon_1 - \epsilon_3} \qquad \textbf{[10.44]}$$

This is the axis at which the maximum stress σ_{max} occurs. A problem arises with the determination of the quadrant for θ since there will be two values obtained as solutions to Eq. (10.44). The angle θ will lie in the first quadrant ($0 < \theta < \pi/2$) if

$$\epsilon_2 > \frac{\epsilon_1 + \epsilon_3}{2} \qquad \textbf{[10.45]}$$

and in the second quadrant if ϵ_2 is less than this value.

Another type of strain-gage rosette in common use is the delta rosette shown in Fig. 10.14. The principal strains in this instance are given by

$$\epsilon_{max}, \epsilon_{min} = \frac{\epsilon_1 + \epsilon_2 + \epsilon_3}{3} \pm \frac{\sqrt{2}}{3}[(\epsilon_1 - \epsilon_2)^2 + (\epsilon_2 - \epsilon_3)^2 + (\epsilon_3 - \epsilon_1)^2]^{1/2} \quad \textbf{[10.46]}$$

The principal stresses are

$$\sigma_{max}, \sigma_{min} = \frac{E(\epsilon_1 + \epsilon_2 + \epsilon_3)}{3(1 - \mu)} \pm \frac{\sqrt{2}\,E}{3(1 + \mu)}[(\epsilon_1 - \epsilon_2)^2 + (\epsilon_2 - \epsilon_3)^2 + (\epsilon_3 - \epsilon_1)^2]^{1/2}$$

$$\textbf{[10.47]}$$

The maximum shear stress axis is calculated from

$$\tau_{max} = \frac{\sqrt{2}\,E}{3(1 + \mu)}[(\epsilon_1 - \epsilon_2)^2 + (\epsilon_2 - \epsilon_3)^2 + (\epsilon_3 - \epsilon_1)^2]^{1/2} \qquad \textbf{[10.48]}$$

The principal stress axis is located according to

$$\tan 2\theta = \frac{\sqrt{3}(\epsilon_3 - \epsilon_2)}{2\epsilon_1 - \epsilon_2 - \epsilon_3} \qquad \textbf{[10.49]}$$

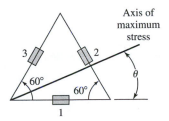

Figure 10.14   Delta strain-gage rosette.

The angle θ will be in the first quadrant when $\epsilon_3 > \epsilon_2$ and in the second quadrant when $\epsilon_2 > \epsilon_3$.

It is worthwhile to mention that resistance strain gages may be sensitive to transverse as well as axial strains. The resistance change produced by a transverse strain, however, is usually less than 2 or 3 percent of the change produced by an axial strain. For this reason it may be neglected in many applications. If the transverse strain is to be considered, the above rosette formulas must be modified accordingly. We shall not present this modification, but refer the reader to Refs. [2] and [17] for more information.

We should note that many complete rosette assemblies and necessary electronic readout equipment are available commercially.

Example 10.8 | **STRAIN-GAGE ROSETTE.** A rectangular rosette is mounted on a steel plate having $E = 29 \times 10^6$ psi and $\mu = 0.3$. The three strains are measured as

$$\epsilon_1 = +500 \ \mu\text{in/in}$$
$$\epsilon_2 = +400 \ \mu\text{in/in}$$
$$\epsilon_3 = -100 \ \mu\text{in/in}$$

Calculate the principal strains and stresses and the maximum shear stress. Locate the axis of the principal stress.

Solution

As an intermediate step we calculate the quantities

$$A = \frac{\epsilon_2 + \epsilon_1}{2} = 200 \ \mu\text{in/in}$$
$$B = [(\epsilon_1 - \epsilon_2)^2 + (\epsilon_2 - \epsilon_3)^2]^{1/2} = 510 \ \mu\text{in/in}$$

Then

$$\epsilon_{\text{max}} = A + \frac{1}{\sqrt{2}} B = 561 \ \mu\text{in/in}$$

$$\epsilon_{\text{min}} = A - \frac{1}{\sqrt{2}} B = -161 \ \mu\text{in/in}$$

$$\sigma_{\text{max}} = \frac{EA}{1-\mu} + \frac{EB}{\sqrt{2}(1+\mu)} = \frac{(29 \times 10^6)(200 \times 10^{-6})}{1 - 0.3}$$

$$+ \frac{(29 \times 10^6)(510 \times 10^{-6})}{\sqrt{2}(1 + 0.3)} = 8280 + 8050 = 16,330 \ \text{psi} \ (112.6 \ \text{MN/m}^2)$$

$$\sigma_{\text{min}} = 8280 - 8050 = 230 \ \text{psi} \ (1.59 \ \text{kN/m}^2)$$

$$\tau_{\text{max}} = \frac{EB}{\sqrt{2}(1+\mu)} = 8050 \ \text{psi} \ (55.5 \ \text{MN/m}^2)$$

We also have

$$\tan 2\theta = \frac{2\epsilon_2 - \epsilon_1 - \epsilon_3}{\epsilon_1 - \epsilon_3} = \frac{(2)(400) - (500) - (-100)}{(500) - (-100)} = 0.667$$

$$2\theta = 33.7 \text{ or } 213.7°$$

$$\theta = 16.8 \text{ or } 106.8°$$

We choose the first-quadrant angle ($\theta = 16.8°$) in accordance with Eq. (10.45).

10.11   THE UNBONDED RESISTANCE STRAIN GAGE

The bonded electrical-resistance strain gage discussed above is the most widely used device for strain measurements. An alternative resistance gage is the unbonded type shown in Fig. 10.15. A spring-loaded mechanism holds the two plates in a close position while the fine-wire filaments are stretched around the mounting pins as shown. The mounting pins must be rigid and also serve as electrical insulators. When plate A moves relative to B, a strain is imposed on these filaments, which may be detected through a measurement of the change in resistance. The allowable displacement of commercial gages is of the order of ± 0.0015 in (0.038 mm), and the wire diameter is usually less than 0.001 in (0.025 mm). The I^2R heating in the unbonded gage can be a problem because the wires have no ready means for heat dissipation other than convection to the surrounding air. The principle of the unbonded gage has been applied to acceleration and diaphragm pressure transducers with good success.

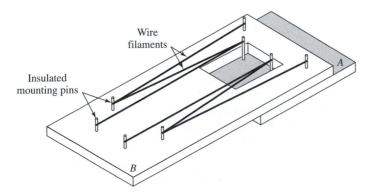

Figure 10.15   Schematic of unbonded resistance strain gage.

10.12   REVIEW QUESTIONS

10.1. Why must a correction be applied for the specific gravity of a sample when a mass balance is used?

10.2. How are elastic elements used for force or torque measurements?

10.3. What is a proving ring?

10.4. What is a Prony brake?

10.5. What does a dynamometer measure?

10.6. Distinguish between stress and strain.

10.7. What is Poisson's ratio?

10.8. What is deformation sensitivity?

10.9. What is strain sensitivity?

10.10. How may a brittle coating be used for strain measurements?

10.11. Define the gage factor for strain gages.

10.12. What is the main advantage of a semiconductor strain gage?

10.13. How are measurements performed with a resistance strain gage?

10.14. How may temperature compensation be performed on resistance strain gages?

10.15. What is meant by a strain-gage rosette? How is it used?

10.13   PROBLEMS

10.1. Many balances use brass weights ($\rho_s = 8490$ kg/m^3) in atmospheric air ($\rho_a = 1.2$ kg/m^3). Plot the percent error in the balance determination as a function of ρ_u/ρ_s.

10.2. Using the conditions given in Example 10.1, calculate the angular sensitivity required if the uncertainty in the weight determination is to be of the order of 1 part in 10^8.

10.3. An LVDT is used to measure the axial deformation of the simple bar of Fig. 10.3. The uncertainty in this measurement is ± 0.025 mm. A steel bar is to be used to measure a force of 1100 N with an uncertainty of 1 percent. Determine a suitable diameter and length of rod for this measurement, assuming the rod dimensions will be known exactly.

10.4. The diaphragm of Fig. 6.10c is to be used as a force-measurement device with the load applied at the center boss. Steel is to be used as the diaphragm material, and the maximum displacement is not to exceed one-third the thickness. A load of 4500 N is to be measured with an uncertainty of 1 percent using an LVDT with an uncertainty of ± 2.5 μm. Determine suitable diaphragm dimensions for this measurement.

10.5. A farmer decides to build a crude weighing device to weigh bags of grain up to 150 lbf. For this purpose he employs a section of 1-in steel pipe (1.315 in OD, 0.957 in ID) as a cantilever beam. He intends to measure the deflection of the beam with a metal carpenter's scale having graduations of $\frac{1}{16}$ in. On the basis of these data, what length of pipe would you recommend for the farmer's application?

10.6. A proving ring is constructed of steel ($E = 30 \times 10^6$ psi) having a cross section with 0.250-in thickness and 1.00-in depth. The overall diameter of the ring is 6.000 in. A micrometer is used for the deflection measurement, and its uncertainty is ±0.0001 in. Assuming that the dimensions of the ring are exact, calculate the applied load when the uncertainty in this load is 1.0 percent. Calculate the percent uncertainty in the load when $y = 0.01$ in and the uncertainties in the dimensions are

$$w_h = w_b = 0.0002 \text{ in}$$
$$w_d = 0.01 \text{ in}$$

where b and h are thickness and depth of the ring, respectively.

10.7. A hollow steel cylinder ($G = 12 \times 10^6$ psi) is subjected to a moment such that ϕ is 1.50°. The dimensions of the cylinder are

$$r_i = 0.500 \pm 0.0003 \text{ in}$$
$$r_o = 0.625 \pm 0.0003 \text{ in}$$
$$L = 5.000 \pm 0.001 \text{ in}$$

The uncertainty in the angular deflection is ±0.05°. Calculate the nominal value of the impressed moment and its uncertainty. Also, calculate the 45° strains.

10.8. In a bridge circuit like that of Fig. 4.25, $R_2 = R_3 = 100 \, \Omega$. The galvanometer resistance is 50 Ω. The strain-gage resistance of zero strain is 120 Ω, and the value of R_4 is adjusted to bring the bridge into balance at zero-strain conditions. The gage factor is 2.0. Calculate the galvanometer current when $\epsilon = 400 \, \mu\text{m/m}$. Take the battery voltage as 4.0 V.

10.9. Calculate the voltage output of the bridge in Prob. 10.8, assuming a detector of very high impedance.

10.10. Show that Eq. (10.41) reduces to

$$\epsilon_{max} = \epsilon_1$$
$$\epsilon_{min} = \epsilon_3$$

when $\theta = 0$. Obtain relations for the principal stresses and maximum shear stress in this instance.

10.11. A delta rosette is placed on a steel plate and indicates the following strains:

$$\epsilon_1 = 400 \ \mu m/m$$
$$\epsilon_2 = 84 \ \mu m/m$$
$$\epsilon_3 = -250 \ \mu m/m$$

Calculate the principal strains and stresses, the maximum shear stress, and the orientation angle for the principal axes.

10.12. Obtain simplified relations for the delta rosette under the conditions that $\epsilon_1 = \epsilon_{max}$; i.e., $\theta = 0$.

10.13. Suppose a rectangular rosette is used to measure the same stresses as in Prob. 10.11. The bottom leg of this rosette is placed in the same location as the bottom arm of the delta rosette. Calculate the strains that would be indicated by the rectangular rosette under these conditions.

10.14. Calculate the percent uncertainty in the maximum and minimum stresses resulting from uncertainties of 2 percent in the strains as measured with a rectangular rosette. For this calculation assume E and μ are known exactly.

10.15. Rework Prob. 10.14 for the delta rosette.

10.16. A rectangular rosette is placed on a steel plate and indicates the following strains:

$$\epsilon_1 = 563 \ \mu m/m$$
$$\epsilon_2 = -155 \ \mu m/m$$
$$\epsilon_3 = -480 \ \mu m/m$$

Calculate the principal strains and stresses, the maximum shear stress, and the orientation angle for the principal axes.

10.17. A strip of steel sheet, $1.6 \times 50 \times 500$ mm, is available for use as a force-measuring elastic element. The strip is to be used by cementing strain gages to its flat surfaces and measuring the deformation under load. The strain gages have a maximum strain of 2000 $\mu m/m$ and a gage factor of 1.90. A battery is available with a voltage of 4.0 V, and the readout voltmeter has a high impedance and an accuracy of $\pm 1 \ \mu V$. The nominal resistance of the strain gages is 120 Ω. Calculate the force range for which the measurement system may be applicable.

10.18. A bridge circuit is used to measure the output of a strain gage having a resistance of 110 Ω under zero conditions. $R_2 = R_3 = 100 \ \Omega$, and the galvanometer resistance is 75 Ω. The gage factor is 2.0. Calculate the galvanometer current when $\epsilon = 300 \ \mu m/m$ for a power voltage of 6.0 V. Also, calculate the voltage output of the bridge for a very high-impedance detector.

10.19. An electrical-resistance strain gage records a strain of 400 $\mu m/m$ on a steel tension member. What is the axial stress?

10.20. A hollow steel cylinder is used as a torque-sensing element. The dimensions are: $r_i = 2.5$ cm, $r_o = 3.2$ cm, $L = 15$ cm. Calculate the angular deflection for an applied moment of 22.6 N · m. Calculate the strain that would be indicated by gages attached as shown in Fig. 10.7.

10.21. A certain automotive engine is reported to produce a torque of 540 N · m at 3000 rpm. What power dynamometer would be necessary to test this engine? Express in both horsepower and kilowatts.

10.22. A stainless-steel hollow cylinder has dimensions of 2.5 cm ID and 2.16 cm OD. The length of the cylinder is 25 cm. This cylinder is to be used for axial and torque force measurements by instrumenting it with appropriate strain gages. Specify the types, number, and location for the gages and the bridge circuit and readout equipment required. Also, specify the range of variables that are suitable for measurement by the device and estimate the uncertainties which may be experienced.

10.23. Something similar to the Prony brake of Fig. 10.8 is to be designed for measurement of a small internal-combustion engine producing a power output of 15 hp. Copper blocks are used instead of wood to dissipate the heat. Present a design for this device with appropriate dimensions and the force measurement system described. Estimate the uncertainty which might be expected in the measurement.

10.24. A bridge circuit is used to measure a strain-gage output. Under zero-strain conditions the gage resistance is 120 Ω. $R_2 = R_3 = 110$ Ω and the galvanometer resistance is 70 Ω. The gage factor is 1.8. Calculate the galvanometer current when $\epsilon = 350$ μm/m for a power supply voltage of 4.0 V. Also, calculate the voltage output for the bridge when used with a high-impedance detector. If the gage is installed on steel, what stress does this measurement represent?

10.25. A small engine is rated to develop 5 hp at 3000 rpm. A solid rod constructed of spring steel is to be used as a torque-measuring device at the rated load. Select two length and diameter combinations to accomplish the measurement assuming an angular deflection of 10°.

10.26. The same measurement as in Prob. 10.25 is to be performed by using a hollow cylinder of spring steel with outside diameter of 12.5 mm and wall thickness of 0.8 mm. What must its length be for an angular deflection of 10°?

10.27. Calculate the percent error caused by buoyancy forces in a balance which uses standard brass weights to weigh 150 g of a material having a density of 100 kg/m^3. The surrounding air is at 25°C and 100 kPa.

10.28. The nominal resistance for a strain gage is 120 Ω and it is placed in an equal arm bridge where all arms also have resistances of 120 Ω. The gage factor is 2.0 and the bridge is operated in a current-sensitive mode with a detector resistance of 75 Ω. The voltage applied to the circuit is 3.7 V. Calculate the current sensitivity of this arrangement.

10.29. The bridge circuit of Prob. 10.28 is operated with a high-impedance detector. Calculate the voltage output for a strain of 2.0 μ in/in.

10.30. A rod having a diameter of 1.6 mm is to be used as a cantilever beam for a force measurement. It is constructed of spring steel. A deflection of 1.0 cm is desired with a measurement uncertainty of ±1 percent. Specify two length and force-loading combinations for these beam parameters.

10.31. A man weighing 200 lb is to be weighed by measuring the elongation of a $\frac{1}{16}$-in-diameter spring-steel wire. What elongation would result for a wire length of 70 cm?

10.32. A Manganin strain gage has a gage factor of 0.47 and a nominal resistance of 120 Ω. The gage is used in the bridge circuit of Prob. 10.29. What will be the voltage output for the specified conditions?

10.33. A strain gage is placed on a spring-steel member and a strain measured as 300 μm/m for the member in tension. Calculate the axial stress.

10.34. A model airplane engine develops 1 hp at 10,000 rpm. Calculate the torque produced at these conditions. Suppose this torque acts at a moment arm of 15 cm and is to be measured with the deflection of a cantilever beam having a nominal deflection of 1.0 cm. Specify suitable design dimensions for a beam to accomplish this measurement.

10.35. The torque of the engine in Prob. 10.34 is to be measured by observing the angular deflection of a hollow cylinder like that shown in Fig. 10.7. An angular deflection of 20° is desired for the 1-hp output. Determine two suitable sets of dimensions (radii and length) to accomplish this objective.

10.36. A strain gage has a nominal resistance of 120 Ω. The gage is installed in a bridge circuit having equal arm resistances of 120 Ω and is operated as a current-sensitive device with a detector resistance of 350 Ω. The voltage source is 4.5 V. Calculate the current sensitivity of the device if the gage factor is 1.9.

10.37. The bridge circuit of Prob. 10.36 is operated with a digital voltmeter having a very high input impedance. What voltage output will result from a strain of 3.1 μm/m?

10.38. A 2.0-cm-diameter rod 50 cm long is subjected to an axial load and the strain measured with the bridge gage arrangement of Prob. 10.37. What axial force would be necessary to produce an axial deformation of 10 μm, and what would be the resulting gage reading in volts?

10.39. A torque is measured by angular deflection of a tube having $r_i = 0.5$ cm, $r_o = 0.5$ cm, and $L = 5$ cm. What relationship must exist between the shear modulus of elasticity and the applied torque to produce an angular deflection of 10°? State units.

10.40. An analytical balance like that of Fig. 10.2 is used to weigh a 15-oz (troy) quantity of platinum in air at 1 atm and 20°C. The density of platinum is 21,380 kg/m^3 and that of the standard brass weights is 8490 kg/m^3. If the price of platinum is $400 per oz, calculate the dollar value of the buoyancy error in weight determination. (*Note:* 1 lbm = 12 troy oz.)

10.41. A hollow steel cylinder is to be used as a torque-sensing device. The maximum torque that will be applied is 30 N-m. If the dimensions of the cylinder are $L = 18$ cm, $r_i = 2.6$ cm, and $r_0 = 2.8$ cm, calculate the maximum angular deflection that will be observed.

10.42. An automotive engine is rated at a power output of 375 horsepower at 4000 RPM. What will be the output torque at the rated conditions?

10.43. A rod, constructed of spring steel, has a diameter of 2.0 mm and is to be used as a force-measuring device by sensing the deflection of the rod when subjected to the loading force. The maximum deflection allowed is 10.0 mm and the length of the rod is 59 mm. Calculate the maximum force the rod can be used to measure.

10.44. A 1.0-m length of spring steel wire having a diameter of 0.8 mm is thread through a pulley arrangement and fixed at one end. The other end is subjected to a force of 800 N. Neglecting any frictional effects of the pulleys, calculate the deflection of the wire, expressed in millimeters.

10.45. An analytical balance which employs brass weights is used to weigh a 200-cm^3 sample of glass wool having an approximate density of 1.5 lbm/ft^3. If the surrounding air is at standard conditions of 1 atm and 20°C, what weight will be indicated by the balance? What percent error do you expect in this measurement?

10.46. Repeat Prob. 10.45 for a measurement employing standard weights constructed of aluminum.

10.14   REFERENCES

1. Dean, M. (ed.): *Semiconductor and Conventional Strain Gages,* Academic Press, New York, 1962.

2. Dove, R. C., and P. H. Adams: *Experimental Stress Analysis and Motion Measurement,* Charles E. Merrill, Columbus, Ohio, 1964.

3. Durelli, A. J., E. A. Phillips, and C. H. Tsao: *Introduction to the Theoretical and Experimental Analysis of Stress and Strain,* McGraw-Hill, New York, 1958.

4. Harris, C. M., and C. E. Crede (eds.): *Basic Theory and Measurements,* vol. 1. *Shock and Vibration Handbook,* McGraw-Hill, New York, 1961.

5. Hetenyi, M. (ed.): *Handbook of Experimental Stress Analysis,* John Wiley & Sons, New York, 1950.

6. Perry, C. C., and H. R. Lissner: *The Strain Gage Primer,* 2d ed., McGraw-Hill, New York, 1962.

7. Herceg, E. E.: *Schaevitz Handbook of Measurement and Control,* Schaevitz Engineering, Pennsauken, NJ.

8. Stein, Peter K.: "Material Considerations for Strain Gages," *Inst. Cont. Syst.,* vol. 37, no. 10, pp. 132–135, October 1964.

9. *Proc. Soc. Exp. Stress Anal.,* published by the Society.

10. Dally, J. W., and W. F. Riley: *Experimental Stress Analysis,* 3d ed., McGraw-Hill, New York, 1991.

11. Juvinall, R. C.: *Engineering Considerations of Stress, Strain, and Strength,* McGraw-Hill, New York, 1967.

12. Beckwith, T. G., and N. L. Buck: *Mechanical Measurements,* 2d ed., Addison-Wesley, Reading, MA, 1969.

13. Omega Engineering Inc.: *Pressure, Strain, and Force Measurement Handbook,* Omega Engineering, Stamford, CT, latest issue.

14. "Measurement of Rotary Speed," ASME Performance Test Codes, ANSI/ASME PTC 19. 13-1961, American Society of Mechanical Engineers, New York, 1961.

15. "Measurement of Shaft Power," ASME Performance Test Codes, ANSI/ASME PTC 19.7-1980, American Society of Mechanical Engineers, New York, 1980.

16. Gindy, S. S.: "Force and Torque Measurement, A Technology Review," *Experimental Techniques,* vol. 9, p. 28, 1985.

17. Budynas, Richard: *Advanced Strength and Applied Stress Analysis,* 2d ed., McGraw-Hill, New York, 1999.

18. Courtney, T. H.: *Mechanical Behavior of Materials,* McGraw-Hill, New York, 1990.

19. Davis, H. E., and G. Troxell: *The Testing of Engineering Materials,* 4th ed., McGraw-Hill, New York, 1982.

20. Wilson, J.: *Experimental Solid Mechanics,* McGraw-Hill, New York, 1993.

21. ———: *Experiments on the Strength of Solids,* McGraw-Hill, New York, 1993.

11

MOTION AND VIBRATION MEASUREMENT

11.1   INTRODUCTION

Measurements of motion and vibration parameters are important in many applications. The experimental quantity that is desired may be velocity, acceleration, or
vibration amplitude. These quantities may be useful in predicting the fatigue failure
of a particular part or machine or may play an important role in analyses which are
used to reduce structure vibrations or noise level.

The central problem in any type of motion or vibration measurement concerns a
determination of the appropriate quantities in reference to some specified state, that is,
velocity, displacement, or acceleration in reference to ground. Ideally, one would like
to have a motion or vibration transducer which connects to the body in motion and
furnishes an output signal proportional to the vibrational input. The ideal transducer
should be independent of its location, that is, functioning equally well whether it is
connected to a vibrating structure on the ground, in an aircraft, or in a space vehicle.
In this chapter we shall see some of the compromises which are necessary to approach
such behavior.

Sound may be classified as a vibratory phenomenon, and we shall indicate some of
the important parameters necessary for specification of sound level. The measurement
and analysis of sound levels are very specialized subjects which are important in
modern building and equipment design [8].

11.2   TWO SIMPLE VIBRATION INSTRUMENTS

Consider the simple wedge shown in Fig. 11.1 which is attached to a vibrating wall.
When the wall is at rest, the wedge appears as in Fig. 11.1a; when the wall is in
motion, the wedge appears as in Fig. 11.1b. An observation is made of the distance
x. At this distance the thickness of the wedge is equal to the double amplitude of the
motion. In terms of x the amplitude is given by

$$a = x \tan \frac{\theta}{2} \qquad \textbf{[11.1]}$$

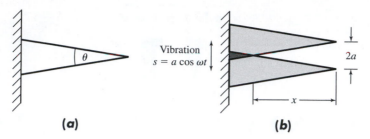

Figure 11.1 Simple wedge as a device for amplitude-displacement measurements. (a) At rest; (b) in motion.

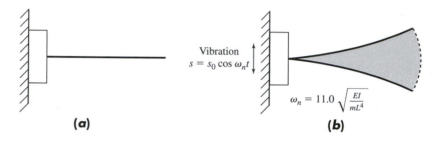

Figure 11.2 Cantilever beam used as a frequency-measurement device. (a) No vibration or vibration other than ω_n; (b) vibration at ω_n.

where θ is the total included angle of the wedge. The wedge-measurement device is necessarily limited to rather large-amplitude motions, usually for $a > 0.8$ mm.

A simple vibration-frequency-measurement device is shown in Fig. 11.2. The small cantilever beam mounted on the block is placed against the vibrating surface, and some appropriate method is provided for varying the beam length. When the beam length is properly adjusted so that its natural frequency is equal to the frequency of the vibrating surface, the resonance condition shown in Fig. 11.2b will result. The natural frequency of such a beam is given by

$$\omega_n = C_n \sqrt{\frac{EI}{mL^4}} \qquad\qquad \textbf{[11.2]}$$

where ω_n = natural frequency, Hz
E = Young's modulus for the beam material, psi or N/m^2
I = moment of inertia of the beam about the centroidal axis in the direction of deflection, in^4 or m^4
m = mass per unit length, lbm/in, or kg/m
L = beam length, in or m
C_n = 11.0 for English or 0.550 for SI units

Example 11.1 | **VIBRATING ROD.** A $\frac{1}{16}$-in-diameter spring-steel rod is to be used for a vibration-frequency measurement as shown in Fig. 11.2. The length of the rod may be varied between 1 and 4 in. The density of this material is 489 lbm/ft^3, and the modulus of elasticity is 28.3×10^6 psi.

Calculate the range of frequencies that may be measured with this device and the allowable uncertainty in L at $L = 4$ in in order that the uncertainty in the frequency is not greater than 1 percent. Assume the material properties are known exactly.

Solution

We have

$$E = 28.3 \times 10^6 \text{ psi}$$

$$I = \frac{\pi r^4}{4} = \frac{\pi(1/32)^4}{4} = 7.49 \times 10^{-7} \text{ in}^4 \qquad \text{(Table A.12)}$$

$$m = \rho\pi r^2 = \frac{\pi(489)(1/32)^2}{(144)}\frac{1}{12} = 8.69 \times 10^{-4} \text{ lbm/in}$$

At $x = 1$ in

$$\omega_n = 11.0 \left[\frac{(28.3 \times 10^6)(7.49 \times 10^{-7})}{(8.69 \times 10^{-4})(1)^4} \right]^{1/2} = 1718 \text{ Hz}$$

At $x = 4$ in

$$\omega_n = 107.5 \text{ Hz}$$

We use Eq. (3.2) to determine the allowable uncertainty in the length measurement in terms of the uncertainty in the frequency measurement.

$$\frac{\partial \omega_n}{\partial L} = 11.0 \sqrt{\frac{EI}{m}} \left(\frac{-2}{L^3} \right)$$

$$w_{\omega_n} = \left[\left(\frac{\partial \omega_n}{\partial L} w_L \right)^2 \right]^{1/2}$$

so that

$$w_L = \frac{w_{\omega_n} L^3}{22.0\sqrt{EI/m}}$$

At $L = 4$ in

$$w_{\omega_n} = (0.01)(107.5) = 1.075 \text{ Hz}$$

$$w_L = \frac{(1.075)(4)^3}{(22.0)(156.2)} = 0.02 \text{ in}$$

At $L = 1.0$ in greater precision is required,

$$w_{\omega_n} = (0.01)(1718) = 17.18 \text{ Hz}$$

$$w_L = \frac{(17.18)(1)^3}{(22)(156.2)} = 0.005 \text{ in}$$

11.3 PRINCIPLES OF THE SEISMIC INSTRUMENT

The seismic instrument is a device that has the functional form of the system shown in Fig. 2.3. A schematic of a typical instrument is shown in Fig. 11.3. The mass

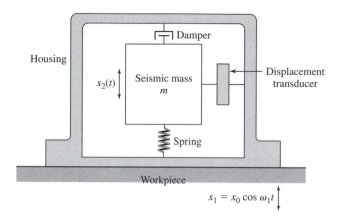

Figure 11.3   Schematic of typical seismic instrument.

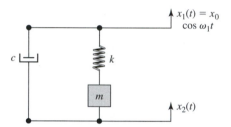

Figure 11.4   Mechanical system for a seismic instrument.

is connected through the parallel spring and damper arrangement to the housing frame. This frame is then connected to the vibration source whose characteristics are to be measured. The mass tends to remain fixed in its spatial position so that the vibrational motion is registered as a relative displacement between the mass and the housing frame. This displacement is then sensed and indicated by an appropriate transducer, as shown in the schematic diagram. Of course, the seismic mass does not remain absolutely steady, but for selected frequency ranges it may afford a satisfactory reference position.

The seismic instrument may be used for either displacement or acceleration measurements by proper selection of mass, spring, and damper combinations. In general, a large mass and soft spring are desirable for vibrational displacement measurements, while a relatively small mass and stiff spring are used for acceleration indications. This will be apparent from the theoretical discussion that follows. We note, of course, that the seismic instrument is a second-order system in the context of the discussion of Chap. 2.

Figure 2.3 is reproduced as Fig. 11.4 to show the mechanical system to be analyzed. Using Newton's second law of motion, we have

$$m\frac{d^2x_2}{dt^2} + c\frac{dx_2}{dt} + kx_2 = c\frac{dx_1}{dt} + kx_1 \qquad \textbf{[11.3]}$$

where it is assumed that the damping force is proportional to velocity. We assume that a harmonic vibratory motion is impressed on the instrument such that

$$x_1 = x_0 \cos \omega_1 t \qquad \textbf{[11.4]}$$

and wish to obtain an expression for the relative displacement $x_2 - x_1$ in terms of this impressed motion. The relative displacement is that which is detected by the transducer shown in Fig. 11.3. Rewriting Eq. (11.3) and substituting Eq. (11.4) gives

$$\frac{d^2 x_2}{dt^2} + \frac{c}{m}\frac{dx_2}{dt} + \frac{k}{m}x_2 = x_0 \left(\frac{k}{m}\cos \omega_1 t - \frac{c}{m}\omega_1 \sin \omega_1 t \right) \qquad \textbf{[11.5]}$$

The solution to Eq. (11.5) is

$$x_2 - x_1 = e^{-(c/2m)t}(A \cos \omega t + B \sin \omega t) + \frac{m x_0 \omega_1^2 \cos(\omega_1 t - \phi)}{\left[\left(k - m\omega_1^2 \right)^2 + c^2 \omega_1^2 \right]^{1/2}} \qquad \textbf{[11.6]}$$

where the frequency is given by

$$\omega = \left[\frac{k}{m} - \left(\frac{c}{2m} \right)^2 \right]^{1/2} \qquad \text{for} \quad \frac{c}{c_c} < 1.0 \qquad \textbf{[11.7]}$$

and the phase angle is given by

$$\phi = \tan^{-1} \frac{c\omega_1}{k - m\omega_1^2} \qquad \textbf{[11.8]}$$

A and B are constants of integration determined from the initial or boundary conditions.

Note that Eq. (11.6) is composed of two terms: (1) the transient term involving the exponential function and (2) the steady-state term. This means that after the initial transient has died out a steady-state harmonic motion is established in accordance with the second term. The frequency of this steady-state motion is the same as that of the impressed motion, and its amplitude is

$$(x_2 - x_1)_0 = \frac{x_0(\omega_1/\omega_n)^2}{\{[1 - (\omega_1/\omega_n)^2]^2 + [2(c/c_c)(\omega_1/\omega_n)]^2\}^{1/2}} \qquad \textbf{[11.9]}$$

where the natural frequency ω_n and the critical damping coefficient c_c are given by

$$\omega_n = \sqrt{\frac{k}{m}} \qquad \textbf{[11.10]}$$

$$c_c = 2\sqrt{mk} \qquad \textbf{[11.11]}$$

The phase angle may also be written

$$\phi = \tan^{-1} \frac{2(c/c_c)(\omega_1/\omega_n)}{1 - (\omega_1/\omega_n)^2} \qquad \textbf{[11.12]}$$

A plot of Eq. (11.9) is given in Fig. 11.5. It may be seen that the output amplitude is very nearly equal to the input amplitude when $c/c_c = 0.7$ and $\omega_1/\omega_n > 2$. For low values of the damping ratio the amplitude may become quite large. The output becomes essentially a linear function of input at high-frequency ratios. Thus, a

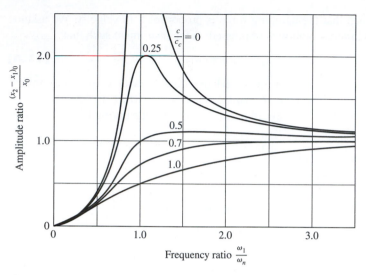

Figure 11.5   Displacement response of a seismic instrument as given by Eq. (11.9).

semismic-vibration pickup for measurement of displacement amplitude should be utilized for measurement of frequencies substantially higher than its natural frequency. The instrument constants c/c_c and ω_n should be known or obtained from calibration. The anticipated accuracy of measurement may then be calculated for various frequencies.

The acceleration amplitude of the input vibration is

$$a_0 = \left(\frac{d^2 x_1}{dt^2}\right)_0 = \omega_1^2 x_0 \qquad\qquad \textbf{[11.13]}$$

We may thus use the measured output of the instrument as a measure of acceleration. There are problems associated with this application, however. In Eq. (11.9) the bracketed term is the one that governs the linearity of the acceleration response since ω_n will be fixed for a given instrument. In Fig. 11.6 we have a plot of

$$\frac{(x_2 - x_1)_0 \omega_n^2}{a_0} \qquad \text{versus} \qquad \frac{\omega_1}{\omega_n}$$

which indicates the nondimensionalized acceleration response. Thus, by measuring $(x_2 - x_1)_0$, we can calculate the input acceleration a_0. Generally, unsatisfactory performance is observed at frequency ratios above 0.4. Therefore, for acceleration measurements we wish to operate at frequencies much lower than the natural frequency, in contrast to the desirable region of operation for amplitude measurements. From the standpoint of instrument construction this means that we wish to have a low natural frequency (soft spring, large mass) for amplitude measurements and a high natural frequency (stiff spring, small mass) for acceleration measurements in order to be able to operate over a wide range of frequencies and still enjoy linear response.

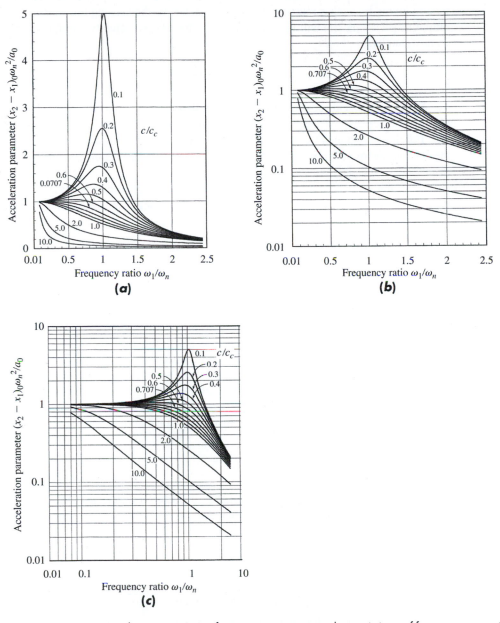

Figure 11.6   Acceleration response of a seismic instrument to a harmonic input of frequency ω_1 as given by Eq. (11.13): (a) linear, (b) semilog, (c) log coordinates.

The seismic instrument may also be used for vibration-velocity measurements by employing a variable-reluctance magnetic pickup as the sensing transducer. The output of such a pickup will be proportional to the relative velocity amplitude, that is, the quantity

$$\left[\frac{d}{dt}(x_2 - x_1)\right]_0$$

From the above discussion it may be seen that the seismic instrument is a very versatile device that may be used for measurement of a variety of vibration parameters. For this reason many commercial vibration and acceleration-pickups operate on the seismic-instrument principle. Calibration methods for the seismic instrument are discussed in Refs. [1], [4], [6], and [7].

The transient response of the seismic instrument is governed partially by the exponential decay term in Eq. (11.6). The time constant for this term could be taken as

$$T = \frac{2m}{c} \tag{11.14}$$

or in terms of the natural frequency and critical damping ratio,

$$T = \frac{1}{\omega_n(c/c_c)} \tag{11.15}$$

The specific transient response of the seismic-instrument system is also a function of the type of input signal, that is, whether it is a step function, harmonic function, ramp function, etc. The linearity of a vibration transducer is thus influenced by the frequency-ratio requirements that are necessary to give linear response as indicated by Eqs. (11.6) and (11.9). The design of a transducer for particular response characteristics must involve a compromise between these two effects, combined with a consideration of the sensitivity of the displacement-sensing transducer and its transient response characteristics.

Example 11.2 | **FREQUENCY LIMIT FOR DISPLACEMENT MEASUREMENT MODE.**   Using Eq. (11.9) and $c/c_c = 0.7$, calculate the value of ω_1/ω_n such that $(x_2 - x_1)_0/x_0 = 0.99$; i.e., the error is 1 percent for a displacement determination.

Solution

We have

$$0.99 = \frac{(\omega_1/\omega_n)^2}{\{[1 - (\omega_1/\omega_n)^2]^2 + [2(0.7)(\omega_1/\omega_n)]^2\}^{1/2}}$$

Rearranging this equation gives the quadratic relation

$$\left[\left(\frac{1}{0.99}\right)^2 - 1\right]\left(\frac{\omega_1}{\omega_n}\right)^4 + 0.04\left(\frac{\omega_1}{\omega_n}\right)^2 - 1 = 0$$

which yields

$$\frac{\omega_1}{\omega_n} = 2.47$$

SEISMIC INSTRUMENT FOR ACCELERATION MEASUREMENT. A small seismic instrument is to be used for measurement of linear acceleration. It has $\omega_n = 100$ rad/s and a displacement-sensing transducer which detects a maximum of ± 0.1 in. The certainty in the displacement measurement is ± 0.001 in. Calculate the maximum acceleration that may be measured with this instrument and the uncertainty in the measurement, assuming ω_n is known exactly.

Example 11.3

Solution

From Fig. 11.6 we see that the maximum attainable value for a_0 is observed when $\omega_1/\omega_n \to 0$ and

$$\frac{\omega_n^2 (x_2 - x_1)_0}{a_0} \to 1.0$$

so that

$$a_0 = \omega_n^2 (x_2 - x_1)_0$$

$$= 100^2 \left(\frac{0.1}{12}\right) = 83.3 \text{ ft/s}^2 \ (25.39 \text{ m/s}^2)$$

The uncertainty is calculated from

$$w_a = \left[\left(\frac{\partial a}{\partial x} w_x\right)^2\right]^{1/2} = \omega_n^2 w_x = (100)^2 \left(\frac{0.001}{12}\right) = 0.833 \text{ ft/s}^2$$

or an uncertainty of 1 percent. It may be noted that larger uncertainties would be present when smaller accelerations were measured.

FREQUENCY RATIO FOR ACCELERATION MEASUREMENT. Calculate the frequency ratio for which the error in acceleration measurement is 1 percent, with $c/c_c = 0.7$; i.e., $[(x_2 - x_1) w_n^2]/a_0 = 0.99$.

Example 11.4

Solution

We have

$$0.99 = \frac{1}{\{[1 - (\omega_1/\omega_n)^2]^2 + [2(0.7)(\omega_1/\omega_n)^2]\}^{1/2}}$$

This gives the quadratic relation

$$\left(\frac{\omega_1}{\omega_n}\right)^4 - 0.04 \left(\frac{\omega_1}{\omega_n}\right)^2 + 1 - \left(\frac{1}{0.99}\right)^2 = 0$$

which yields

$$\frac{\omega_1}{\omega_n} = 0.306$$

DECAY OF TRANSIENT TERM. Calculate the time required for the exponential transient term in Eq. (11.6) to decrease by 99 percent with $\omega_n = 100$ rad/s and $c/c_c = 0.7$. Compare this time with the period of an acceptable frequency ω_1 for displacement and acceleration amplitude measurements as calculated from the 99 percent relationships of Examples 11.2 and 11.4.

Example 11.5

Solution

We have

$$e^{-(c/2m)t} = 0.01 \quad \text{and} \quad \frac{ct}{2m} = 4.61$$

$$4.61 = \omega_n \left(\frac{c}{c_c}\right) t$$

$$t = 0.0658 \, \text{s}$$

For Example 11.2 $\omega_1/\omega_n = 2.47$ and the corresponding period for ω_1 in the displacement measurement is

$$P = \frac{2\pi}{\omega_1} = \frac{2\pi}{247} = 0.0257 \, \text{s}$$

From Example 11.4 $\omega_1/\omega_n = 0.306$ and the corresponding period for ω_1 in the acceleration measurement is

$$P = \frac{2\pi}{\omega_1} = \frac{2\pi}{30.6} = 0.205 \, \text{s}$$

Comment

No specific conclusions may be drawn from this comparison because the transient response depends on the manner in which the instrument is subjected to a change.

11.4   PRACTICAL CONSIDERATIONS FOR SEISMIC INSTRUMENTS

The previous paragraphs have shown the basic response characteristics of seismic instruments to an impressed harmonic vibrational motion. Let us now consider some of the ways that these instruments might be constructed in practice.

In Fig. 11.7 a seismic instrument is illustrated that uses a voltage-divider potentiometer for sensing the relative displacement between the frame and the seismic mass. To provide the damping for the system, the case of the instrument might be filled with a viscous liquid which would interact continuously with the frame and the mass. Because of the relatively large mass of the potentiometer, such systems have rather low natural frequencies (less than about 100 Hz) and, as such, are limited to acceleration measurements at frequencies less than about 50 Hz. Such acceleration transducers may weigh 500 g or more.

The linear variable differential transformer (LVDT) described in Sec. 4.21 offers another convenient means for measurement of the relative displacement between the seismic mass and accelerometer housing. Such devices have somewhat higher natural frequencies than potentiometer devices (270 to 300 Hz) but are still restricted to applications with low-frequency-response requirements. The differential transformer, however, has a much lower resistance to motion than the potentiometer and is capable of much better resolution. In addition, the seismic accelerometer using an LVDT can be considerably lighter in construction than one with

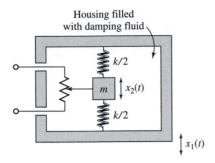

Figure 11.7   Seismic instrument using a voltage-divider potentiometer for sensing relative displacement.

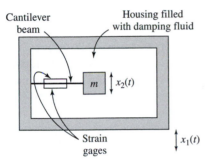

Figure 11.8   Seismic instrument utilizing an electrical-resistance strain gage for sensing relative displacement.

a potentiometer. Table 4.4 gives a comparison of the characteristics of the LVDT and potentiometer.

The electrical-resistance strain gage discussed in Chap. 10 may also be used for a displacement-sensing device in a seismic instrument. Consider the schematic in Fig. 11.8. The seismic mass is mounted on a cantilever beam. On each side of the beam a resistance strain gage is mounted to sense the strain in the beam resulting from the vibrational displacement of the mass. Damping for the system is provided by the viscous liquid, which fills the housing. The outputs of the strain gages are connected to an appropriate bridge circuit, which is used to indicate the relative displacement between the mass and housing frame. The natural frequencies of such systems are fairly low and roughly comparable to those for the LVDT systems. The low natural frequencies result from the fact that the cantilever beam must be sufficiently large to accommodate the mounting of the resistance strain gages.

For high-frequency measurements the seismic instrument frequently employs a piezoelectric transducer (Sec. 4.23), as shown in Fig. 11.9. The natural frequency of such instruments may be as high as 100 kHz, and the entire instrument may be quite small and light in weight. A total weight of 50 g for a piezoelectric accelerometer

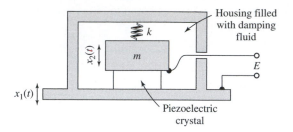

Figure 11.9   Seismic instrument utilizing a piezoelectric transducer.

is not uncommon. Piezoelectric devices have rather high electric outputs but are not generally suitable for measurements at frequencies below about 10 Hz. Electrical impedance matching between the transducer and readout circuitry is usually a critical matter requiring careful design considerations.

Seismic-vibration instruments may be influenced by temperature effects. Devices employing variable-resistance-displacement sensors will require a correction to account for resistance changes due to temperature. The damping of the instrument is affected by changes in viscosity of the fluid surrounding the seismic mass. Silicone oils are frequently employed in such applications, and their viscosity is strongly dependent on temperature. It would not be uncommon for such oils to experience a change in viscosity by a factor of 2 for a temperature change of only 50°F (28°C). The viscosity-temperature effect could be eliminated, of course, if a fluid with a small change in viscosity with temperature were used. Unfortunately, fluids which have this desirable behavior, viz., gases, have low viscosities as well so that they are generally unsuitable for use as damping agents in the seismic instrument. One way of alleviating the viscosity-temperature problem is to install an electrical-resistance heater in the fluid to maintain the temperature at a constant value regardless of the surrounding temperature.

Obviously, vibration measurements are not always performed on large, massive pieces of equipment. It is frequently of interest to measure the vibration of a thin plate or other small, low-mass structures. The question immediately arises about the effect of the presence of the vibration instrument in such measurements. If the mass of the instrument is not small in comparison with the mass of the workpiece on which it is to be installed, the vibrational characteristics may be altered appreciably. In such cases a correct interpretation of the output of the vibrational instrument may only be obtained by analyses beyond the scope of our discussion. A simple experimental technique, however, will tell the engineer whether the presence of the instrument has altered the vibrational characteristics of the structure under test. First, a measurement is made with the vibration instrument in place. Then, an additional mass is mounted on the structure along with the instrument and a second measurement is taken under the same conditions as the first. If there is no appreciable difference between the two measurements, it may be assumed that the presence of the vibrometer has not altered the vibrational characteristics. If there is an appreciable difference in the measurements, an analysis will be necessary to determine the "free" vibrational characteristics in terms of the experimental measurements.

11.5 SOUND MEASUREMENTS

Sound waves are a vibratory phenomenon and for this reason are appropriately discussed in this chapter. They might also be discussed in the chapter on pressure measurements because acoustic effects are usually measured in terms of harmonic pressure fluctuations that they produce in a liquid or gaseous medium. They are also characterized by an energy flux per unit area and per unit time as the acoustic wave moves through the medium. A mathematical description of different acoustic waves will be given later in this section.

It is standard practice in acoustic measurements to relate sound intensity and sound pressure to certain reference values I_0 and p_0, which correspond to the intensity and mean pressure fluctuations of the faintest audible sound at a frequency of 1000 Hz. These reference levels are

$$I_0 = 10^{-16} \text{ W/cm}^2 \qquad \textbf{[11.16]}$$

$$p_0 = 2 \times 10^{-4} \text{ dyn/cm}^2$$
$$= 2 \times 10^{-5} \text{ N/m}^2 (0.0002 \text{ } \mu\text{bar})$$
$$= 2.9 \times 10^{-9} \text{ psi} \qquad \textbf{[11.17]}$$

Sound intensity and pressure levels are measured in decibels. Thus,

$$\text{Intensity level (dB)} = 10 \log \frac{I}{I_0} \qquad \textbf{[11.18]}$$

$$\text{Pressure level (dB)} = 20 \log \frac{p}{p_0} \qquad \textbf{[11.19]}$$

When pressure fluctuations and particle displacements are in phase, such as in a plane acoustic wave, these levels are equal, $10 \log I/I_0 = 20 \log p/p_0$.

The intensity or pressure level is usually called the *sound-pressure level* (SPL).

The magnitudes of the particle velocity and pressure fluctuations created by a sound wave are small. For example, a plane sound wave having an intensity of 140 dB generates a maximum oscillation velocity of approximately 2.4 ft/s (0.73 m/s) and a root mean square pressure fluctuation of about 0.029 psi (200 Pa). A sound intensity of 90 dB is considered the maximum permissible level for extended human exposure.

In many circumstances we shall be interested in the sound intensity that results from several sound sources. This calculation could, of course, be performed with Eqs. (11.18) and (11.19), but the chart in Fig. 11.10 is frequently used to advantage in such computations. The following examples illustrate the use of the figure.

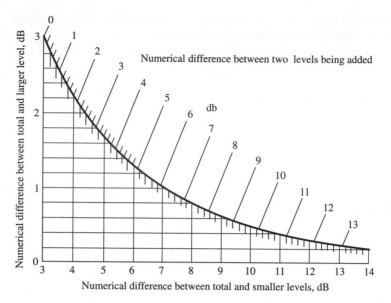

Figure 11.10   Calculation aid for adding or subtracting sound levels measured in decibels.

Example 11.6 | **ADDITION OF SOUND SOURCES.**   Calculate the total sound intensity resulting from two sound sources at 60 and 70 dB.

Solution

The numerical difference between the two levels being added is 10 dB. Consulting Fig. 11.10, we find that the numerical difference between the total and larger level is about 0.4 dB. Therefore, the total sound intensity is $70 + 0.4 = 70.4$ dB.

This same result could also be obtained by using Eq. (11.18)

$$\log\left(\frac{I_1}{I_0}\right) = \frac{60}{10} = 6 \qquad \log\left(\frac{I_2}{I_0}\right) = \frac{70}{10} = 7$$

$$I_1 = 10^6 I_0 \qquad I_2 = 10^7 I_0 \qquad I_{\text{total}} = I_1 + I_2 = 1.1 \times 10^7 I_0$$

$$\log\left(\frac{I_{\text{total}}}{I_0}\right) = 7.0414$$

$$\text{Total intensity level (dB)} = 10\log\left(\frac{I_{\text{total}}}{I_0}\right) = 70.41 \text{ dB}$$

Example 11.7 | **ADDITION OF THREE SOUND SOURCES.**   Calculate the total sound intensity resulting from three sound sources at 50, 55, and 60 dB.

Solution

We first combine the two lower levels to find their equivalent intensity. This intensity is then combined with the highest level to obtain the total sound level. Figure 11.10 is used for the calculations.

50- and 55-dB levels:

Difference = 5 dB
Amount to be added to larger level = 1.2 dB
Equivalent level = 55 + 1.2 = 56.2 dB

56.2- and 60-dB combination:

Difference = 3.8 dB
Amount to be added to larger level = 1.5 dB
Total sound level = 60 + 1.5 = 61.5 db

The human ear responds differently to different sound frequencies and to certain groups of frequencies, which are called "noise." The acoustical engineer is responsible for designing rooms or systems in such a way that unwanted frequencies or noise are filtered out while clarity of transmission of certain other sounds is still retained. An auditorium, for example, should be designed so that very little outside noise is transmitted through the walls, while the inside structure and fixtures should allow clear transmission of audio performances on the stage. The walls should have good sound insulation, while the inside surfaces should be properly designed so that unwanted reverberations are eliminated without causing undue attenuation of sound from the stage. Almost all acoustical design problems involve the exercise of considerable judgment based on practical experience as well as a knowledge of some basic sound-level measurements.

Sound-level measurements are performed with some type of microphone, which may be considered a type of seismic-vibration instrument. The electric output of the microphone is proportional to sound-pressure level, which may be used to calculate the sound intensity according to Eq. (11.18). Appropriate power amplifiers and readout meters or recorders indicate the sound level. In general, the microphone must be calibrated in a test facility with a source of known frequency and intensity. Even with careful calibration, accuracies of better than ±1 dB may not be expected in sound-pressure-level measurements. The characteristics of some typical microphones are given in Table 11.1.

A typical practical application of sound-level measurements may call for an analysis of the noise spectrum in a certain sound source. For this purpose a noise analyzer with bandpass-filter circuits is used. Several sound-pressure-level measurements are taken to determine the sound intensity in various wavelength bands. Based on experience, the acoustical engineer may then take steps to attenuate whatever frequencies he or she feels are necessary to achieve the design objective. For many commercial applications, noise and vibration meters calibration may be achieved with a simple

Table 11.1 Characteristics of microphones according to Ref. [9]

Mechanism	Frequency Range	Approximate Open-Circuit Sensitivity, dB, Below 1 V/dyn · cm^2	Approximate Impedance, Ω	Application and Remarks
Electrodynamic	Up to 20 kHz	−85	10	Field measurements, communications, etc.
Electrostatic	Up to 50 kHz	−50	500,000	Precision measurements, standards
Piezoelectric:				
Rochelle	Audio region ultrasonic	−50	100,000	Temperature-dependent, hygroscopic
ADP crystal		−50		Use in underwater sound, solids, etc.
Quartz	Mainly ultrasonic	−90 to −100	High	
Barium titanate		−90	Low	High-intensity work in air; water; measurements with small probes
Magnetostrictive	Mainly ultrasonic	−100	Low	Underwater sound
Ribbon microphone	Audio region	−100	1	Directive

whistle or tuning fork in a quiet room. Sound-level readings will generally follow the inverse square law with respect to distance.

Example 11.8 | **MICROPHONE SENSITIVITY AND OUTPUT.** A certain microphone has an open-circuit sensitivity of −80 dB referenced to 1 V for sound-pressure excitation of 1 dyn/cm^2. Calculate the voltage output when exposed to a sound field of (*a*) 100 dB and (*b*) 50 dB.

Solution

We first calculate the sound pressure from Eq. (11.19). We have, for the 100-dB level,

$$100 = 20\log \frac{p}{2 \times 10^{-4}}$$
$$p = 20 \text{ dyn/cm}^2$$

The reference voltage for an excitation of 1 dyn/cm^2 is calculated as

$$-80 \text{ dB} = 20\log \frac{E}{1.0}$$
$$E = 10^{-4} \text{ V}$$

Because we assume the output voltage varies in a linear manner with the impressed sound field, the output voltage at 100-dB sound level is

$$E = 20 \times 10^{-4} = 2 \text{ mV}$$

For a sound level of 50 dB the sound pressure is obtained from

$$50 \text{ dB} = 20 \log \frac{p}{2 \times 10^{-4}}$$

$$p = 6.24 \times 10^{-2} \text{ dyn/cm}^2$$

The output voltage is therefore

$$E = (6.24 \times 10^{-2})(10^{-4}) = 6.24 \times 10^{-6} \text{ V}$$

A good summary of acoustical principles and materials for use in sound-attenuation applications is given in Refs. [2], [3], and [8]. The following paragraphs present a summary discussion of some of the types of acoustical phenomena that are encountered in practice.

TRAVELING PLANE WAVES

A plane acoustic wave may be described in terms of the displacement of a particle at a distance x from the wave source and at a time t. The analytical relation expressing the particle displacement as a function of distance and time for a harmonic wave is

$$\xi = \xi_m \cos \frac{2\pi}{\lambda}(ct - x) \qquad \textbf{[11.20]}$$

where ξ = particle displacement
$\quad\quad \xi_m$ = particle-displacement amplitude
$\quad\quad c$ = sound velocity
$\quad\quad \lambda$ = wavelength

An observer located at some particular distance x_1 from a plane wave source will experience a periodic variation of the particle displacement and velocity. Another observer located at a distance x_2 from the source will experience exactly the same relative periodic motion; however, the absolute time at which the maximum and minimum points in the motion occur will depend on the distance $(x_2 - x_1)$ and the wave velocity c. The amplitude of the oscillatory motion will decrease with the distance from the wave source owing to viscous dissipation in the fluid. The plane wave described by Eq. (11.20) is called a "traveling wave" because the particle displacement is dependent on time and the distance from the wave source. Traveling waves may be described in a similar manner for cylindrical wave propagation.

The fluctuations in pressure due to the passage of a plane sound wave may be described in terms of the amplitude of the particle displacement through the relation

$$p = -\beta \frac{\partial \xi}{\partial x} = -\beta \frac{2\pi}{\lambda} \xi_m \sin \frac{2\pi}{\lambda}(ct - x) \qquad \textbf{[11.21]}$$

where β is the adiabatic bulk modulus of the fluid defined by

$$\beta = -V\frac{dp}{dV}$$ [11.22]

The intensity of a sound wave is defined as the flux of energy per unit time and per unit area. Several equivalent expressions for the intensity of a plate wave are given in Eq. (11.23):

$$
\begin{aligned}
I &= \tfrac{1}{2}\rho_a c(\dot{\xi}_m)^2 \\
&= \tfrac{1}{2}\rho_a c\omega^2\,\xi_m^2 \\
&= \rho_a c(\dot{\xi}_{rms})^2 \\
&= p_{rms}\xi_{rms}\omega \\
&= \frac{p_{rms}^2}{\rho_a c}
\end{aligned}
$$ [11.23]

where
ρ_a = air density
ω = frequency
$\dot{\xi}_m$ = velocity amplitude of the wave
$\dot{\xi}_{rms}$ = rms value of the particle velocity
p_{rms} = rms pressure fluctuation

It is worthwhile to note that the intensity expressions containing the pressure terms do not apply for the case of cylindrical or spherical waves unless the radius of curvature of these waves is large enough so that the wave front may be approximated by a plane.

STANDING WAVES

Standing waves, like traveling waves, are described in terms of periodic particle displacements and the corresponding periodic particle velocities. For the case of traveling waves the amplitude of the particle displacement is the same regardless of the position x provided that viscous dissipation effects are neglected. For standing waves the amplitude of the particle displacement will follow a periodic variation with the distance x from the sound source. At even quarter-wavelengths the amplitude will be zero, and at odd quarter-wavelengths the particle-displacement amplitude will take on its maximum value. Consequently, the term "standing wave" is derived from the fact that the amplitude of the particle displacement has a periodic variation which is independent of time. Thus, we have the idea that the wave "stands" in a certain position. The particle motion in a standing wave may be represented by the relation

$$\xi = 2\xi_m \sin\left(\frac{2\pi x}{\lambda}\right)\cos\left(\frac{2\pi ct}{\lambda} + \text{const}\right)$$ [11.24]

	180	Rocket engine
	160	Supersonic boom
	140	Threshold of pain
Hydraulic press (3 ft)	130	
Large pneumatic riveter (4 ft)		
		Boiler shop (maximum level)
Pneumatic chipper (5 ft)		
	120	
Overhead jet aircraft, four-engine (500 ft)		
		Jet-engine test-control room
Unmuffled motorcycle		
	110	Construction noise (compressors and
Chipping hammer (3 ft)		hammers) (10 ft)
		Woodworking shop
Annealing furnace (4 ft)		
		Loud power mower
Rock and roll band	100	
Subway train (20 ft)		
Heavy trucks (20 ft)		Inside subway car
Train whistles (500 ft)		Food blender
	90	
10-hp outboard (50 ft)		
		Inside sedan in city traffic
Small trucks accelerating (30 ft)		
		Heavy traffic (25 to 50 ft)
	80	Office with tabulating machines
Light trucks in city (20 ft)		
Automobiles (20 ft)		
	70	
Dishwashers		Average traffic (100 ft)
		Accounting office
Conversational speech (3 ft)		
	60	
	50	Private business office
		Light traffic (100 ft)
		Average residence
	40	
	30	
		Broadcasting studio (music)
	20	
	10	
	0	

Figure 11.11   Typical overall sound levels: decibels RE 0.0002 μbar, the minimum audible sound level at 1000 Hz.

CONSTANT-PRESSURE SOUND FIELD

Constant-sound-pressure waves are distinctly different from either traveling waves or standing waves and are encountered in reverberant chambers whose walls reflect the major portion of the wave energy striking them. For the ideal case the energy of the sound source is distributed uniformly throughout the reverberant room so that the fluid will experience a uniform compression and expansion in all directions.

Traveling waves and standing waves represent a vector field; that is, they depend on the direction in which the wave is propagating. Reverberant waves, for the ideal case, are independent of either distance or direction from the wave source; hence, they represent a scalar field. There is evidently no simple correlation between the pressure fluctuations and the particle displacements in the free space of a reverberant room.

PSYCHOACOUSTIC FACTORS

The sound-pressure scale for typical sound sources is indicated in Fig. 11.11, ranging from the threshold of hearing to the intensive sound of a rocket engine. We have already noted that the human ear responds differently to different frequencies; therefore, in order to interpret the response of the ear to a particular sound source we must know the distribution of the sound over the entire frequency spectrum. The nonlinear response of the ear has been determined, resulting in the Fletcher-Munson [11] curves for equal loudness shown in Fig. 11.12. According to these curves, the ear is much more sensitive to midfrequencies than to low or high frequencies. Based on

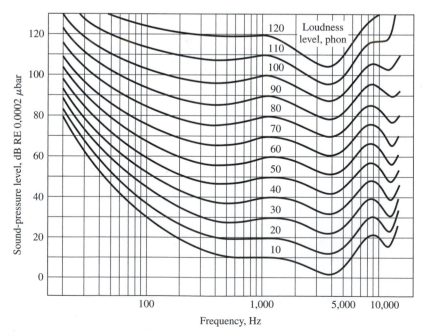

Figure 11.12   Equal loudness contours for pure tones. (*American Standard for Noise Measurement Z24.2–1942.*)

the above concepts, it is possible to define the *loudness level* as a relative measure of sound strengths judged by an average human listener. The unit of loudness level is the *phon,* which is defined so that the loudness in phons is numerically equal to the sound-pressure level at a frequency of 1000 Hz. The loudness levels are noted on the curves of Fig. 11.12. Inspecting the 40-phon curve we find that a signal at 20 Hz requires a SPL of 94 dB to sound as loud as a signal having a frequency of 1000 Hz and SPL of 40 dB. Likewise, a signal at 10 kHz requires a SPL of 46 dB to sound as loud as a signal at 1000 Hz and 40 dB.

VELOCITY AND PRESSURE FLUCTUATIONS IN SOUND FIELD.   Verify that the | **Example 11.9**

velocity amplitude and root-mean-square (rms)-pressure fluctuation are 0.7 m/s and 202 Pa for a 140-dB plane sound wave in air at 20°C and 100 kPa.

Solution

From Eq. (11.18)

$$140 \text{ dB} = 10 \log \frac{I}{I_0}$$

where $I_0 = 10^{-16} \text{ W/cm}^2 = 10^{-12} \text{ W/m}^2$, so that

$$I = (10^{-12})(10^{14}) = 100 \text{ W/m}^2$$

The air density is obtained from the ideal-gas law

$$\rho_a = \frac{p}{RT} = \frac{10^5}{(287)(293)} = 1.189 \text{ kg/m}^3$$

and the velocity of sound from Eq. (7.49) is

$$c = \sqrt{\gamma g_c RT} = [(1.4)(1.0)(287)(293)]^{1/2}$$
$$= 343 \text{ m/s}$$

The velocity amplitude $\dot{\xi}$ is obtained from Eq. (11.23) as

$$\dot{\xi} = \left(\frac{2I}{\rho_a c} \right)^{1/2}$$

$$= \left[\frac{(2)(100)}{(1.189)(343)} \right]^{1/2} = 0.70 \text{ m/s}$$

The rms-pressure fluctuation is also obtained from Eq. (11.23) as

$$p_{\text{rms}} = (\rho_a c I)^{1/2}$$
$$= [(1.189)(343)(100)]^{1/2}$$
$$= 202 \text{ Pa} = 0.002 \text{ atm}$$

SOUND POWER TRANSMITTED BY LOUDSPEAKER.   A 15-in-diameter "woofer" | **Example 11.10**

loudspeaker produces a SPL of 105 dB 1 m from the speaker for a power input to the speaker of 1.0 W. The air conditions are 100 kPa and 20°C. Assuming the SPL occurs over the area of the speaker, calculate the ratio of sound energy to the energy input to the speaker.

Solution

We first calculate the sound intensity from Eq. (11.18):

$$105 = 10 \log \frac{I}{10^{-12}}$$

and

$$I = 0.0316 \text{ W/m}^2$$

The area for a 15-in diameter is

$$A = \frac{\pi(15/12)^2}{4} = 1.227 \text{ ft}^2 = 0.114 \text{ m}^2$$

The sound power is, therefore,

$$\text{Sound power} = IA = (0.0316)(0.114) = 0.0036 \text{ W}$$

The desired ratio is

$$\frac{\text{Sound power}}{\text{Power to speaker}} = 0.0036$$

Comment

We note, of course, that all the sound power leaving the surface of the speaker cone does not pass through the area used in the above calculation. Much of the energy goes to other parts of the enclosing space. If one assumes all the energy goes through a hemisphere area with a radius of 1.0 m, the area would be

$$A = \tfrac{1}{2}(4\pi r^2) = 2\pi = 6.283 \text{ m}^2$$

and the sound power would be

$$\text{Sound power} = IA = (0.0316)(6.283) = 0.199 \text{ W}$$

The output/input ratio, then, would be 0.199, indicating a rather good conversion of electric power to sound power.

MEASUREMENT SYSTEMS

The typical sound-measurement system is shown in Fig. 11.13. An appropriate microphone is used to sense the sound wave. Before measurements begin an acoustic calibrator is placed over the microphone to effect a calibration. The calibrator consists of a special cavity for the particular microphone and a transducer that will produce a known sound-pressure level inside the cavity for given driving voltage inputs. The response of the microphone to the calibration signals is noted on the sound-level meter, and thereby furnishes reference levels for use in comparing the signals produced by the unknown sound field.

In practice, the microphone may be mounted directly on the sound-level meter or separated with a connecting cable in order to accommodate some particular measurement objective. The sound-level meter incorporates the functions shown in Fig. 11.14. An input preamplifier amplifies the signal that feeds the attenuator switch, which may

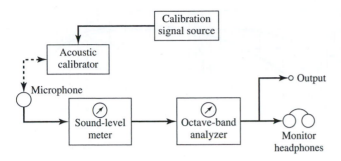

Figure 11.13   Schematic of instrument arrangement for sound measurements.

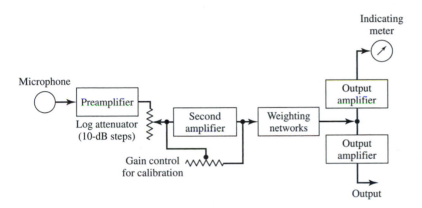

Figure 11.14   Schematic of sound-level meter.

be varied in 10-dB steps. The signal is then amplified further, and three standardized weighting networks modify the signal in accordance with approximations to the standard loudness contours. The frequency-response characteristics of these networks are shown in Fig. 11.15. The A scale approximates the 40-phon contour and is used for levels below 55 dB. The B scale approximates the 70-phon contour and is used for levels between 55 and 85 dB. Finally, the C scale is used above 85 dB and provides a fairly flat response curve. The output from the weighting networks is either observed on the self-contained indicating meter or recorded with some type of external digital device. Checking the A curve we find that it attenuates the signal by about 54 dB at 20 Hz, which corresponds to the 54-dB boost one would need in the 40-phon curve ($94 - 40$ dB) to cause the 20-Hz signal to sound as loud as the 1000-Hz signal. Similar behavior is evident for the B curve in approximating the 55-phon curve.

The entire sound spectrum is of importance because of the influence of different frequencies on human response. For this reason the output of the sound-level meter is usually fed to an octave-band analyzer, which is an instrument containing amplifiers, standard filter circuits, and an indicating-meter circuit to measure the signal strength in

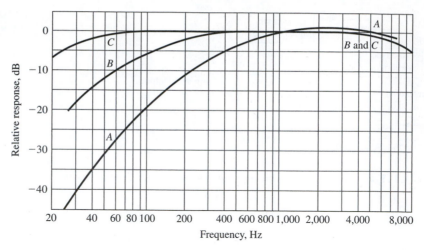

Figure 11.15   Standardized weighting networks for sound-level meters for random sound incidence upon the microphone. *A* curve approximates inverse of 40-phon loudness, *B* curve approximates inverse of 55-phon loudness, *C* curve approximates flat response. (*American Standard Specification for General Purpose Sound-Level Meters, SI-4-1961.*)

Table 11.2   Standard octave-band-analyzer filters

Lowpass to 75 Hz
75–150 Hz
150–300 Hz
300–600 Hz
600–1200 Hz
1200–2400 Hz
2400–4800 Hz
Highpass from 4800 Hz

each filter passband. The standard octave passbands are given in Table 11.2. The output from the microphone may also be captured in a digital memory device and a Fourier analysis performed to determine the frequency content when complex waveforms are measured. Such analysis can be quite useful in examining speech disorders and may suggest appropriate therapeutic measures. Digital recorders offer the opportunity to record sound sources for later display and computer analysis.

ACOUSTIC PROPERTIES OF MATERIALS

One of the important objectives of acoustic measurements is a specification of the sound-absorption properties of materials. With such information available, the acoustical engineer may then choose the proper material to accomplish environmental objectives in noise control or architectural design. There are many complicating factors

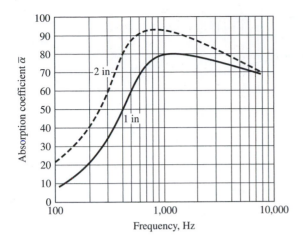

Figure 11.16 Increase in sound absorption resulting from a doubling in thickness of a typical homogeneous porous sound-absorbing material, according to Ref. [3].

to be considered in the selection of sound-insulation materials, and we shall discuss only two parameters that are of common interest. For further information the reader should consult Refs. [3] and [10].

The sound-absorption coefficient $\bar{\alpha}$ is defined by

$$\bar{\alpha} = \frac{\text{sound energy absorbed}}{\text{sound energy incident upon surface}}$$

The absorption coefficient is strongly frequency-dependent and influenced by such factors as porosity of the surface, thickness of material, rigidity of material, etc. Figure 11.16 illustrates the frequency dependence for a typical porous sound-absorbing material. The material is clearly most effective in absorbing sound fields at mid- and high frequencies. The absorption characteristics of some different types of materials are shown in Fig. 11.17. The more rigid wood panels are substantially less effective as absorbers than the porous material. In general, rigid materials are poor sound absorbers because they have the ability to vibrate and reemit sound. A highly damped, nonrigid material like the porous sheet or, to a lesser degree, the assembly represented by curve B in Fig. 11.17 is more effective in absorbing an incident sound wave. A thin sheet of steel mounted in an undamped fashion would be a very poor absorber because of the ease with which it may be set into vibration.

Fourier analysis of the frequency spectrum for transmission and absorption can point the way to improved performance of various materials.

NOISE-REDUCTION COEFFICIENT

Since the sound-absorption characteristics of materials are so strongly frequency-dependent, some average parameter must be selected which represents the material

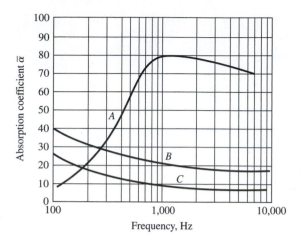

Figure 11.17   Absorption coefficients for (a) 1-in-thick porous acoustical material; (b) two sheets of $\frac{1}{8}$-in-thick plywood, separated by tufts of randomly glued felt causing $\frac{1}{8}$-in spacing between sheets; (c) one sheet of $\frac{3}{8}$-in-thick randomly braced plywood, according to Ref. [3].

behavior over the entire frequency spectrum. One could choose an integrated average of the absorption coefficient, but the accepted practice is to define a noise-reduction coefficient (NRC) as

$$\text{NRC} = \sum \frac{\text{sound-absorption coefficients of materials at 250, 500, 1000, and 2000 Hz}}{4}$$

Thus, the NRC may be determined by measuring the absorption coefficients at the four test frequencies and by taking their arithmetic average. It should be noted that this definition does not take into account the absorption characteristics at a low frequency, say, 125 Hz. For this reason comparison of materials on the basis of NRC alone may be misleading if the absorption characteristics at low frequencies are of particular interest. Certain inconsistencies are present in published data of noise-reduction co-efficients because of the influence of sample mounting and test procedures on the final reported measurements. For further information the interested reader should consult Refs. [3] and [12].

11.6 REVIEW QUESTIONS

11.1. Describe the basic concept of the seismic instrument.

11.2. Under what conditions is a seismic instrument suitable for (a) amplitude measurements and (b) acceleration measurements?

11.3. Describe some alternative constructions of a seismic instrument. When would each construction be used?

11.4. When would a piezoelectric transducer be used to advantage in a seismic instrument?

11.5. What are the reference-energy flux and pressure levels for sound fields?

11.6. What is the maximum permissible sound level for human exposure?

11.7. What is meant by the open-circuit sensitivity of a microphone?

11.8. Distinguish among traveling, standing, and constant-pressure sound fields.

11.9. Why is an octave-band analyzer used in sound measurements?

11.10. What is meant by loudness level?

11.11. Is the ear more or less sensitive to low frequencies than to midfrequencies (i.e., 1000 Hz)?

11.12. How is a sound-level meter calibrated?

11.13. Define the sound-absorption coefficient.

11.14. Define the noise-reduction coefficient (NRC).

11.7 PROBLEMS

11.1. The vibrating wedge shown in Fig. 11.1 is used for an amplitude measurement. The length of the wedge is 15 cm $\pm$ 0.5 mm and the thickness is 2.5 cm $\pm$ 0.2 mm. The x distance is measured as $x = 5.6$ cm $\pm$ 1.3 mm. Calculate the vibration amplitude and its uncertainty in percent.

11.2. A small cantilever vibrometer is available for measurement of vibration frequency, but the specification sheet is lost, so that the properties of the device are not known. The instrument is calibrated by placing it on a large compressor in the laboratory which is rotating at 300 ± 2.0 rpm. The measured length for resonance conditions is 5.6 cm $\pm$ 0.2 mm. Calculate the frequency that the instrument will indicate when $L = 10$ cm $\pm$ 0.5 mm. Also, calculate the uncertainty in the measurement in this length.

11.3. Consider the cantilever beam of Example 11.1. Suppose the uncertainties in the material properties and dimensions are

$$w_E = 2\%$$
$$w_r = 0.0005 \text{ in}$$
$$w_p = 5 \text{ lbm/ft}^3$$
$$w_L = 0.01 \text{ in}$$

Calculate the resulting uncertainty in ω_n at $L = 1.0$ and $L = 4.0$ in.

11.4. A cantilever beam is to be used for the measurement of frequencies between 100 and 1000 Hz. The beam is to be made from a rod of spring steel having the properties given in Example 11.1. The maximum length of the rod is to be 12.5 cm, and the uncertainties in material dimensions and properties are those given in Prob. 11.3. Calculate the nominal diameter of the rod and the uncertainty in the frequency at 100 and 1000 Hz.

11.5. A seismic accelerometer is to be used to measure linear acceleration over a range from 30 to 300 m/s^2. The natural frequency of the instrument is 200 Hz, and this value may vary by ± 2 Hz owing to temperature fluctuations. Calculate the allowable uncertainty in the relative displacement measurement in order to ensure an uncertainty of no more than 5 percent in the acceleration measurement.

11.6. A seismic instrument is to be used to measure velocity. Show that the input velocity amplitude is

$$v_0 = \left(\frac{dx_1}{dt}\right)_0 = \omega_1 x_0$$

Subsequently show that this velocity amplitude may be expressed in terms of the steady-state relative displacement as

$$(x_2 - x_1)_0 = \frac{v_0(\omega_1/\omega_n)}{\omega_n\{[1 - (\omega_1/\omega_n)^2]^2 + [2(c/c_c)(\omega_1/\omega_n)]^2\}^{1/2}}$$

11.7. Calculate the value of the time constant for the instrument in Prob. 11.5 if $c/c_c = 0.65$.

11.8. Plot the error in acceleration measurement of a seismic instrument for $c/c_c = 0.7$ versus frequency ratio, that is,

$$1 - \frac{(x_2 - x_1)_0\omega_n^2}{a_0} \quad \text{versus} \quad \frac{\omega_1}{\omega_n}$$

11.9. A large seismic instrument is constructed so that $m = 45$ kg and $c/c_c = 0.707$. A spring with $k = 2.9$ kN/m is used so that the instrument will be relatively insensitive to low-frequency signals for displacement measurements and relatively insensitive to high-frequency signals for acceleration measurements. Calculate the value of linear acceleration that will produce a relative displacement of 2.5 mm on the instrument. Calculate the value of ω_1/ω_n such that

$$\frac{(x_2 - x_1)_0}{x_0} = 0.99$$

11.10. A seismic accelerometer is to be designed so that the time constant calculated from Eq. (11.15) is equal to the period of the maximum acceptable frequency for a 1 percent error in measurement; i.e., $(x_2 - x_1)_0/a_0 = 0.99$. Plot ω_n versus c/c_c in accordance with this condition.

11.11. Calculate the phase angle ϕ for the condition of Example 11.2.

11.12. Calculate the phase angle ϕ for the conditions of Example 11.3.

11.13. Calculate the energy flux and rms-pressure variation corresponding to a sound intensity of 100 dB.

11.14. The estimated accuracy of a sound-intensity measurement is usually expressed in decibels. Calculate the uncertainties in intensity in watts per square centimeter corresponding to ± 1, ± 2, and ± 3 dB. Express these as percentage values at sound levels of 50, 80, and 120 dB.

11.15. A standing sound wave is created in room air (20°C, 1 atm) with a frequency of 1000 Hz. The peak sound intensity occurring in the wave is 120 dB. Plot the sound intensity and pressure fluctuations as functions of x for one wavelength. Over what fractional portion of a wavelength could the intensity be considered constant, consistent with an estimated uncertainty of ± 1 dB in the intensity measurement?

11.16. Calculate the equivalent sound intensity of a combination of two, three, and four 50-dB sources.

11.17. Calculate the total sound intensity of a combination of 63- and 73-dB sources.

11.18. Four sound sources of 62, 58, 65, and 70 dB are added together. Calculate the total sound intensity in two different ways showing that the same result is obtained.

11.19. A microphone has an open-circuit sensitivity of -55 dB. Calculate the voltage output for sound levels of 60 and 80 dB.

11.20. Two microphones having sensitivities of -55 and -70 dB are available. What amplification factor must be applied to the signal of the least sensitive unit in order to make its output the same as the other microphone?

11.21. Calculate the noise-reduction coefficient (NRC) for the materials shown in Figs. 11.16 and 11.17.

11.22. Calculate the attenuation in decibels for the 2-in porous material of Fig. 11.16 within each octave band of Table 11.2. For this calculation, choose some average absorption coefficient for each range of interest. Assume that all energy not absorbed is transmitted through the material.

11.23. A 1000-Hz tone is generated with an intensity of 70 dB. What must be the intensities of tones at 50, 100, and 10,000 Hz in order for them to be detected with the same loudness by the ear?

11.24. Suppose that the playback of a certain musical passage in a high-fidelity system produces an intensity of 80 dB at 1000 Hz when the reproduction is judged to be most authentic to the original source. For various reasons a person may wish to play the music at a lower level. When the 1000-Hz tones are reduced to 45 dB, how much must the 60- and 100-Hz tones be boosted in order to retain the same sense of authenticity to the original source?

11.25. Compare the sound-level-meter weighting-characteristic curves with the loudness curves of 40, 70, and 90 phon, corresponding to the A, B, and C scales, respectively.

11.26. You are asked to design a cantilever device which will have a natural frequency of 440 Hz to be used in music applications. Present a suitable design and specify the uncertainty which might be expected in the natural frequency.

11.27. Four sound sources of 63, 78, 89, and 92 dB are to be added together. Calculate the total sound intensity in two different ways showing that the same result is obtained.

11.28. A seismic instrument is to be designed to measure frequencies between 100 and 400 Hz for obtaining both displacement and acceleration measurements. Specify some mass-spring-damper arrangements to accomplish this and estimate the uncertainty.

11.29. A condenser microphone has an open-circuit sensitivity of -40 dB. Calculate the voltage output for sound levels of 100 and 120 dB.

11.30. Calculate the attenuation in decibels for material B in Fig. 11.17 within each octave band of Table 11.2.

11.31. Three sound sources at 75, 89, and 100 dB are added together. What is the resultant sound level?

11.32. Calculate the rms-pressure fluctuations for sound waves at levels of 100, 110, and 130 dB.

11.33. A microphone having an open-circuit sensitivity of -55 dB is said to have a noise level of 28 dB. For a sound level of 90 dB, what is the voltage output referenced to the noise level?

11.34. Suppose noise levels of a tape recorder are stated as -55 dB, while another figure is given as -60 dB "A-weighted." Explain the two statements. Why would one specification be preferred over the other?

11.35. Suppose all the electric-power input to a speaker is delivered as sound power through a hemisphere having a radius of 1.0 m enclosing the speaker. What SPL would result from a 1-W input to the speaker?

11.36. A steel rod having a diameter of 0.8 mm is to be designed to have a natural frequency of 500 Hz. What should be the length in centimeters?

11.37. A certain loudspeaker produces a sound-pressure level of 90 dB at a distance of 1.0 m from the speaker cone for a power input to the speaker of 1 W. Calculate the ratio of total sound power to electric-power input, assuming all the sound power passes through a 1-m radius hemisphere in front of the speaker.

11.38. A microphone having an open-circuit sensitivity of -60 dB referenced to 1 V is used to measure the sound level from the speaker in Prob. 11.37. Determine the voltage output from the microphone.

11.39. It is possible to experience "feedback distortion" when a record turntable is placed too close to a loudspeaker. The problem is caused mainly by low-frequency notes because of their larger displacements. For a turntable mass of 8 kg, design an appropriate seismic-type system which will attenuate frequencies below 180 Hz. Specify two suitable sets of design parameters.

11.40. Calculate the particle displacement amplitudes for a 90-dB tone at 100, 1000, and 10,000 Hz.

11.41. Calculate the velocity amplitudes of the waves in Prob. 11.40.

11.42. A small seismic instrument is to be used for measurement of linear acceleration. The natural frequency for the device is specified as 20 Hz and the displacement-sensing device connected to the instrument can detect a maximum of ± 3.0 mm with an uncertainty of ± 0.02 mm. Determine the maximum acceleration which may be measured with the instrument and the uncertainty resulting from the displacement measurement.

11.43. For what frequency would the measurement in Prob. 11.42 be in error by 1 percent?

11.44. Two loudspeakers each produce 1000-Hz tones at 1 m of 90 and 100 dB, respectively. What SPL would be sensed if both speakers are operated together?

11.45. A cantilever beam having a thickness of 0.8 mm and a width of 25 mm is to be used for frequency measurements between 200 and 1200 Hz. Specify the lengths that are necessary for these two frequencies.

11.46. The sound-intensity levels in a rocket nozzle can approach 160 dB. For frequencies of 300 and 1000 Hz, determine the rms-pressure fluctuations, particle-displacement amplitudes, and particle-velocity amplitudes.

11.47. The material of Fig. 11.16 is to be used to attenuate a 1000-Hz source at 100 dB. Estimate the thickness of material required to lower the level to 90 dB.

11.48. What would be the attenuation in decibels of a 1000-Hz tone at 90 dB for each of the materials in Fig. 11.17?

11.49. A seismic-type instrument has a natural frequency of 60 Hz. What is the maximum percent overshoot in acceleration response for damping ratios of 0.2 and 0.3? At what frequencies do these maximum values occur?

11.50. The instrument in Prob. 11.49 is operated under a critically damped situation. Approximately what will be the acceleration attenuation at the natural frequency, expressed in dB?

11.51. Using Fig. 11.6 as an approximate calculation tool, estimate the range of values of the damping ratio that will yield a flat acceleration response of ± 1 dB for $0.5 < \omega_1/\omega_n < 1.5$.

11.52. Repeat Prob. 11.51 for $0.2 < \omega_1/\omega_n < 1.0$.

11.53. Four sound sources having SPL values of 82, 76, 95, and 111 dB are added together. What is the resultant sound pressure level? Perform the calculation using Eq. (11.19) and also with Fig. 11.10.

11.54. In checking a stereo speaker system for balance, the left and right channels are found to differ by 3 dB at SPL $= 80$ dB and a frequency of 1000 Hz. Calculate the intensity of sound for each speaker expressed in W/m^2 and comment on the results.

11.55. A subwoofer for a home theater system is stated to produce a SPL of 110 dB at 16 Hz. Calculate the resulting rms pressure fluctuation and rms particle displacement amplitude.

11.56. In a home theater system the main two front speakers deliver sound levels of 95 and 90 dB while the front center speaker delivers 78 dB. The rear two speakers each deliver 75 dB. Assuming all measurements are made at the same listener location, calculate the total SPL delivered to the listener. What percent of the total sound intensity (absolute value) is delivered by each speaker?

11.57. The output from the microphone of Example 11.8 is to be connected to an amplifier which must deliver an output voltage of at least 0.1 V for the maximum SPL condition (100 dB). The lowest level of sound that will be detected by the microphone is estimated as 25 dB SPL, and a S/N ratio of 30 dB is desired for the amplifier circuit at that condition. What must be the design S/N ratio for the amplifier, referred to full-voltage output of 0.1 V?

11.58. A seismic-type instrument is to be used for displacement measurements with an accuracy of ± 1 percent. What natural frequency will be necessary for $c/c_c = 0.7$ and measurements of $\omega > 100$ Hz?

11.59. Calculate the voltage output of a dynamic microphone when exposed to sound-pressure levels of 70 and 90 dB if the open-circuit sensitivity of the microphone is -37 dB.

11.60. A 100-Hz tone is generated with a SPL of 80 dB. What must be the SPL of a 10-kHz tone to be detected with the same loudness by the ear?

11.61. Four sound sources of 60, 70, 80, and 90 dB are excited in a room. What is the overall SPL?

11.62. A microphone having a noise level of 30 dB has an open-circuit sensitivity of -60 dB. Calculate the voltage output referenced to the noise level when the microphone is exposed to a SPL of 85 dB.

11.63. Calculate the length of a steel rod with diameter of 1.0 mm such that its natural frequency will be 600 Hz.

11.64. What would be the attenuation in dB of a 500-Hz tone for each of the materials in Fig. 11.17?

11.65. A seismic-type instrument is to be used for displacement measurements with an accuracy of ± 2 percent. What natural frequency will be necessary for $c/c_c = 0.7$ and measurements with $\omega > 120$ Hz?

11.8 REFERENCES

1. Beckwith, T. G., R. D. Maranguni, and J. H. Lienhard: *Mechanical Measurements,* 5th ed., Addison-Wesley, Reading, MA, 1993.

2. Beranek, L. L.: *Acoustics,* McGraw-Hill, New York, 1954.

3. ———: *Noise Reduction,* McGraw-Hill, New York, 1960. Reprinted by permission of the author.

4. Dove, R. C., and P. H. Adams: *Experimental Stress Analysis and Motion Measurement,* Charles E. Merrill, Columbus, OH, 1964.

5. Gross, E. E.: "Noise Measuring and Sound Control," *Refrig. Eng.,* vol. 66, p. 49, 1957.

6. Harris, C. M., and C. M. Crede (eds.): *Basic Theory and Measurements,* vol. 1, *Shock and Vibration Handbook,* McGraw-Hill, New York, 1961.

7. Hetenyi, M.: *Handbook of Experimental Stress Analysis,* John Wiley & Sons, New York, 1950.

8. *ASHRAE Guide and Data Book,* Am. Soc. Heat, Refrig. Air Cond., Atlanta, GA, latest edition.

9. Condon, E. U. (ed.): *Handbook of Physics,* McGraw-Hill, New York, 1958.

10. Harris, C. M. (ed.): *Handbook of Noise Control,* McGraw-Hill, New York, 1957.

11. Fletcher, H., and W. A. Munson: "Loudness, Its Definition, Measurement, and Calculation," *J. Acous. Soc. Am.,* vol. 5, pp. 82–108, 1933.

12. Broch, J. T.: *Acoustic Noise Measurements,* 2d ed., Bruel and Kjaer Instrument Company, 1971.

13. Wilson, J.: *Experiments in the Dynamics of Solids,* McGraw-Hill, New York, 1993.

14. Meirovitch, L., and M. Modest: *Elements of Vibration Analysis,* 2d ed., McGraw-Hill, New York, 1986.

15. Kelly, S. G.: *Fundamentals of Mechanical Vibrations, IBM,* McGraw-Hill, New York, 1993.

12

THERMAL- AND NUCLEAR-RADIATION MEASUREMENTS

12.1 INTRODUCTION

In Chap. 8 we mentioned the importance of thermal-radiation measurements in temperature determinations. In this chapter we shall examine the physical principles and operating characteristics of some of the more important thermal-radiation detectors and indicate their range of applicability. For a rather complete survey of the field of infrared engineering and thermal-radiation detectors the reader should consult the references at the end of this chapter.

Radioactivity measurements are a broad field of activity performed by engineers and are important in many applications. In this chapter we shall discuss some of the measurements that find wide application and indicate the devices used for performing these measurements.

12.2 DETECTION OF THERMAL RADIATION

The measurement of thermal radiation is basically a measurement of radiant energy flux. The detection of this energy flux may be accomplished through a measurement of the temperature of a thin metal strip exposed to the radiation. The strip is usually blackened to absorb most of the radiation incident upon it and is constructed as thin as possible to minimize the heat capacity and thereby bring about the most desirable transient characteristics. A schematic of the general thermal-radiation detector is given in Fig. 12.1. The temperature attained by the element is not only a function of the radiant energy absorbed, but is also dependent on the convection losses to the surroundings and conduction to the mounting fixtures. Convection losses from the element may be reduced by enclosing the detector in an evacuated system with an appropriate window for transmission of the radiation. The infrared-transmission characteristics of several substances employed as window materials are given in Fig. 12.2 according to Ref. [1]. Conduction losses may be reduced with suitable insulating materials.

Figure 12.1   Schematic of the general thermal-radiation detector.

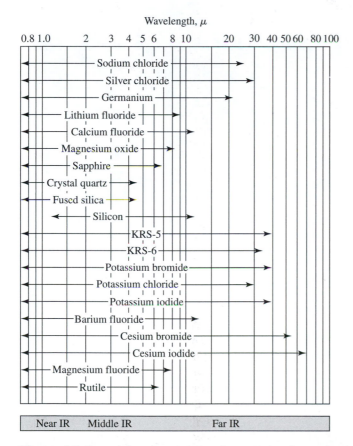

Figure 12.2   Infrared transmission characteristics of several optical materials, according to Ref. [1] as arranged by Hackforth [7]. Sample thickness = 2 mm, long-wavelength cutoff at 10 percent transmission.

Either thermocouples or thermopiles may be used for detecting the temperature of the blackened radiation-sensitive element. Thermopiles offer the advantage that they produce a higher voltage output. Many ingenious methods of construction have been devised for such thermopiles, a commercial example of which is given in Fig. 12.3. The expanded view of the thermopile shows the blackened junction pairs surrounded

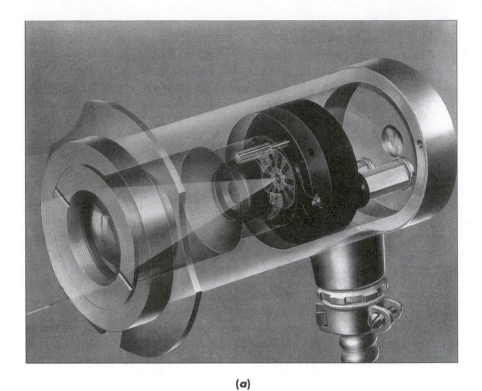

(a)

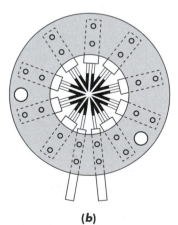

(b)

Figure 12.3   A commercial radiometer utilizing a thermopile sensor. (a) Cutaway of instrument; (b) detail of thermopile sensor. (*Courtesy of Honeywell, Inc.*)

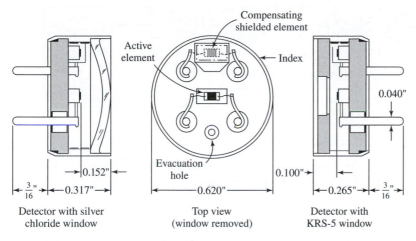

Figure 12.4 A commercial thermistor radiation detector. (*Courtesy of Barnes Engineering Co., Stamford, CT.*)

by an annular ring of mica that serves as both electrical and thermal insulation. The thermopile registers the difference in temperature between the hot junctions and the ambient temperature surrounding the detector. The lens at the front of the device focuses the radiation on the thermopile junctions. Special circuitries are used to provide for compensation of ambient temperatures between 50 and 250°F (10 to 121°C).

Thermal radiation may also be sensed by a metal bolometer, which consists of a thin strip of blackened metal foil such as platinum. The temperature of the foil is indicated through its change in resistance with temperature. An appropriate bridge circuit is used to measure the resistance.

Thermistors are widely used as thermal-radiation detectors, and a cutaway drawing of a commercial detection device is given in Fig. 12.4. Two thermistors are enclosed in the detector case, which is covered by an appropriate glass window having satisfactory transmission characteristics. One thermistor element is exposed to the incoming radiation that is to be measured, while the other element is shielded from this radiation. The shielded element is connected in the circuit so that it furnishes a continuous compensation for the temperature of the detector enclosure. A schematic of a commercial radiometer using thermistor detectors is given in Fig. 12.5. The incoming radiation is focused by the mirror system onto the thermistor detector. A motor-driven chopper periodically interrupts the radiation so that an alternating signal is produced which may be amplified more easily. The amplified signal is subsequently rectified to produce an output voltage proportional to the radiant flux incident on the thermistor detector. In this system mirrors may be adjusted so that the optical system can be focused on an area as small as 1 mm^2. The movable mirror and focusing lamp are used for this purpose. The system is exceedingly sensitive and may even be used for detection of radiation from sources near room temperatures.

Relatively inexpensive handheld and battery-powered radiometers are available which can sight on areas of about 1.5 cm in diameter (see Ref. [33]). These instruments

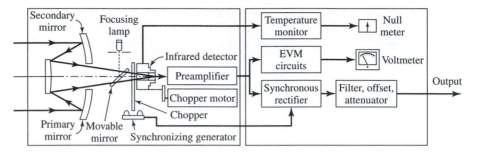

Figure 12.5   Schematic of a commercial radiometer utilizing a thermistor detector. (*Courtesy of Barnes Engineering Co., Stamford, CT.*)

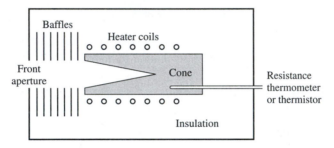

Figure 12.6   Typical construction of a blackbody source.

have internal circuits for setting an emissivity value for use in Eq. (8.25). The output is then a *calculated* temperature based on the assumed emissivity and the measured incoming radiant flux. We should note that the radiant flux leaving a surface (radiosity) may not give a true indication of the surface temperature. Suppose, for example, that a cool surface at the bottom of a cavity has a high reflectivity (low emissivity) and is surrounded by cavity walls at higher temperatures. One might imagine sighting on a cool bottom of a tin can with sides at a higher temperature. The radiometer will sense all radiation leaving the bottom surface which will be due to (1) emission from the bottom and (2) reflection of energy arriving from the side walls. Substantial errors can result in these circumstances.

Thermal-radiation detectors are calibrated directly by obtaining the output as a function of the known radiation from a blackbody source at various temperatures. A typical blackbody source is constructed as shown in Fig. 12.6. The conical cavity is constructed of some high-conductivity material such as aluminum or copper, and the inside surface is blackened. An electric heater maintains the cavity at a desired temperature, which is indicated and controlled through a sensitive electric-resistance thermometer or thermistor. Baffles near the opening prevent stray radiation from the surroundings from influencing the radiation output of the cavity. The detailed construction of blackbody standards is discussed by Marcus [9]. It may be noted that cavities like the one discussed above may give effective emissivities within 1 percent of blackbody conditions.

Table 12.1 Characteristics of thermal receivers[1]

Type	Response Time	Energy Threshold, W
Thermocouple	0.2 s	3.0×10^{-6}
Thermistor bolometer	3.0 ms	7.2×10^{-8}
Metal bolometer	4.0 ms	3.3×10^{-8}

| [1]According to Ref. [2].

A comparison of the approximate transient and energy-threshold characteristics of three types of thermal-radiation receivers is given in Table 12.1 according to Ref. [2]. The energy-threshold values are the minimum energy which may adequately be detected by such a device.

Photoconductive-radiation detectors have been discussed in Sec. 4.25 and the response characteristics of a number of practical sensors noted. The lead-sulfide cell is widely used for detection of radiation in the 1- and 4-μm range.

Various electronic devices may be used for readout of the signals and coupled with microprocessors for control.

CALCULATION OF SIGNAL NOISE. A certain thermistor-radiation-bolometer detector has characteristics such that the minimum input power necessary to produce a signal-to-noise (S/N) ratio of unity is 10^{-8} W. Suppose such a detector is located 1.0 m from a blackbody radiation source and is perfectly black. Calculate the temperature of the source necessary to produce an S/N ratio of 10, assuming that the detector and surroundings are maintained at 20°C. The source is a black sphere, 5.0 cm in diameter, and the detector area is 1 mm². | **Example 12.1**

Solution

The net energy absorbed by the detector is

$$q = \sigma A \left(T_s^4 - 293^4 \right) \left(\frac{2.5}{100} \right)^2$$

where A is the detector area. For an S/N ratio of 10, the input power must be $10 \times 10^{-8} = 10^{-7}$ W $= q$. Thus,

$$T_s^4 - 293^4 = \frac{(10^{-7})(100/2.5)^2}{(5.669 \times 10^{-8})(1 \times 10^{-6})} = 2.82 \times 10^9$$

$$T_s = 317.7 \text{ K} = 44.7°\text{C}$$

For an S/N ratio of 100, the corresponding value of T_s would be 161°C. From the results of this example we see why optical systems like those shown in Figs. 12.3 and 12.5 are used to collect radiation over a fairly large area and to focus it on the small surface area of the detector.

12.3   MEASUREMENT OF EMISSIVITY

An apparatus for the measurement of total normal emissivity has been described by Snyder, Gier, and Dunkle [11]. The apparatus uses a thermopile radiometer and is constructed as shown in Fig. 12.7. An electric heater is used to maintain the temperature of the sample, while thermocouples embedded in the sample furnish an indication of its temperature. A detailed drawing of the thermopile receiver is shown in Fig. 12.7b. It is constructed of 160 junctions of silver constantan mounted in a cylindrical housing, which is blackened on the inside. Two blackened aluminum-foil strips are attached to the junctions. The rear shield has a narrow slot which allows the hot-junction strip to be exposed to the radiant flux from the sample, while the cold-junction strip is

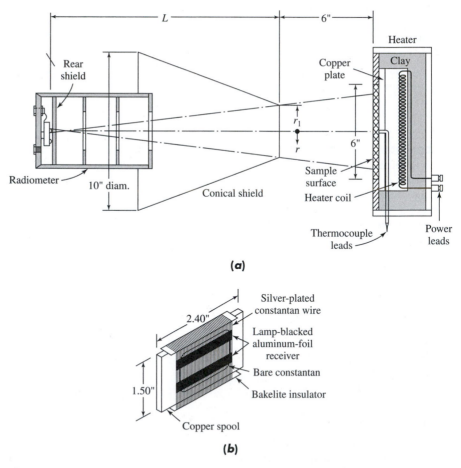

(a)

(b)

Figure 12.7   Apparatus for measurement of normal emissivity, according to Ref. [11]. (a) Schematic; (b) detail of thermopile construction.

exposed to the temperature of the thermopile enclosure. The temperature difference between the hot and cold junction is thus taken as an indication of the radiant energy flux, which, in turn, is related to the emissivity of the sample. The conical shield on the front of the device ensures a proper concentration of energy from the sample. According to Ref. [11], the device furnishes essentially a linear output of 0.0404 mV per W/m^2 radiant flux. If the temperatures of the hot and cold junctions are assumed to be essentially the same as the inside temperature of the radiometer (this apparently is a good assumption because of the small total energy absorbed in the radiometer) and the sample is assumed gray, the total normal emissivity of the sample ϵ_s is

$$\epsilon_s = \frac{7.84E}{F_{ts}\sigma\left(T_s^4 - T_R^4\right)} \qquad \text{[12.1]}$$

where E = voltage output of the thermopile, mV

T_s = sample temperature, °R

T_R = radiometer temperature, °R

The view factor F_{ts} is calculated from

$$F_{ts} = \frac{r_1^2}{r_1^2 + L^2} \qquad \text{[12.2]}$$

In order for measurement to be valid the temperature of the radiometer must be maintained at very nearly that of the surroundings.

Spectral characteristics of surfaces are discussed in Refs. [2], [20], [22], and [36].

We have already discussed in Sec. 8.6 how a radiation pyrometer may be used to obtain estimates of surface emissivities. The basic technique is as follows:

1. Provide a source to heat the surface.

2. Measure the surface temperature T with a thermocouple or other device.

3. Read the apparent temperature T_a with the radiometer, assuming

$$\epsilon = 1.0$$

4. Calculate the apparent emissivity from

$$\epsilon_a = \left(\frac{T_a}{T}\right)^4$$

5. Set the electronics of the radiometer for the apparent emissivity and repeat the radiation measurement of temperature. If the thermocouple and radiation measurements agree, the value of ϵ_a may be assumed to be reasonable. If they disagree by an unacceptable amount, one may assume the measurement is influenced by reflected radiation from surrounding surfaces. This problem arises more with low-emissivity ($\epsilon \sim 0.1$) surfaces than with high-emissivity ($\epsilon \sim 0.9$) surfaces.

The reader should recognize that surface emissivities vary widely with wavelength, temperature, and surface conditions. A polished copper plate, for example, may have an emissivity on the order of 0.05. If the plate is allowed to oxidize, the emissivity may be as large as 0.75, an increase of *1400 percent*. The visual appearance of a surface can be very deceiving insofar as emissivity is concerned. Snow white enamel at room temperature emits about 90 percent of radiation of a perfect blackbody. An abbreviated table of emissivities for several surfaces is given in the appendix. Some further information is given in the commercial publication of Ref. [33].

Example 12.2 | **UNCERTAINTY IN EMISSIVITY MEASUREMENT.**   A measurement of total emissivity is made with the apparatus shown in Fig. 12.7. $L = 14.25$ in, and $r_1 = 1.50$ in. The output is 0.823 ± 0.005 mV, $T_s = 703 \pm 1.0°$F, and $T_R = 70 \pm 0.5°$F. Calculate the nominal value of the emissivity and estimate the uncertainty in the measurement.

Solution

We calculate the view factor F_{ts} as

$$F_{ts} = \frac{(1.50)^2}{(1.50)^2 + (14.25)^2} = 0.011$$

The emissivity is then calculated with the use of Eq. (12.1).

$$\epsilon_s = \frac{(7.84)(0.823)}{(0.011)(0.1714 \times 10^{-8})(1163^4 - 530^4)}$$

$$= 0.195$$

We use Eq. (3.2) to calculate the uncertainty. The appropriate parameters are

$$\frac{\partial \epsilon}{\partial E} = \frac{7.84}{F_{ts}\sigma\left(T_s^4 - T_R^4\right)} = 0.237 \text{ mV}^{-1}$$

$$w_E = 0.005 \text{ mV}$$

$$\frac{\partial \epsilon}{\partial T_s} = \frac{-(7.84)(4)T_s^3 E}{F_{ts}\sigma\left(T_s^4 - T_R^4\right)^2} = -70 \times 10^{-4} \text{ °R}^{-1}$$

$$w_{T_s} = 1.0°\text{R}$$

$$\frac{\partial \epsilon}{\partial T_R} = \frac{(7.84)(4)T_R^3 E}{F_{ts}\sigma\left(T_s^4 - T_R^4\right)^2} = 0.663 \times 10^{-5} \text{ °R}^{-1}$$

$$w_{T_R} = 0.5°\text{R}$$

and the uncertainty in the emissivity is calculated as

$$w_\epsilon = \{[(0.237)(0.005)]^2 + [(-7.0 \times 10^{-4})(1.0)]^2 + [(0.663 \times 10^{-5})(0.5)]^2\}^{1/2}$$

$$= 0.00137 \quad \text{or} \quad 0.7\%$$

If the uncertainty in the source temperature had been $\pm 5°$F, the resulting uncertainty in the emissivity would be 0.0037 or 1.9 percent.

12.4   REFLECTIVITY AND TRANSMISSIVITY MEASUREMENTS

In Chap. 8 we described the nature of thermal radiation and the properties which describe surface characteristics. Emissivity is one important property, and Sec. 12.3 has described one method for its measurement. Reflectivity and transmissivity are other important surface properties, and we shall now describe a technique for their measurement through the use of an integrating sphere. For completeness we shall describe the technique for determining monochromatic properties, that is, properties over a range of thermal-radiation wavelengths.

Let us assume that a suitable *monochromator* is available to produce monochromatic radiation. This radiation might be produced by some kind of prism arrangement or through the use of narrowband optical filters. We wish to determine the fraction of some incident radiation that is either reflected or transmitted by a surface. For this purpose the integrating sphere of Fig. 12.8 is employed. The monochromatic radiation is chopped to produce an ac source to a detector that may be easily amplified. Consider first the measurement of reflectivity. The pivoted mirror permits a divergence of the monochromatic beam either directly to the specimen inside the sphere or to some location on the inner surface of the sphere. The inside of the sphere is coated with a thick layer (about 2 mm) of magnesium oxide to produce a very high and diffuse reflectance of about 0.99. Thus, radiation incident on the inner surface will be diffusely reflected throughout the sphere, producing a uniform irradiation of the inner surface. Around the surface radiation detectors are placed to sense the reflected radiation. The general measurement technique consists of a comparison of the detector signal for direct impingement of the reference beam with that which results from reflected radiation from the sample. The ratio of these two signals is the reflectivity of the sample. It must be noted that a chopped (ac) input beam is essential for this measurement technique because the detectors in the side of the sphere sense all radiation incident upon them, including the continuous signal resulting from thermal emission of the wall of the sphere. Appropriate electronic circuits filter out all but the chopped signals so that the output will be indicative of the input chopped radiation.

A measurement of transmissivity is made by placing a sample over the input port to the sphere. After transmission through the specimen the signal is then compared to that obtained for direct impingement on the inner surface of the sphere.

The integrating-sphere technique is regarded as the most reliable method for obtaining reflectance and transmissivity measurements and may properly be described as an absolute measurement standard. It is applicable only up to about 3-μm wavelength because of the spurious properties of magnesium oxide at higher wavelengths. The use of the integrating sphere for radiation measurements has been given a very complete exposition in the literature, and the interested reader should consult Refs. [13], [15], and [5] for additional information. A good discussion of radiation-property measurements in both gases and solids is given in Refs. [14], [16] to [20], [30], [31], and [36]. A complete survey of thermal-radiation-property data is available in Refs. [21] and [24].

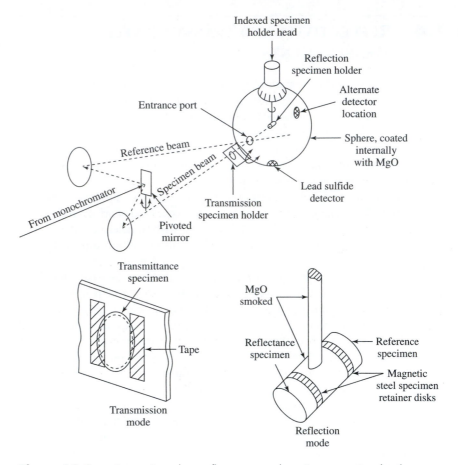

Figure 12.8   Integrating-sphere reflectometer and specimen mounting details.

12.5   SOLAR RADIATION MEASUREMENTS

The sun does not radiate as a blackbody. Estimates of the temperature of the sun vary widely but are generally in the billions of kelvins. Measurements indicate that the radiation spectrum of the sun has a shape similar to that of a blackbody, as shown in Fig. 12.9. Most of the radiation is contained between 0.2 and 2.0 μm, and the peak occurs at about 0.5 μm. If this were a true blackbody spectrum and Wien's law [Eq. (8.21)] applied, the equivalent blackbody temperature would be

$$T \approx \frac{5215.6}{0.5} = 10{,}431°\text{R} = 5795 \text{ K}$$

Figure 12.9 indicates the strong absorption of solar radiation in the atmosphere because of carbon dioxide and water-vapor concentrations.

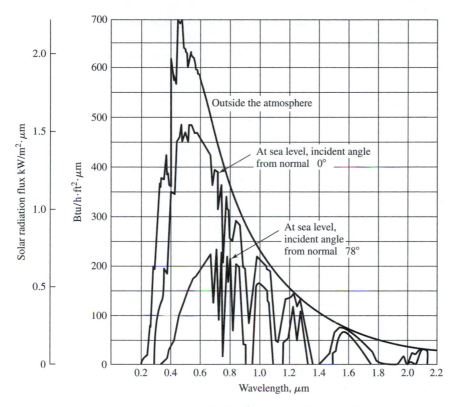

Figure 12.9   Spectral distribution of solar radiation as a function of atmospheric conditions and angle of incidence, according to Ref. [32].

Measurements of solar radiation are important because of the increasing numbers of solar heating and cooling applications and the need for accurate solar irradiation data to predict performance.

Two basic types of instruments are employed for solar radiation measurements: (1) a *pyrheliometer,* which collimates the radiation to determine the beam intensity as a function of incidence angle, and (2) a *pyranometer,* which measures the total hemispheric solar irradiation. The pyranometer measurements are the most common. A typical instrument is the Eppley pyranometer,[1] which employs two concentric rings serving as hot and cold junctions for a 50-element thermopile. The rings are coated black and white to produce a temperature difference when exposed to solar radiation. Barium sulfate is used as the whitening agent, and the entire assembly is covered with a hemispherical glass cover which transmits radiation from about 280 to 2800 nm. A typical sensitivity for the device is 7.5 mV · cm^2 · min/cal or 107.5 mV · cm^2/W. For

---

| [1] The Eppley Laboratory Inc., Newport, RI 02840.

a solar irradiation of 800 W/m² the output of the device would be

$$E = (107.5)(800 \times 10^{-4}) = 8.6 \, \text{mV}$$

The *langley* unit is frequently used in solar measurements and is defined as

$$1 \, \text{langley} = 1 \, \text{cal/cm}^2 = 4.186 \, \text{J/cm}^2 = 3.687 \, \text{Btu/ft}^2$$

The total solar radiation arriving at the outer edge of the atmosphere is called the *solar constant* and has a value of approximately 1395 W/m².

The Robitsch pyranometer operates on the principle of differential expansion of bimetallic strips exposed to solar radiation. Because the operation is mechanical, readout can be accomplished with a simple mechanical linkage, and no external electric-power source is required. Photovoltaic cells may also be employed for solar measurements. Silicon cells, cadmium sulfide cells, and selenium cells have all been used in various applications. Calibration of solar pyranometers is made with the Abbot water flow and silver disk pyrheliometers as primary and secondary standards; see Refs. [26] and [28]. Rather complete reviews of solar measurements are given in Refs. [22], [23], [27], and [29].

12.6 NUCLEAR RADIATION

In the following sections we shall be concerned with methods for detecting nuclear radiation. A detailed consideration of the origins of such radiation is quite beyond the scope of this discussion, but a few introductory remarks are appropriate in order to categorize the types of nuclear radiation and to indicate some of their particular characteristics. A detailed discussion of the principles of nuclear radiation is given in Refs. [8] and [38]; Price [10] presents a comprehensive appraisal of the subject of nuclear-radiation detection.

We shall be concerned with the detection of four kinds of nuclear radiation: (1) alpha (α) particles, (2) beta (β) particles, (3) neutrons, and (4) gamma (γ) rays. An *alpha particle* is a helium nucleus that has a positive charge of 2 and a relative mass of 0.000549. Beta particles are electrons that carry a negative charge and have a very small mass. The *neutron* has a zero charge and a relative mass of unity. *Gamma rays* are high-energy electromagnetic waves that result from nuclear transformations and, as such, do not have mass or charge in the classical sense. They may, however, produce ionizing effects in their interaction with matter. Alpha particles are absorbed rather readily in many materials, beta particles are usually more penetrating, and neutrons and gamma rays are the most penetrating types of radiation because they usually have higher energies and do not interact with coulomb force fields when entering a material. The interaction of these nuclear radiations with particular materials is a very complicated subject and forms the basis for nuclear-shielding applications [35]. Our concern is with the detection of these types of radiation.

12.7   DETECTION OF NUCLEAR RADIATION

Nuclear radiation is detected through an interaction of radiation with the detecting device, which produces an ionization process. The degree of ionization may be measured with appropriate electronic circuitry. Two types of detection operations are normally performed: (1) a measurement of the *number* of interactions of nuclear radiation with the detector and (2) a measurement of the total effect of the radiation. The first type of operation is a counting process, while the second operation may be characterized as a mean-level measurement. The counting operation frequently ignores the energy level of the irradiation. The popular Geiger-Müller counter is typically used for nuclear counting operations, while ionization chambers and photographic plates are used for energy-level measurements. We shall discuss these different types of detectors and indicate their range of applicability.

12.8   THE GEIGER-MÜLLER COUNTER

A typical cylindrical-tube arrangement for a Geiger-Müller (G-M) tube is shown in Fig. 12.10. The anode is a tungsten or platinum wire, while the cylindrical tube forms the cathode for the circuit. The tube is filled with argon with perhaps a small concentration of alcohol or some other hydrocarbon gas. The gas pressure is slightly below atmospheric. The ionizing particle or radiation is transmitted through the cathode material, or some window material is installed therein, and through interaction with the gas molecules produces an ionization of the gas. If the voltage E is sufficiently high, each particle will produce a voltage pulse. The counting performance of the tube is indicated in Fig. 12.11. The plateau region typically slopes slightly upward at a rate of from 1 to 10 percent per 100 V. The tube must be operated in the plateau region, which has a width of approximately 200 V for commercial tubes. When the particle causes a discharge or pulse, there is a time delay before the tube can detect another particle and register another pulse. This delay is roughly the time required to recharge the anode and cathode system, that is, to establish a new space charge in

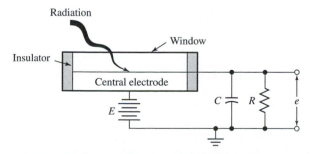

Figure 12.10   Typical cylindrical-tube arrangement for a Geiger-Müller counter.

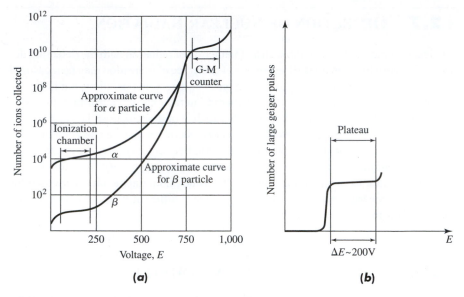

Figure 12.11   (a) Counting performance of the system in Fig. 12.10 as a function of impressed voltage; (b) detail of Geiger-counter region.

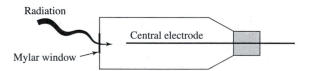

Figure 12.12   Construction of an end-type Geiger counter.

the gas. The counting rate of the G-M tube is thus limited by this delay time. The maximum counting rates are of the order of 10^4 count/s.

An end-type Geiger-counter tube may be constructed as shown in Fig. 12.12. This type of tube is used for counting α and β particles and low-energy γ rays. For these low-energy radiations the window is usually a thin sheet of mica or Mylar.

12.9   IONIZATION CHAMBERS

An ionization chamber may be constructed in basically the same way as the Geiger counter shown in Fig. 12.10 except that the tube is operated at a much lower voltage. This region of operation is indicated in Fig. 12.11. The arrangement may be modified, however, to accommodate specific applications. A schematic of a typical parallel-plate ionization chamber is shown in Fig. 12.13, according to Price [10]. In operation, the

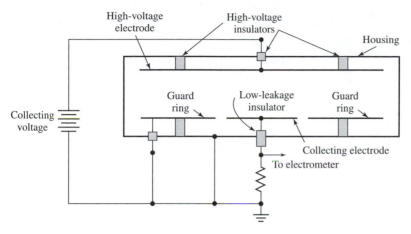

Figure 12.13   Parallel-plate ionization chamber, according to Ref. [10].

chamber is charged with a voltage that is high enough to ensure that electrons produced by the ionizing radiation will be collected on the anode. The voltage is not so high that it triggers excitation of other molecules by electrons produced in the ionization process. This is in contrast to the operation of the G-M counter, which has a sufficiently high voltage that once ionization is effected by the incoming radiation, subsequent ionization of other molecules also takes place in an avalanche fashion, which produces the voltage pulse used for counting. To measure the energy level of the incoming radiation, a measurement is made of the output current of the device or a determination is made of the charge released in the chamber over a period of time. The ionization chamber may also be used as a pulse-type device for the measurement of the number and energies of high-energy alpha particles. In this application a measurement of the pulse height (strength) is made in order to determine the energy levels of the alpha particles.

12.10   PHOTOGRAPHIC DETECTION METHODS

When certain types of photographic films are exposed to nuclear radiation and subsequently developed, the opacity of the print may be taken as an indication of the total amount of radiation incident on the film during the time of exposure. Many specialized films are available for nuclear-radiation measurement, and this method is used for measuring the total radiation exposure for workers in atomic-energy installations. The photographic film badge may be used for detecting α and β particles, γ rays, and neutrons. In order to use a single badge for measurement of the exposure rates to the different types of radiation, several openings, or windows, may be constructed in a single badge. A different type of filter is placed over each window so that only one type of radiation is permitted to strike the photographic emulsion under each window. Thus, the opacity of the developed film under each of the windows gives an indication of the different radiation-exposure rates.

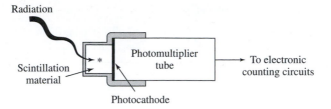

Figure 12.14   Schematic of a scintillation counter.

12.11   THE SCINTILLATION COUNTER

The scintillation counter is the most versatile of the conventional devices for the measurement of nuclear radiation. A schematic of the device is shown in Fig. 12.14. The ionizing radiation strikes the scintillation crystal, which is enclosed in the thin aluminum housing, and produces a flash of light. The light flashes are detected by the photomultiplier tube, the output of which is fed to appropriate electronic circuits. If the number of pulses is to be counted, the output circuit will include appropriate electronic counters. The scintillation counter may also be used for measuring the energy of the incoming radiation since, for certain crystals, the intensity of the flash of light is proportional to the energy of the radiation.

A number of scintillation materials are used, including both solids and liquids. The selection of the proper type of material for detection of specific radiations is discussed by Price [10] and Birks [3].

12.12   NEUTRON DETECTION

The three detection methods described above depend on an ionization process caused by an interaction of the α, β, or γ radiation with a gas or scintillation material. These detectors may not be used for neutrons because neutrons do not produce an ionizing effect. For this reason the measurement of neutron flux is usually an indirect process which involves the utilization of an intermediate reaction to produce some type of ionizing radiation (α, β, or γ rays), which may then be used to indicate the incoming flux. A variety of neutron-measurement devices are available, and the type which is used in practice depends strongly on the energy level of the neutrons to be measured and the total neutron flux. We shall discuss the principles of operation of only one type of detector which is applicable for the measurement of thermal neutron fluxes ($E \sim 0.025$ eV, 1 electron volt $= 1.6 \times 10^{-19}$ J).

A typical reaction which is used for neutron detection is the interaction of neutrons with B^{10} to produce Li^7 and an alpha particle. The alpha particles may subsequently produce ionizations in a gas and corresponding voltage pulses. The reaction is utilized in two ways: (1) The boron is present in the form of borontrifluoride (BF$_3$) gas and

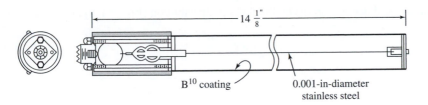

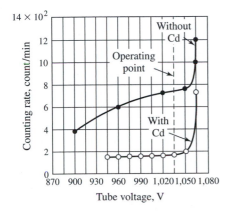

Figure 12.15   Diagram of a B^{10}-lined neutron counter. Filling gas is helium plus 5 percent ether at 10-cmHg pressure. (*From Ref. [12].*)

placed in a chamber very similar to the G-M counter shown in Fig. 12.10. The potential difference between the central wire and the cylindrical shell is about 1300 V. The wire diameter is about 0.002 in, while the cylinder diameter is about 1 in, with a length of about 6 in. (2) An alternative method of utilizing the B^{10} (n, α) reaction is shown in Fig. 12.15. The inside surface of the cylinder is coated with B^{10}, while the volume is filled with helium and 5 percent ether at 10 cmHg pressure. The sensitivity of a typical BF_3 detector is about 1.0 count/s per unit neutron flux. The sensitivity of the boron-coated chamber is about 5 to 10 counts/s per unit neutron flux. Neutron flux is measured in neutrons per square centimeter per second.

   If sufficiently high neutron fluxes are encountered, the detector may be operated as a current-sensitive device. The current output is then proportional to the incoming flux. Currents of about 0.1 mA may be obtained for neutron fluxes of 10^{10} neutrons/cm^2 · s.

12.13   STATISTICS OF COUNTING

We have already seen that many nuclear-radiation measurements rely on a counting operation. In this section we wish to consider the statistics associated with such operations and indicate the possible errors involved.

   In a given sample of radioactive material there are a large number of atoms, and the probability that each of these atoms will decay is quite small. The probability that

n atoms will decay is given by the Poisson distribution [Eq. (3.10)] as

$$p(n) = \frac{n^n e^{-n}}{n!}$$ [12.3]

where $Np = n$

N = total number of atoms

p = probability that each atom will decay

Equation (12.3) is generally valid for $N > 100$ and $p < 0.01$, and the standard deviation of the distribution may be shown to be[2]

$$\sigma = \sqrt{n}$$ [12.4]

The quantity n represents the average number of radioactive decays recorded in a series of observations. The radioactivity of a certain sample of material might be studied by observing the number of particles or gamma rays given off in specified increments of time. The counting operation could be performed with a G-M counter or other device. In general, the number of counts in a fixed interval of time, say, 1 min, might vary considerably. Several counts would be taken, and the arithmetic average of the number of counts in each time interval is given by $\bar{n}$. The standard deviation of the measurement would be given by Eq. (12.4). In the above and following equations n is taken to represent the average number of counts.

The parameter which is of interest in an experimental determination is the *counting rate,* defined by

$$r = \frac{n}{t}$$ [12.5]

where t is the elapsed time for the average number of counts n. The uncertainty of the determination of the counting rate depends on the number of counts and the time t.

When a counting measurement is made to determine the counting rate, the detector senses not only the signal from the radioactive source, but also some background signal which is the result of spurious radioactivity of the environment. A separate measurement must be made of the background counting rate and this information used to determine the counting rate resulting from the radioactive source. The counting rate and its associated uncertainty are written as

$$r = r \pm w_r$$ [12.6]

The counting rate for the source is

$$t_s = r_T - r_b$$ [12.7]

where r_T is the total counting rate and r_b is the counting rate due to the background.

| [2]See Ref. [10], p. 56.

The uncertainty in the source counting rate is obtained from Eq. (3.26) as

$$w_s = \left(w_T^2 + w_b^2\right)^{1/2} \qquad \textbf{[12.8]}$$

where w_T and w_b are the uncertainties in the total and background counting rates, respectively. The uncertainties for these counting rates may be estimated though the use of the standard deviation for the Poisson distribution given by Eq. (12.4) if the nuclear radiation satisfies the conditions for this distribution. The accuracy of the counting-rate measurement is clearly dependent on the length of time available for the measurement. A longer time affords the opportunity for observation of a large number of counts and hence the establishment of a better average counting rate. Example 12.3 illustrates the use of the above relations for determining counting rates.

OPTIMUM COUNTING TIMES. A certain radioactive sample is observed, and the total and background counting rates are approximately 800 and 35 counts/min, respectively. If a total time of 45 min is available for both counting measurements, estimate the apportionment of this total time such that there is minimum uncertainty in the source counting rate. | **Example 12.3**

Solution

We may assume that both the source and background counts follow the Poisson distribution so that the uncertainty is proportional to the standard deviation; that is,

$$\sigma_s = n_s^{1/2}$$

$$\sigma_b = n_b^{1/2}$$

$$\sigma_T = n_T^{1/2}$$

Thus, $\qquad w_s = \dfrac{n_s^{1/2}}{t_s} \qquad w_b = \dfrac{n_b^{1/2}}{t_b} \qquad w_T = \dfrac{n_T^{1/2}}{t_T} \qquad \textbf{[a]}$

where t_s, t_b, and t_T are the counting times for the source, background, and total, respectively. We have also

$$n_s = r_s t_s \qquad n_b = r_b t_b \qquad n_T = r_T t_T \qquad \textbf{[b]}$$

so that $\qquad w_s^2 = \dfrac{r_s}{t_s} = \dfrac{r_b}{t_b} + \dfrac{r_T}{t_T} \qquad \textbf{[c]}$

in accordance with Eq. (12.8). We wish to optimize w_s by proportioning the times t_b and t_T in accordance with

$$t_b + t_T = \text{const} = 45 \text{ min} \qquad \textbf{[d]}$$

Thus, we optimize w_s by setting $dw_s/dt_b = 0$ and obtain

$$2w_s \frac{dw_s}{dt_b} = 0 = -\frac{r_b}{t_b^2} - \frac{r_T}{t_T^2}\frac{dt_T}{dt_b} \qquad \textbf{[e]}$$

From Eq. (d)

$$\frac{dt_T}{dt_b} = -1$$

so that the optimum condition is

$$\frac{t_b}{t_T} = \left(\frac{r_b}{r_T}\right)^{1/2}$$

[f]

For the given counting rates

$$t_b + t_T = 45$$

$$\frac{t_b}{t_T} = \left(\frac{35}{800}\right)^{1/2} = 0.2091$$

and $t_T = 37.2$ min; $t_b = 7.8$ min. In the above calculations it has been assumed that the time interval is accurately determined and that the uncertainty for this determination is zero.

Example 12.4 | **DETERMINATION OF OPTIMUM TIME.**   For the estimated counting rates given in Example 12.3, calculate the optimum total time necessary to give a standard deviation of the source counting rate of 1 percent. Repeat for total and background counting rates of 400 and 100, respectively.

Solution

The nominal value of the source counting rate is

$$r_s = r_T - r_b = 800 - 35 = 765 \text{ counts/min}$$

We have

$$\sigma_s^2 = \frac{r_b}{t_b} + \frac{r_T}{t_T} = [(0.01)(765)]^2 = 58.5$$

Also, from relation (f) of Example 12.3

$$t_b = t_T \left(\frac{r_b}{r_T}\right)^{1/2}$$

so that

$$t_T = \frac{1}{\sigma_s^2}\left(r_b^{1/2} r_T^{1/2} + r_T\right)$$

$$= \frac{1}{58.5}[(35)^{1/2}(800)^{1/2} + 800] = 16.5 \text{ min}$$

The total counting time is $t_b + t_T$, or

$$t_{\text{tot}} = (1.209)(16.5) = 20 \text{ min}$$

For total and background counting rates of 400 and 100

$$r_s = 400 - 100 = 300 \text{ counts/min}$$

$$\sigma_s^2 = [(0.01)(300)]^2 = 9$$

$$t_T = \tfrac{1}{9}[(100)^{1/2}(400)^{1/2} + 400] = 66.7 \text{ min}$$

The total counting time is

$$t_{\text{tot}} = 66.7\left[\left(\frac{100}{400}\right)^{1/2} + 1\right] = 100 \text{ min}$$

DETERMINATION OF SOURCE UNCERTAINTY.   A radioactive sample is observed | **Example 12.5**
over a period of 30 min, and a total of 21,552 counts is recorded. The sample is then removed, and
the background radiation is observed for another 30 min. The background produces 1850 counts.
Calculate the source counting rate and its uncertainty.

Solution

We have

$$r_T = \frac{21,552}{30} = 715.1 \text{ counts/min}$$

$$r_b = \frac{1850}{30} = 61.7 \text{ counts/min}$$

so that

$$r_s = 715 - 62 = 653 \text{ counts/min}$$

The standard deviation of the source counting rate is calculated from Eq. (*c*) of Example 12.3.

$$w_s = \sigma_s = \left(\frac{r_b}{t_b} + \frac{r_T}{t_T} \right)^{1/2} = 5.08 \text{ counts/min}$$

12.14   RADIATION EFFECTS IN HUMANS

Nuclear radiation interacts with human tissue by producing ionization, and thereby
leaving in its wake deteriorating material. Both the type of radiation and its energy
level influence the amount of ionizing energy that will be absorbed in human tissue.
The irradiation, or amount of radiation incident per unit mass of material, is designated
with the unit sievert (Sv) in SI units of J/kg. The actual energy absorbed per unit mass
is measured in the SI unit of gray (Gy), also designated in J/kg. The relationship
between the irradiated energy in sievert and the actual energy absorbed in gray is
given by a multiplying factor W_R called the radiation-weighting factor, which depends
on the type of radiation and its energy level. Table 12.2 gives an abbreviated list of
approximate values for W_R. The relationship is:

Absorption in gray (Gy) $= W_R \times$ Irradiation in sievert (Sv)

EFFECTS OF RADIATION ON HUMANS

As would be expected the effect of radiation dose on humans depends on the dose level
and the duration of exposure. Table 12.3 lists some typical doses and their effects.

MEASUREMENT UNITS FOR RADIATION SOURCES

The sievert and gray are units of measure for irradiation and absorption of radiation.
For radiation sources that decay with time, the SI unit employed for emission from

Table 12.2   Radiation weighting factors ($1.0 \, J = 6.241 \times 10^{18}$ eV)

Type of Radiation	Energy Level, MeV	W_R
X-rays and γ-rays	Variable	1.0
β particles	Variable	1.0
Protons	Variable	2.0
α particles and other atomic nuclei	Variable	20.0
Neutrons	<0.01	5.0
	0.01 to 0.1	10
	0.1 to 2.0	20
	2.0 to 20	10
	>20	5

Table 12.3   Radiation dosage effects on humans

Exposure	Effect
Whole-body exposure > 5 Gy	Usually results in death
Local exposure of hair-bearing skin > 1 Gy	Loss of hair
Local treatment of lymphoma, 20 to 40 Gy	
Abdominal X-ray, 1.5 Gy	
1 Sv sudden exposure	5 percent risk of cancer
0.1 Sv sudden exposure	0.5 percent risk of cancer

Long-term exposure limits:

Threshold limit value annual dose for radiation workers in 1 year = 0.05 Sv
Threshold limit value, annual dose, 5-year average = 0.02 Sv
General public recommended annual dose = 1.0 mSv

the radioactive source is the becquerel (Bq), defined as

$$1 \, Bq = 1 \text{ radiation emission per second}$$

A non-SI unit that is still employed is the curie (Ci), which is related to the becquerel through

$$1 \, Ci = 37 \, GBq$$

Note that the becquerel and curie are measures of the rate of emissions from a source, and not the energy. The energy emitted would be expressed by

$$\text{Energy emitted per second} = \text{energy per unit} \times \text{emission rate in Bq}$$

OTHER UNITS

The sievert and gray are the accepted SI units for radiation exposure and their designation follows the standard prefixes given in Table 2.7. Regretfully, a number of holdover units are still in use, and it worthwhile to familiarize the reader with their definitions and conversions.

The roentgen (R) is the amount of γ-radiation to produce ionization of

$$1 \text{ roentgen (R)} = 258 \text{ micro coulomb/kg } (\mu\text{C/kg})$$

This amount of radiation will produce approximately 1 radiation absorbed dose (rad). A related unit is the roentgen equivalent man (rem) that is akin to the gray in that it measures the equivalent effect on humans. The rad and rem are non-SI units and their usage is generally discouraged. Should they be encountered, however, the appropriate conversion factors are

$$1 \text{ rad} = 0.01 \text{ Gy}$$
$$1 \text{ rem} = 0.01 \text{ Sv}$$

12.15   REVIEW QUESTIONS

12.1. How is thermal radiation distinguished from other EM radiation?

12.2. How is thermal radiation detected?

12.3. Why is it sometimes necessary to chop thermal radiation before it strikes the detector?

12.4. What is an integrating sphere? How is it used for reflectivity and transmissivity measurements?

12.5. What are the four types of nuclear radiation that are normally measured?

12.6. Describe the principle of operation of the Geiger counter.

12.7. How does an ionization chamber work?

12.8. How does a scintillation counter operate?

12.9. What types of instruments are used for detection of (*a*) alpha particles, (*b*) beta particles, (*c*) gamma rays, and (*d*) neutrons?

12.10. Describe the techniques for measuring neutron irradiation.

12.11. Why is a Poisson distribution an adequate representation of nuclear counting statistics?

12.16   PROBLEMS

12.1. Consider the thermistor detector and radiation source of Example 12.1. Instead of the bare detector, an optical system collects the radiation over a circular area of radius r and focuses it on the detector, which is maintained at 20°C. Plot the source temperature necessary to maintain a signal-to-noise (S/N) ratio of 200 as a function of the radius of the collecting optical system. Assume that the distance between the source and detector is 1.0 m, as in Example 12.1. Neglect losses in the optical system for this calculation.

12.2. Repeat Prob. 12.1 for distances of 6 and 12 ft between the source and detector.

12.3. A thermocouple detector has characteristics such that the minimum input power necessary to produce an S/N ratio of unity is 3×10^{-6} W. A 2-in-diameter optical system collects the input thermal radiation and focuses it on the thermocouple. The optical-thermocouple system is to be used to measure the temperature of a 2-in black sphere 6 ft away. The thermocouple bead is maintained at a temperature of 70°F. Estimate and plot the uncertainty in the temperature measurement as a function of the temperature of the sphere.

12.4. The apparatus of Example 12.2 is used for an emissivity determination between the temperature limits of 150 and 540°C. Plot the percent uncertainty in emissivity as a function of T_s for nominal emissivities of 0.2 and 0.8. Take the uncertainties in radiometer temperature and output voltage as

$$w_{T_R} = \pm 0.3°C$$
$$w_R = 0.004 \text{ mV}$$

The uncertainty in the source temperature is taken as ±0.5°C.

12.5. A radioactive sample is observed, and the total and background counting rates are 500 and 100 counts/min, respectively. Estimate the total time necessary to produce a standard deviation of 2 percent in the nominal source counting rate. Repeat for counting rates of 100 and 22.

12.6. The following counting data are collected on a radioactive sample. Each of the number of counts is taken in an exact time interval of 1 min. The background counting rate is known accurately as 23 counts/min. Calculate the nominal value of the source counting rate and the uncertainty in the rate. Apply Chauvenet's criterion to eliminate some of the data if necessary. Assume that the Poisson distribution applies.

Interval Number	No. of Counts
1	523
2	410
3	342
4	595
5	490
6	611
7	547
8	512

12.7. After a set of radioactive counting measurements has been made, it is found that the timing device has been spurious in operation. A set of calibration measurements on the timer has given the standard deviation of the time-interval measurements as σ_t. Using the approach taken in Example 12.3, derive an expression for the standard deviation of the source counting rate in terms of the standard deviations of the number of background and total counts and the standard deviation of the time interval.

12.8. Derive an expression for the optimum time t_T under the conditions that $r_b = r_s$. Express in terms of σ_s and r_s.

12.9. Plot the optimum counting time from Prob. 12.8 as

$$\frac{(t_T)_{\text{opt}}\sigma_s^2}{r_s} \qquad \text{versus} \qquad \frac{r_b}{r_s}$$

12.10. Under the condition that the total number of counts for both the background and sample measurements is limited, i.e.,

$$n_b + n_T = K = \text{const}$$

show that the minimum error condition is

$$\frac{n_b}{n_T} = \frac{r_b}{r_T}$$

and the total counting time for the background and sample measurements is

$$t_b + t_T = \frac{2K}{2r_b + r_s}$$

Plot $\dfrac{(t_b + t_r)r_s}{K}$ versus $\dfrac{r_b}{r_s}$

12.11. The emissivity-parameter procedure described in the paragraphs following Eq. (12.2) is used to determine the emissivity of a surface at 300°C. The surrounding radiation temperature is only 25°C and thus is expected to have little influence on the measurement. With $\epsilon = 1.0$, the apparent temperature is indicated as 231°C. The uncertainty in the radiometer indication of temperature is ±3°C and the uncertainty in the surface-temperature measurement is ±1°C. What surface emissivity is indicated by these measurements? What is its uncertainty?

12.12. Assuming the internal emissivity compensation of the radiometer in Prob. 12.11 is exact, what temperature will be indicated when the instrument is set for the nominal value obtained in Prob. 12.11?

12.13. Repeat Prob. 12.11 for an indicated temperature of 151°C when ϵ is set equal to 1.0.

12.14. Assume that a surface follows gray body behavior as described in Sec. 8.6. The peak in the radiant emissive power curve of E_λ vs. λ will follow Wien's law described by Eq. (8.24). A device is proposed that will scan the radiant emission from a surface and determine the maximum point and therefore enable a determination of the temperature and emissivity of the surface. The calculation is made through the equations

$$E_\lambda = \epsilon E_{b\lambda}$$
$$\lambda_{\max} T = 2897.6 \ \mu\text{m} \cdot \text{K}$$

$E_{b\lambda}$ is expressed in the Planck blackbody formula of Eq. (8.18). The two measurement outputs of the device are $\lambda_{\max}$ and the corresponding value of $E_\lambda(\max)$. Assuming each parameter is measured with a percent uncertainty of ± 4 percent, determine the percent uncertainty in the calculated values of ϵ and T at temperature levels of 400, 600, and 1000 K.

12.15. What percent uncertainties in measurements of $\lambda_{\max}$ and E_λ (max) are needed with the device of Prob. 12.14 to obtain uncertainties of 5 percent in the calculated values of T and ϵ at temperature levels of 400, 600, and 1000 K?

12.16. An alternative to the procedure of Prob. 12.14 is proposed. The device will scan over all wavelengths and, with an internal integration circuit, perform the calculation

$$E = \int E_\lambda \, d\lambda$$

over all wavelengths scanned. In addition, the maximum point $\lambda_{\max}$ will be measured. Then,

$$E = \epsilon E_b = \epsilon \sigma T^4$$

and

$$\lambda_{\max} T = 2897.6 \ \mu\text{m} \cdot \text{K}$$

may be solved for ϵ and T.

Comment on this alternative measurement proposal.

12.17. A sample of radioactive material is observed to emit radiation at the rate of 600 counts/min while the background is observed at 100 counts/min. Estimate the total counting time necessary to ensure a standard deviation of 3 percent in the source counting rate.

12.18. Repeat Prob. 12.11 for an indicated temperature of 200°C when ε is set equal to 1.0.

12.19. Assuming the internal emissivity compensation for the radiometer in Prob. 12.18 is exact, what temperature will be indicated when the instrument is set for the nominal value obtained in Prob. 12.18?

12.17 REFERENCES

1. Ballard, S., and W. Wolfe: "Optical Materials in Equipment Design, A Critique, ONR," *Proc. Infrared Inform, Symposia,* vol. 4, no. 1, p. 185, March 1959.

2. Billings, B. H., E. E. Barr, and W. L. Hyde: "An Investigation of the Properties of Evaporated Metal Bolometers," *J. Opt. Soc. Am.,* vol. 37, p. 123, 1947.

3. Birks, J. B.: *Scintillation Counters,* McGraw-Hill, New York, 1953.

4. Fusion, N.: "The Infrared Sensitivity of Superconducting Bolometers," *J. Opt. Soc. Am.,* vol. 38, p. 845, 1948.

5. Gier, J. T., R. V. Dunkle, and J. T. Bevans: "Measurement of Absolute Spectral Reflectivity from 1.0 to 15 Microns, *J. Opt. Soc. Am.,* vol. 44, p. 558, 1954.

6. Glaser, P. E., and H. H. Blau: "A New Technique for Measuring the Spectral Emissivity of Solids at High Temperatures," *Trans. ASME,* vol. 81C, p. 92, 1959.

7. Hackforth, H. L.: *Infrared Radiation,* McGraw-Hill, New York, 1960.

8. Kaplan, I.: *Nuclear Physics,* 2d ed., Addison-Wesley, Reading, MA, 1963.

9. Marcus, N.: "A Blackbody Standard," *Instr. Autom.,* vol. 28, March 1955.

10. Price, W. J.: *Nuclear Radiation Detection,* 2d ed., McGraw-Hill, New York, 1964.

11. Snyder, N. W., J. T. Gier, and R. V. Dunkle: "Total Normal Emissivity Measurements on Aircraft Materials between 100 and 800°F," *Trans. ASME,* vol. 77, p. 1011, 1955.

12. *Reactor Handbook,* vol. 2, p. 951, U.S.A.E.C. Document, McGraw-Hill, New York, 1955.

13. Edwards, D. K., J. T. Gier, K. E. Nelson, and R. D. Roddick: "Integrating Sphere for Imperfectly Diffuse Samples," *J. Opt. Soc. Am.,* vol. 51, no. 11, pp. 1279–1288, November 1961.

14. Cox, R. L.: "Emittance of Translucent Materials: A Qualitative Evaluation of Governing Fundamental Properties and Preliminary Design of an Experimental Apparatus to Measure These Properties to 6000°F," M.S. thesis, Southern Methodist University, Dallas, TX, 1961.

15. Birkebak, R. C., and S. H. Cho: "Integrating Sphere Reflectometer Center-Mounted Blockage Effects," ASME Paper 67-HT-54, August 1967.

16. *Precision Measurement and Calibration: Optics, Metrology, and Radiation,* Natl. Bur. Std., Handbook 77, vol. 3, 1961.

17. Richmond, J. C.: "Measurement of Thermal Radiation of Solids," NASA Sp. 31, 1963.

18. Katzoff, S.: "Symposium in Thermal Radiation of Solids." NASA Sp. 55, 1965.

19. Smith, R. A., F. E. Jones, and R. P. Chasmar: *The Detection and Measurement of Infra-Red Radiation,* Oxford University Press, New York, 1957.

20. Richmond, J. C., S. T. Dunn, D. P. Dewitt, and W. D. Hayes: "Procedures for the Precise Determination of Thermal Radiation Properties," *Air Force Materials Lab., Tech. Doc. Rept.* ML-TDR-64-257, pt. 2, April 1965.

21. Gubareff, G. G., J. E. Jansen, and R. H. Torborg: *Thermal Radiation Properties Survey,* Honeywell Research Center, Honeywell, Inc., Minneapolis, MN, 1960.

22. Duffie, J. A., and W. A. Beckman: *Solar Energy Thermal Processes,* 2d ed., John Wiley & Sons, New York, 1991.

23. Yellott, J. I.: "Low Temperature Application of Solar Energy," in *The Measurement of Solar Radiation,* ASHRAE, New York, 1967.

24. Touloukian, Y. S., et al.: *Thermophysical Properties of Matter,* vol. 7, *Thermal Radiative Properties—Metallic Elements and Alloys* (1970), vol. 8, *Thermal*

Radiative Properties—Non-Metallic Solids (1972), and vol. 9: *Thermal Radiative Properties—Coatings* (1972), Plenum Data Corporation.

25. Siegel, R., and J. R. Howell: *Thermal Radiation Heat Transfer,* 3d ed., McGraw-Hill, New York, 1992.

26. Abbot, C. G., F. E. Fowle, and L. B. Aldrich: "Improvements and Tests of Solar Constant Methods and Apparatus," *Ann. Astrophys. Observ.*, vol. 3, p. 39, 1913.

27. Aldrich, L. B., and W. B. Hoover: *Pyrheliometry, Ann. Anstrophys. Observ.*, vol. 7, p. 99, 1954.

28. Abbot, C. G.: *The Silver Disk Pyrheliometer, Smithsonian Misc. Collections,* vol. 56, no. 19, p. 1, 1911.

29. Morikofer, W.: "On the Principles of Solar Radiation Measuring Instruments," *Trans. Conf. on Use of Solar Energy,* vol. 1, p. 60, University of Arizona Press, Tucson, 1958.

30. Nutter, G. D.: "Recent Advances and Trends in Radiation Thermometry." ASME paper 71 WA/Temp-3, November 1971.

31. Androulakis, J. G.: "The Universal High Temperature Emissiometer," ASME paper 72-HT-1, August 1972.

32. Threlkeld, J. L., and R. C. Jordan: "Direct Solar Radiation Available on Clear Days," *ASHAE Trans.* vol. 64, p. 45, 1958.

33. *Omega Temperature Measurement Handbook,* Omega Engineering Company, Stamford, CT, 1999.

34. Modest, M.: *Radiative Heat Transfer,* McGraw-Hill, New York, 1993.

35. Levi, Hans W.: *Nuclear Chemical Engineering,* 2d ed., McGraw-Hill, New York, 1981.

36. Hamilton, J., and J. H. F. Yang: *Modern Atomic and Nuclear Physics,* 2d ed., McGraw-Hill, New York, 1996.

37. Benedict, M., T. H. Pigford, and H. W. Levi: *Nuclear Chemical Engineering,* 2d ed., McGraw-Hill, New York, 1981.

13

AIR-POLLUTION SAMPLING AND MEASUREMENT

13.1 INTRODUCTION

Most citizens are acutely aware of the need for air-pollution control because of continual information from government sources and the press. The lay public, however, is not aware that accurate measurements are needed to bring about effective air-pollution control. Measurements are important, first, to establish acceptable levels of contamination of the atmosphere taking into account meaningful biological data in both human and animal subjects. Second, measurements must be performed in different locales to determine the offending sources and degree of control required. On the basis of such measurements control standards can be formulated and an enforcement mechanism set in motion. Finally, periodic sampling and measurement are required for use in proper legal proceedings against violators.

Our presentation of air-pollution measurement techniques certainly cannot be complete because this is an extensive and complicated field involving many disciplines in engineering and science. We shall, however, be able to discuss general sampling techniques for both particulate and gaseous contaminants. We shall also discuss air-pollution standards in a general way, keeping in mind that specific standards will change from time to time. A number of books discuss air-pollution control techniques in detail, for example, Refs. [1] to [6].

13.2 UNITS FOR POLLUTION MEASUREMENT

Quantities of pollutants may be expressed on either a volumetric or a mass basis. For the mass basis appropriate units would be grams per cubic meter or pounds-mass per cubic foot. The volumetric unit normally employed is parts per million (ppm), which is defined as

$$1 \text{ ppm} = \frac{1 \text{ volume of gaseous pollutant}}{10^6 \text{ volumes (air + pollutant)}} \qquad \textbf{[13.1]}$$

Table 13.1   Mass concentrations for 1 ppm of common pollutants at 1 atm

Pollutant	Mass Concentration ($\mu g/m^3$)	
	0°C	25°C
Carbon monoxide (CO)	1250	1145
Nitric oxide (NO)		1230
Nitrogen dioxide (NO_2)		1880
Ozone (O_3)	2141	1962
PAN [$CH_3(CO)O_2NO_2$]	5398	4945
Sulfur dioxide (SO_2)	2860	2620

or 0.0001 percent by volume $= 1$ ppm.

To convert the volumetric unit to a mass basis, one must obviously know the molecular weight of the pollutant in order to calculate its volume at a given temperature and pressure. Assuming ideal-gas behavior, the conversion can be made with

$$\frac{m_p}{V} = (\text{ppm})\frac{M_p p}{\Re T} \times 10^{-6}$$ **[13.2]**

where $m_p/V =$ mass concentration of pollutant

$M_p =$ molecular weight of pollutant

$p =$ total pressure of the air and pollutant mixture

$\Re =$ universal gas constant

$T =$ absolute temperature of the mixture

The units for m_p/V will be kilograms per cubic meter in the SI system. A more common unit is micrograms per cubic meter, in which case we can note

$$1 \text{ kg/m}^3 = 10^9 \, \mu g/m^3$$

Equivalent volumetric and mass concentrations for some common pollutants are given in Table 13.1.

13.3   AIR-POLLUTION STANDARDS

Local, state, and federal agencies may set air-pollution standards or allowable levels on two bases:

1.   Pollutant concentration from the source of pollution

2.   Ambient pollution concentration from all sources

Thus, the particulate discharge from a smokestack can be controlled by specifying either an allowable discharge concentration level at each stack or a maximum allowable concentration level in the particular locale. One can argue for both types

Table 13.2 Ambient-air quality standards

Pollutant	Averaging Times	California Standard		Federal Standard		
		Concentration, $\mu g/m^3$	Method of Measurement	Primary, $\mu g/m^3$	Secondary, $\mu g/m^3$	Method of Measurement
Particulates	Annual geometric mean	60	High-volume sampler	75	60	High-volume sampler
		100		260	150	
	24 h					
Carbon monoxide	1 h	46,000 (40 ppm)	Nondispersive infrared spectroscopy	46,000 (40 ppm)	Same	Nondispersive spectroscopy
	8 h	—		10,000 (9 ppm)		
Sulfur dioxide	Annual average	—	Conductimetric	80 (0.03 ppm)	60 (0.02 ppm)	Pararosaniline or flame photometric
	24 h	105 (0.04 ppm)		365 (0.14 ppm)	260 (0.1 ppm)	
	3 h	—		—	1300 (0.5 ppm)	
	1 h	1310 (0.5 ppm)		—	—	
Hydrocarbons (nonmethane)	3 h (6–9 A.M.)	—	—	160 (0.24 ppm)	Same	Flame ionization
Photochemical oxidants	1 h	200 (0.1 ppm)	Neutral buffered KI	160 (0.08 ppm)	Same	Buffered KI or chemiluminescence
Nitrogen dioxide	Annual arithmetic mean	—	Saltzman method	100 (0.05 ppm)	Same	Saltzman or colorimetric method with NaOH
	1 h	470 (0.25 ppm)				

of standards. Measuring the concentration at the source clearly marks the emission of a prospective offender and thus provides a ready enforcement mechanism. On the other hand, it is the level of pollution in the ambient air which eventually influences public health and thus may be cited as the most meaningful type of standard.

The ambient standard is not as easy to enforce as a source standard. In an industrial area with many smokestacks, how shall the measurement engineer say *which* stack(s) are causing ambient pollution levels that are too high? In Los Angeles it would be impossible to say which automobiles were causing the most ambient pollution without performing measurements on each.

Arguments for both types of standards are discussed by Stern [4], and the interested reader may consult this reference for more detailed information. Our purpose in this section is to emphasize the need for both source and ambient measurements so that later sections on specific measurement techniques can be considered in their proper perspective.

It should be clear to the reader that the local environment should be taken into account in formulating allowable air-pollution emission levels. Thus, the stringent auto emission standard which may be required for Los Angeles may be totally unnecessary in the wide-open space of Lubbock, Texas. As another example, the ambient

Table 13.3   Typical source emission standards (Federal Register, December 23, 1971)

Incinerators, particulates	0.08 grains/std. ft^3 (1 lbm = 7000 grains)
Cement plants	
Particulates	
Kilns	0.3 lb/ton feed
Clinker cookers	0.1 lb/ton feed
Visible emissions	
Kilns	10% opacity
Others	Less than 10% opacity
Fossil fuel-fired steam	
generators greater than	
250×10^6 Btu/h heat input	
Particulates	0.1 lb/10^6 Btu (max. 2 h avg.)
SO$_2$	
Coal-fired	1.2 lb/10^6 Btu (max. 2 h avg.)
Oil-fired	0.8 lb/10^6 Btu
Nitrogen oxides	
Coal-fired	0.7 lb/10^6 Btu (max. 2 h avg. NO$_2$)
Oil-fired	0.3 lb/10^6 Btu
Gas-fired	0.2 lb/10^6 Btu
Visible emissions	Less than 20% opacity, except that 2 min in any 1-h emission can be as great as 40%
Sulfuric acid plants	
SO$_2$	4 lb/ton acid
Acid mist	0.15 lb/ton acid
Visible emissions	Less than 10% opacity

standards to be used around coal-burning facilities on the east coast may be unrealistic in the southwest where natural gas and oil are commonly used as fuels.

Imposition of strict air-pollution standards may cause severe economic dislocations on a local, state, or national level. Local unemployment may result if a business cannot afford to install pollution-control equipment and therefore must close a plant. Energy requirements and availability at all levels may be influenced by the air-pollution-control measures required. These political and economic factors are beyond our consideration, but we mention them because they are important in setting standards, some samples of which we give in Tables 13.2 and 13.3. Table 13.2 gives a summary of ambient standards for California and the federal basis. The California standards are generally more restrictive because that state has more severe air-pollution problems than the nation as a whole. Table 13.3 lists the federal source standards for some typical sources. The methods of measurement mentioned in these tables will be discussed in later sections.

We should caution the reader that air-pollution standards are subject to revision *up* or *down* as better legal and technical experience is gained with enforcement and control measures.

13.4   GENERAL AIR-SAMPLING TRAIN

Figure 13.1 illustrates the general air-sampling train. Not all elements of the train must be employed in each application, but we include them for generality. The sampling train has four basic sections:

1. Sample-conditioning devices
2. Sample-collection equipment
3. Metering devices
4. Vacuum source

Detailed discussion of different elements of these sections is the basis for this entire chapter, but we shall indicate the general nature of each in the following paragraphs.

Sample-conditioning devices alter the incoming air sample so that it is easier to collect the specific gaseous or particulate component desired. The conditioning can consist of

Filtration

Drying to remove moisture

Saturation with water vapor to provide a later reference level

A variety of chemical reactions to remove unwanted pollutants or gases which may interfere with later measurements.

In the sample-collection section the gaseous or particulate pollutant is removed from the airstream by one or more of a variety of methods such as the following:

GAS COLLECTION

Adsorption onto solid surface

Absorption with chemical reaction into a liquid

Condensation in a freeze-out trap

PARTICULATE COLLECTION

Filtration in single or multiple stages

Removal by wet or dry impingers

Removal by sedimentation

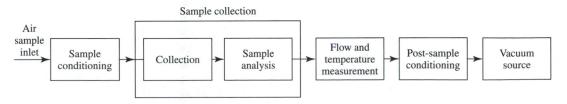

Figure 13.1   General air-sampling train.

Once the sample is collected, it is measured quantitatively by appropriate methods. After sample collection, the flow rate, temperature, and pressure of the airflow stream are measured in accordance with methods we have described in previous chapters. The vacuum source provides the means to pull the air sample through the train. To protect the vacuum source, filtration or moisture removal may be necessary after the flow-measurement section. We should remark that all connecting tubing must be nonreactive with the pollutant(s) to be measured.

13.5   GAS SAMPLING TECHNIQUES

We now consider specific methods for collecting gaseous samples from the overall air sample. We shall also describe some of the analysis techniques employed with the sample.

ABSORPTION TECHNIQUES

In absorption techniques the air sample is bubbled through a liquid so that the pollutant is absorbed by chemical reaction. The absorption column can be an impinger, an atomizing scrubber, a fritted glass scrubber, or some type of packed column. The absorption rate is influenced by the rate of gas flow through the absorber, concentration of the absorbing solution, and overall contact time. After absorption the liquid reactant is analyzed to determine the concentration of the component proportional to the original gaseous pollutant.

ADSORPTION TECHNIQUES

In the adsorption process the gaseous molecules to be analyzed are placed in contact with a solid surface, to which they adhere. Typical solids employed for the process are activated carbon, activated alumina (AlO_2), silica gel (SiO_2), and molecular sieves. The solid can then be washed with a reagent to absorb the pollutant, and an analysis can be performed. Obviously, the adsorption process requires periodic regeneration of the solid.

CONDENSATION AND FREEZE-OUT TECHNIQUES

If the air sample is passed through cooling chambers which are maintained at very low temperatures, it is possible to cause various gaseous components to condense or freeze out. In such an operation it may be necessary to employ multiple freeze-out sections in series to remove water vapor or other constituents before extracting the desired pollutant. Table 13.4 gives a summary of typical low-temperature solutions.
We shall examine sulfur dioxide and combustion products measurements in more detail later. For now Table 13.5 gives a convenient summary of some of the methods employed for gas sampling measurements.

Table 13.4   Summary of low-temperature solutions

Coolant	Temperature, °C	Coolant	Temperature, °C
Distilled water and ice	0	Toluene slush	−95
Ice and NaCl	−21	Carbon disulfide slush	−111.6
Carbon tetrachloride slush	−22.9	Methyl cyclohexane slush	−126.3
Chlorobenzene slush	−45.2	N-Pentane slush	−130
Chloroform slush	−63.5	Liquid air	−147
Dry ice and acetone	−78.5	Isopentane slush	−160.5
Dry ice and Cellosolve	−78.5	Liquid oxygen	−183
Dry ice and isopropanol	−78.5	Liquid nitrogen	−196
Ethyl acetate slush	−83.6		

NOTE: The various slushes are prepared by adding small quantities of liquid nitrogen to the respective solvents in a Dewar vessel until a thick slush consistency is obtained.

13.6   PARTICULATE SAMPLING TECHNIQUES

The range of particulate-matter sizes encountered in air-pollution applications is extremely broad, as illustrated in Fig. 13.2. The particles range from about 0.001 to 500 μm in diameter, with most atmospheric particles falling between 0.1 and 10 μm. For very small particles below 0.1 μm motion and transport are governed mainly by molecular collisions, and particles above about 20 μm tend to settle out of the atmosphere and to accumulate in their local surroundings. The rate at which particles will settle out depends on both size and density. For a particle density about equal to that of water (1.0 g/cm^3) the approximate settling velocities are given in Table 13.6.

SETTLING AND SEDIMENTATION

Because of the settling phenomenon, one mechanism for collecting samples consists of a simple container placed in the appropriate locale and observed over a sufficient length of time. Thus, the particulate pollution from an iron foundry can be measured by placing a container near the foundry and measuring the accumulation over a period of several days. Such simple methods are not without problems, however. Typically, the container is filled with water to trap the particles once they fall inside, and some protective cover is installed to prevent leaves and twigs from invalidating the measurement. The container is also placed at a sufficient height above the ground to minimize the collection of dust raised by people or motor vehicles moving in its immediate vicinity.

MECHANICAL COLLECTION

When a gaseous sample is taken over a very short time interval compared to the total time of the experiment, it is said to undergo a *grab sampling* procedure. The sample

Table 13.5   Summary of gas analysis methods

Gas	Name of Method	Sampling Method	Chemical Reaction	Concentration Range	Analysis Method
CO	Iodine pentoxide	Heated I_2O_5 tube preceded by purification train	$5CO + I_2O_5 \rightarrow 5CO_2 + I_2$	1–10 ppm	Titrimetric, photometric, or coulometric determination of I_2
CO	Infrared absorption	Pressurized nondispersive IR spectrometer with 40-in cell	—	0–150 ppm	Determined from IR absorption; suitable for continuous monitoring
CO_2	Infrared absorption	Same as CO	—	0–150 ppm	Same as CO (IR method)
CO_2	Titrimetric	Bubbler	$CO_2 + Ba(OH)_2 \xrightarrow{BaCl_2} BaCO_3 \downarrow + H_2O$	0–150 ppm	Titration of remaining $Ba(OH)_2$ with standard oxalic acid
H_2S	Cadmium sulfide	Two bubblers in series	H_2S reacts with Cd salt in ammoniacal solution to form insoluble CdS	0–0.5 ppm	CdS dissolved in HCl and sulfide determined by iodimetric titration
H_2S	Methylene blue	Greenburg-Smith impinger	H_2S reacts with alkaline $CdSO_4$; sulfide ion then reacts with ρ-aminodimethylaniline and finic and chloride ions to give methylene blue	$<20 \times 10^{-3}$ ppm	Colorimetric determination of methylene blue
NO_2	Saltzman	Fritted glass bubbler or freeze-out	Dye complex formed between sulfanilic acid, nitrite ion, and α-naphthylamine in an acid medium	0–5 ppm	Colorimetric determination of dye complex
NO	Saltzman	Same as NO_2	Oxidize NO to NO_2 by passing through potassium permanganate, then same as NO_2	0–5 ppm	Colorimetric determination
C_2H_2	Cuprous acetylide	Silica-gel detector tube in acetone-dry ice bath	Acetylenic hydrocarbons absorbed on silica gel, then reacted with ammoniacal cuprous chloride to produce red-colored copper acetylide	0–0.01 ppm	Hydrocarbon concentration determined by color intensity of detector tube
CH_2O (formaldehyde)	Chromotropic acid	Glass midget impinger with fritted disk inlet	CH_2O reacts with chromotropic acid-sulfuric acid solution to form purple chromogon	0.01–200 ppm	Spectrophotometric determination of chromogon
SO_2	West-Gaeke	Midget impinger	SO_2 reacts with sodium tetrachloromercurate to form dichorosulfito-mercurate ion	0.005–5 ppm	Addition of acid-bleached pararosaniline and formaldehyde to form pararosaniline methylsulfonic acid; colorimetric determination
SO_2	Hydrogen peroxide	Impinger or fritted bubbler	$SO_2 + H_2O_2 \rightarrow H_2SO_4$	0.01–10 ppm	Titration of H_2SO_4 with standard alkali
SO_2	Conductimetric	Sampler probe and scrubber	$SO_2 + H_2O_2 \rightarrow H_2SO_4$	0.01–20 ppm	Change in conductivity of H_2SO_4
SO_2	Lead peroxide candle	PbO_2 paste on cylinder or plate	$PbO_2 + SO_2 \rightarrow PbSO_4$	0.02 mg SO_2/ 100 cm^2/day	Gravimetric determination of $PbSO_4$

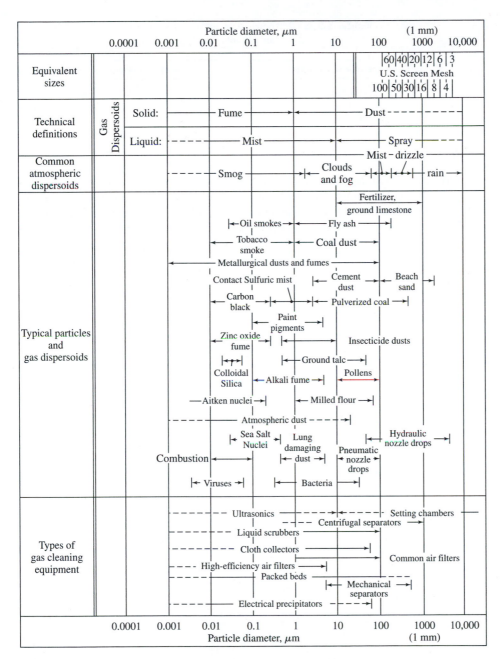

Figure 13.2   Characteristics of particulate matter. (*From Ref. [12].*)

Table 13.6   Approximate settling velocities for particles with density of 1.0 g/cm^3

Particle size, μm	Settling Velocity
0.1	0.4 nm/s
1.0	40 nm/s
10	30 mm/s
100	0.3 m/s

may be collected in a container which has previously been carefully evacuated and sealed in the laboratory. Once collected, the sample can be returned to the lab for analysis. When an evacuated container is not available, a clean glass grab sampler must be purged for a sufficient length of time to remove air contained in it. Consider the sampler shown in Fig. 13.3. The volume of the sampler is V, and the concentration of sample gas in the sampler at any instant of time is C. If we assume complete mixing in the volume, the input minus the output flow rate of the sample must equal the accumulation, or

$$QC_i - QC = \frac{d(VC)}{d\tau}$$ [13.3]

where Q is the volume flow rate and C is now exit concentration; i.e., $C_e = C$. For constant Q, V, and C_i, we write

$$\int_0^C \frac{dC}{C_i - C} = \frac{Q}{V} \int_0^\tau d\tau$$

or

$$\ln\left(\frac{C_i - C}{C}\right) = -\frac{Q\tau}{V}$$

Rearranging, we obtain

$$\frac{C}{C_i} = 1 - e^{-Q\tau/V}$$ [13.4]

and we see that the concentration in the sampler approaches C_i for τ sufficiently large; the buildup is illustrated in the following tabulation:

$Q\tau/V$	C/C_i
1	0.632
2	0.865
3	0.950
4	0.982
5	0.993
10	0.999955

Thus, by proper monitoring of the grab sampling time, the purity of the sample can be made to approach the desired level.

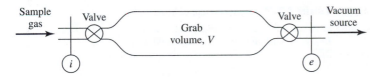

Figure 13.3   Purge of a grab sampler.

FILTRATION TECHNIQUES

Using the vacuum source of the sampling train, a volume of air may be drawn across an appropriate filter which collects the particulate matter. The filter is then removed after a known period of time and taken to the laboratory for examination.

One filtration-collection technique employs a high-volume sampler with a motor similar to that of a vacuum cleaner to pull the sample across a large filter of glass-fiber paper. Such filters typically operate for 24 h with a volume flow rate of about 1.5 m^3/min. Ordinary analytical filter paper can be used for particles between 0.5 and 1.0 μm in size.

For smaller-volume samples, porous ceramic, fritted glass, or granular filters can be employed for particles above 1.0 μm in size. Membrane or molecular filters of cellulose esters can be used down to 0.1 μm. For detailed time-varying measurements a tape sampler device is used which has a cellulose tape filter placed over a sampling nozzle connected to the source. The tape is automatically moved at preset intervals so that concentrations can be observed as a function of time. The tape is then examined in the laboratory to determine the percent of light transmitted through each collection spot. The results of such measurements are reported as a *coefficient of haze* (Coh), expressed as Coh/1000 linear ft of air sample. This measurement is especially valuable in determining the soiling effects of various particulate matter. The following measurements are performed:

$$I_0 = \text{optical transmission of clean tape}$$
$$I = \text{optical transmission of exposed tape}$$

Then, we define

$$\text{Opacity} = \frac{I}{I_0} \qquad \textbf{[13.5]}$$

$$\text{Optical density (OD)} = \log_{10} \frac{I}{I_0} \qquad \textbf{[13.6]}$$

$$\text{Coefficient of haze (Coh)} = \frac{\text{OD}}{0.1} \qquad \textbf{[13.7]}$$

$$= \frac{\log_{10}(\text{opacity})}{0.1}$$

$$= 100 \log_{10}(T_{\text{frac}})^{-1} \qquad \textbf{[13.8]}$$

where, now, T_{frac} is the fractional transmittance of the dirty filter. For clean air the coefficient of haze is zero.

Table 13.7   Coefficient of haze

Degree of Air Pollution	Coh/1000 ft
Light	0–0.9
Moderate	1–1.9
Heavy	2–2.9
Very heavy	3–3.9
Extremely heavy	4–4.9

To express the Coh in terms of lineal feet of air sample, we observe

$$\text{No. of 1000-ft lengths in sample} = \frac{V}{A} \times \frac{1}{1000} \qquad \textbf{[13.9]}$$

where V is the sample volume in cubic feet and A is the sample area on the tape in square feet. By combining Eqs. (13.8) and (13.9), we obtain

$$\frac{\text{Coh}}{1000 \text{ ft}} = \frac{10^5 A}{V} \log_{10}(T_{\text{frac}})^{-1} \qquad \textbf{[13.10]}$$

$$= \frac{10^5}{u_g \tau} \log_{10}(T_{\text{frac}})^{-1} \qquad \textbf{[13.11]}$$

where u_g is the gas velocity in feet per second and τ is the sample time in seconds. Table 13.7 gives values of the Coh for different degrees of air pollution.

Example 13.1 | **CALCULATION OF HAZE COEFFICIENT.**   A sample is collected at a rate of 0.05 ft³/s for a period of 45 min through a filter area having a diameter of 1.0 in. As a result of the sample collection, the light transmission of the filter paper is reduced from 80 to 55 percent. Calculate the Coh/1000 ft.

Solution

The fractional transmittance is

$$T_{\text{frac}} = \frac{0.55}{0.80} = 0.6875$$

From Eq. (13.8) the Coh is

$$\text{Coh} = 100 \log_{10}\left(\frac{1}{0.6875}\right) = 16.272$$

and with Eq. (13.11) we find

$$\text{Coh/1000 ft} = \frac{(10^5)\pi(0.5)^2}{(144)(0.05)(45)(60)} \log_{10}\left(\frac{1}{0.6875}\right)$$

$$= 0.657$$

In this case only a light degree of air pollution is observed.

IMPINGEMENT AND PRECIPITATOR COLLECTORS

In impingement collection devices the air sample is first accelerated to high velocity and then forced to undergo an abrupt change in direction. For particles above about 2 μm their inertia causes them to impact with the surface causing the change in flow directions. In some devices the surface is wetted and particles are collected in the runoff liquid. In dry devices a cascade arrangement is typically used, with jets of progressively higher velocities to remove progressively smaller particles.

Electrostatic precipitators are very efficient means of particulate collection. The air sample is drawn across a wire grid which is charged to 12 to 30 kV. The particles become charged and are subsequently collected on collector plates carrying an opposite electrical charge. When the device is shut off, the particles can be removed from the plates for laboratory analysis. Precipitators can operate effectively with a broad range of particle sizes, from 0.01 to 10 μm. We might observe that electrostatic precipitators are widely used in homes to remove particulate matter resulting from cooking or cigarette smoke.

ISOKINETIC SAMPLING

Up to this point we have described techniques for removing particulate matter from an air sample for subsequent laboratory analysis. But we have not discussed one important sampling technique which must be used to ensure that a representative sample reaches our collection device. With a high-volume sampling device, the particles are collected almost immediately on the filter paper. However, when a sample must be collected from a stack, by either grab or continuous methods, more care must be exerted. Typically, a probe is inserted into the flow at various points and a sample is withdrawn with a vacuum source. Because the particles are being transported by the flow stream, they are influenced by the viscosity of the gas, and their motion depends on their size and inertia. If the suction velocity in the probe is much slower than the flow velocity in the stack, much of the flow approaching the probe must be diverted around it. Particles with a large inertia may not be able to follow the change in direction and will enter the probe in a disproportionately high concentration. When the suction velocity is much larger than the flow velocity, large particles again may not be able to make the direction change and thus pass on by the probe. In this case the sample will have too low a concentration of particles. Thus, the most desirable technique is to draw the sample into the probe at a velocity equal to the flow velocity in the stack. This is accomplished by measuring the stack velocity with a Pitot tube and then adjusting the sampling flow rate until a like velocity is produced in the sampling probe. Watson [11] has developed a semiempirical equation to describe this effect.

$$\frac{C}{C_0} = \frac{U_0}{U}\left\{1 + f(p)\left[\left(\frac{U}{U_0}\right)^{1/2} - 1\right]\right\}^2 \qquad \textbf{[13.12]}$$

where C = indicated particulate concentration
          C_0 = true particulate concentration in stack
          U_0 = stack flow velocity, ft/s or m/s
          U = mean velocity at sampling orifice, ft/s or m/s
          D = diameter of sampling orifice, ft or m
          d = particle diameter, ft or m
          ρ = density of particles, ft³/lbm or m³/kg
          μ = viscosity of gas, lbm/ft · s or kg/m · s
          $p = d^2 \rho U_0 / 10 \mu D$

The function $f(p)$ is given approximately by

$$f(p) = 0.0426 p^2 - 0.407 p + 1.0 \qquad\qquad \textbf{[13.12a]}$$

and the overall effect on concentration measurement is illustrated in Fig. 13.4 for the flow conditions indicated. Clearly, large errors can result from nonisokinetic sampling when larger particle sizes are involved. The error is small for small particles.

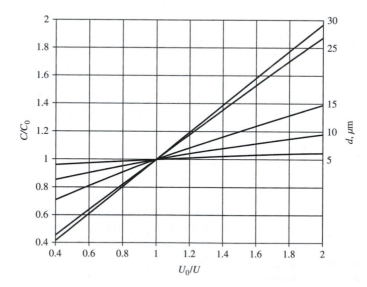

Figure 13.4   Effect of nonisokinetic sampling as calculated with Eq. (13.12) for spheres with density of 1.0 g/cm³, D =1.0 mm and U_0 =1.0 m/s in air at 1 atm and 300 K.

13.7   SULFUR DIOXIDE MEASUREMENTS

Table 13.5 has already indicated some of the techniques employed for SO_2 measurement. Sulfur dioxide arises from the combustion of hydrocarbon fuels with a substantial sulfur content and represents a major source of air pollution. Because of the large number of industries which burn coal and fuel oil with a substantial sulfur content, the measurement techniques are particularly broad in application. We shall discuss three basic measurement techniques and refer the reader to specific manufacturers for details of equipment operation. The control measures which must be employed to remove SO_2 are beyond the scope of our discussion but are examined very thoroughly in Refs. [1] to [6].

COLORIMETRY ANALYSIS FOR SO_2

Figure 13.5 illustrates the basic apparatus for colorimetric determinations of SO_2. A vacuum pump draws the air sample through an inlet flow measurement device and then into a rotary-disk scrubber. An absorbing agent wets the surface of the disks to provide a large surface for the gas contact area, which allows for rapid absorption

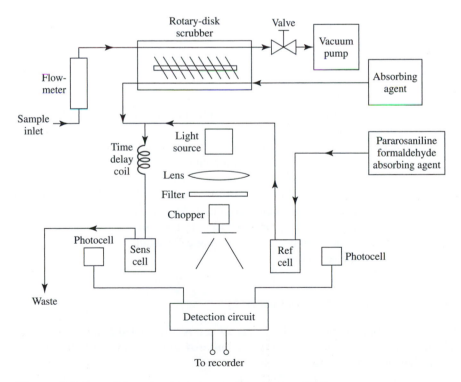

Figure 13.5   Schematic of colorimetric determination of SO_2.

and response. The solution containing the pollutant is next mixed with the reagent, bleached pararosaniline, and allowed to pass vertically through a time-delay coil where the color-development reaction occurs.

A light source is filtered and passed through a motor-driven chopper to provide an alternating source which is adaptable to electronic amplification. Two matched cadmium sulfide photocells are employed to sense the light transmission from a sensing cell containing the sample which has undergone the colorimetric reaction and also from a reference cell containing only the reagent. The electronic detection circuit compares the outputs of the two cells and sends an output voltage to an appropriate recorder. Practical systems may be programmed to take samples and perform the analysis at predetermined time intervals.

ELECTROCONDUCTIVITY ANALYSIS FOR SO_2

The electroconductivity measurement relies on oxidation of SO_2 to produce sulfate ions, which change the electrical conductivity of the solution in proportion to the amount of SO_2 present. With water as the reagent, we have the reactions

$$H_2O + SO_2 \rightarrow H_2SO_3$$
$$H_2SO_3 + \tfrac{1}{2}O_2 \rightarrow H_2SO_4$$

[13.13]

As an alternative we can use H_2O_2; we will then experience less interference from acid gases like CO_2. Thus,

$$H_2O_2 + SO_2 \rightarrow H_2SO_4$$

[13.14]

The electrical conductivity is a function of temperature so that the experimental apparatus must be constructed with a temperature-control system. The apparatus is indicated in the schematic of Fig. 13.6.

The reagent is passed first through a reference cell and then into an absorbing column, where the air sample is mixed in a countercurrent-flow process. The SO_2 is absorbed in the reagent, and the new solution is then passed through a measuring conductivity cell. An ac voltage is impressed on the conductivity cells and the current is measured to enable a calculation of the resistances. Zero-level calibration is achieved by passing the air through a soda-lime absorber to remove all SO_2. For this condition the conductivities in the reference and measuring cells should be the same and thus may be used to set the zero-output level.

COULOMETRIC ANALYSIS FOR SO_2

To perform a coulometric analysis, a detector cell is constructed as illustrated in the schematic of Fig. 13.7. The cell contains a buffered solution of KI, which generates I_2 at the anode according to

$$2I^- \rightarrow I_2 + 2e^-$$

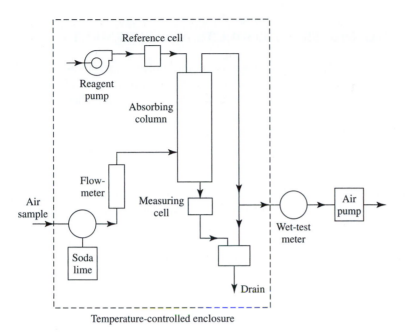

Figure 13.6   Schematic of electric conductivity analysis of SO$_2$.

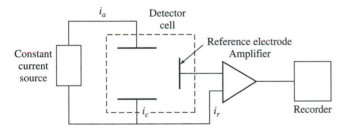

Figure 13.7   Schematic of coulometric analysis of SO$_2$.

When the air sample containing SO$_2$ is drawn in the cell, it reacts with I$_2$ to remove it from the cell. The unreacted portion of the I$_2$ is then reduced to I$^-$ at the cathode. As a result of the reaction, a reference electrode will detect a difference in the anode and cathode currents as shown:

$$i_r = i_a - i_c$$

This current may subsequently be amplified to provide an output proportional to the SO$_2$ concentration.

13.8   COMBUSTION PRODUCTS MEASUREMENTS

The analysis of products of combustion is important not only for air-pollution-control applications but also for maintenance of most efficient burning rates and energy utilization. We shall discuss three methods applicable to such measurements.

ORSAT APPARATUS

Figure 13.8 illustrates the simple Orsat apparatus, which is employed to analyze products of combustion. It consists of a measuring burette and three reagent pipettes which are used successively to absorb carbon dioxide, oxygen, and carbon monoxide from the mixture. First, a sample of the combustion products is drawn into the measuring burette. Next, the sampling manifold is shut off and the sample is forced into the first reagent pipette, where carbon dioxide is absorbed. The sample is then brought back into the measuring burette, and the reduction in volume is noted. The procedure is then repeated for the other two pipettes, which absorb the O_2 and CO successively. In the measuring process the combustion products are contained over water in the burette and remain saturated with water vapor. As a result, the volumetric proportions of the products are obtained on a so-called dry basis, that is, exclusive of the water vapor which is present.

Potassium hydroxide is normally used as the reagent to absorb carbon dioxide. A mixture of pyrogallic acid and solution of potassium hydroxide is employed as the reagent for absorption of oxygen. Cuprous chloride is used to absorb carbon monoxide. Precautions must be taken to ensure that the reagents are fresh in order to minimize errors. The cuprous chloride must be changed after 10 absorptions or so if the carbon monoxide content is above 1 percent. The Orsat device is mainly used for portable remote applications which do not require continual monitoring.

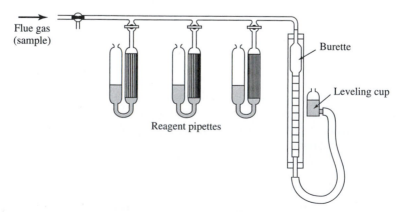

Flue gas (sample)

Burette

Leveling cup

Reagent pipettes

Figure 13.8   Schematic of Orsat apparatus.

EXCESS AIR CALCULATION.   An operator reports the following raw volume measure- | **Example 13.2**
ments on an Orsat device in measuring the combustion products for methane fuel (CH_4). Based
on these data, calculate the air-fuel ratio and percent excess air for the combustion process:

	Volume (cm^3)
Initial sample	95
After CO_2 absorption	83
After O_2 absorption	79
After CO absorption	78

Solution

The analysis of the "dry products" is

$$n_{CO_2} = 95 - 83 = 12$$
$$n_{O_2} = 83 - 79 = 4$$
$$n_{CO} = 79 - 78 = 1$$

On a mole-fraction basis we obtain

$$x_{CO_2} = \frac{12}{95} = 12.6\%$$

$$x_{O_2} = \frac{4}{95} = 4.2\%$$

$$x_{CO} = \frac{1}{95} = 1.05\%$$

We now write the combustion equation for 95 mol of dry products.

$$aCH_4 + xO_2 + 3.76xN_2 \rightarrow 12CO_2 + 4O_2 + (1)CO + bH_2O + 3.76xN_2$$

The carbon balance yields

$$a = 12 + 1 = 13$$

For the hydrogen $4a = 2b$ so that $b = 26$, and the oxygen balance gives

$$2x = (12)(2) + (4)(2) + 1 + 26 = 59$$
$$x = 29.5$$

The air-fuel ratio for dry air is

$$\text{AF ratio} = \frac{m_a}{m_{CH_4}}$$

$$= \frac{(29.5)[32 + (3.76)(28)]}{(13)(16)}$$

$$= 19.47$$

For stoichiometric combustion (no excess oxygen) the reaction would be

$$13CH_4 + 26O_2 + (3.76)(26)N_2 \rightarrow 13CO_2 + 26H_2O + (3.76)(26)N_2$$

The percent excess air is therefore

$$\frac{29.5 - 26}{26} \times 100 = 13.5\%$$

GAS CHROMATOGRAPH

Another technique for measuring products of combustion, and other gas samples as well, is *gas chromatography,* indicated in the schematic diagram of Fig. 13.9. The sample gas is typically collected in a small glass syringe and introduced into the chromatographic column. The column can vary in length from 1 to 20 m and in diameter from 0.25 to 50 mm. Inside the column either a solid adsorbent, like granular silica, alumina, or carbon, or a liquid adsorbent is contained to retard the flow of different constituents of the gas sample. The liquid can either coat the wall when the column is a thin capillary or be distributed over an inert granular material like diatomaceous earth in larger packed columns. Hundreds of liquids are available as adsorbents, depending on the particular application.

Once the gas sample is impressed on the column, an inert carrier gas (typically helium) transports the sample through the column. Different components will be retained in the column for different lengths of time so that they will appear in the discharge stream from the column at different times. The device may be calibrated to predict the retention time for various gas-liquid combinations. The components in the output flowstream may be detected in various ways. A thermal-conductivity detector is most common, but ultrasonic and electron-capture may also be used. Special detectors are available for specific elements.

The output of the chromatographic detector can be processed electronically and displayed on a chart recorder. The display takes the form of a series of spikes. The time of appearance of the spike indicates the particular component, while the height of the spike indicates the quantity present. The chromatographic column may require heating or cooling to achieve the proper separation times so that a temperature control system is required. A flow regulator is also employed to maintain a constant flow rate of the carrier gas.

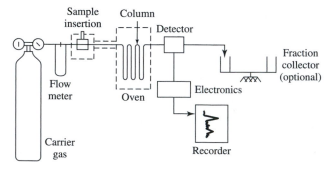

Figure 13.9   Schematic of gas chromatograph.

NONDISPERSIVE INFRARED ABSORPTION

Nondispersive infrared (IR) measurements may be used to determine concentration of gaseous pollutants as illustrated in the schematic diagram of Fig. 13.10. The absorption of IR radiation occurs in narrow-wavelength bands, with each gas exhibiting its own peculiar characteristics. Two infrared sources are provided, as shown, with a chopper to impose an ac signal, which is handled more easily by subsequent electronic detection circuits. Filters may be placed in the path of the sources so that only the absorption-wavelength band for the particular gas to be studied is investigated. Two cells are exposed to the IR radiation: (1) a reference cell containing an inert gas (usually nitrogen) and (2) a cell which admits passage of the sample gas containing the pollutant to be investigated. No IR absorption occurs in the reference cell, while the absorption in the sample cell is proportional to the concentration of the component of interest.

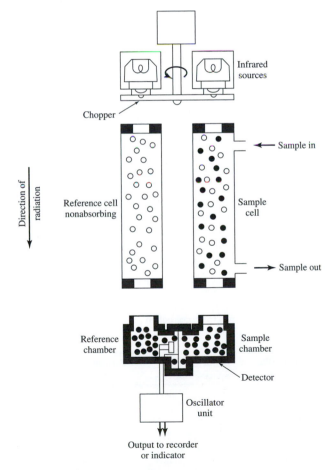

Figure 13.10   Schematic of nondispersive infrared analyzer.

Beyond the reference and sample cells there are two detector chambers which absorb the IR radiation transmitted. These chambers are sealed so that absorption of the radiation causes the temperature of the gas inside to rise. The two chambers are separated by a thin diaphragm. Because more radiation is transmitted through the reference cell, there will be a greater temperature rise in the reference detector chamber than in the sample detector. As a result, a pressure differential is created which causes movement of the diaphragm; this can be detected with a capacitance pickup. The resulting signal is amplified and transmitted to an appropriate readout device. Both the chromatograph and infrared absorption devices may be used for continuous monitoring of combustion products.

13.9   OPACITY MEASUREMENTS

Most lay people measure air pollution by visual observation. If a discharge stack at a plant emits heavy black smoke, this is judged to be harmful pollution. The accumulation of smog around expressways on still days is evidenced by decreased visibility.

The opacity of the gas stream discharging from a stack is indeed a measure of the concentration of particulate matter in the gas. The *color* of the discharge gas, however, depends on the specific particulate matter involved in the process. Unburned carbon is black, while very harmful asbestos or lead particles might be gray in color. Air-pollution-control officials cannot provide continuous stack gas monitoring of all plants because this would involve excessive costs in instrumentation and personnel, and so they frequently rely on visual observation of the discharge gases to indicate probable air-pollution offenders.

To aid in standardizing visual observations, the Ringleman smoke-chart system has been devised. Five charts are constructed in even increments of blackness as indicated in Table 13.8. Observers are then trained to compare gas emissions with the charts. In the training process the chart is placed at this distance of 50 ft from the observer and in a line of sight with the smoke. At this distance the charts take on an appearance of different degrees of gray. In observations the observer should stand facing away from the sun. For training purposes optical transmission measurements

Table 13.8   Spacing of lines for construction of Ringleman chart

Ringleman Chart No.	Width of Black Lines, mm	Width of White Spaces, mm	Percent Black
0	All white		0
1	1	9	20
2	2.3	7.7	40
3	3.7	6.3	60
4	5.5	4.5	80
5	All black		100

may also be performed on the smoke to correlate with the Ringleman charts. A person who has undergone an intensive training period is then permitted to make observations in the field strictly on the basis of visual acuity. Such persons are appropriately called "smoke readers"; when they are employed by air-pollution-control groups, they must be retrained about every 6 months. The trained smoke observer can obviously make many measurements in a region at very low cost. Typically, a plant is not allowed to operate at greater than a Ringleman number 2 discharge opacity. The trained observer can read the opacity within $\pm 1/2$ Ringleman number [14].

The opacity of a stack discharge can be easily changed without changing the total emission of pollutants. Because opacity is a function of optical path length, one simple way to decrease the opacity is to reduce the stack diameter and to increase the flow velocity. Another method is to increase the number of stacks employed for the discharge. Air-pollution-control regulations frequently do not allow the use of such techniques once a citation has been issued for excess discharge opacity.

All visual observations can be in serious error if the stack gases contain significant quantities of water vapor which may condense out. In such cases a violator may be excused if engineering data clearly show that the opacity could be the result of the water discharge in the process.

13.10 ODOR MEASUREMENT

We have all encountered odor problems at one time or another. Because odors must be detected by human beings, quantitative measurements and standards are very difficult to establish. Different people react to various odors in different ways. A person who eats garlic likes the taste and odor, while a person sitting nearby may find the garlic smell quite objectionable. We can become so accustomed to odors that they are no longer objectionable. A person who works at a stockyard does not notice the smell, but a newcomer visiting such a facility may find the odor very offensive.

Ideally, we could like to specify acceptable odor levels in terms of odorant concentration in parts per million or micrograms per cubic meter. The technique normally employed is to expose a group of individuals to varying concentration levels of an odorant and to ask them to make a rating on the following scale:

5   overpowering perception

4   strong odor

3   easily perceived odor

2   faint perception

1   very faint perception

0   no perception

This rating scale is applied to the intensity of the odor, its pervasiveness, and the acceptability or degree of like or dislike of the odor. When these data are collected, it

Table 13.9 Odor thresholds in air, from Ref. [10]

Odorant	Odor Threshold, ppm
Acetic acid	1.0
Acetone	100
Amine trimethyl	0.0021
Ammonia	46.8
Aniline	1.0
Carbon disulfide	0.21
Chlorine	0.31
Diethyl formamide	100
Diphenyl sulfide	0.21
Formaldehyde	1.0
Hydrogen sulfide	0.00047
Methanol	100
Methylene chloride	214
Nitrobenzene	0.0047
Phenol	0.047
Trimethyl amine	0.00021

is generally found that they follow a relationship of the form,

$$P = K \log S \qquad [13.15]$$

where P is the human sensor response, or intensity (from 0 to 5), and S is the ratio of the odor concentration in parts per million to that for 0 response. K is a constant which varies from 0.3 to 0.6, depending on odorant. Using such techniques it is possible to determine odor thresholds for a number of substances, as shown in Table 13.9. Other values for petrochemicals are given in Ref. [9]. A cursory examination of this table indicates the extremely broad range of odor thresholds (over six orders of magnitude) and therefore the wide variety of instrumentation which may be required for measurements.

In some air-pollution problems it is convenient to define an *odor unit* as

1 odor unit = quantity of odorant that will contaminate 1 ft³ of clean
air to the mean threshold level

From Table 13.9 we see that 1 odor unit of acetone would be the quantity necessary to produce a concentration of 100 ppm, while 1 odor unit of formaldehyde would be that necessary to produce a concentration of 1.0 ppm.

13.11 REVIEW QUESTIONS

13.1. Distinguish between source and ambient air-pollution standards. What are the advantages of each?

13.2. Name three techniques for particulate measurement.

13.3. What is the range of sizes of most atmospheric particulate matter?

13.4. What is meant by "grab sampling"?

13.5. Define the coefficient of haze (Coh).

13.6. What is the principle of operation of an electrostatic precipitator?

13.7. What is meant by "isokinetic sampling"?

13.8. Describe the colorimetry analysis process for SO_2.

13.9. How does an Orsat apparatus work?

13.10. How does a gas chromatograph work?

13.11. What is a Ringleman chart?

13.12. What easy measures can be taken to reduce the opacity of a stack discharge?

13.12   PROBLEMS

13.1. Verify the mass concentrations of carbon monoxide and sulfur dioxide given in Table 13.1.

13.2. A glass grab sampler has a volume of 100 cm^3. Calculate the time required to effect a 90 percent purge of the sampler with a flow rate of $125 \text{ cm}^3/\text{min}$.

13.3. A sample is collected at a rate of $10 \text{ ft}^3/\text{min}$ for a period of 1 h through a filter area having a diameter of 1.4 in. As a result of the sample collection, the light transmission of the filter paper is reduced from 80 to 40 percent. Calculate the Coh/1000 ft.

13.4. An analysis of the dry products of combustion for CH_4 yields

$$CO_2 = 10 \text{ percent}$$
$$CO = 0.7 \text{ percent}$$
$$O_2 = 2.0 \text{ percent}$$

Calculate the air-fuel ratio and percent excess air.

13.5. Air at 1 atm and 20°C is burned with gaseous butane (C_4H_{10}). Calculate percentage compositions of the dry products of combustion for 25 percent excess air.

13.6. Calculate the mass concentrations of gaseous butane (C_4H_{10}) and propane (C_3H_8) for 1 ppm in air at 1 atm and 20°C. Express in $\mu g/m^3$.

13.7. Calculate the concentration of water vapor in atmospheric air at 80°F and 50 percent relative humidity. Express in units of ppm and $\mu g/m^3$.

13.8. Consult standard thermodynamic tables and determine the mass concentrations for 1 ppm of Freon-12 and ammonia in atmospheric air at 20°C. Consult the references at the end of this chapter (or others) to determine acceptable tolerance limits for these two gases.

13.9. Air at 1 atm and 20°C is burned with gaseous octane and 30 percent excess air. Calculate the percentage compositions of the dry products of combustion. Repeat for 100 percent excess air.

13.10. A glass grab sampler has a volume of 125 cm³. Assuming a flow rate of 100 cm³/min, calculate the times required for 80, 90, 95, and 99 percent purge of the sampler.

13.11. A grab sampler has an initial concentration C_0 of a certain gaseous substance. The inlet flow stream has a concentration C_i of the same species. Derive an expression for the concentration of the species in the sampler volume as a function of time.

13.12. Using the results of Prob. 13.11 and an initial concentration $C_0 = 0.1$ and inlet concentration $C_i = 0.5$ for the size and flow rate of Prob. 13.10, calculate the time to obtain concentrations of the gas of 0.2, 0.3, 0.4, and 0.49. Comment on the results of the calculation.

13.13. A grab sampler is used to collect a sample of pure gas. If the initial concentration of the gas in the sample volume is 0.5, and the volume and flow conditions are those of Prob. 13.10, calculate the time required to obtain a sample that is 99 percent pure.

13.14. An analysis of the dry products of combustion for propane (C_3H_8) yields

$$CO_2 = 12 \text{ percent}$$
$$CO = 0.5 \text{ percent}$$
$$O_2 = 3 \text{ percent}$$

Calculate the air-fuel ratio and the percent excess air.

13.15. Calculate the mass concentrations of gaseous butane and gaseous propane for 1.5 ppm in air at standard atmospheric pressure in Denver, Colorado (altitude = 5000 ft), and a temperature of 25°C. Express in units of $\mu g/m^3$.

13.16. Air at 1 atm and 20°C is burned with gaseous propane. Calculate the percentage compositions of the dry products of combustion for 30 percent excess air.

13.17. The grab sampler of Prob. 13.10 has an initial concentration of the pure gas to be sampled of 0.1. Calculate the time required to obtain a sample that is 95 percent pure.

13.18. An airstream at 1 atm and 27°C contains small spherical particles having a density of approximately 1000 kg/m³. The flow velocity of the airstream is about 0.9 m/s, and the mean velocity at a sampling device inserted in the airstream is about 0.75 m/s. The diameter of the spherical particles is about 16 μm. Calculate the error in the particle concentration as a result of the sampling process.

13.13 REFERENCES

1. Wark, K., and C. F. Warner: *Air Pollution,* IEP Publishers, New York, 1976.

2. Seinfeld, J. H.: *Air Pollution,* McGraw-Hill, New York, 1975.

3. Perkins, H. C.: *Air Pollution,* McGraw-Hill, New York, 1974.

4. Stern, A. C.: *Air Pollution,* vols. I, II, III, Academic Press, New York, 1968.

5. Magill, P. L., F. R. Holden, and C. Ackley: *Air Pollution Handbook,* McGraw-Hill, New York, 1956.

6. *Air Pollution Engineering Manual,* EPA, NAPCA, AP-40, 1967.

7. Moncrieff, R. W.: *The Chemical Senses,* Leonard Hill Books, London, 1967.

8. Duffee, R. A.: "Appraisal of Odor Measurement Techniques," *J. Air Pollution Cont. Assoc.,* vol. 18, p. 472, 1968.

9. Hellman, T. M., and F. H. Small: "Characterization of the Odor Properties of 101 Petrochemicals Using Sensory Methods," *J. Air Pollution Cont. Assoc.,* vol. 24, p. 979, 1974.

10. Leonardos, G., D. Kendall, and N. Bernard: "Odor Threshold Determinations of 53 Odorant Chemicals," *J. Air Pollution Cont. Assoc.,* vol. 19, p. 91, 1969.

11. Watson, H. H.: "Errors Due to Anisokinetic Sampling of Aerosols," *Am. Ind. Hyg. Assoc. Quart.,* vol. 15, p. 21, 1954.

12. Lapple, C. E.: "Characteristics of Particles and Particle Dispersoids," *Stanford Res. Inst. J.,* vol. 5, p. 94, 1961.

13. Ringleman, M.: "Methode d'Estimation des Fumes Produites par les Foyer Industriels," *La Revue Technique,* 268, June 1898.

14. Conner, W. D., and J. R. Hodkinson: *Optical Properties and Visual Effects of Smoke Stack Plumes,* Publication 999-AP-30, USHEW, 1967.

15. de Nevers, Noel: *Air Pollution Control Engineering,* McGraw-Hill, New York, 1995.

16. Corbitt, R. A.: *Standard Handbook of Environmental Engineering,* McGraw-Hill, New York, 1999.

chapter

14

DATA ACQUISITION AND PROCESSING

14.1 INTRODUCTION

All the previous chapters have illustrated, in one way or another, the manner in which an experiment can be planned and executed to produce meaningful data. Selection of proper instrumentation and consideration of experimental uncertainties are all necessary parts of the planning process. But very little has been said about the manner of collecting data and the later processing of these data to produce the desired results of the experiments. The acquisition of data might consist simply of several people (or perhaps only one person) reading a number of instruments and recording the observations on a data sheet. The subsequent processing of data could be done in many ways, from simple hand calculations to complicated computer routines.

There are systems available today for rapidly collecting a great bulk of data, processing it, and displaying the desired results in printed form. The purpose of this chapter is to present a brief qualitative description of such systems and the functions of the elements that go together to make up an overall data acquisition and processing installation. These systems change rapidly. So, this discussion takes a rather general approach.

14.2 THE GENERAL DATA ACQUISITION SYSTEM

The essential element in a data acquisition system is the instrument transducer, which furnishes an electrical signal that is indicative of the physical variable being measured. The signal may be an analog voltage, current, resistance, or frequency, or a digital representation of any of these quantities in the form of a series of electric pulses. The present discussion assumes that suitable transducers are available to convert the physical variables of interest into electrical signals. As examples of such transducers we may recall that a thermocouple gives a voltage representation of temperature, a strain gage gives a resistance representation of strain, and so forth. These types of transducers have been discussed in previous chapters.

The object of any data acquisition and processing system is to collect the data, process them in the desired fashion, and record the results in a form suitable for storage, presentation, or additional subsequent processing. Thus, a recording potentiometer is a simple data acquisition system that may be used for collecting temperature data from thermocouples. In this case the data points must be read from the recorder chart. A more complicated system would convert the analog voltage signal from the thermocouple to an equivalent digital signal, which could be used to operate a printing recorder so that the numerical value of the temperature is printed on a sheet of paper. Such a system is much more complicated than the simple recorder because of the analog-to-digital conversion process. It is easy to see, however, that the digital output has many advantages.

The major elements of any data acquisition and processing system are shown in block diagram form in Fig. 14.1. This figure divides the system into three major parts. The first is the input stage, which consists of appropriate transducers and an input circuit and additional signal conditioning circuits as needed (amplifiers, filters, etc.). The second is the signal-conversion stage, in which the information is readied for transmission, if such is required (as in situations in which the transducers are physically remote from the location at which data display is desired), as well as the transmission and receiving equipment and any necessary data processors. An example of the latter would be the conversion of a signal from analog-to-digital (A/D) form. The last is an output stage which provides two primary functions: data display and data storage. Examples include data display and storage in printed form on a sheet of paper or in graphical form on suitable paper and storage on magnetic tapes or disks, or in the internal memory of a computer.

The output stage must include suitable coupling circuits to convert the data to a form which may be used to drive a printer, graphics device, or computer processing unit.

It is a rare circumstance when data involving only one experimental variable are to be collected. The data acquisition and processing system, then, should include provision for collecting and analyzing multiple channels of data input. This collection process could be accomplished by having a channel like the one shown in Fig. 14.1 for each variable to be studied. The cost of such a system, however, would be quite

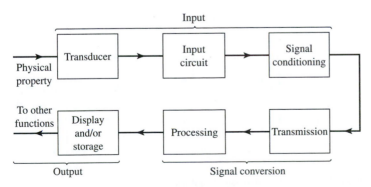

Figure 14.1   General data acquisition system.

high because of equipment duplication so that a scanner/programmer is normally employed for multiple-channel work. The scanner is a device that samples the data channels in rapid sequence so that only one conversion and output stage is necessary. Equipment available today makes possible any particular sequencing of a large number of data channels at the discretion of the laboratory personnel. Thus, the system may be programmed to collect any desired range of variables in any order, and the term "scanner/programmer" is quite appropriate. In many systems computational elements are provided in order that the data may be manipulated immediately, without external computer processing.

Many experimental situations require the collection of data at regular time intervals or in some particular time sequence. The acquisition system may perform this timing function automatically by incorporating a digital clock and time standard in the scanner and/or conversion stages.

Finally, it may be advantageous to apply signal conditioning to the output of the device functioning as the scanner/programmer. This conditioning might be amplification, analog-to-digital conversion for only some of the channels, filtering, distortion or harmonic analysis of the waveform of some of the signals, and so forth. When all the above elements are combined, a very flexible data acquisition and processing system results, as shown in Fig. 14.2. It should be noted that the programmable feature of the scanner is normally used in the converter stage as well. This is essential since some channels may require signal conditioning while others may not.

The use of a flexible system like the one described above depends on many factors, not the least of which is cost. Progress in the development of microprocessors and digital signal processing (DSP) [1, 4, 6, 8] has greatly reduced the cost of such systems. Data acquisition processing systems which use microprocessors to perform the scanner/programmer function are widespread.

In examining Fig. 14.2, we note that the general system may be used to feed a computer, which, in turn, has a software input to process the data or cause the

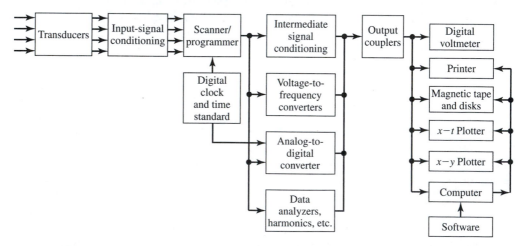

Figure 14.2   Schematic of a multichannel data acquisition system.

computer to execute some process control operation. The output from the computer can also be fed to a printer for either text or graphical displays. Many standard software packages are available to perform the computer operations.

There are two basic types of interfaces to connect the measurement system to a personal computer: stand alone and plug in. The plug-in systems are designed for a particular type of computer, usually the IBM PC or Apple Macintosh. The plug-ins consist of special terminal boards for connection of thermocouples, RTDs, or other sensors and outputs. Menu-driven software is included which enables a setup of scale ranges, units, digital input and output, data logging, and process control for any of the channels. Hundreds, or even thousands, of analog inputs can be allowed for a single PC. The stand-alone interface does not depend on the type of computer to be used. All that is required is a standard serial communications port on the computer. The stand-alone interfaces can incorporate a single-variable channel, multiple thermocouples, or many channel data loggers with different analog inputs. Stand alones may or may not be supplied with software capability. If not supplied, then the user must provide his or her own software. In general, a variety of software packages are available from many manufacturers.

Some data acquisition systems have on-board microprocessors for data analysis and storage. Depending on the complexity of the system, it may be possible to analyze some data channels while storing the output of others for later manipulation. Because of the reduction in cost of memory devices, we may expect very elaborate data acquisitions systems in the future which incorporate digital oscilloscope features in one central processing unit. Cost will be a factor in decisions to utilize such systems.

14.3   S<small>IGNAL</small> C<small>ONDITIONING</small> R<small>EVISITED</small>

As was pointed out in Chap. 4, noise is present in all physical situations in which measurements are attempted or information is conveyed. A consideration of noise and its effects on experiments led us, in Sec. 4.12, to consider filtering electrical signals. We considered circuits which would transmit or "pass" only certain bands or ranges of frequencies which are part of an input signal. It was pointed out that some filters are composed only of passive elements (resistors, capacitors, and inductors), while others contain amplifiers. Circuits of the first type are called *passive filters,* while those in the second category are termed *active filters* [5, 6].

Table 4.2 displayed four very simple passive filters. In order to illustrate active filters and the methods of analysis applied to such filters, we will analyze an active version of the first filter shown in this table, a lowpass *RC* filter. For the passive form, the maximum ratio of output voltage to input voltage is unity. As we shall see, the addition of an operational amplifier allows the realization of output-to-input-voltage ratios which are larger than 1. The gain of the operational amplifier serves to improve the overall performance of the filter. It should be mentioned that we will be analyzing only a very simple circuit; there are commercial design manuals which present the analysis of more complex active filters.

The basic ideal operational amplifier is presented in Fig. 14.3, along with the relationship between the output voltage v_o and the two input voltages v_1 and v_2. The amplifier gain A will be taken to be infinite, and the input resistances of both the positively and the negatively labeled terminals of the amplifier will be assumed to be infinite. These approximations greatly simplify the analysis [3].

The lowpass, single-section active filter to be analyzed is shown in Fig. 14.4. As was the case for each circuit containing an operational amplifier which was analyzed in Chap. 4, we are seeking a relation between the output voltage at v_o and the input voltage v_i. It is simplest to assume that all voltages, and consequently all currents, in the circuit are single-frequency sinusoids; that is,

$$v_1 = V_1 e^{j\omega t} \qquad v_i = V_i e^{j\omega t}$$

$$i_1 = I_1 e^{j\omega t} \qquad i_2 = I_2 e^{j\omega t}$$

[14.1]

Using the reference directions shown in Fig. 14.4, we can write the following four expressions for the currents shown:

$$i_1 = \frac{v_i - v_1}{R} \qquad i_2 = \frac{v_2}{R}$$

$$i_3 = -C\frac{dv_1}{dt} \qquad i_4 = \frac{v_o - v_2}{(K-1)R}$$

[14.2]

In these equations the ground voltage has been taken to be zero.

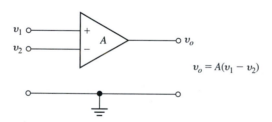

$$v_o = A(v_1 - v_2)$$

Figure 14.3   The ideal operational amplifier.

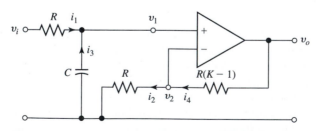

Figure 14.4   Lowpass, single-section active filter.

Since the input resistances of the two operational amplifier terminals $(+, -)$ are infinite, it follows that the sum of the two currents entering each terminal must be zero. We write

$$i_1 + i_3 = 0$$
$$-i_2 + i_4 = 0$$

Substituting the sinusoidal forms given in Eqs. (14.1) and the current expression in Eqs. (14.2) results in

$$\frac{V_i - V_1}{R} e^{j\omega t} - C\frac{d}{dt}(V_1 e^{j\omega t}) = 0 \qquad \textbf{[14.3a]}$$

$$-\frac{V_2}{R} e^{j\omega t} + \frac{V_o - V_2}{(K-1)R} e^{j\omega t} = 0 \qquad \textbf{[14.3b]}$$

After carrying out the derivative with respect to time and canceling common factors, we arrive at

$$V_i - V_1 - j\omega RC V_1 = 0 \qquad \textbf{[14.4a]}$$

$$-V_2 + \frac{V_o - V_2}{K-1} = 0 \qquad \textbf{[14.4b]}$$

The relationship between the output voltage and the two input voltages of the operational amplifier is

$$v_o = A(v_1 - v_2)$$

or, upon substituting the sinusoidal time dependencies, we have

$$V_o = A(V_1 - V_2)$$

Since the amplifier gain is approximated by infinity, it is necessary for V_1 to equal V_2 if the output voltage V_o is to be finite.

Solving for V_1 in Eq. (14.4a) and V_2 in Eq. (14.4b) and equating the results yield

$$\frac{V_i}{1 + j\omega RC} = \frac{V_o}{K}$$

The ratio between the output and input voltages is given by the expression

$$\frac{V_o}{V_i} = \frac{K}{1 + j\omega RC}$$

or, in terms of amplitudes,

$$\left|\frac{V_o}{V_i}\right| = \frac{K}{\sqrt{1 + (\omega RC)^2}}$$

This result should be compared with the equation given for the lowpass RC passive filter in Table 4.2. The two results differ only by the factor of K in the numerator. Since this may be selected by the designer (refer to Fig. 14.4), the effect of the operational amplifier in the active filter is to allow a choice of the overall gain displayed by the active filter compared with the passive filter. The frequency response of the two circuits is identical. The only difference is the constant K appearing in the relation between output and input voltages for the active filter.

This example indicates that using active filters rather than passive filters often makes adding more amplifier circuits to a given data acquisition and processing system unnecessary.

The other major advantage of active filters is based on a unique property of integrated circuits. Simply stated, there are no integrated circuit forms of inductors. To this day, the inductor remains a large element. In order to fabricate passive filters containing inductors it is necessary to sacrifice some degree of compactness. Active filters can be designed to avoid totally the need for inductors. That is, the combination of resistors, capacitors, and operational amplifiers will replace and outperform any passive filter containing inductors. This means that circuit compactness and integration can always be obtained. The proof of these assertions is complex and well beyond the scope of this chapter. It is mentioned because this avoidance of inductors is a major advantage of active filters.

Although we have analyzed only a lowpass filter, both bandpass and highpass active filters exist and can often be used in place of passive filters. In many situations operational amplifiers allow the circuit designer to use smaller values of resistance and capacitance than would be needed for passive filters. This implies an associated reduction in cost and circuit size. There are also available designs for filters containing more than one operational amplifier. Such multistage circuits exhibit sharper cutoffs, and hence more useful filtering behavior, than the single-stage circuit analyzed in this chapter.

Once the signal has been filtered, it is necessary to consider the means available to transmit it if the transducer is at a remote location. This is the subject of the next section. It must be mentioned that in some situations there are advantages to filtering the signals *both* before and after transmission, while in other cases filtering *either* before *or after* transmission will be adequate. More information on active filters is given in Refs. [3], [5], and [7].

14.4 DATA TRANSMISSION

Given a transducer in one location and the need to display the transducer's output in another location, one is immediately faced with choosing the channel over which the necessary transmission will occur. There are many possible choices, ranging from current and voltage on copper wires to the propagation of electromagnetic waves through space to the transmission of light signals through an optical fiber. In selecting a communication channel, the system designer must consider two paramount issues, cost and noise immunity. A detailed examination of these two issues would fill several additional books, but we can consider some of the important characteristics of various transmission channels.

Coaxial cable provides a very secure means of transmitting a signal from one location to another. It consists of a central conductor through which the signals of interest are sent surrounded by another conductor which is held at zero voltage. Any stray electromagnetic noise is attenuated by this outer shield before it can corrupt the signal present on the inner conductor. Coaxial cable is often the communication medium of choice in laboratory situations where low voltage levels are being transmitted. Of course, if the transducer is on a probe heading toward Venus, it is not possible to string a coaxial cable between the transducer and the earth. In these situations it is necessary to consider communication channels which do not depend on a material link between the transmitter and receiver. This is needed not only in communicating from space, but also in any situation in which the cost of a physical link between transmitter and receiver is prohibitive.

It is obvious that empty space is free and available as a channel over which signals can be transmitted. Of course, there may be additional expense in the receiver and transmitter, depending on the frequencies used, but the communication link itself is quite inexpensive. The major drawback is the susceptibility of signals propagating through space to noise. For example, a lightning discharge generates a large amount of noise over a wide range of frequencies. If a signal is propagating in the vicinity of a lightning strike, it will be mixed with the noise produced by the lightning. This illustrates the basic trade-off between cost and noise immunity. Of course, the effect noise has on a given signal depends on the frequency being used and the degree of protection the designer builds in. The details, again, are beyond our immediate interest, but a common example of such designed protection is given by AM and FM radio systems. A comparison of the quality of reception of an AM signal and an FM signal during a thunderstorm will immediately indicate that noise is much less significant on the FM channel. This is a result of both the frequency used and the more complex electronics involved in FM radio transmission.

We shall discuss the relative advantages of digital signals in the next section, but it is convenient to examine a simple technique which can be used to provide some degree of noise protection when the signals being transmitted are in digital form. The science of information theory has developed around the topics of transmitting information over channels and protecting the information against noise. Such systematic attempts to build noise immunity into a data transmission link are called *codes*. We present the simplest possible example of such a code to conclude this section.

Consider a case in which a value of temperature is being transmitted from a remote location in the form of a four-bit binary word. The set of possible four-bit words and the associated temperature ranges are shown below:

0000	0–5°C	1000	40–45°C
0001	5–10	1001	45–50
0010	10–15	1010	50–55
0011	15–20	1011	55–60
0100	20–25	1100	60–65
0101	25–30	1101	65–70
0110	30–35	1110	70–75
0111	35–40	1111	75–80

The meaning of this table is quite simple. If, for example, the voltage produced by the transducer indicates that the temperature is between 15 and 20°C, the binary signal 0011 will be sent by the transmitter. This process of transmitting temperatures could occur as frequently as desired.

The receiver would then "see" a series of four-bit binary words. If the experiment is to produce valid results, the four-bit binary word received must be identical to the one sent. This may or may not be the case in a given sending-receiving event since the signal received depends not only on the signal sent, but also on the noise present along the channel while transmission was occurring.

A very simple technique can be used to provide a limited degree of protection against noise. It allows the receiver to determine whether one of the bits in the received four-bit word is incorrect. In such a case the receiver can automatically request a second transmission. The technique is called *parity checking*. As each four-bit word is readied for transmission, the signal-conditioning circuits add a fifth binary bit such that the total number of 1s in the five-bit word is even.

Imagine that the temperatures at three successive times are 57, 52, and 48°. The set of temperatures generates 3 four-bit binary words, 1011 1010 1001.

The parity check bit (the fifth bit) would be added to each of these words so that the total number of 1s would be even. The resulting set of 3 five-bit words would be 10111 10100 10010.

If the following set of words arrived at the receiving station, 10111 10110 10010, the presence of an odd number of 1s in the second word would initiate a request for retransmission. This request could be generated by part of the receiving circuit, or it could depend on an individual monitoring the received signals.

It should be noted that this simple coding technique will not protect against an even number of errors (i.e., 10111 sent and 10001 received) since in such cases the number of 1s in the received word remains even. In addition, the technique will not be able to discriminate between one error and three errors occurring in a single word. In short, if an even number of errors occurs (and this includes zero errors), the parity check will be satisfied. If an odd number of errors occurs, the parity check will generate a request for retransmission.

Much more sophisticated coding techniques exist. In fact, codes exist which will not only indicate the presence of errors in received data but contain, within the received data, the means to correct the errors without requiring a retransmission. This is an important result and serves as a tribute to the ingenuity of those individuals who have developed so-called error-correcting codes.

Data in digital form are a prerequisite if coding techniques are to be used. Transducers, of course, produce signals which are necessarily analog in nature. In the next section we shall discuss the means used to convert signals from analog-to-digital (A/D) form and from digital-to-analog (D/A) form.

Of course, digital signal transmission and storage has advantages for ordinary laboratory use as well as in applications where long-distance data transmission is required. Some relative values for storage capability for digital media are digital memory "sticks" (approximately 13×50 mm): 128 GB; writable CD: 700 MB; writable DVD: 9 GB; and digital hard drives in excess of 3 TB.

14.5   ANALOG-TO-DIGITAL AND DIGITAL-TO-ANALOG CONVERSION

The basic concepts of digital and analog representations of electrical signals were discussed in Sec. 4.2. In analog representations the physical variables of interest are treated as continuous quantities, while in digital representations the quantities are restricted to discrete, noncontinuous "chunks." A further refinement of the digital representation was presented in the last section, where a range of temperatures of interest was broken into separate segments, each 5° wide. Then a separate four-bit binary word was used to represent each segment. This is a two-step digital representation; the first step involves discretizing the analog quantity of interest (the temperature in this case), and the second assigns a different four-bit word to each range. It is clear that 2^n different ranges can be designated by means of an n-bit binary word. For example, the selection of an eight-bit word allows the physical signal of interest to be discretized into 2^8 or 256 segments. The phrase *analog-to-digital conversion* is used to describe this kind of signal processing. The immediate question of interest concerns the value of such conversions.

The first reason for converting signals from analog to digital form involves noise immunity during transmission. The problem of determining whether or not a single digital pulse (either a binary 1 or binary 0) is present at a given time is relatively simple compared with the task of determining whether a voltage is, for example, 48 V or 4.81 V or 4.79 V, and so forth. In crude terms, it is easier to "see electrically" the presence or absence of a pulse than to discern the precise value of an analog signal in the presence of noise induced along the transmission path.

In addition, the techniques of coding, which were discussed in the last section, have been developed only for digital signals. This means that in order to take advantage of the error-detection and error-correction capabilities of these codes, it is essential to convert all the data signals of interest to digital form.

The third major reason for the widespread use of digital representations centers on the data processing capabilities of the digital computer. After all, the digital computer, by its nature, manipulates data only in digital form. Therefore, an A/D conversion must be made if one wants to carry out signal processing using digital computers. One example should suffice to illustrate the attractiveness of digital computers for many types of signal analysis.

In statistical studies it is often necessary to evaluate expressions of the form

$$\overline{y(t)\,y(t - \tau)}$$

where $y(t)$ is some physical signal of interest, the overscore indicates a time average, and τ is a time delay. Such so-called correlation functions are important in analyzing the statistical properties of physical signals. It is necessary that the time average be calculated for different values of the time delay. In some cases such correlation functions are evaluated between two different physical signals, that is,

$$\overline{x(t)\,y(t - \tau)}$$

In order to carry out the evaluation of the time average in analog form, it is necessary to take the signal $y(t)$ and delay it by the quantity τ. How can this be accomplished for a typical electromagnetic signal? The most obvious and direct method is physically to delay the signal by propagating it over a sufficiently long path. Suppose, for concreteness, that a time delay of 1 s was required in some application. Since electromagnetic signals travel with the velocity of light, a time delay of 1 s can be obtained only by using a signal path which is 300 million m in length (a distance not easily available in the normal laboratory!). Even a time delay of 1 μs requires a rather cumbersome 300 m. The reason for this difficulty is the high velocity with which electromagnetic waves travel. There are two basic ways out of this dilemma.

The first involves converting the analog signal of interest, $y(t)$, to an equivalent digital form. The digital values of the signal are then stored in a large computer memory, with a value for each discrete time stored in a different memory location. Once this storage has been completed, the computer can be programmed to carry out the appropriate time averaging. For example, it could "fetch" $y(0)$ and $y(1)$, multiply these two numbers together, and continue this "fetching" and multiplying process until the time average had been calculated. What we have done is to replace a very long delay line with a computer. Complicated data processing is frequently easily performed on desktop personal computers. In order to take advantage of the computer a signal conversion from analog to digital form must be carried out. Although we cannot go into other applications, it must be mentioned that the calculation of correlation functions is only one of the useful processing steps which can be accomplished using digital representations of signals and a computer.

SAMPLE RATES AND ALIASING

In performing an A/D conversion, one must sample the waveform at some periodic rate. The faster the sampling rate, the more closely the analog waveform may be described. If the sampling rate is too low, substantial errors may be experienced. Consider a simple periodic voltage signal expressed by

$$E(t) = A \sin (2\pi f t + \phi)$$

where f is the frequency in hertz and ϕ is the phase shift. Suppose that this signal is sampled in time increments of δt over one waveform such that there are N time increments. At each integral sample the voltage would then be

$$E(n\delta t) = A \sin (2\pi f n\delta t + \phi) \qquad \textbf{[14.5]}$$

where $n = 0, 1, 2, \ldots, (N - 1)$. Because of the periodic waveform, it would be possible to obtain the same amplitude at some different frequency, such that

$$\sin (t) = \sin (t + 2\pi a) \qquad \textbf{[14.6]}$$

where a is an integer. Writing Eq. (14.5) in this form, we obtain

$$E(n\delta t) = A \sin (2\pi f n \delta t + 2\pi a + \phi)$$

or
$$E(n\delta t) = A \sin \left[2\pi n \delta t \left(f + \frac{m}{\delta t} \right) + \phi \right]$$ **[14.7]**

where, now, $m = a/n$, another integer. So, we obtain the same amplitude at frequencies of f or $f + m/\delta t$. The latter frequency is called an *alias frequency*.

Aliasing may or may not be a problem. It may be alleviated by adhering to a sampling theorem which stipulates that the sampling rate f_s be at least twice the highest frequency component of the analog signal. For mechanical systems the frequency components of signals are relatively low (less than 100 kHz) so that electronic sampling at twice the signal frequency is an easy matter.

An alternate method of avoiding aliasing is to filter the signal before sampling. Suppose a complex signal is to be sampled which contains frequency components as high as 30 kHz, but one is only interested in the components up to 10 kHz. A filter could be installed to filter out components above 10 kHz (see Chap. 4) and then sampling could take place at as low a rate as 20 kHz (twice the 10-kHz signal). If the entire 30-kHz signal were preserved, sampling would have to take place at a rate of at least 60 kHz to avoid aliasing.

The two principles which must be observed in the A/D sampling process follow:

1. Sampling must take place at a rate at least twice the highest frequency component in the signal.

2. If the frequency content of the signal is unknown, frequencies above the largest frequency of interest should be filtered out and the highest frequency of interest must be less than half the sampling rate. This action is particularly important when unknown higher harmonics may be present in the signal. For example, a complex sound source may contain harmonics above 20 kHz, but for hearing purposes people cannot hear above this frequency, so they may be filtered out without the loss of significant information.

DIGITAL-TO-ANALOG (D/A) CONVERTER

Figure 14.5 illustrates a schematic for an eight-bit digital-to-analog (D/A) converter. An operational amplifier is connected to the resistance network to act as an inverting summer or adder (see Sec. 4.9, Fig. 4.32h). A stable reference voltage E_{ref} is applied to the network and provision is made to switch the resistors into the amplifier. Note that the resistors are progressively increased in value by factors of 2.

Switch 7 controls the most significant bit (MSB) because it presents the largest voltage to the amplifier input. In contrast, switch zero represents the least significant bit (LSB) because it presents the lowest voltage to the amplifier input. The switches are integrated circuit devices which are activated by the incoming digital signal to allow input to the operational amplifier.

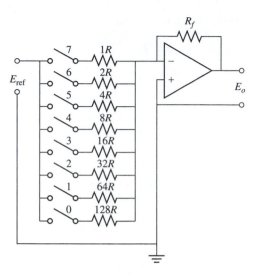

Figure 14.5   Eight-bit digital-to-analog converter.

The output voltage of the amplifier, E_o, is thus

$$E_o = -E_{\text{ref}} R_f \left(\frac{1}{R} + \frac{1}{2R} + \cdots + \frac{1}{128R} \right) \qquad \textbf{[14.8]}$$

The resistor R_f may be selected to produce the desired output voltage. To see how the device operates, suppose we have an incoming eight-bit word 11000001. The sum of the inverse resistances above would be

$$\left(\frac{1}{R} \right) \left(1 + \frac{1}{2} + 0 + 0 + 0 + 0 + 0 + \frac{1}{128} \right) = \left(\frac{1}{R} \right)(1.5078125)$$

The digital equivalent to the binary word is

$$2^7 + 2^6 + 0 + 0 + 0 + 0 + 0 + 2^0 = 193$$

which is also equal to

$$193 = 2^7 \times 1.5078125$$

Both the reference voltage and resistor R_f may be adjusted (designed) to produce a desired output range for the device. Typical practical D/A devices produce output full-scale readings of 0 to 10 V or ±5 V. Obviously, greater resolution may be obtained by using a larger number of bits. Thirty-two-bit and sixty-four bit devices are very common.

ANALOG-TO-DIGITAL (A/D) CONVERTER

Two types of analog-to-digital (A/D) converters will be described: (1) the successive approximation comparator type and (2) the dual-slope integrator type.

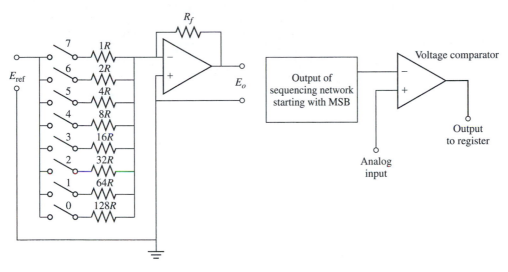

Figure 14.6   Successive-approximation analog-to-digital converter.

A successive-approximation A/D converter is illustrated in Fig. 14.6 for an eight-bit configuration. A stable reference voltage E_{ref} is fed to the resistance and switch network as in the D/A converter. The output of the network is fed to an operational amplifier as before, but now the output from the op-amp is fed to a voltage comparator which compares the signal with a given analog input. An integrated circuit switch sequences through the resistor network beginning with the smallest resistor (MSB), which produces the largest input to the op-amp. The output is compared with the analog input signal—if it is too large, a zero is recorded, if it is too small, a unity bit is recorded. The process is continued through to the least significant bit and the sequence of 0s and 1s represent the value of the analog signal. This digital information may then be placed in memory or manipulated for whatever purposes are desired.

The successive-approximation A/D converter has the advantage of rapid conversion time on the order of $2\,\mu s$ for an eight-bit device. Sixteen-bit devices have conversion times on the order of $30\,\mu s$. This converter has the disadvantage that if the input signal changes during the bit-by-bit process, an error will be experienced in the final representation. To alleviate this difficulty, one may employ sample-hold amplifiers (SHA) which quickly sample the input signal and hold it during the conversion process. With SHA devices, scan rates exceeding 10 MHz may be achieved in commercial applications.

A dual-slope integrator A/D converter is illustrated in Fig. 14.7. In this device the incoming analog signal is digitized in time steps which increase by factors of 2 up to 2^M, where M is the number of bits. The signals are then fed to an integrator (Sec. 4.9, Fig. 4.32k) and operated for a predetermined time interval T. An accurate reference voltage of opposite polarity is then impressed on the same integrator and the time Δt measured for the output of the integrator to return to zero. Because the

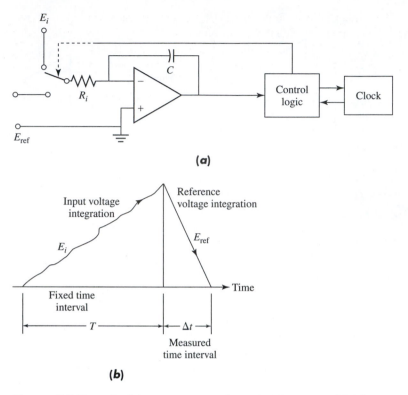

(a)

(b)

Figure 14.7   Dual-slope integrator analog-to-digital converter. (a) Schematic; (b) voltage integration.

resistor-capacitor components of the integrator are the same in both processes, the input voltage E_i is related to the reference voltage E_{ref} by

$$E_i = E_{ref}\left(\frac{\Delta t}{T}\right)$$ **[14.9]**

Dual-slope converters have the advantages of high accuracy and the ability to reject 60-Hz line frequency noise when the time period T is an even multiple of this frequency. The conversion rates for dual-slope converters are slow, of the order of 30/s. This conversion rate is quite satisfactory, however, in applications which change slowly with time, as in many thermal and fluid systems.

RESOLUTION AND QUANTIZATION

The *resolution* of the converters discussed above is expressed as the fraction of the LSB to the largest number represented in terms of the full-scale voltage of the device.

$$\text{Resolution} = Q = \frac{E_{FS}}{2^M}$$ **[14.10]**

where M is the number of bits. Thus, an eight-bit converter has a resolution of one part in 256. The *quantization* of the device is defined as $\pm\frac{1}{2}$LSB. So, an eight-bit A/D converter with an E_{FS} of 10 V could detect a minimum signal of $10/256 = 0.0391$ V. If the device receives an input signal less than this value, it will indicate a value of zero.

If an A/D converter receives an input signal greater than the full-scale E_{FS} for which it is designed, it will be incapable of indicating the correct value. In this case we say that a *saturation error* results. In other words, once the full scale is reached, all the bits are filled and the device cannot indicate a higher value. For an $E_{FS} = 10$ V an input signal of 12 V would be indicated as only 10 V, producing a rather substantial error. To alleviate this difficulty, one employs input conditioning circuits to adjust the range of voltage(s) presented to the converter.

The above discussion indicates, of course, that signal amplification may be necessary in some applications. The eight-bit converter with $E_{FS} = 10$ V mentioned above would be totally unsuitable for thermocouple applications because its resolution is only 39 mV, a value far above the output of most thermocouples.

In practice, both D/A and A/D converters are constructed on single integrated circuit chips, and are available in a myriad of configurations for different applications.

When numbers are represented by the binary code of successive powers of 2, it is said to be a straight binary code. An alternate scheme is to represent each digit of a decimal number by a separate four-bit binary code. (Four bits are required to represent the largest digit, 9.) This representation is called a *binary coded decimal* (*BCD*) and is frequently used for data transmission. The four-bit digit codes are listed below:

Digit	Four-Bit Binary Code
0	0000
1	0001
2	0010
3	0011
4	0100
5	0101
6	0110
7	0111
8	1000
9	1001

The decimal number 846_{10} would thus have a BCD code of 1000 0100 0110.

To express both positive and negative numbers in binary form a convention is used which employs the MSB as a sign indicator. Zero designates positive numbers and 1 designates negative numbers. This convention is termed a *bipolar* or *complementary* code. Thus, -9 would be expressed as 11001 and $+9$ would be expressed as 01001 in this system.

In specifying the performance of instruments, manufacturers frequently will state that the A/D converter is a 16-bit, bipolar, 15-bit resolution, indicating that one bit is used for sign designation.

Example 14.1

A/D CONVERTER FOR THERMOCOUPLE.   A 12-bit A/D converter is to be employed with a chromel-alumel thermocouple such that resolution of 1°F is obtained at 100°F. What is the maximum full-scale voltage that may be accommodated for this application?

Solution

From Table 8.3 we have the following data for chromel-alumel thermocouples:

$T(°F)$	$E(mV)$
50	0.412
150	2.667

The sensitivity at 100°F is therefore

$$S = \frac{(2.667 - 0.412)}{(150 - 50)} = 0.02254 \ mV/°F$$

Therefore, for the 1°F resolution the A/D converter must have a resolution of 0.02254 mV; so, we set

$$0.02254 \ mV = Q = \frac{E_{FS}}{2^{12}}$$

and obtain

$$E_{FS} = (0.02254)(2^{12}) = 92.324 \ mV$$

The maximum recommended temperature for this type of thermocouple is about 2500°F, which corresponds to an output of 54.8 mV (from Table 8.3). Thus, the 12-bit A/D converter with $E_{FS} = 92$ mV should provide a 1°F resolution over the entire range of the thermocouple. An instrumentation amplifier, of course, would most likely be employed before the signal is presented to the A/D converter input. Input conditioning circuits to adapt to the thermocouple characteristic may also be used.

Example 14.2

RESOLUTION OF A/D CONVERTER.   For the thermocouple in Example 14.1, what temperature resolution would be expected with a 16-bit A/D converter operating with a full-scale voltage of 100 mV?

Solution

The resolution for this converter is

$$Q = \frac{E_{FS}}{2^{16}} = \frac{100 \ mV}{2^{16}} = 0.00153 \ mV$$

The thermocouple sensitivity was 0.002254 mV/°F, so this resolution corresponds to

$$\Delta T = \frac{0.00153 \ mV}{0.02254 \ mV/°F} = 0.068°F$$

Example 14.3

NUMBER OF BITS FOR SPECIFIED RESOLUTION.   An A/D converter is to operate with a full-scale voltage of 10 V. How many bits should be employed to obtain a resolution of 0.01 percent?

Solution

We set the resolution equal to 0.01 percent of the full-scale voltage and obtain

$$\frac{Q}{E_{\text{FS}}} = 0.0001 = \frac{1}{2^M}$$

Solving for M, we obtain

$$M = \frac{\ln(1/0.0001)}{\ln 2}$$
$$= 13.29$$

which should be rounded up to 14 bits. For the 14-bit converter the resolution would be

$$\frac{Q}{E_{\text{FS}}} = \frac{1}{2^{14}} = 0.000061 = 0.0061\%$$

14.6   DATA STORAGE AND DISPLAY

Data storage and data display do not always require different apparatus. The relatively simple methods of data display often provide simultaneous storage. Examples include the xy plotter, the strip-chart recorder, paper tape output from a calculator or a computer, and even the recording of experimental data by pen in a notebook. The data are both displayed and stored in all these cases. A storage oscilloscope also provides a dual function because in such a device there is at least temporary storage of the data for as long as the instrument is left on.

There are also devices which are intended to provide only a storage function. Such storage media are most useful if the data stored are easily retrievable for display or subsequent processing. As an example, let us consider the problem NASA has in transmitting television pictures to earth from deep-space probes. Normally, the images "seen" by the cameras are stored in digital form in memory devices on board the craft. When a transmission is made to earth, the stored data are sent, bit by bit, using appropriate coding techniques, to earth, where the received information is again stored on magnetic devices. This information is digitally filtered to reduce the effects of noise, and the resulting image is printed. In addition to magnetic tape, other forms of more or less permanent storage are CDs, magnetic cores, magnetic disks, and various kinds of semiconductor memories. Semiconductor memories are of both the permanent and volatile type. Stored information in the latter type may be lost if the power is removed.

Many technologies and associated devices compete for dominance in the display area. We mention only two. The first is the light-emitting-diode (LED) technology, in which semiconductor diodes which emit light under positive voltages are used as light sources. These devices have the significant advantage of having no filaments or other parts that can burn out; as a consequence, they have very impressive operating lives. They do require significant amounts of power compared with other technologies and tend to "wash out" in bright ambient lighting. The power requirement is the reason that digital watches with LED displays are not constantly on. The user must press a button to cause a brief time display. If the displays were left on constantly, batteries would last only a few days or weeks.

The second technology is based on liquid crystals. These are materials which display different light-reflecting properties when voltages are applied to them. By sandwiching a liquid crystal between two thin transparent metal electrodes, a display can be easily fabricated. This display requires less power than a LED array, but since it depends on light reflection, it is not easily visible in dimly lit surroundings and not at all visible in the dark. Advances in portable computer and video display technology offer a broad range of opportunities for use in instrumentation output devices.

14.7 THE PROGRAM AS A SUBSTITUTE FOR WIRED LOGIC

The general data acquisition system which we have been discussing often serves many functions. In addition to gathering, processing, storing, and displaying data, it often is required to perform a control function. In the case of a nuclear power plant an indication in the data acquisition system that a nuclear core temperature is exceeding some operational limit will result in an automatic shutdown of the nuclear reaction. The entire data acquisition and control system must be able to make logical judgments. It is expected that this system will be backed up by human judgment as well, but for maximum safety the electronic system itself must be capable of such decisions. How is the control function fulfilled? A simple schematic answer is presented in Fig. 14.8. This sketch shows the familiar elements of a transducer and the subsequent data acquisition and display functions. In addition, there is a new element in which a decision is made electronically. In this module the question, "Is the temperature T

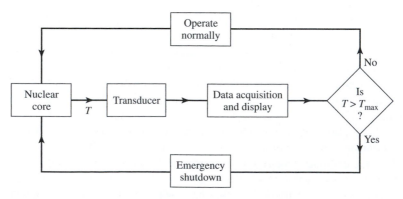

Figure 14.8   Program substitute for wired logic.

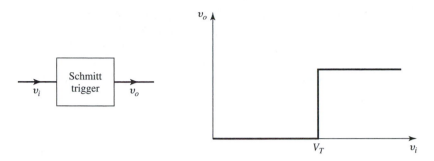

Figure 14.9   The Schmitt trigger.

greater than the safe operating temperature T_{max}?" is electronically answered. If the answer is "yes," the system generates an electronic signal which begins an automatic emergency shutdown of the nuclear reaction. If "no," the system indicates that all is well and allows the core to continue to operate.

In order to implement this logical decision-making process a single integrated circuit called a Schmitt trigger may be used. The output-input characteristic of such a device is shown in Fig. 14.9. The threshold voltage V_T can be electrically controlled. This voltage is set to be the voltage produced by the transducer in Fig. 14.8 when the nuclear core temperature is equal to T_{max}. As long as the actual temperature is below this value, the voltage output of the Schmitt trigger will be zero. Once the safe operating temperature limit is exceeded, the Schmitt trigger will have a positive voltage at its output, and this voltage may be fed, through appropriate interface circuitry, to the equipment which initiates the emergency shutdown procedure. This example illustrates a segment of electrical engineering known as *logic design,* the purpose of which is to develop circuits which can perform logical operations and comparisons. Subsequent action, or the lack of it, depends on the outcome of these logical operations. Although the example given above is analog in nature, circuit designers have tended to concentrate in the area of digital logic circuits. Again, this preference for digital rather than analog representations results from the higher noise immunity of digital circuits. After all, in many industrial situations logic circuits must operate in quite noisy electrical environments.

As an example of a reasonably simple logic design problem, let us consider the task of preparing a circuit which will monitor a traffic control signal and generate an error message when an illegal combination of lights appears. This signal could be routed to maintenance headquarters to alert the staff to the existence of a traffic signal problem. In the state in which this is being written legal light combinations are red only, yellow only, green only, and red-yellow together. All others are illegal. Figure 14.10 illustrates the solution to this problem in the form of a block diagram. It should be noted that error messages occur whenever three or no lamps are lit as well as when any two lamps other than red-yellow are lit. In any of these cases an error signal must be generated as indicated in the diagram. The logical processing shown in this problem can be carried out by means of digital logic circuits. To illustrate this

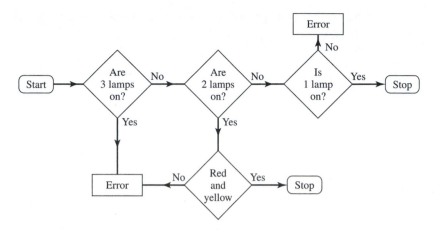

Figure 14.10   Logic for lamp circuit.

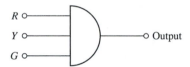

Figure 14.11   Logical AND gate.

briefly, imagine that we have available three electrical wires; the first of these will be at a positive voltage if the red lamp is lit and 0 V if the red lamp is out, the second line is tied to the yellow lamp, and the third is tied to the green lamp in analogous fashion. In order to decide if all three lamps are on, a logical AND gate is used as sketched in Fig. 14.11. This gate will produce a positive output voltage only if each of the inputs is positive. In this case an output will occur only if all three of the lamps are simultaneously lit. Since this is an error condition, this output signal could be used to generate the necessary error message. In an analogous fashion, other gates are used to check if any other error conditions exist. In addition to AND gates, there are other classes of digital logic circuits, and, indeed, the designing of circuits containing these decision-making gates is a separate area of study.

Our purpose in discussing this topic here is not to educate the reader thoroughly in the area of logic design. This would be impossible in the space of one section within one chapter of a text devoted to experimental methods. The salient point is that logical decisions can be made by wiring together various gates and connecting them in the appropriate way to the input and output circuits. Such systems are called *hard-wired* logic arrays. In order to change the function of a given logic system it is necessary physically to rearrange and reconnect the gates. In situations in which many different logical functions must be performed this leads to a great proliferation in the types of logic arrays which are manufactured. Microprocessors now provide universal logic circuits that may be programmed to perform needed logic functions for a particular application [2].

Microprocessors which are available today are in the form of one circuit containing many leads. Some of these leads provide interconnection to the required dc power supplies and ground, while others serve as interfaces over which data can be fed into the system and results fed out of the system. An external digital clock must also be connected to the unit to provide the timing necessary for the step-by-step digital processing. The chip itself contains a memory in which data *and* a user-generated program may be stored. In addition, the microprocessor has available a set of registers in which arithmetic operations may be performed.

Within limits set by the available instructions, the size of the memory, the interface circuits, and the cleverness of the designer, the microprocessor is a universal logic circuit. It may be programmed to perform a particular desired set of logical procedures. Rather than having to manufacture a wide array of custom-designed circuits, the semiconductor manufacturers are free to concentrate on producing only one microprocessor design.

14.8 SUMMARY

This chapter began by considering the entire data acquisition, conditioning, transmitting, storage, and display process. We developed some additional concepts in the areas of filtering and signal conditioning and surveyed some of the technologies available to carry out signal processing in digital form.

Because of the rapid advances made in digital data acquisition systems, the experimentalist will be well advised to consult manufacturers' catalogs and representatives for the latest information. To aid in understanding these materials, we give here a brief glossary of terms used to describe such systems. Although some of these terms have been mentioned before, it is convenient to assemble them in one place.

14.9 GLOSSARY

A/D or ADC This term refers to analog-to-digital converters discussed previously.

ASCII This is the abbreviation for American Standard Code for Information Interchange which sets standard binary codes for various symbols to be used for transmission of data.

asynchronous communication In this mode of communication data are sent when ready, in contrast to being sent on the basis of a predetermined signal from a time clock.

baud This term is used to designate data transmission speed in bits per second (BPS).

buffer When input and output (I/O) operations operate at different speeds, a buffer is used which stores the data until they are ready for use or transmission.

bus This term refers to the parallel lines which are used to transfer signals between components.

byte This term refers to a character in binary; equal to eight bits.

CPU This abbreviation designates the central processing unit of the acquisition system or the computer.

DMA This term designates direct memory access which means the data may be stored directly in the memory of the computer, such as an IBM PC. This is a high-speed storage mode because the data do not have to go through intermediate transmission and/or storage operations. With DMA, manipulation of the data may begin immediately in accordance with the software supplied.

EPROM This abbreviation designates erasable programmable read only memory.

handshake This term refers to an organized interface procedure which enables orderly transfer of data and data-status information on a predetermined basis. This type of transmission is in contrast to the asynchronous mode.

IEEE standard 488–1981 This term refers to the Institute for Electrical and Electronic Engineers interface bus which is also called a general-purpose interface bus (GPIB) or the Hewlett-Packard interface bus (HP-IB). This standard bus allows bidirectional transmission of 8 eight-bit data lines as well as eight bidirectional control lines. Many instrument manufacturers use this standard to ensure compatibility, *but not all.*

LLEM This is a low-level expansion multiplexer. See multiplexer (analog) below.

multiplexer (analog) The analog multiplexer allows multiple analog signals to be fed to a single input port. It is analogous to a single-pole, multithrow, mechanical, electrical switch.

parallel transmission This term means that all data bits are sent simultaneously. The mode is typically used for communications between computers and printers. Contrasted to serial transmission.

resolution This term refers to the number of bits used to quantize an input signal:

$$8 \text{ bit} = 2^8 \ = 1 \text{ part in } 256$$
$$12 \text{ bit} = 2^{12} = 1 \text{ part in } 4096$$
$$16 \text{ bit} = 2^{16} = 1 \text{ part in } 65{,}536$$
$$32 \text{ bit} = 2^{32} = 1 \text{ part in } 4{,}294{,}967{,}296$$

RS-232 standard This is a serial transmission interface standard which may be used to communicate over telephone lines between data terminal equipment (DTE) and data communication equipment (DCE). The standard involves up to 25 lines, but many are never used.

RS-232 to IEEE 488 controller This is a controller which converts the RS-232 output of a computer into a GPIB.

serial transmission This term refers to the transmission of data one bit at a time as with the RS-232 standard.

In summary, the choices for data acquisition systems are almost endless and are limited only by cost.

14.10 REVIEW QUESTIONS

14.1. What are the major elements of a data acquisition and processing system?

14.2. In what kinds of situations would data transmission be necessary?

14.3. What are the differences between active and passive filters?

14.4. What are the functions of a lowpass active filter?

14.5. Why are digital representations of signals preferred for data transmission?

14.6. Why does coaxial cable exhibit higher noise immunity than signal propagation through free space?

14.7. What is coding?

14.8. What is the function of a microprocessor?

14.9. What is meant by "aliasing"? How may it be alleviated?

14.10. How is resolution for an A/D converter defined?

14.11. What is meant by "saturation error"?

14.12. Distinguish between successive approximation and dual-slope A/D converters. Which type provides for more rapid sampling rates?

14.13. How is bipolar (+ or −) coding performed with binary number representation?

14.14. What is meant by BCD?

14.11 PROBLEMS

14.1. Find the ratio of V_o to V_i for the following circuit, where

$$v_o = V_o e^{jwt} \qquad v = V_i e^{jwt}$$

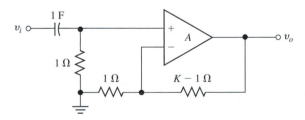

Figure Problem 14.1

14.2. At what frequency ω_0 will the ratio of output to input (V_o/V_i) in the above circuit be 50 percent of the low-frequency ($\omega \to 0$) value?

14.3. Determine V_o/V_i for the following circuit.

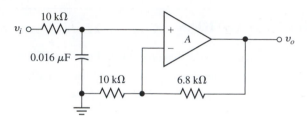

Figure Problem 14.2

14.4. Express the following decimal numbers in binary form: 932, 10721, 36, 9.

14.5. Express the following decimal numbers in BCD form: 932, 10721, 36, 9.

14.6. An iron-constantan thermocouple (Table 8.3) is to be used to measure temperatures from 32 to 250°F. A suitable instrumentation amplifier is employed to provide an input to an eight-bit A/D converter ranging from 0 to 10 V full scale. What temperature increment resolution will be obtained and what amplifier gain is required?

14.7. Using the concepts of decibels defined in Eq. (4.51), express the resolution for 8-, 12-, and 16-bit A/D converters as decibels below full-scale voltage.

14.8. How many bits are required to achieve a resolution of 0.001 percent of full scale?

14.9. How many bits are required to represent a decimal number of 10^{15}?

14.10. An A/D converter is to be designed for a special instrument that will accommodate type N thermocouples (see Table 8.3a) operating over the range of 0 to 1000°C. A resolution of 0.5°C at 25°C is desired. Determine the number of bits required and the maximum voltage operating range.

14.11. If the A/D converter designed in Prob. 14.10 is employed with a type S thermocouple (see Table 8.3a), what will be the resulting temperature resolution?

14.12. A 16-bit A/D converter is used with the thermocouples in Probs. 14.10 and 14.11 as well as a type E thermocouple over the temperature range of 0 to 1000°C. What will be the temperature resolution for each couple at 25°C for a full-scale voltage matching the maximum need?

14.13. Repeat Prob. 14.10 for a resolution of 0.05°C at 25°C.

14.14. The voltage output of a certain device has a dynamic range of 50 dB with a maximum voltage of 100 mV. Determine the number of bits for an A/D converter to ensure a resolution of ±0.1 dB at the 0-dB condition.

14.15. Repeat Prob. 14.14 for a dynamic range of 80 dB.

14.16. A set of standard loudness curves is shown in Fig. 11.12. For the 20-phon, level an A/D converter is to be designed to provide a resolution of ±1 dB at the minimum SPL condition while also providing a voltage output of 1.0 V at the 80-dB sound pressure level condition. Determine the number of bits necessary for this converter.

14.17. An A/D converter is to be designed to accommodate sound sources resembling the *A* curve of Fig. 11.15. What number of bits would be required to provide a resolution of ±0.5 dB at a frequency of 30 Hz and a voltage output of 1.0 V at the maximum point of the curve?

14.18. An A/D converter is to be supplied for an experimental transducer that has a dynamic range of 45 dB and a maximum voltage of 90 mV. Determine the number of bits for the converter to ensure a resolution of ±0.2 dB at the 0 dB condition.

14.19. A 20-bit A/D converter is employed with the thermocouple in Problem 14.10 over the temperature range of 0 to 800°C. What will be the temperature resolution at 25°C for a full-scale voltage matching the temperature requirements?

14.20. For the 40 phon sound level shown in Fig. 11.12 an A/D converter is to be provided to assure a resolution of ±1 dB at the minimum SPL condition while also providing a voltage output of 0.4 V at the 80-dB SPL condition. Specify the number of bits required for the converter.

14.21. Repeat Prob. 14.10 for a resolution of 0.1°C at 25°C.

14.12 REFERENCES

1. Krutz, R.: *Interfacing Techniques in Digital Design,* Wiley, New York, 1988.

2. Money, S. A.: *Microprocessors in Instrumentation and Control,* McGraw-Hill, New York, 1985.

3. Wait, J., L. P. Huelsman, and G. Korn: *Introduction to Operational Amplifiers,* 2d ed., McGraw-Hill, New York, 1992.

4. Mitra, S. K.: *Digital Signal Processing: A Computer Based Approach,* McGraw-Hill, New York, 1998.

5. Huelsman, L. P.: *Active and Passive Analog Filter Design: An Introduction,* McGraw-Hill, New York, 1993.

6. Gopalan, K. G.: *Introduction to Digital Electronic Circuits,* McGraw-Hill, New York, 1996.

7. Antoniou, A.: *Digital Filters: Analysis, Design and Applications,* 2d ed., McGraw-Hill, New York, 1993.

8. De Micheli, G.: *Synthesis and Optimization of Digital Circuits,* McGraw-Hill, New York, 1994.

9. Northrop, R. B.: *Introduction to Instrumentation and Measurements,* CRC Press, Boca Raton, FL, 1997.

chapter

15

REPORT WRITING AND PRESENTATIONS

15.1   INTRODUCTION

Previous chapters have discussed a large variety of instruments and techniques for obtaining and analyzing experimental data. At some point the experimentalist will find it necessary to prepare a written report describing an experimental program and may also need to make a verbal presentation to supervisory personnel or other interested parties. The purpose of this chapter is to furnish some guidelines for assembling such reports and presentations.

First, we will give some general comments on style and overall philosophy of reports. Next, we will discuss the types of reports a person may be called upon to write, followed by a discussion of the different elements of a report. Most engineering reports require some graphical displays, so a section on this subject is in order. Reports are processed in different ways, from simple typed statements to elaborate typeset productions. A brief section on this subject may aid the reader in the report construction phase.

Next, a section on oral presentations is given, with a discussion of audiovisual aids. Finally, some guidelines are offered for conducting meetings, planning sessions, and conferences—a source of a huge amount of wasted time for many engineers and managers.

15.2   SOME GENERAL COMMENTS

Many books have been written on the subject of report writing, and the author has no intention of competing with these works. The importance of good report writing and data presentation cannot be overemphasized. No matter how good an experiment or how brilliant a discovery, it is worthless unless the information is communicated to other people.

Very good advice for those engaged in writing or speaking activities is contained in the aphorism:

A man who uses a great many words to express his meaning is like a bad marksman who instead of aiming a single stone at an object takes up a handful and throws at it in hopes he may hit.—*Samuel Johnson*

Another cogent expression in the same vein is ascribed to Abraham Lincoln:

I must apologize for this long letter, but I didn't have time to write a short one.

In other words, say what you have to say and then shut up! Be succinct but not laconic. When graphs or tables will present the idea clearly, use them, but do not also include a wordy explanation which tells the reader what can be plainly seen by careful inspection of the graph.

Third-person present tense is generally accepted as the most formal grammatical style for technical reports, and it is seldom incorrect to use such a style. In some instances first person may be used to emphasize a point or to stress the fact that a statement is primarily the opinion of the writer. The usual scientific writing style is also in the passive mode. Examples of the two styles are:

Third person. Equation (5) is recommended for the final correlation in accordance with the limitations of the data as discussed above.

First person. We (I) recommended Eq. (5) for the final correlation in accordance with the limitations of the data presented in our discussion above.

In the first-person statement the writer is making the recommendation on a much more personal basis than in the third-person statement. The selection of the proper statement is a matter of idiom, which depends on many factors, including consideration of the person(s) who will read the report. For a formal paper in a scientific journal the third-person statement might be preferable, whereas the first-person usage might be desirable in an engineering report to an individual.

Engineers could learn a few lessons from politicians and business executives. Have you ever noticed the difference in writing style in the annual reports of corporations? If the earnings situation is good, the point is brought forth very quickly and very clearly. Wordiness is avoided in the presentation of good points so that the reader will not miss them. When the profit statement is poor, the discussion becomes more circumlocutious and clear graphical or tabular presentations are seldom employed. The point is that a writer should make sure that the strong points of a report are not buried in excess words. The degree to which weak points are submerged depends on the audience for which the report is intended and the circumstances under which the report is written.

The above comments should not be misconstrued. The engineer should not bury bad data in a presentation in hopes that no one will notice! On the other hand, one should recognize that some engineering reports are written for lay persons, and it is extremely important for the presentation not to give conflicting or confusing results which may be misinterpreted. Of course, a paper for publication in the technical or scientific literature should be completely objective. The purpose of these brief paragraphs has been to illustrate the fact that engineers may, in *some circumstances,* find it necessary to do a bit of selling in their writing. Such writing requires a great amount of skill and experience in order to be effective.

Be specific if you have something to be specific about. Consider the following two statements:

1. An analysis of the experimental data showed that the average deviation from the theoretical values was less than 1 percent. Uncertainties in the primary data were shown to account for a deviation of 0.5 percent. In view of this excellent agreement between Eq. (42) and the experimental data this relation is believed to be an adequate representation of the physical phenomena and is recommended for calculation purposes.

2. The experimental data are in good agreement with the theoretical development. In view of this favorable comparison the assumptions pertaining to the derivation are verified.

Note the difference between the two statements. The first statement is quite specific and leaves the reader with a feeling of confidence in the experimental data and the-oretical analysis. Upon examining the second statement, the reader will immediately ask: How good is "good"; how favorable is "favorable"? The author has seen such statements applied to experimental data that differed from theoretical values by such a large factor that the veracity of the writer might be questioned by a careful reader.

A brief, general procedure for report writing might take the following form:

1. Make a written outline of the report with as much detail as possible.

2. Let the outline "cool" for a period of time while you direct your thoughts to other matters.

3. Go back to the outline and make whatever changes you feel are necessary.

4. Write the report in rough draft form as quickly as possible.

5. Let the report cool for a period of time, preferably a week or so.

6. Go back and make corrections on your draft. You will probably find that you did not say things quite the way you would like to, did not include as much information as you wanted to, or made several stupid mistakes in some of the data analysis.

   Use a hard-copy, expanded-line printout for this proofreading stage. Except for the automated spell-check and grammar-check, most persons will find that a report is easier to edit on a printed page than on a computer screen.

7. If possible, have a colleague scrutinize the report carefully. This person should be one whose competence you respect.

8. Consider your colleague's comments very carefully, and rewrite the report in its final form.

9. During the outline and preliminary writing process be careful to set units and symbols for equations and labels for graphical displays in accordance with the guidelines given in later sections in this chapter.

Once a draft of the report is available, corrections must be made; Table 15.1 shows the accepted nomenclature for proofreader's marks. It is a good idea to get in the habit of using these standard marks.

Table 15.1   Standard proofreader's marks

Delete	lower case Word
Dele_Ate and close up	Capital letter
Quad (one em) space	SMALL CAPITAL LETTER
Move down	**Boldface** type
Move up	*Italic* type
[Move to left	Roman type
Move to right	Wrong font
Equalize spacing between words	Insert space
Broken letter	Close up
Begin a new paragraph	Turn letter
[No new paragraph	Period
Let type stand as set	Comma
Verify or (supply) information	Apostrophe
Transpose letters or marked words	Quotation marks
Spell out ((abbrev.) or 7)	Semicolon
Push space % down	Colon
Straighten type	Question mark
Align type	Exclamation mark
Run in material on same line	Hyphen
Change (x/y) to built-up fraction	En dash
	Em dash
Change x/y to shilling fraction	Two-em dash
Set s as subscript	Parentheses
Set s as exponent	Brackets

Of course, the objective of a report is to communicate the ideas and information gained in the experimental work. The care and skill with which the discussion and conclusions are drawn will determine the overall success of the report.

The above remarks on report writing can be considered as general ones which may be applicable to the broad range of reports, papers, or monographs the engineer will be called upon to write. A carefully constructed outline, of course, is always the best starting point and will help the writer to make sure that all pertinent information is included.

15.3   TYPES OF REPORTS

INFORMAL REPORTS

Informal reports usually follow a letter or interoffice memo format and are brief and to the point. Two examples are given below.

MEMO FORMAT
TO: J. J. Brown
FROM: B. R. Smith
DATE:
SUBJECT: Test of TV Satellite Uplink Quality

In accordance with your request of _____, we conducted tests with the TV satellite uplink on _____. Signals were monitored and videotaped at locations in Boulder, Colorado; Andover, Massachusetts; and Dayton, Ohio. All three sites reported satisfactory video quality at transmission levels below 125 watts. This value is well below the maximum power level of 300 watts for the uplink. Difficulties were experienced with the audio signal caused by 60-Hz interference from the on-desk monitor. The interference was eliminated by better location of the desk microphone or using an alternate lapel mike.

The three receiving sites are returning videotapes of all the tests to my office where they will be available for your inspection. Based on our verbal communications with the sites, we believe that the entire uplink operation is now delivering satisfactory video and audio signals.

LETTER FORMAT
Mr. J. R. Marshall
Acme Development Corp.
501 Main Street
Dallas, Texas 75201

Dear Mr. Marshall:

Following our meeting of _____ we conducted preliminary tests of the HVAC systems in your two-story building at 10123 North Road. The purpose of these tests was to determine if the main unit was performing in accordance with specifications. We made the following measurements:

1. Dry-bulb and wet-bulb temperatures for all inlet and exit airstreams of the main unit

2. Air velocities for these airstreams

3. Leakage from the outside air damper

4. Power consumption of the main unit

Calculations based on these measurements indicate that the main unit is capable of achieving its rated cooling capacity; however, measurements on the outside air damper indicate a leakage of 15 percent when the damper is fully closed. This leakage places an additional load on the cooling unit so that some deterioration in performance is experienced in very hot and humid weather.

We recommend that the outside air damper be replaced; therafter, we would expect entirely satisfactory operation of the system.

Please let us know if we can be of further service.

Sincerely,
R. W. Smith, P.E.
Consulting Engineer

FORMAL REPORTS

Formal reports are usually organized to include several of the elements described in the following section. Lengths obviously vary with the complexity of the report. The format for the report may be specified by the person or organization to receive the document without much discretion on the part of the writer.

PAPERS AND JOURNAL ARTICLES

Papers and journal articles written for the professional engineering or scientific community may certainly be classified as "formal" reports. In many cases the journal or literature of the sponsoring organization will include a section entitled "Information for Authors" which specifies the format to be followed. The writer should pay close attention to such information because nonadherence to the required format may result in rejection of the paper even if it is judged by the referees to have good technical merit.

TUTORIAL REPORTS

A tutorial report may also be classified as "formal," but—in contrast to papers or journal articles—it is not directed at a knowledgeable professional in the field. The purpose of the tutorial report is to teach someone who is not well informed in the subject matter. For example, a manufacturer of flowmeters might ask a member of its engineering staff to prepare a report which describes how the flowmeters work, the difficulties to be encountered in installation, proper selection techniques, and accuracies which may be expected. Such a report would probably be issued as a company brochure directed at possible users of the product. The report must be factual and helpful to the reader, but it does not convey "new" information as would a paper or a journal article.

BOOKS AND MONOGRAPHS

We mention books and monographs mainly for the sake of completeness. Obviously, some large reports are long enough to be called "books," and, indeed, some are eventually published as books. Book publishers frequently have their own standards for format and style.

15.4 CONTENTS OF A REPORT

We have mentioned that a good outline is always the starting point for a good report. To aid in this outline construction, we now give a brief discussion of different report elements. Not all these elements will be included in every report, and the writer must choose those appropriate to the audience.

FRONT MATTER

Front matter includes the title page, with author affiliation(s), sponsor of report activity (if any), table of contents, list of nomenclature, list of figures, preface (if any), and a letter of transmittal if required.

In many cases the letter of transmittal will be a simple interoffice memo for the company. For a major report to a project sponsor the letter may be more elaborate.

A variety of schemes are used for organizing the table of contents. Many authors like to number every paragraph, thereby creating many subheadings. A technical paper for a journal would use no *numbered* headings and would not include a table of contents. Most book publishers do not use more than one subheading in each chapter; that is, Chap. 6 might have Sects. 6.1, 6.2, 6.3, etc., but *not* 6.1.1, 6.1.2, and so on.

Nomenclature lists are normally required for both journal articles and comprehensive technical reports. They may not be necessary in brief reports for a limited audience.

Common practice is to list terms or variables in alphabetical order, followed by Greek or foreign symbols in order, and then followed by subscripts and superscripts. *The units of all terms should be given with the nomenclature list.* If more than one set of units may be used, as with English and SI, both should be stated. Standard abbreviations should be employed for stating the units. Some examples of nomenclature listings are as follows:

a	Local velocity of sound, m/s or ft/s
A_c	Flow cross-sectional area, m^2 or ft^2
k	Thermal conductivity, $W/m \cdot °C$ or $Btu/h \cdot ft \cdot °F$
$Re = \rho u d / \mu$	Reynolds number, dimensionless
μ	Dynamic viscosity, $kg/m \cdot s$ or $lbm/h \cdot ft$
ρ	Density, kg/m^3 or lbm/ft^3

Subscripts

i	Inside
o	Outside
∞	Free-stream conditions

When using built-up units, as with thermal conductivity k above, several choices of style are available, such as:

1. $\dfrac{Btu}{(h)(ft)(°F)}$

2. $Btu/(h)(ft)(°F)$

3. $Btu/h \cdot ft \cdot °F$

4. $Btu/h\text{-}ft\text{-}°F$

5. $Btu/h/ft/°F$

The first four styles all mean the same thing. Styles 1 and 2 are more work for the typist, style 4 is easiest for the typist, while style 3 may be more common in typeset publications. Style 5 does appear from time to time, but the multiple slash marks can be confused for multiple divisions unless the reader knows the subject. In general, style 5 should be avoided. Inclusion of units in the nomenclature list is very important. A reader who is unfamiliar with the terms will find the absence of units very annoying.

Lists of tables and figures are frequently included in large reports but are never used in papers for journal publication. Inclusion of a preface is optional in reports: it is normally used only in monographs (books).

ABSTRACTS

Someone stated an old rule for army methods of communication: Tell them what you are going to tell them; tell them; and then tell them what you told them. The abstract should attempt to accomplish the first objective in a very short, succinct format, without mathematical formulations. It should tell what was done, and the conclusions which resulted from the work. Keep in mind that there are many people who will read *only* the abstract of a report because of heavy demands on their time. This is especially true of managers who must review a large number of reports on a broad range of subjects. A well-written abstract becomes particularly useful for those people, and may arouse their interest in examining the report in more detail.

Some writers may choose to use the terms "summary" or "executive summary" instead of abstract, but the purpose of the section is the same.

INTRODUCTION

In some reports the introduction will take the place of a preface. In a very extensive report or book, both a preface and an introduction may be employed, while in brief reports neither may be used. The purpose of the introduction is to lay the groundwork for the more detailed discussions in the body of the report. Sometimes a review of previous work is given in the introduction.

Sometimes the introduction section is used to state clearly the motivation for performing the work, i.e., to define the problem.

BACKGROUND AND PREVIOUS WORK

A survey of the literature may be appropriate and can be included under this title. It could also be included in an introduction.

This section can vary widely in length, depending on what one assumes for the knowledge level of the intended reader. The length also depends on what the reader *needs to know* in order to understand and appreciate the remainder of the report. It is rather easy to go off the deep end in this section because with most technical subjects there are great volumes of background and previous work. The writer would be wise to keep this section focused so that it both acknowledges previous work and points to the need for the current study.

Because of space limitations, most technical papers and journal articles try to minimize the length of this section.

THEORETICAL PRESENTATION(S)

In some reports a large section will be devoted to development of theoretical information applicable to the subject. This section enables the reader to understand the implications of the experimental work and aids in proper interpretation of the data.

Of course, some reports are strictly theoretical in nature, and thus this section forms the main body for those reports. To encourage more efficient use of the reader's time, some writers may use this section as a vehicle to summarize theoretical presentations and relegate long detailed derivations to appendices. In this way the reader may get to the heart of the presentation more quickly while retaining the option of examining details later.

In the theoretical presentation(s) one has the option of defining units for the various terms as they are introduced or specifying the units only in the nomenclature list. The latter practice is preferable for papers because it conserves space. For unusual variables the units might be defined when they are introduced.

Display of mathematical formulas should follow standard practice in the field at the discretion of the writer. In many cases formulas will be programmed with variables for computer use. For these cases the standard for the particular computer language should be followed.

PRESENTATION OF EQUATIONS

The technical writer has several choices of format for presentation of equations. When the equations are intended for direct input to a computer, one may wish to write the equation in computer format. The following compound interest formula

$$S = \frac{(1 + I)^N - 1}{I} \qquad \textbf{[15.1]}$$

could thus be written as a computer statement

$$S = ((1 + I)^N - 1)/I \qquad \textbf{[15.2]}$$

but this would be awkward to read in a text. Equation (15.1) could also be written as

$$S = ((1 + I)^N - 1)/I \qquad \textbf{[15.3]}$$

or

$$S = [(1 + I)^N - 1]/I \qquad \textbf{[15.4]}$$

where Eq. (15.3) uses sequential parentheses (( )), while Eq. (15.4) uses both brackets and parentheses. For text purposes Eqs. (15.3) and (15.4) are usually preferable to Eq. (15.2). Equation (15.3) is easier for the typist. The general rule is to move outward from parentheses to brackets to { } signs. So, if Eq. (15.4) were to be multiplied by some factor P, the expression might appear as

$$S = P\{[(1 + I)^N - 1]/I\}$$

If more than the ( ), [ ], or { } notations are required, it is usually best to define some new variables. A multiplication sign ($\times$) may or may not be used after the P.

We say that Eq. (15.1) uses a *built-up* process for presentation of division and that Eq. (15.3) uses a shilling *fraction* form (/) for division. For complex equations both techniques are usually employed. For example, the following equation from Chap. 2 uses both built-up and shilling fractional forms:

$$x = \frac{(F_0/k) \cos(\omega_1 t - \phi)}{\{[1 - (\omega_1/\omega_n)^2]^2 + [2(c/c_c)(\omega_1/\omega_n)]^2\}^{1/2}} \qquad \textbf{[15.5]}$$

If the built-up form were used for the denominator, the equation would appear as

$$x = \frac{\left(\dfrac{F_0}{k}\right) \cos(\omega_1 t - \phi)}{\left\{\left[1 - \left(\dfrac{\omega_1}{\omega_n}\right)^2\right]^2 + \left[2\left(\dfrac{c}{c_c}\right)\left(\dfrac{\omega_1}{\omega_n}\right)\right]^2\right\}^{1/2}} \qquad \textbf{[15.6]}$$

Either form is correct, but Eq. (15.5) is easier to type and appears less clumsy. On the other hand, the following radiation heat-transfer formula presented in built-up form

$$q = \frac{A_1\left(T_1^4 - T_2^4\right)}{\dfrac{A_1 + A_2 - 2A_1 F_{12}}{A_2 - A_1(F_{12})^2} + \left(\dfrac{1}{\epsilon_1} - 1\right) + \dfrac{A_1}{A_2}\left(\dfrac{1}{\epsilon_2} - 1\right)}$$

would look awkward in fractional form. The writer must exert discretion in these matters.

EXPERIMENTAL APPARATUS AND PROCEDURE

Sufficient information must be supplied on the apparatus and experimental procedures for the reader to understand what was done. If the report is concerned with research and new knowledge, a rather detailed discussion of the apparatus may be necessary. If test results are being reported in accordance with standard procedures (ASME, ASTM, etc.), then the appropriate procedures can be cited without furnishing details in the report. Any deviations from the standard procedures should be noted.

Great variations in length of this section can be used. An extensive report might include all details of the apparatus and instrumentation, while a technical paper or journal article would give only a brief summary.

RESULTS OF EXPERIMENTS

One normally will include a separate section in the report to give results of the experiments in a form which is consistent with the needs of the intended audience.

Clear tabular and graphical presentations should be used as much as possible to conserve words in the report. Of course, some verbal discussions of the graphs and tables must be given, but such discussion should focus the reader's attention on salient features of the data, not just recite numbers or parameters which are obvious upon inspection. In other words, the written and graphical presentations should be complementary so the reader's time is conserved.

Section 15.5 on graphical presentations gives suggestions for good practice.

INTERPRETATION OF RESULTS

Once the experimental results have been presented in a clear form, the author has a responsibility to interpret the results in the light of the theoretical presentation and the work of others in the same general subject area. In this section the background, theoretical presentation, and experimental results are all brought together to lead the reader to the conclusions of the study.

Of course, there are many instances where one is concerned only with the presentation of results and no interpretation is required. For instance, a calibration test for a thermocouple calls only for a calibration curve, table, or polynomial relation. The results speak for themselves.

CONCLUSIONS AND RECOMMENDATIONS

By the time the reader has reached this section most of the conclusions of the work should already have been drawn. The object of the conclusion *section* is to collect all the important results and interpretations in clear summary form. This is the section which tells the reader what was covered in the body of the report. There will be many readers who will read only the abstract or conclusion sections of a report, so it is especially important that they be carefully written.

Some writers like to include recommendations for action or further study in this section. A typical action statement might be:

> Calibration of the model 802 flowmeter has now been completed and production runs may begin immediately.

A further-study recommendation could be:

> The calibration procedure for the model 802 flowmeter has now been established and reliable data obtained for the temperature range of 50 to 100°F. It is recommended that further data be obtained up to a temperature of 300°F. Once these data are obtained, production runs may begin.

ACKNOWLEDGMENTS

Many people may contribute to a project who are not listed as formal authors of the report. An acknowledgment section may be used to recognize these contributions. For technical papers it is normally placed after the conclusions section, while for other reports it may appear as a part of front matter.

APPENDIX MATERIALS

Wide latitude is available to the writer in the types of material which may be placed in appendices. Of course, the writer may choose to have no appendix at all and include all information in the body of the report. Otherwise one or more of the following might appear in the appendix:

1. Detailed mathematical derivations which are summarized in the body of the report, as described under "Theoretical Presentation(s)" above
2. Tabulations of raw experimental data which are summarized or correlated in the body of the report
3. Calibration information for instruments or sensors employed in the experiments
4. Uncertainty analysis of the experiments, the results of which may be stated or summarized in the body of the report
5. Tabulations or graphs of material properties used in the report
6. In some cases calculation charts or materials obtained from other sources
7. Detailed computer programs used in the work, which may be referenced in the report

We can see that the general purpose of the appendix is to remove detailed clutter from the body of the report so that the reader may appreciate the work and conclusions in a shorter period of time. For those interested, the details are still available for study. Most technical papers and journal articles do not have much appendix material (if any at all) because the very object of these papers is to present just the essence of the work.

REFERENCES AND BIBLIOGRAPHIES

Most technical reports require references to the work of others. References should be cited when a work was used in writing the report. For example, the statement:

Cheesewright [1] discusses turbulent natural convection ...

with a citation:

1. Cheesewright, R.: Turbulent Natural Convection from a Vertical Plane Surface, *J. Heat Trans.,* vol. 90, p. 1, 1968.

is a valid reference citation. On the other hand, a standard textbook on heat transfer might have general information on free convection and yet not be employed as a specific reference for the report. It is common practice to list such sources under the title of "Bibliography." Normally, only extensive works like textbooks or monographs have separate bibliographies. Because of the extra space consumed, bibliographies are almost never used in papers for journal publication.

CITATION FORMATS

The citation listing shown above, with a numbering of the references, is an acceptable format. For technical papers the numbering of the references should usually be in the order in which they are cited in the body of the paper. For a report, this sequence may also be desirable but is not required. The advantage of this citation format is that it conserves space if the writer chooses to leave off the author's name(s) in the citation, particularly so when citing multiple references. As an example, the citation

> Turbulence effects are described in Refs. [1, 3, 5] . . .

saves considerable space by leaving out several author names. A second advantage of this style of reference citation is the quick visual access afforded the reader by a numerical listing.

Some latitude is available in the manner of literature citation. The above reference could be listed without its title, i.e.,

> 1. Cheesewright, R.: *J. Heat Trans.,* vol. 90, p. 1, 1968.

Although such citations may be accepted by various journals, and do conserve space, they should be avoided whenever possible because the absence of the article title is very annoying to a reader. The object of a report is to communicate. A procedure which impedes communication is to be shunned.

An alternate citation technique is to list references in alphabetical order of the first author, then second author, and then year. No specific reference number is given and citation in the body of the report refers to the year of publication. When the same author has more than one citation in one year, the year is followed by a, b, c, and so on. For one or two authors both names are cited; for more than two authors only the first author followed by "et al." is cited. The following examples illustrate this technique.

1. Lee, Y., Korpela, S. A., and Horne, R. N., 1982, "Structure of Multi-Cellular Natural Convection in a Tall Vertical Annulus," *Proceedings, 7th International Heat Transfer Conference,* U. Grigull et al., eds., Hemisphere Publishing Corp., Washington, D.C., vol. 2, pp. 221–226.
2. Kwon, O. K., and Pletcher, R. H., 1981, "Prediction of the Incompressible Flow Over a Rearward-Facing Step," Technical Report HTL-26, CFD-7, Iowa State Univ., Ames, Iowa.
3. Sparrow, E. M., 1980*a*, "Fluid-to-Fluid Conjugate Heat Transfer for a Vertical Pipe-Internal Forced Convection and External Natural Convection," *ASME Journal of Heat Transfer,* vol. 102, pp. 402–407.
4. Sparrow, E. M., 1980*b*, "Forced-Convection Heat Transfer in a Duct Having Spanwise-Periodic Rectangular Protuberances," *Numerical Heat Transfer,* vol. 3, pp. 149–167.

The corresponding citations in the report would appear as:

1. Lee et al. (1982) discovered . . .
2. Kwon and Pletcher (1981) predicted . . .

3. Sparrow (1980*a*) studied . . .

4. Sparrow (1980*b*) observed . . .

The format described above is required by several technical journals because their editorial advisers consider it important that the author's name be included in the body of the paper along with each citation.

The reader will notice that some of the above citations list the journal title in full (*Journal of Heat Transfer*) while only an abbreviation (*J. Heat Trans.*) is used in others. Either is correct, but some journals and/or publishers follow their own standard format.

15.5 GRAPHICAL PRESENTATIONS

Most engineering reports include some form of graphical presentation: a simple plot of raw data, a statistical distribution of data, correlation of data, and perhaps a comparison of experimental data with analytical predictions.

We have already noted in Sec. 3.17 that the majority of graphs are of the x-y plot variety, and illustrated some of the different uses for this type of graphical display. The x-y chart, and others, are part of the overall graphics assembly which may be required for a good report. Some of the graphics are:

1. General curves or "trends"

2. Detailed plots of experimental data

3. Accurate graphs which are used for calculation purposes

4. Schematic diagrams

5. Scale diagrams for experimental apparatus

6. Graphs in a format to be used for presentation

STYLE

Regardless of the purpose of the graph, there are certain elements of good style which should be common to all.

1. Label the graph. Make sure all coordinates are labeled consistently; i.e., don't use all uppercase on one coordinate and mixed upper- and lowercase on the other.

2. If numerical values of variables are graphed, make sure that each coordinate has scale markings and that each label for a variable includes the *units* for that variable. Try to maintain the same unit system for all variables; i.e., do not mix English and SI units.

   One must recognize that accepted practice in the field may dictate the use of mixed units. It would not be unusual at all, e.g., to use mixed units in the energy conservation field. For example, the energy-efficiency ratio (EER) for air conditioners is defined by

$$EER = \frac{\text{cooling in Btu/h}}{\text{power input in watts}}.$$

When a coordinate is dimensionless, i.e., has no units in a consistent system; it is *not* necessary to enter "dimensionless" on the axis or legend label. One may expect that a reader familiar with the subject area will recognize dimensionless parameters when they appear. When ambiguity or uncertainty may arise regarding the definition of a variable, that definition might be included on the chart. For example, one could write either Re or $Re = \rho u_m d/\mu$ as the Reynolds number label for a chart concerned with flow in a circular channel.

3. Recognize that some graphs may be reduced in size later because of space limitations. This is particularly true for those to be published in technical papers or symposia volumes. If any reduction is anticipated, the size of the labels and unit specifications on the graph must be increased so that they will be legible in their eventual format. For graphs prepared on computers this will be an easy matter to arrange.

Figure 15.1 illustrates some problems and solutions for a graphical presentation. In (a) an exponential graph is presented with legends that are *readable,* but just barely, in a reduced format for book or technical paper publication. Since most technical paper graphics are produced directly from author's copy, it is important that the author furnish graphics material that will not detract from the overall readability of the paper. Figure (b) gives a more suitable display with legends that are increased in size so they are easily readable when the figure is reduced in size.

Even the legend sizes in (b) may not be satisfactory for an oral presentation of the paper using visual-projection devices. Authors and presenters are notorious for flashing graphical displays on a screen with legends so small that they seem to vanish, or strain a person with exceptional visual acuity. This practice is not only annoying, but obviously detracts from the information transfer process. The solution is to use a still larger legend format as shown in Fig. 15.1c, which would give a visual appearance less favorable than (b) insofar as the written paper is concerned. You may be assured that the larger legends will be appreciated by the audience attending the presentation. The actual graphic from which the visual will be prepared is typically larger in physical size than that shown in (c). A cropped section from a graphic about twice the size of (c) is shown in (d), which looks out of balance on a sheet of paper but works fine for a visual presentation.

Summarizing, Fig. 15.1a would do quite well if presented in an enlarged size for a technical report, (b) is better for the reduced size of a technical paper, and (c) is superior for visual presentations. The report writer-presenter should realize that it may be desirable to prepare graphs in all three legend formats. With computer graphics that is not a cumbersome task.

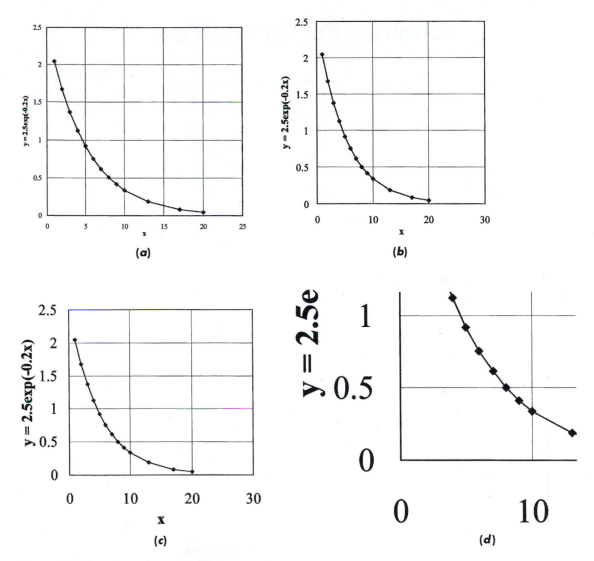

Figure 15.1 Legend sizes for different purposes.

4. The purpose of graphs is to convey information. Sometimes this is best accomplished by plotting several curves on one graph; sometimes not. For example, suppose that six pressure traverses are taken along the length of a wind tunnel for six different flow rates. In this case it would probably be best to include all six sets of data on one graph with each curve clearly labeled. A use of six *separate* graphs would not communicate as well and would certainly consume more space. On the other hand, suppose a heated plate is placed in the tunnel and heat-transfer data are collected for the same six flow rates. It would not be prudent to present

Cooling Performance

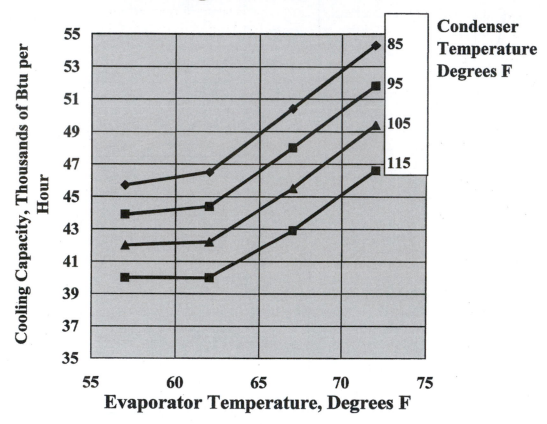

Figure 15.2   Example of *G* and *A* chart.

these data on the same graph as the pressure data because the display would be too cluttered. A second graph would be in order. There is no hard rule to apply in these circumstances. Common sense usually works fine.

5. There is a class of presentations called *G&A* graphs (for *G*enerals and *A*dmirals). They are usually broad-brush and never loaded with a lot of data points or complicated mathematical expressions. The purpose of these graphs is to present general trends without precise data points, and they are normally employed for readers or viewers who are not very familiar with the subject. A sample of a G&A chart is given in Fig. 15.2. Note that the legends are displayed in larger type fonts and terms are defined without mathematical symbols.

6. Careful thought should be given to the selection of coordinates for a graph. Even with plots of raw data one may wish to choose a logarithmic plot over a linear system if an exponential or power variation of the data is expected. The reader

will find that a review of Sec. 3.17 and Examples 3.29 through 3.32 may be in order at this point. In that section consideration was given to the types of coordinates that should be employed for graphical displays, and their relation to trendlines and data correlations.

LEGENDS FOR GRAPHS

Some of the choices available for establishing the legends for graphs are:

1. Placing the legends for the curve(s) directly on the graph itself.
2. Labeling the curves with numerals (1, 2, 3, etc.) or letters (a, b, c, etc.) which are then defined or explained in the figure title.
3. The same as 2, but definitions and explanations are given in the body of the text material.

   Style 1 should be used for all simple plots. If labeling of curves clutters the overall graph, then style 2 may be selected as preferable. Although style 3 is indeed used by some writers, it is generally inferior because the reader cannot obtain information directly from the graph. For oral presentations with visual aids even style 2 may be hazardous unless the figure title is reproduced in large enough type. A very common comment at technical presentations is:

   I don't understand what you have plotted there . . .

or

   What do the labels a, b, c, on the curves mean . . .?

Although the presenter may have given a verbal statement regarding labels, the audience may miss the comment while studying the graph. Our point, of course, is that one may need to choose a different legend style for an audiovisual presentation than for a written report.

GRIDLINES AND SCALE MARKINGS

Some discretion is required of the writer in display of gridlines and scale markings on graphs. Figure 15.3a shows a loglog plot with display of both major and minor gridlines and scale nomenclature for the major logarithmic cycles. This graph would be suitable for a report where some detail is desired for presentation of the data points. For presentation in a journal article or conference paper the figure may be reduced in size considerably, as described previously, and inclusion of the gridlines may clutter the display. In such cases the gridlines are frequently omitted entirely as shown in Fig. 15.3b. To retain the sense of the logarithmic scales, tick marks are placed on the coordinate axes to indicate the minor scale divisions, in this case, on the inside of the axes. The tick marks could also be displayed on the outside of the axes, or across the axes, but the inside position is usually preferred when gridlines are omitted from the chart.

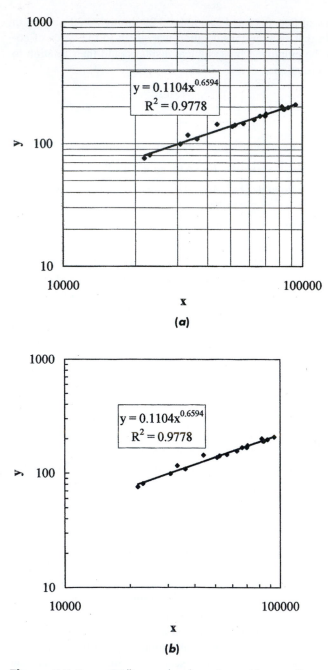

Figure 15.3   Gridlines and scale markings. (*Continued*)

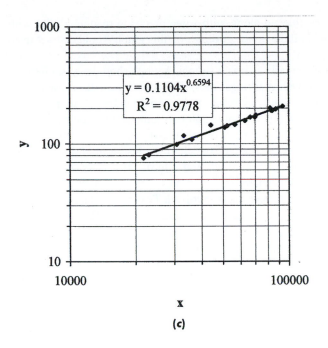

(c)

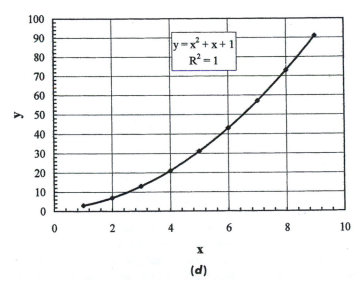

(d)

Figure 15.3 *(Continued)*

A modification of (*a*) is shown in Fig. 15.3*c* with both gridlines and outside tick marks indicating the scale divisions. Inclusion of the tick marks in this case adds a marginal amount of information, and might be better left off. An example of a graphical display on linear coordinates is shown in Fig. 15.3*d*, where major division gridlines are employed along with inside tick marks for the minor scale divisions. An uncluttered appearance is achieved along with some precision reading ability with the tick marks. Inclusion of all the minor gridlines on this graph would distract from the presentation.

The point of the above remarks is rather simple. A graph should not be cluttered with excessive gridlines and scale markings unless they contribute to the presentation. For formal papers, the format that is indicated in Fig. 15.3b is frequently preferred. There is an interaction between the amount of information contained on a graph and the need for gridlines. In Fig. 15.4a four sets of calculated points are presented on the chart, with labels to the right of the body of the chart. Inclusion of all the gridlines does not distract from the presentation at all, and affords the reader some ease in reading numerical values from the chart.

In Fig. 15.4b a different circumstance is involved. Two sets of data for the variables fv and fvp are presented as functions of the same variable I. The two variables have markedly different numerical values, so they must be plotted with different scales for their respective coordinates. The presentation here places the ordinates for fv and fvp on the right and left sides of the graph, respectively. Tick marks are used to indicate the scale divisions for I, fv, and fvp. There is a lot of information on this one graph, so much that addition of gridlines will clutter the display. Figure 15.4c adds the gridlines for I and only *one* ordinate, fvp. The chart becomes too cluttered. If one were to add the gridlines for fv also, the appearance would be terribly congested. Projections of a chart like that in Fig. 15.4c for an oral presentation would be a waste of time for presenter and audience alike. The point of this discussion is that one must be careful to avoid presenting so much information on a chart that the display becomes cluttered. Exclusion of gridlines provides an option for alleviating the clutter.

THREE-DIMENSIONAL CHARTS

Sometimes, the writer may choose three-dimensional charts for presentation of experimental or calculated variables. Such choices are subject to conjecture, but two sample displays are given in Fig. 15.5 for illustration purposes. Figure 15.5a shows a 3-D mesh chart of the response of a second-order system to a step input (also shown earlier as Fig. 2.9), while (b) gives the same kind of display for the Bessel function $J_n(x)$ for $n = 1$ to 4. Some of the labels have purposely been left off the charts to avoid cluttering already complex graphics. What is accomplished here? The reader will have to answer this question for himself or herself. The answer depends on the objectives of the report or presentation, and also on the space visualization capabilities of the reader. The availability of computer graphics makes 3-D presentations very easy to execute. Sometimes a 3-D presentation will add to understanding, sometimes not.

COSMETICS

The availability of computer graphics software packages provides an almost limitless number of possibilities for addition of cosmetic effects to graphs and charts. Variation

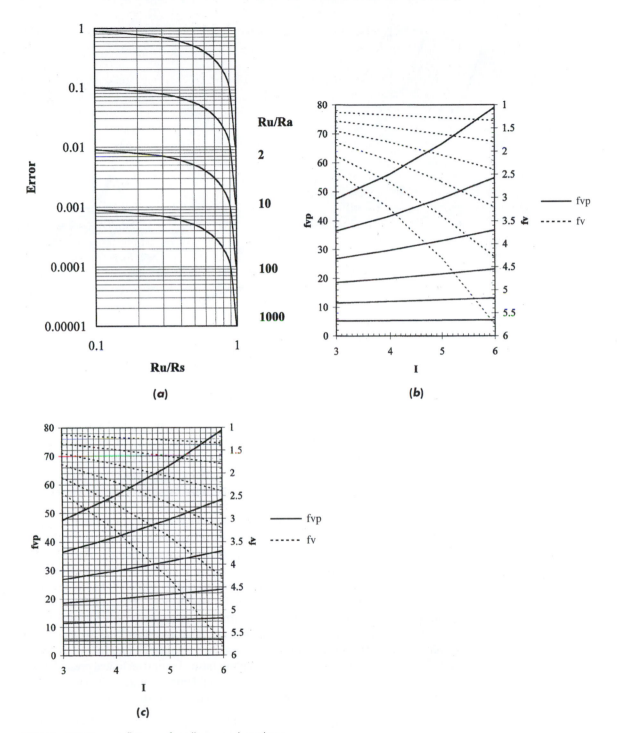

(a)

(b)

(c)

Figure 15.4   Influence of gridlines on chart clutter.

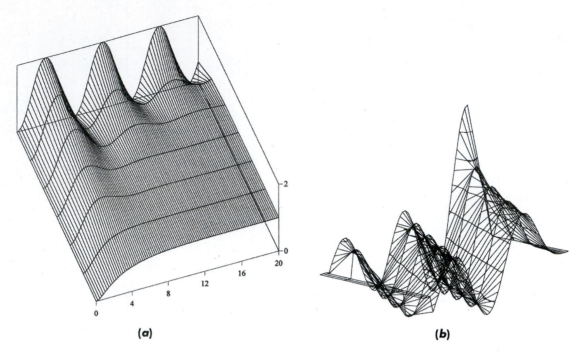

(a)   **(b)**

Figure 15.5   3-D charts: (a) Response of second-order system to step input; (b) Bessel function $J_n(x)$ for $n = 0$ to 4.

of color, shadings and fill, textures, and the like can be accomplished with the click of a button. For the most part, such cosmetics are unnecessary, except for *G & A* charts. When cosmetics are added, they should be executed with discretion. Two examples will be given: shading, or fill, either in the background or internal to a chart, and use of color to designate scale divisions. The writer must be careful not to *detract* from the chart with *too much* effect.

The 3-D charts of Fig. 15.5 are already somewhat startling in a visual sense. In Fig. 15.6*c* and *d* a gradient shading effect has been added as a background to "enhance" the presentation. The display in (*c*) clearly involves too much shading. The display in (*d*) is better. Whether it is better than the original in Fig. 15.5 is a matter of opinion. In Fig. 15.6*a* different colors (printed here in grayscale) have been used to designate scale divisions on the vertical coordinate axis. The display of the damping effects is somewhat obscured by this addition, but not in an objectionable manner. In (*b*) considerably more colored scale divisions have been added and the display of the damping effect is almost totally obscured. Again, one must be careful not to do *too much*. The present writer has seen some displays flashed at oral technical presentations that were so doctored with cosmetics that they would mimic modern abstract art. And, there was perhaps an analogous amount of information transfer as well.

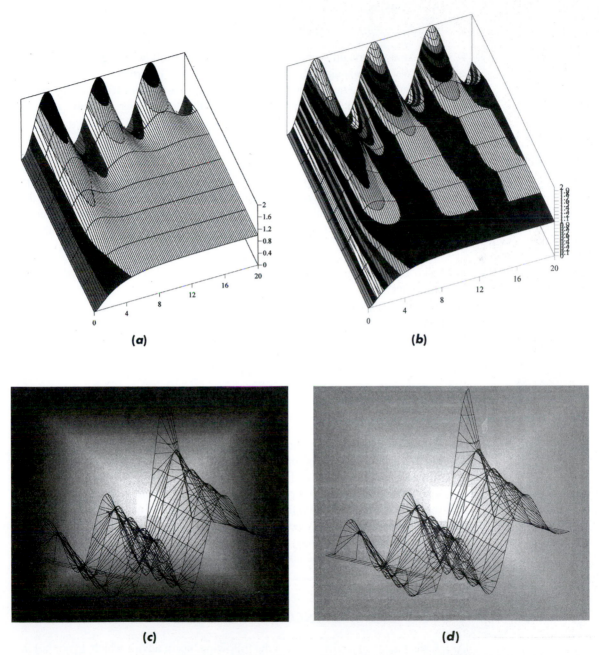

Figure 15.6   Cosmetic or enhancement effects added to 3-D charts of Fig. 15.5. (*a*) and (*b*) Color divisions for vertical coordinate axis (printed in grayscale); (*c*) and (*d*) use of gradient shading or fill.

An old saying that goes along with army methods of instruction is:

KISS: Keep It Simple Stupid.

A simple straightforward graphical presentation which follows the guidelines given above, as well as the suggestions of Sec. 3.17, is seldom incorrect.

The graphics that have been presented in Figs. 15.1 through 15.6 were prepared in Microsoft Excel [6], but could be generated with other software as well.

TABULAR MATTER

Sometimes it may be necessary or desirable to present information in tabular form instead of a graphical display. An example might be a table which summarizes the range of experimental variables investigated or tables which give data to a greater precision than can be read from a graph. The general rules for tables are similar to those for graphs:

1. Give the table a number and title. This information is usually placed at the top of the table.

2. The name of a variable listed in the table may or may not be necessary. Sometimes a symbol is satisfactory if the variable is easily recognized (like viscosity, density, etc.). On the other hand, the units for the variables listed should always be given in a format which is consistent with that used in the body of the report. See the above comments for ways to express units.

3. When numbers less than unity are presented, they should *always* be preceded by a zero and a decimal point; that is, the number 0.3432 should *not* be listed as .3432.

4. Tables used in a report may not be suitable for a presentation because of the small type size. If there is sufficient time, they should be redone in a large type. In general, complicated tables should be avoided in oral presentations. If the table involves so much information that the type size will be too small to read, the table should be omitted from the presentation. If the audience can't read the visual, don't present it. When tabular material is important, compromise is in order. Prepare an abbreviated summary table in larger type size for the visual. See the comments offered above regarding label sizes for graphs.

15.6   MISCELLANEOUS HELPFUL HINTS

As a further aid to the writer we give the following bits of advice which do not fall in the categories of other sections on report writing.

1. Always give great attention to the audience for which the report is written.

2. Set up a list of symbols to be used in the report before starting to write. Then be sure to use these symbols consistently as the report is written. *This is not an*

insignificant item. Many inexperienced writers make the mistake of switching symbols for a physical quantity in different sections of the report. Such practice can be very confusing to a reader and is not easily detected in proofreading. Even an experienced technical typist will have difficulty catching such mistakes. Be particularly careful about switching between uppercase and lowercase letters for the same variable.

3. In choosing symbols to be used in a report, try to follow the practice used in standard textbooks or technical journals in the field. Don't go off and invent new symbols which will make the report hard for the reader to understand. As simple as this advice may sound, it is violated over and over again, even by experienced writers.

4. Once writing is started, write the report as if you had to make it correct on the first draft. Take your time. This will pay off later. Of course, unless you are very good, or very lucky, changes will have to be made, but they usually can be minimized with this approach. Avoid the procedure followed by some persons of assembling a draft consisting of notes or comments in the form of incomplete sentences. You will just have to go back and rewrite the whole thing and must still go through a correction stage. It is an old saying but still appropriate: *Try to do it right the first time.*

5. Be kind to the typist or computer operator. In many cases the preparation of a report draft is a hybrid affair. The writer types the text on a computer, while leaving blanks for insertion of handwritten equations that involve Greek or math symbols. The writer should be careful with handwriting so there is no confusion between English and Greek symbols. Here are some specific suggestions:

   a. Use block letters for uppercase: viz.,

   U, not $\mathcal{U}$;

   E, not $\mathcal{E}$; etc.

   b. Use script letters for lowercase: viz.,

   u, not U or μ (mistaken for Greek "mu")

   v, not Y or ν (mistaken for Greek "nu")

   k, not K (mistaken for Greek "kappa")

   h, not h

   p, not ρ (mistaken for Greek "rho")

   w, not w (mistaken for Greek "omega")

   c. Be sure to make the "tails" very definite on Greek "eta" (η) and "mu" (μ).

   d. Put "flag" to left on Greek "nu" (ν) and to right on lowercase "vee" (v).

15.7 WORD PROCESSORS AND COMPUTERS

A number of word processors and word processing software systems are readily available in the market. Almost all are suitable for preparing technical reports. For those systems where preparation of complicated mathematical expressions is a problem, the situation may be alleviated with addition of an equation software package like MathType.[1]

No matter how simple or elaborate the word processor, information must be fed in by a human. This person may be the author (experimentalist) or someone who takes the author's write-up and prepares the report draft. If the report is highly technical and mathematical, the input person obviously must have experience and skill in handling such material.

We must note a behavioral pattern that applies to many persons. A document, letter, or report looks different in the form of a hard copy printout than it does on a computer or word processor screen. Mistakes that somehow went unnoticed on the screen seem to appear out of nowhere when hard copy is examined. For this reason, it is advisable to perform at least the final proofreading from a printed version of the document or report.

15.8   **PROCESSING OF REPORTS**

A knowledge of the eventual use for a report and the reproduction or processing required may save considerable time for both the writer and personnel who must do the typing (processing). Clearly, simple informal reports require only simple typing. In many cases equations with Greek or unusual symbols can be written in by hand. For more formal reports which will receive wide distribution, one would normally expect all equations to be typed.

Professional societies, and some publishers, require technical papers to be submitted in a specific digital or "camera-ready" format. They are subsequently printed and bound as either individual papers or part of a symposium volume. In many cases, publication is in the form of a CD or DVD containing all the papers for the symposium. As above, they should be typed and carefully checked for errors. Papers which are to be typeset for journal publication offer more flexibility than camera-ready copy. In these cases minor handwritten corrections can be placed in a double-spaced manuscript, as they will be incorporated in the final copy during the typesetting process. Writers usually have opportunity to correct the typeset copy when proofs are returned for their approval.

For company reports or manuals which are to be typeset a professional copy editor marks the copy by hand to establish the final style. Hand corrections are easily accommodated in the copyediting process. The point is that once it is known that the final report will be typeset, considerably less perfection is required in the original manuscript.

---

[1] MathType, Design Science, Inc., Long Beach, CA.

Table 15.2 gives an abbreviated summary of our discussion of report writing in the form of a checklist. An author can check off the items that apply and note whether the respective task has been performed. Obviously, there will be additional items which will come to mind for comprehensive reports, and there are too many listings for brief simple reports. The checklist is a starting point, though, and may help the writer avoid important mistakes or omissions.

Table 15.2   Sample report checklist

REPORT CHECKLIST

Title of report _____
_____
Author(s) _____
Date _____

1. Overall format
   (*a*) Consistency, section and paragraph numbers, etc. _____
   (*b*) Matches specs of sponsor or publisher _____
2. Title page
   (*a*) Conforms to format specs _____
   (*b*) Special title page or cover _____
3. Abstract or (executive) summary
   (*a*) Included _____
   (*b*) Conforms to length specs _____
4. Nomenclature
   (*a*) Included _____
   (*b*) Matches with equations, figures, and tables _____
5. Introduction, purpose, objective _____
6. Background, previous work, and/or review of literature _____
7. Theory
   (*a*) Included _____
   (*b*) Part referred to appendix _____
8. Description of experiment (if appropriate)
   (*a*) Apparatus _____
   (*b*) Procedure _____
9. Results
   (*a*) Included _____
   (*b*) Some referred to appendix _____
   (*c*) Reported in abstract _____
10. Discussion of results _____
11. Conclusions
    (*a*) Included _____
    (*b*) Match with abstract _____
12. References format
    (*a*) Consistent _____
    (*b*) Match with text _____
13. Appendix
    (*a*) Included _____
    (*b*) References in body of text _____

Table 15.2   (Continued)

14. Equations
    (*a*) Numbered _____
    (*b*) Consistent symbols _____
    (*c*) Match text references _____
15. Tables
    (*a*) Titled _____
    (*b*) Numbered _____
    (*c*) Format _____
    (*d*) Units included _____
16. Figures
    (*a*) Titled _____
    (*b*) Numbered _____
    (*c*) Coordinate labels _____
    (*d*) Coordinate units _____
    (*e*) Data series labels _____
    (*f*) Line sizes _____
    (*g*) Gridlines _____
    (*h*) Scale markings _____
17. Margins _____
18. Page numbers _____
19. Grammar and write-up
    (*a*) Style _____
    (*b*) Spelling _____
    (*c*) Punctuation, capitalization _____
    (*d*) Other _____
    _____
20. Comments

15.9   ORAL PRESENTATIONS

There are many times when an oral presentation must be made of the results of the experiments. Sometimes the persons attending this presentation will have copies of a written report and sometimes not. It is usually best to plan the presentation on the assumption that the attendees do *not* have the report or that, if they have it, they will not read the report until later.

Some persons choose to present just highlights of the report and furnish attendees with printed copies of the visuals employed for the presentation. The attendees are given the printed material at the time of the presentation and may add their own annotations and markings during the talk. The attendees may or may not read the complete report. With or without handout materials some guidelines for planning the presentations are given below.

1. Determine the audience for the presentation. Are they likely to be familiar with background information on your topic? If not, then you should include enough remarks to emphasize the importance of the work.

2. Determine the alloted time for the presentation and *plan to stay within this time limit*. If there are to be questions or discussions, then the formal presentation should end before the overall time limit is reached. It is probably not an overstatement to say that a majority of presenters do *not* stay within their time limit when one is imposed. Typically, when several presentations are scheduled, the presenters at the end get squeezed for time by early speakers who overtalk. For this reason some stern meeting management may be required by the person chairing the session. Of course, if no particular time limit is set for the presentation, the author should just set a reasonable time independently, taking into account the complexity of the report and the audience. One should neither bore the audience with too much detail nor leave them sketchily informed with a presentation which is too telegraphic.

3. Use slides, overhead view graphs, or computer visuals for the presentation. If a report has been prepared with good graphics, the slides can be made from figures in the report. For less formal presentations some blackboard work may be in order. Be sure to heed the advice given in the selection on graphics presentations when preparing the figures. Pay particular attention to the size of labels for coordinates and data. Video displays are used increasingly to demonstrate experimental behavior. The general advice to be offered on this matter is that professional audiovisual personnel be consulted in the preparation, editing, and presentation stages. Video presentations are usually unnecessary except when motion must be displayed. In general, higher illumination and finer resolution can be achieved with slides and viewgraphs than with video displays.

4. Avoid the use of complicated mathematical relations. If such relations are used in the report, present only summary comments which will enable the listener to understand what was done, and for what reason, but without seeing the details.

5. Don't leave the audience hanging. Be sure to summarize the results and conclusions with some clear, concise statements.

6. In the question-and-answer session some questioners tend to speak softly and cannot be heard by other people in the audience. It is a good idea to quickly repeat such a question, particularly if you are using a microphone, so that everyone will understand your response.

7. Finally, a courteous "thank you" is in order at the close of the presentation.

15.10   PLANNING SESSIONS AND CONFERENCES

Seldom does an experiment involve only one person. In many cases a very large number of people will be involved, and it will be necessary to get everyone together in some kind of conference. It is not an overstatement to say that a great amount of time is wasted in conferences because the sessions are not carefully planned or administered. The following remarks should provide some help in making more efficient use of the time devoted to conferences or meetings.

1. Establish the *need* for the meeting, and tell the attendees in advance. There can be many purposes for the conference, such as to brief all persons on the overall scope of the project, to come to a decision on equipment purchases, to reconcile experimental data, to divide and assign individual responsibilities, etc. The important point is that the purpose(s) be established in advance of the meeting. Don't just get everyone together "to have a meeting."

2. Decide in advance *who* should attend the meeting. Don't involve people needlessly and waste their time. For example, you don't need anyone from the computer center to decide on a pump purchase, but you may want to consult some experienced shop people. On the other hand, a meeting to discuss data reduction and analysis might properly involve some computer people.

3. Schedule the meeting and anticipate the total time needed. *Plan to stay within that time limit.*

4. If the meeting is to be rather long, establish an agenda with a certain amount of time allotted for each item. Be reasonable in setting these time limits. If there are five agenda items, this does not mean that each will consume one-fifth of the total meeting time. The idea is to *schedule* discussion. Don't let it just happen. In a political sense, the organizer of the meeting will be well advised to schedule noncontroversial items early in the meeting so that they may be disposed of in an expeditious manner. This accomplishes two objectives:
   a. It gives a sense of accomplishment and movement to the attendees early-on.
   b. It reserves time for discussions of more controversial/uncertain issues during the overall time which may be available.

5. Have action items to be accomplished, such as assignment of responsibilities either on an individual or subgroup basis, time schedules for reporting, or other specific tasks.

6. If follow-up meetings are necessary, schedule a time for these meetings which fits the schedule of all participants.

7. As in a report, it is a good idea to summarize the results of the meeting at the close in case there have been any misunderstandings on someone's part. In addition to a verbal summary at the end of the meeting, the chairperson or the person who called the meeting should issue a written summary to the participants at some later time.

The purpose of the above remarks has not been to complicate overly the task of "holding a meeting." There is some skill to the task, though. The present writer has seen a skillful organizer accomplish more in a one-hour session than some blunderer might achieve in an all-day meeting.

15.11   REVIEW QUESTIONS

15.1. Why is it important to include units on legends for graphs?

15.2. Why is a nomenclature list necessary?

15.3. How should one go about selecting nomenclature for physical variables?

15.4. Give an acceptable format for listing references.

15.5. Name some precautions to be taken in preparation of figures.

15.6. What is the purpose of an abstract?

15.7. How should one make use of appendices?

15.8. How does a conclusion section differ from an abstract?

15.9. Construct a nomenclature list for the flow-measurement chapter of this book.

15.10. Construct a nomenclature list for the temperature-measurement chapter of this book.

15.11. Construct a nomenclature list for Chap. 3 of this book.

15.12. List some action items for planning and conducting meetings.

15.13. Discuss alternate ways to label multiple curves on a graph. What are the advantages and disadvantages of each?

15.14. Write Eq. (7.4) in (*a*) all fraction format and (*b*) spreadsheet format.

15.15. Write Eq. (7.4) in fraction format using (*a*) all parentheses (( )) and (*b*) successive parentheses and brackets [( )].

15.16. Write Eq. (7.14) in (*a*) all fraction format with all parentheses and (*b*) spreadsheet format. In the latter case, devise some new variables which may simplify the presentation.

15.17. Write Eq. (9.11) in (*a*) all fraction format and (*b*) spreadsheet format.

15.18. Write spreadsheet statements for the correlation-coefficient equations of Chap. 3. Define symbols as necessary.

WRITING PROJECTS

The following projects ask the reader to present some of the subject matter in this book in a formal report format. Each project may include:

1. A title page

2. An abstract or executive summary

3. An introduction

4. A presentation of theory

5. Results and/or conclusions

6. A nomenclature list

7. A list of references in an acceptable format

8. Figures and tables as appropriate

A. A tutorial report on zeroth-, first-, and second-order systems.

B. A tutorial report on the use of uncertainty analysis in the design of experiments. Numerical examples should be included.

C. A tutorial report on the use of manometers for pressure measurements, including numerical examples.

D. A tutorial report on the use of Pitot tubes for velocity measurements, including numerical examples.

E. A tutorial report on the use of thermocouples for temperature measurements, including numerical examples.

F. A report which discusses the seismic instrument and its application for measurements of displacement, velocity, and acceleration. Be sure to show all steps in any theoretical discussion.

G. List the references in Chap. 3 of this book in alphabetical order following the last format of the "References and Bibliographies" section of this chapter.

15.12   REFERENCES

1. Olsen, L., and T. Huckin: *Technical Writing and Professional Communication,* 2d ed., McGraw-Hill, New York, 1991.

2. Wieder, S.: *Introduction to MathCAD for Scientists and Engineers,* McGraw-Hill, New York, 1992.

3. Universal Technical Systems: *TK Solver Plus—College Edition,* McGraw-Hill, New York, 1988.

4. Eisenberg, Anne: *Effective Technical Communication,* 2d ed., McGraw-Hill, New York, 1992.

5. Bertoline, G. R., E. Wiebe, C. Miller, and L. Nasman: *Microcomputer Systems Engineering Graphics Communication,* McGraw-Hill, New York, 1995.

6. Holman, J. P.: *What Every Engineer Should Know about EXCEL*, CRC Press, Taylor and Francis, Boca Raton, FL, 2006.

7. *Mathcad 8,* Mathsoft, Inc., Cambridge, MA, 1999.

8. *TK Solver,* Universal Technical Systems Inc., Rockford, IL, 1999.

chapter

16

DESIGN OF EXPERIMENTS

16.1 INTRODUCTION

In previous chapters we have indicated several elements of the experimental design process through discussions of experiment planning, uncertainty analysis applied to selection of instrumentation, methods of measurement, and reporting of experimental results. The purpose of this chapter is to bring some of this information together to form an overall protocol for execution of experiment design. The protocol is far from complete and, as we will see, is applicable mainly to certain types of experimental programs. After formulating the protocol, we will apply the information to specific examples.

16.2 TYPES OF EXPERIMENTS

Different types of experiments have already been described briefly in Chap. 2 and their reporting requirements discussed in Chap. 15. At this point we choose to expand the categorization of experiments into nine types for purposes of discussion. In addition, we construct a listing of experiment characteristics which may or not be applicable to each category. The categories of experiments and the applicable characteristics are shown in Table 16.1. This table is rather involved and includes a number of entries titled as "variable" or "possibly." The intent is not to be vague but rather to indicate the wide range of possibilities for the particular characteristic(s).

Type 1 and 2 experiments involving basic research are very specialized and require execution by people expert in the field. The *design* of such experiments is therefore a procedure that is highly variable. Although the protocol we shall outline is generally applicable to such projects, we defer serious consideration to the experts in their respective fields. This comment should not be interpreted to brush aside the importance of information from previous research and the need for careful design and selection of measurement methods.

Table 16.1   Characteristics of different types of experiments

A. Type of Experiment	1. Fundamental Research, Company Proprietary, or Government Classified	2. Fundamental Research, Open Results	3. Developmental Research, Company Proprietary	4. Developmental Research, Open Results
B. Type of output publication or reports	Internal reports, with portions possibly for journals	Conference papers, journal articles	Internal reports, some highly restricted	Open results unlikely until patented or legally protected
C. Presentation requirements, special meetings, etc.	Internal and external	Professional society meetings	Internal restricted	Internal
D. Outcome specified, design not specified			Yes, within limits	Yes, within limits
E. Experimental apparatus designed, outcome or objective not specified	Sometimes	Sometimes	Sometimes	Sometimes
F. Outcome known or anticipated	No	No	Sometimes	Sometimes
G. Method or procedure known, outcome unknown or surprises anticipated	Sometimes	Sometimes		Sometimes
H. Analysis and theoretical requirements	Variable	Variable	Variable	Variable
I. Uncertainty analysis required in design execution	Yes	Yes	Yes	Yes
J. Personnel require special background	Yes	Yes	Possibly not	Possibly not
K. Significant involvement of other groups of people	Not necessarily	Not necessarily	Yes, management, finance, legal	Yes, management, finance, legal
L. Novel experimental design required	Variable, mostly no	Variable, mostly no	Probably yes	Probably yes
M. Special instruments required or off-the-shelf	Usually off-the-shelf	Usually off-the-shelf	Some special may be developed	Some special may be developed
N. Experiment designed, execute only	No	No	Sometimes	Sometimes
O. Expense limits, buget restraints	Usually modest	Modest	Highly variable	Highly variable
P. Time constraints	Usually relaxed	Relaxed	Usually rushed	Usually rushed
Q. Space/facility requirements	Variable	Variable	Variable	Variable
R. Environmental constraints	Yes	Yes	Yes	Yes
S. Safety requirements	Yes	Yes	Yes	Yes
T. Example of this type of experiment	Study of laser or infrared transmission in exotic materials / Studies of genetic engineering and cloning	Study of boiling of fluorocarbons / Study of combustion products for engines	Semiconductor chip growth/manufacturing methods / Laser cutting methods in manufacturing	Semiconductor chip growth/manufacturing methods / Laser cutting methods in manufacturing

5. Testing According to Code or Specified Standards	6. Testing According to Accepted Method, but Not Code	7. Testing for Commercial Promotion Purposes	8. Just Want to Know	9. Preliminary to Design or Execution of One or More of Items 1–8
Internal report or report to regulatory agency	Internal report	Product literature, advertising material	Informal report	Internal report
Minimal	Internal	Special audiovisual presentations	Verbal to supervisor	Minimal internal
Yes, within limits	Yes, within limits	Yes	Possibly	Depends on type
Yes	No		Possibly	Depends on type
Yes	Sometimes	Usually	Possibly	Depends on type
Yes	Yes	Surprises not anticipated	Possibly	Depends on type
Minimal	Minimal	Minimal	Minimal or none	Depends on type
No	Yes	Probably not	Minimal	Depends on type
Usually trained for tests	Usually not	Usually not	No	Depends on type
Usually not	Probably not	Publications and promotional persons	No	Depends on type
No	No	No	No	Depends on type
Off-the-shelf	Off-the-shelf	Off-the-shelf but may need special effects	Off-the-shelf but may need special effects	Depends on type
Yes	May need construction	Usually not	Possibly	Depends on type
Usually well defined	Usually well defined	Controlled	Low budget	Variable
Variable	Usually well defined	Usually well defined	Short time	Variable
May use external facility	Variable	Variable	No special	Variable
Yes	Yes	Yes	Yes	Yes
Yes	Yes	Yes	Yes	Yes
Calorific value of foods by ASTM test	Forced convection in a tube	Video demonstration/ test of strength of paper towels	Anything, e.g., How long does it take to boil a cup of water in a microwave?	Anything
Viscosity index of oils by ASTM test	Sound absorption in solid materials	Comfort test of auto seating		
Air-conditioner performance by ASHRAE standards	Performance of a stereo amplifier	Consumer magazine test of small appliances		

Type 3 and 4 experiments involve developmental work and may frequently be assigned to the average engineering professional. Our design protocol should be applicable to such projects. Type 5 experiments that require testing according to "code" usually require very little "*design*" but may involve a significant amount of coordination to ensure that the results match the code requirements. We will assume that our design protocol discussion is applicable to type 6 through 9 experiments.

In preparing to conduct or design an experiment, one may view the process as an activity that works backward from the reporting requirements indicated in row B of Table 16.1. This is not an insignificant point. The design of the experiment must always be responsive to the output specified by management or the client customer. In the case of an "open" basic research project, the eventual client may be the pertinent scholarly journal in the field. Even basic research projects have sponsors, though, and their requirements and desires must be respected. For a development project the eventual output may be a legal patent, or even a process that is maintained as a trade secret. The handling of the results for such projects is obviously different from that for results that will be published in the open literature.

The *Just Want to Know* category of experiment is not as trivial as it may seem. A curious inquiry by an astute engineer may frequently lead to a profitable product development or modification. This might also be described as a What If . . . scenario. It could lead to implementation of the final category of experiment which then may develop into a substantial experimental program. Obviously, the Just Want to Know category cannot be employed willy-nilly in an industrial environment that is profit-oriented. There must be a profit light at the end of the curiosity tunnel. The example noted in Table 16.1 of "How long does it take to boil a cup of water in a microwave" might suggest the use of certain cookware for such purposes, specific object placement in commercial units, or possible improvements in design of the heating cycles for the unit.

Where uncertainty analysis is indicated as an essential element of the experimental project, we imply the simple type of analysis executed with application of Eqs. (3.2) of Chapter 3. For some experiments, more elaborate statistical design might be employed, but we will not consider such projects in our design protocol setup.

Environmental and safety requirements are listed as factors for all the experiment categories. To some readers these listings may seem obvious and not worthy of mention. On the other hand, they are items frequently overlooked in the early stages of experiment planning, only to arise later at embarrassingly high costs. A project involving use of hazardous wastes must be concerned not only with safety requirements, but also with the regulations which govern disposal of such wastes. A project to study the heat-transfer characteristics of ammonia might not involve a serious waste disposal problem but could find restrictions imposed because of toxicity-safety considerations. The selection of fluid handling components for such an experiment might require special attention.

It is rather obvious that a person well trained and experienced in a particular field will find it easier to design experiments pertaining to that field than a novice with little experience. In these cases the expert may jump to the conclusion and state: "Oh yes, it's done this way. . . ." The novice, but reasonably competent person, should be able to achieve results comparable to the expert with exercise of an experimental design protocol. The experienced person may also benefit by using the procedure as a checklist. In that way he or she may avoid omitting an important factor with a hasty jump to a final design result.

16.3   EXPERIMENT DESIGN FACTORS

With the above discussion of types of experiments in mind, we now consider factors that will enter into the experiment design process. These factors may be common to many or all of the experiments. The design factors are summarized in Table 16.2. Fifteen items are presented with a short descriptive title for each item.

The Source of Item Description in column B indicates the person(s), literature information, or other item in the table that will be used by the experiment designer to obtain the specified information. The Action and Results of column C indicate the results which the experiment designer should obtain to satisfy the item requirement in column A. Finally, column D indicates what influence the results or action will have on other factors; for example, the results for item 5 serve as the input or "source" for item 6.

The design factors in Table 16.2 are listed in an approximate sequence for execution in an experiment design process. For an experienced design engineer, the table may serve as a checklist in conjunction with the information in Table 16.1. Otherwise, it is employed as part of the design protocol which we now describe.

16.4   EXPERIMENT DESIGN PROTOCOL
AND EXAMPLES

The experiment design protocol we propose is as follows:

A. Determine the type or category of experiment by consulting Table 16.1. Consider overlapping categories.

B. Examine each of the characteristic entries in Table 16.1 pertaining to the experiment type selected. Write down a preliminary checklist of Things to Do based on this examination.

C. Begin working through the sequence of tasks in Table 16.2. Write down known information for as many items as possible. Prepare a list of needed activities in accordance with the known information and items yet to be determined.

D. To the extent possible, prepare a written schedule for accomplishing the tasks in Table 16.2.

E. Refine the design by working through all the tasks in Table 16.2 in detail.

This protocol may appear extremely brief or overly simplified; however, an examination of Tables 16.1 and 16.2 will reveal many steps that are anything but simple. In a sense, the protocol is compressed by using the tabular presentations. It would be possible to expand the discussion with an elaboration for each cell of the tables, but

Table 16.2 Experiment design factors

Item Number	A. Item Description	B. Source of Item Description	C. Action and/or Results Required	D. Influence on Other Factors
1	Overall objectives	Management or client/customer	Refine objectives to obtain specific information of item 2 description	Essential for proper design of experiment
2	List of specific results needed including ranges(s) of variables, accuracy, and uncertainties desired	Coordinated between management and experimental personnel	As noted in description	Required for design of size and scope of experiment and selection of instrumentation
3	Sample presentation format for results	Coordinated between management and experimental personnel	As noted	Indicates connection between item 2 and results expected. Serves as check on specification of results needed
4	Method/technique for overall experiment	Code specification for standard test, various literature for established techniques, novel experiment design in other cases	Locate and review appropriate literature and testing procedures	Necessary for selection of instrumentation and measurement methods
5	Parameters needed to calculate/determine results	Experimental personnel and literature	Review literature and theoretical calculation procedures	Indicates information needed for item 6
6	Measured quantities needed to calculate above parameters	Item 5	Determine	Indicates information needed for item 8
7	Ranges expected for measured quantities	Items 2 and 5	Determine	Indicates information needed for item 8
8	Method/techniques(s) for individual measurements	Items 6 and 7	Determine	Required for instrumentation selection/purchase
9	Anticipated uncertainties in individual measurements	Manufacturers literature, information in other sources such as chapters of this book	Estimate based on information available	Needed for items 13, 14, and 15
10	Apparatus preliminary design	Results of above determinations	Execute shop drawing and specifications as necessary	Needed for construction and/or purchases
11	Sample calculations based on apparatus design	Above factors	Perform	May suggest modifications in design or measurement techniques
12	Calculations needed to check consistency of measurements; energy, mass, force balances, etc.	Items 4, 5, and 6	Determine	Needed in execution of experiment
13	Decision on measurement, methods for individual parameters	All information above	Determine	Needed for construction and/or purchases
14	Estimate of uncertainties of final results	Calculation methods of Chap. 3	Calculate	May possibly indicate need for modification of experimental design or selection of instrumentation
15	Modifications to apparatus design and measurement techniques based on one or more of the above factors	Above items	Perform minimal modifications to meet overall objectives	Excessive modifications may increase cost

in most cases that would tend to obscure the overall view of experiment design we would like to convey here. Most readers will find the tabular displays relatively easy to follow.

FORCED-CONVECTION MEASUREMENTS FOR A NEW REFRIGERANT. The | **Example 16.1**
manufacturer of a new refrigerant desires to determine the forced-convection heat-transfer coefficients of their fluid in terms of the conventional parameters used to describe such a performance. They contract with an independent testing laboratory to perform the measurements. After consultation between laboratory personnel and the manufacturer-client, the preliminary specifications for the test are established as:

Fluid properties for saturated liquid conditions from 0°C to 40°C

Density $\approx$ 1200 kg/m^3

Saturation pressure: 300 to 1000 kPa

Dynamic viscosity $\approx 3 \times 10^{-4}$ kg/m · s

Thermal conductivity $\approx$ 0.075 W/m · °C

Specific heat $\approx$ 1.4 kJ/kg · °C

Prandtl number $= 5.6$

Desired range of Reynolds number: 20,000 to 150,000

Fluid (liquid) temperature range: Same as that given above

Flow geometry, smooth circular tube:

    Tube diameters: 2.0 to 35 mm
    Tube lengths: Sufficient for developed flow
    Temperature differences between tube and fluid: 5 to 15°C

Heat fluxes: As determined in experiment design

Flow rates: As determined in experiment design

Desired uncertainty in determination of heat-transfer coefficients: How good can you get?

Anticipated results: Correlate with conventional forced-convection relations available in standard heat-transfer literature and handbooks. No surprises are anticipated, and management will view with some alarm significant deviations from these expected results.

The experimental team is asked to come up with a suitable plan for design of the experiment that will be acceptable to the client in terms of meeting the above preliminary specifications. The team is expected to propose modifications to the specifications if some of the parameters appear unreasonable or inconsistent. Based on the design and/or proposed modifications, a proposal will be presented to the client for execution of the experiment along with appropriate cost and time schedules.

Design Protocol

A number of the elements in the proposed design protocol of Sec. 16.4 have been executed to develop the specifications and data parameters presented above. As a first step in the protocol we consult Table 16.1 and check off the pertinent items.

The type of experiment is selected as number 6 because there are accepted procedures for determining forced-convection heat-transfer coefficients in smooth tubes. How do we know that? There is an accepted general apparatus description in Sec. 9.7 of this book which we may use a guide for development of the experimental design. We could arrive at the same conclusion by consulting the literature pertaining to previous experimental determinations of forced-convection heat-transfer coefficients for liquids and gases.

Checking the rows in Table 16.1 for the type 6 experiment, same comment as in table:
Rows B, C, D, E, H, I, J, K, L, M

Row F: The outcome-correlation may be anticipated from the appropriate relation in Table 8.6 or other references.

Row G: No; surprises are not anticipated, and will be unwelcome.

Row N: Must design the experimental apparatus and instrumentation to match the specifications.

Row O, P: Budget and time schedules will be proposed to the client when the technical design is established.

Row Q: Space and appropriate facilities are expected to be available.

Row R: Must obtain manufacturers specifications for fluid toxicity and hazardous waste disposal requirements.

Row S: Standard safety measures are assumed to apply.

Preliminary Calculations

The experimental apparatus is illustrated in Fig. 9.17, which we reproduce here for convenience as Fig. Example 16.1a. Assuming negligible heat loss through the insulation, the energy balance is

$$q = \dot{m}c_p \, \Delta T_{\text{fluid}} = hA(T_w - \overline{T}_{\text{fluid}}) = EI$$
$$= \dot{m}c_p(T_2 - T_1)$$

where
$$A = \pi dL$$
$$\dot{m} = \rho u_m A_c$$
$$A_c = \frac{\pi d^2}{4}$$

This energy balance may be used as a check on the experimental measurements, provided the uncertainties in T_1 and T_2 do not obscure the calculation.

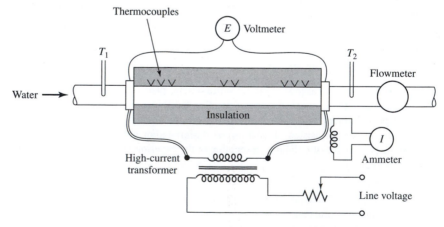

Figure Example 16.1a   Schematic of apparatus for determination of forced-convection heat-transfer coefficients in smooth tubes.

Estimate of Range of Values for Heat-Transfer Coefficients

For this calculation we employ the following equation from Table 8.6:

$$\text{Nu} = 0.023 \, \text{Re}^{0.8} \, \text{Pr}^{0.4} = \frac{hd}{k}$$

With the property values specified by the client, we have

$\text{Re} = 20{,}000, \, d = 35 \text{ mm}$

$$h = \frac{(0.023) \, (20000)^{0.8} \, (5.6)^{0.4} \, (0.075)}{(0.035)} = 271 \text{ W/m}^2 \cdot {}^\circ\text{C}$$

$\text{Re} = 20{,}000, \, d = 2.0 \text{ mm}$

$$h = 4740 \text{ W/m}^2 \cdot {}^\circ\text{C}$$

$\text{Re} = 150{,}000, \, d = 2.0 \text{ mm}$

$$h = \frac{(0.023)(150{,}000)^{0.8}(5.6)^{0.4}(0.075)}{(0.002)} = 23{,}760 \text{ W/m}^2 \cdot {}^\circ\text{C}$$

$\text{Re} = 150{,}000, \, d = 35 \text{ mm}$

$$h = \frac{(0.023)(150{,}000)^{0.8}(5.6)^{0.4}(0.075)}{(0.035)} = 1360 \text{ W/m}^2 \cdot {}^\circ\text{C}$$

Estimate of Flow Rate Ranges

$$\dot{m} = \rho u_m \, A_c:$$

$$A_c = \frac{\pi d^2}{4}$$

$$\text{Re} = \rho u_m \frac{d}{\mu} = \frac{d(\dot{m}/A_c)}{\mu}$$

$\text{Re} = 20{,}000, \, d = 2.0 \text{ mm}, \, A_c = \pi d^2/4 = \pi(0.002)^2/4 = 3.14 \times 10^{-6} \text{ m}^2$

$$\frac{\dot{m}}{A_c} = \text{Re} \left(\frac{\mu}{d}\right) = \frac{(20{,}000) \, (3 \times 10^{-4})}{(0.002)} = 3000$$

$$\dot{m} = 9.42 \times 10^2 \text{ kg/s}$$

$\text{Re} = 20{,}000, \, d = 35 \text{ mm}, \, A_c = \dfrac{\pi(0.035)^2}{4} = 9.62 \times 10^{-4} \text{ m}^2$

$$\dot{m} = 2.89 \text{ kg/s}$$

$\text{Re} = 150{,}000, \, d = 2.0 \text{ mm}, \, A_c = 3.14 \times 10^{-4} \text{ m}^2$

$$\frac{\dot{m}}{A_c} = \text{Re} \left(\frac{\mu}{d}\right) = \frac{(150{,}000)(3 \times 15^{-4})}{0.002} = 22{,}500$$

$$\dot{m} = 0.071 \text{ kg/s}$$

$\text{Re} = 150{,}000, \, d = 35 \text{ mm}, \, A_c = 9.62 \times 10^{-4} \text{ m}^2$

$$\frac{\dot{m}}{A_c} = 22{,}500$$

$$\dot{m} = 21.6 \text{ kg/s}$$

The range of flow rates is very broad, ranging from 0.0094 to 22 kg/s. This suggests that multiple experimental setups may be required to cover the range, at excessive cost to the client.

Heat-Transfer Rates

Let us examine the corresponding heat-transfer rates, for the maximum suggested temperature difference of 15°C. A tube length of 2.0 m is selected, so that

For $d = 2.0$ mm $A = \pi dL = \pi(0.002)(2) = 0.0126 \text{ m}^2$

For $d = 35$ mm $A = \pi(0.035)(2) = 0.22 \text{ m}^2$

Using the calculated values of h, and $T_w - T_{\text{fluid}} = 15°C$ we obtain

Re = 20,000, $d = 2$ mm

$$q = hA(T_w - T_{\text{fluid}}) = (4740)(0.0126)(15) = 896 \text{ W}$$

Re = 20,000, $d = 35$ mm

$$q = (271)(0.22)(15) = 894 \text{ W}$$

Re = 150,000, $d = 2$ mm

$$q = hA(T_w - T_{\text{fluid}}) = (23,760)(0.0126)(15) = 4491 \text{ W}$$

Re = 150,000, $d = 35$ mm

$$q = (1360)(0.22)(15) = 4488 \text{ W}$$

Change in Fluid Temperature

The fluid temperature rise may be computed from

$$q = \dot{m}c_p \Delta T_{\text{fluid}}$$

At Re = 20,000, $d = 2$ mm and $L = 2$ m, we have

$$\Delta T_{\text{fluid}} = \frac{896}{(9.42 \times 10^{-3})(1400)} = 67.9°C$$

Similarly, for other conditions

$$
\begin{array}{lll}
\text{Re} = 20,000, & d = 35 \text{ mm} & \Delta T_{\text{fluid}} = 0.22°C \\
\text{Re} = 150,000, & d = 2 \text{ mm} & \Delta T_{\text{fluid}} = 45.2°C \\
\text{Re} = 150,000, & d = 35 \text{ mm} & \Delta T_{\text{fluid}} = 0.148°C
\end{array}
$$

Suggested Modifications to Design Parameters

The above calculations indicate the need for a very broad range of flow rates. If the results follow the suggested correlation, as expected, then the broad range of diameters may not be necessary; that is, a single tube diameter that operates over the required Re range should be satisfactory. We therefore propose a single intermediate tube diameter of 12 mm for all the tests. The calculation for a 12-mm-diameter tube with $L = 2.0$ m gives

$$T_1 = 0°C$$

$$\overline{T}_w - T_{\text{fluid}} = 15°C$$

Re	20,000	150,000
$\dot{m}$	0.0566 kg/s	0.424 kg/s
h	790 W/m^2 · °C	3961 W/m^2 · °C
q	894 W	4480 W
$T_2 - T_1$	11.13°C	7.55°C

Note that the flow rates are now within a factor of 10, which may be accommodated with a single flow device like the turbine meter of Sec. 7.6.

Primary Measurements Summary

Based on the preliminary design calculations described above we now list the primary measurements to be performed, their range of values, and an estimate of the uncertainties that may be expected as shown in Table Example 16.1a. These uncertainties will be employed to make a preliminary estimate of the accuracy (uncertainty) of the results for h, the convection heat-transfer coefficient. Note that a direct measurement of the temperature difference $T_2 - T_1$ is proposed using a thermopile. This direct measurement facilitates a better energy balance calculation than one based on a calculated value of $T_2 - T_1$ from the individual measured values of T_1 and T_2.

Calculation Checks to Determine Satisfactory Experiment Operation

The only calculation check we propose is an energy balance on the tube test section. Assuming no heat loss through the insulation (this may be checked by calculation or measurement), we have

$$q = \dot{m}c_p(T_2 - T_1) = (?)P = EI$$

Note that the calculation is independent of the determination of h. First, we estimate the uncertainty in the determination of q by the two methods; electric power, and flow energy balance. Using the design data at the Reynolds number limits, and Eq. (3.2a) for a product function uncertainty:

$Re = 20,000$

Flow energy   Electric power

$$\frac{W_q}{q} = \left[(0.005)^2 + \left(\frac{0.3}{11.13}\right)^2\right]^{1/2} \qquad \frac{W_P}{P} = [(0.003)^2 + (0.003)^2]^{1/2}$$

$$= 0.027 \qquad\qquad = 0.0042$$

Table Example 16.1a Proposed measurement methods and associated ranges

Variables	Method of Measurement	Range	Estimated Uncertainty (with Calibration)	Comments
T_1, T_2	Type E thermocouple	0–40°C	±0.5°C	High output thermocouple
$T_2 - T_1$ (direct)	Type E thermocouple-thermopile	0–15°C	±0.3°C	
T_w	Type E thermocouple	0–60°C	±0.5°C	
$\dot{m}$	Turbine meter	0.05–0.5 kg/s	±0.5%	
E	Voltmeter	2–20 volts	±0.3%	
I	Ammeter	100–400 amp	±0.3%	
pressure (for reference)	Bourden gage & transducer	As specified	—	

$$\text{Re} = 150,000$$

$$\frac{W_q}{q} = \left[(0.005)^2 + \left(\frac{0.3}{7.55}\right)^2\right]^{1/2} \qquad \frac{W_P}{P} = [(0.003)^2 + (0.003)^2]^{1/2}$$

$$= 0.041 \qquad\qquad\qquad\qquad = 0.0042$$

The above calculations are very optimistic, indicating that the energy balance might be expected to check within 4 percent. Ten percent may be more realistic. The calculation also shows that the electric power measurement is clearly preferred for calculating the best transfer coefficient from

$$q = EI = hA(\overline{T}_w - \overline{T}_{\text{fluid}})$$

Heating Device Design Calculations

The heating mechanism proposed is an electrically heated tube with an impressed alternating current. We select a stainless-steel tube with dimensions

$$\text{Inside diameter} = d_i = 12 \text{ mm}$$

$$\text{Wall thickness} = 0.8 \text{ mm}$$

$$\text{Length} = L = 2.0 \text{ m}$$

The resistivity of stainless steel is about $\rho = 70\ \mu\Omega \cdot \text{cm}$, so the tube resistance is

$$R = \frac{\rho L}{A} = \frac{(70 \times 10^{-6})(200)(4)}{\pi(1.36^2 - 1.2^2)}$$

$$= 0.0435\ \Omega$$

The electric power is $P = I^2 R$ or E^2/R, so, for the design range of power from 894 to 4480 W we have

P	894	4480	W
I	143	321	A
E	6.22	14	V

These are electric parameters that may be accommodated with an off-the-shelf variac and transformer.

Estimate of Uncertainty in Determination of h

The previous calculation indicated an uncertainty in the determination of electric power input as about ±0.4 percent. Let us assume a value of 1.0 percent for use in calculating h. We have

$$h = \frac{q}{A(T_w - T_f)} = \frac{q}{A\Delta T}$$

where we set $\Delta T = T_w - T_f$. We take two conditions $\Delta T = 5°C$ and $15°C$ along with $W_T = ±0.5°C$. Then

$$W_{\Delta T} = [0.5^2 + 0.5^2] = 0.71°C$$

h is a product function of q and ΔT so, from Eq. (3.2a),

$$\frac{W_h}{h} = \left[(1)^2 \left(\frac{W_q}{q}\right)^2 + (-1)^2 \left(\frac{W_{\Delta T}}{\Delta T}\right)^2\right]^{1/2}$$

We assume $q = P = EI$, so that $W_q/q = 0.01$.

At $\Delta T = 5°C$

$$\frac{W_h}{h} = \left[(1)^2(0.01)^2 + (-1)^2 \left(\frac{0.71}{5}\right)^2\right]^{1/2} = 0.142 = 14.2\%$$

And at $\Delta T = 15°C$

$$\frac{W_h}{h} = \left[(1)^2(0.01)^2 + (-1)^2 \left(\frac{0.71}{15}\right)^2\right]^{1/2} = 0.0483 = 4.83\%$$

Thus, small temperature differences between the heating tube and the fluid should be avoided. Note that the precision of the electric power (voltage and current) measurements is not critical in the determination of h.

Expected Results

The expected results will take the form of a nondimensional plot of $Nu/Pr^{0.4}$ vs. Re on loglog coordinates. An *excellent* presentation would resemble the fictitious data set shown in Fig. Example 16.1b. The plot should include a correlation trendline and correlation coefficient r^2 obtained from Eq. (3.40). Data points that appear particularly out of line may be eliminated at the discretion of the experimentalist, if justified. See Examples 3.29 and 3.31. Preliminary to calculation and presentation of such a correlation, there should be intermediate plots of temperatures along the surface of the tube as illustrated in Fig. Example 16.1c. These plots should be smooth curves, to assure no anomalies in the wall temperature measurements.

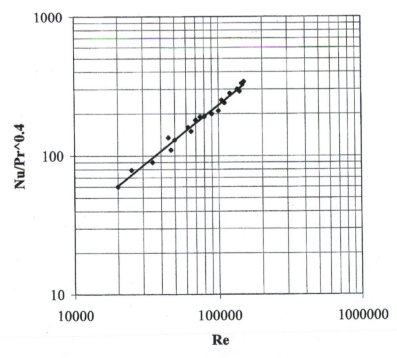

Figure Example 16.1b   Sample data correlation.

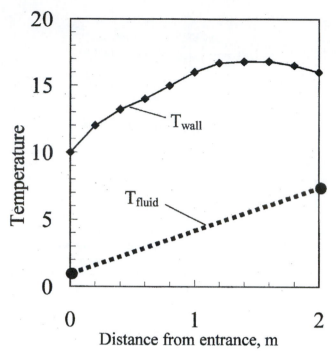

Figure Example 16.1c   Sample wall temperature profiles.

The electrically heated tube generates a uniform heat flow so the bulk fluid temperature is assumed to increase linearly. The arithmetic average of the temperature measured along the tube is usually employed as the value for calculating $\overline{T}_w - T_{fluid}$. T_{fluid} is the average between inlet and exit conditions, or $(T_1 + T_2)/2$.

Data Collection and Calculation of Results

Table Example 16.1b provides a summary of the information and data to be collected in the experiment and the calculation formulas that may be employed to obtain the expected results described above. The data may be collected manually or with a data acquisition system. Rapid transients are not involved, so there should be ample time for data collection. If a data acquisition system is employed, the output should be suitable for direct entry into the data analysis program. The calculation and presentation of graphical results will be prepared with the computer and software facilities available.

Environmental and Safety Considerations

With fluorinated hydrocarbon refrigerants there are likely restrictions on the leakage (toxicity) aspects of the apparatus, and regulations governing disposal of the fluids. We assume that the manufacturer of the fluid will furnish all information necessary to conform to these requirements. The specific details of piping, valves, pumps, filters, for example, are left as a mechanical design problem that will be executed by other persons.

Table Example 16.1*b*   Proposed data sheet and calculated results

---

A. General observations and records
    1. Ambient temperature and pressure
    2. Material of construction and electrical properties of heated test section
    3. Calibration information for instruments
    4. Observations of unsteady operation or unusual behavior

B. Primary measurements
    1. Inlet and exit fluid bulk temperatures for test section T_1 and T_2
    2. Tube wall temperatures T_w, at least eight
    3. Voltage impressed on test section E, volts
    4. Current through test section I, amperes
    5. Flowmeter readout × factor to convert to mass flow rate m in kg/s
    6. Direct measurement of fluid $\Delta T = T_2 - T_1$
    7. Temperature at outside of insulation (check on effectiveness of insulation)
    8. Pressure at inlet to test section
    9. Dimensions of test section tube, inside and outside diameters d_i and d_o, and tube length L

C. Estimates of uncertainty in primary measurements at high and low limits of Reynolds number and
    high and low limits of temperature difference between tube wall and fluid. Estimate for quantities
    listed in B.

D. Calculated results
    1. Flow rate $= m =$ meter reading × factor to convert to kg/s
    2. Flow cross-sectional area $A_c = \pi d_i^2/4$
    3. Surface area for heat transfer $A = \pi d_i L$
    4. Mean fluid temperature $Tm = (T_1 + T_2)/2$
    5. Fluid viscosity, thermal conductivity, and Prandtl number evaluated at Tm
    6. Mean wall temperature $Twm = \Sigma Tw/n$ for uniform spacing
    7. Reynolds number $\mathrm{Re} = (m/A_c)d_i/\mu$
    8. Mean wall to fluid temperature difference $\Delta Tm = Twm - Tm$
    9. Electric power input $P = EI$
    10. Convection heat-transfer coefficient

$$h = P/A\,\Delta Tm, \mathrm{W/m}^2 \cdot {}^\circ\mathrm{C}$$

    11. Nusselt number $= hd_i/k$
    12. Parameter $\mathrm{Nu}/\mathrm{Pr}^{0.4}$
    13. Energy balance check for high and low values of Re:

$$P = EI = (?) = mc_p(T_2 - T_1) \text{ with } T_2 - T_1 \text{ values from direct measurement}$$

    14. Check of electric resistance of tube:

$$R_{\mathrm{elec}} = \frac{\rho_{\mathrm{elec}} L}{A_{\mathrm{tube}}} = \frac{4\rho_e L}{\pi\left(d_o^2 - d_i^2\right)} = (?) = \frac{E}{I}$$

E. Calculated uncertainties of results for high and low values of Reynolds number Re, and high and low
    wall to fluid temperature differences. Assume fluid properties and tube dimensions are exact.
    1. Power: $w_P/P = [(w_E/E)^2 + (w_I/I)^2]^{1/2}$
    2. Average wall to fluid temperature difference $\Delta Tm = Twm - Tm$

$$w_{\Delta Tm} = [(w_{Twm})^2 + (w_{Tm})^2]^{1/2}$$

(Continued)

Table Example 16.1b   (Continued)

---

3. Heat-transfer coefficient:

$$w_h/h = [(w_P/P)^2 + (w_{\Delta Tm}/\Delta Tm)^2]^{1/2}$$

4. Reynolds number: $w_{Re}/Re = w_m/m$
5. Nusselt number: $w_{Nu}/Nu = w_h/h$

F. Graphical presentation of results
   1. Sample wall temperature profiles, T_w vs. x along tube length
   2. $Nu/Pr^{0.4}$ vs. Re on loglog coordinates with trendline fit to data
   3. Other correlation parameters if needed
   4. Bounding lines on chart in No. 2 indicating calculated limits of uncertainties

---

Summary Comments

This example has made use of the experiment categories and design factors tables that form the basis for a design protocol presented in this chapter. We may note that the engineer executing this experiment design must have some prior knowledge of instrumentation and the general subject matter to be explored with the experimental program. The person is not expected to start from a zero knowledge level, and should be conversant with the terms that appear in Tables 16.1 and 16.2. Of course, one purpose of previous chapters is to furnish such background information.

   In this example, we note that no surprises are expected in the results and the desire of the client-manufacturer is that the heat-transfer coefficient will follow accepted correlations currently used for design calculations in industry. If that is the expectation, then one may ask why the experiments are being conducted at all. The answer to the question has two parts. First, surprises do occur from time to time, and a prudent course of action is to back up expectations with firm experimental data. Second, the backup experimental data provide useful marketing and promotional information that may be furnished to skeptical persons who demand such data before construction of equipment employing any new fluids. The method of publishing results of the experiments has been left open. The manufacturer may want to retain the information for limited disclosure to interested customers, or may choose to seek publication in a technical society forum. Whatever choice is made, the results should be reported in a complete fashion with appropriate comments regarding uncertainties, and so forth.

   No time schedule has been specified in this experiment design. Such a schedule should arise out of the consultations between the fluid manufacturer and the testing laboratory, taking into account specific information regarding the instruments and space available at the local facility.

Example 16.2 | **PERFORMANCE TEST OF A SMALL "DRONE" ENGINE.**   The manufacturer of small radio-controlled aircraft drones used for military target practice requests the engine suppliers for the drones to furnish performance specifications for their engines. The information received is less than satisfactory and there seems to be a noticeable difference in performance from one engine manufacturer to another, even though they state the same power output. The aircraft manufacturer decides to make his own performance measurements of the small two-stroke cycle engines and assigns a member of his small engineering staff to design and conduct the experiments. Limited budget and facilities are available for the tests.

After extended discussions, the manager and engineer assigned to the test arrive at preliminary specifications for the tests:

Approximate power output of engines: 3 to 5 hp

Speed range: 1000 to 10,000 rpm

Power loading device: Propellers used in the application

Speed control device: Throttle and carburetor employed in application

Fuel consumption: To be determined

Measures of performance:

    a. Torque vs. rpm for different propellers
    b. Power output vs. rpm for different propellers
    c. Fuel consumption vs. rpm for different propellers
    d. Qualitative observations of operating characteristics of engine; smoothness of operation, response to throttle changes, etc.

Because of the limited budget, a decision is made to keep the test as simple as possible with minimum expenditures for any data acquisition systems or elaborate instrumentation.

Design Protocol

Consulting Table 16.1, we see that this experiment falls into a category 8 with possible implications for a category 4 in the event that the results lead to design changes in the drone aircraft or their radio control system. The measurement technique for the test is not established, and one would not expect to find a "code" specification.

Working through the rows in Table 16.1, we construct the following comments:

Same as table: Rows B, C, D, H, I, J, K, M, O, P, Q

Row E: Apparatus not designed

Row F: Outcome not known

Row G: Procedure not known, no surprises anticipated

Row L: Because of low budget some novel design of experiment may be necessary

Row N: Experiment not designed

Row R: Space for test must be ventilated for engine exhaust

Row S: Normal safety measures

Following the design protocol of Sec. 16.4, we find that the "Things to Do" list is very short: Perform the technical design for the experiment.

Consulting Table 16.2, we find that the output type of information has been specified but the ranges of variables remain to be determined. The presentation format has also been specified but the anticipated results are not yet known. One would expect the power output of the engine to generally increase with speed (rpm) and possibly reach peak values that depend on the propeller-loading device. The fuel consumption would be expected to increase with power output, but not in a linear fashion.

The next step in the design process is a calculation of the anticipated ranges of the output data to complete cell 2A of Table 16.2.

Torque and Power

Only two measurements are required to determine power through an application of Eq. (10.21).

$$\text{Power} = \frac{2\pi TN}{33,000} \text{ hp}$$

These measurements are torque T and speed N (rpm). Assuming a power output of 5 percent of engine rating at 1000 rpm, and full power at 10,000 rpm, the anticipated values of torque are

3-hp engine:

$$N = 1000 \, \text{rpm} \qquad T = 0.79 \, \text{lbf} \cdot \text{ft}$$
$$N = 10{,}000 \, \text{rpm} \qquad T = 1.58 \, \text{lbf} \cdot \text{ft}$$

5-hp engine:

$$N = 1000 \, \text{rpm} \qquad T = 1.32 \, \text{lbf} \cdot \text{ft}$$
$$N = 10{,}000 \, \text{rpm} \qquad T = 2.63 \, \text{lbf} \cdot \text{ft}$$

The power output will likely peak at a lower speed, so the maximum torque should be higher than these values.

Speed Measurement

The engine speed may be measured directly by an indicator tachometer connected to the front of the propeller shaft. An alternate method would be a strobe light directed at the propeller.

Torque Measurement

We propose an engine mounting on a platform that is free to rotate like the cradled dynamometer of Sec. 10.4. The torque measurement may then be performed by measuring the force at the end of a predetermined moment arm. Assuming that the maximum torque will be greater than the calculated values above, we choose a measurement range of 0.25 to 5 lbf · ft. If the engine mounting cradle has a 3-in moment arm, the corresponding force measurement to be performed ranges from about 1.0 to 20 lbf, or 4.5 to 89 N. Force transducers are available for this range.

Fuel Consumption

Assuming 15 percent overall efficiency for the engine, which may be too high, full power output at 10,000 rpm, and 10 percent output at 1000 rpm, the input fuel energies will be:

1000 rpm:

$$\text{Fuel energy} = (3 \text{hp})(2545 \, \text{Btu/hp} \cdot \text{h})(0.1)/0.15 = 5090 \, \text{Btu/h}$$

10,000 rpm:

$$\text{Fuel energy} = 50{,}900 \, \text{Btu/h}$$

The heating value for gasoline is about 140,000 Btu/gal, so these fuel energy inputs are equivalent to

$$5090 \, \text{Btu/h} = 0.036 \, \text{gal/h} = 0.0382 \, \text{cm}^3/\text{s}$$

$$50{,}900 \, \text{Btu/h} = 0.36 \, \text{gal/h} = 0.382 \, \text{cm}^3/\text{s}$$

For an overall efficiency of 5 percent the fuel consumption rates would be three times these values. A discussion concluded that a few preliminary tests should be conducted to determine approximate values for the actual fuel consumption rates. These values may then be employed to set the ranges for the flow measurement instruments.

Fuel Flow Measurement

Two methods are appropriate for a low-cost fuel flow measurement:

a. Direct measurement of the volume consumed in a period of time.

b. Flow rate measurements with a set of rotameters (see Sec. 7.6) to cover the anticipated flow rate range. The rotameters may be calibrated using the direct volumetric flow rate technique.

At the low flow rate calculated above, the time to consume 10 cm^3 of fuel is

$$\Delta \tau = \frac{10}{0.038} = 262 \text{ s}$$

or a reasonable time period for measurement.

Anticipated Measurement Problems

The small engines to be tested will likely exhibit some unsteady speed operation for a given throttle setting. As a result, the torque and speed measurements recorded should be the best average values obtainable by the operator. One should not be surprised at a variation of 5 percent in T and N. If these values are taken as the uncertainties in T and N, the uncertainty in the power determination would be

$$\frac{w_P}{P} = \left[\left(\frac{w_T}{T} \right)^2 + \left(\frac{w_N}{N} \right)^2 \right]^{1/2}$$

$$= (0.05^2 + 0.05^2)^{1/2} = 0.071 = 7.1\%$$

For the low flow rate range the fuel consumption may be measured by recording the time required to consume a given volume of fuel. This measurement automatically averages over any speed variation. At the higher flow rates, the operator should record the best estimate of the reading on the rotameters. As mentioned above, final selection of the rotameters should be made after test runs to determine the actual operating ranges for fuel consumption.

Expected Results

The expected output results would be in the form of curves that may have the appearances shown in Fig. Example 16.2a and b for each propeller tested. An efficiency parameter may also be defined as

$$\text{Efficiency parameter} = \eta = \frac{\text{Power output}}{\text{Fuel consumption}}$$

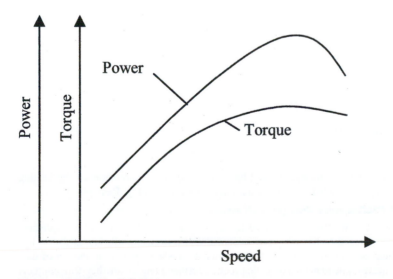

Figure Example 16.2a

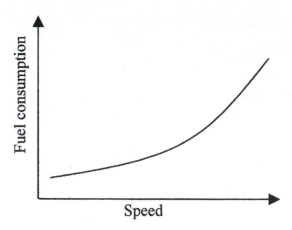

Figure Example 16.2*b*

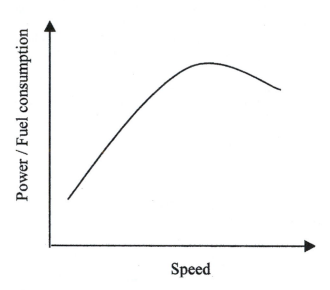

Figure Example 16.2*c*

This parameter may be nondimensionalized by multiplying the denominator by the heating value of the fuel, and a sample plot of the parameter is indicated in Fig. Example 16.2*c*.

Summary of Preliminary Design Activities

Consulting Table 16.2, we find that most of the actions in col. C have been accomplished. Provision for recording data has not been assigned, and a decision whether to use a data acquisition system or manual collection of data must be made. Computer reduction of data must also be indicated. The items in col. C that require further attention are therefore numbers 10, 11, and 15. A few crude tests to determine the range of fuel consumption are also needed.

Table Example 16.2   Data sheet and calculated results

A. General information
    1. Ambient temperature and pressure
    2. Engine description
    3. Propeller type and description
    4. Observations of steadiness of operation, fluctuations in speed or torque, difficulties
       with throttle settings, etc.
B. Primary measurements
    1. Throttle setting (describe)
    2. Speed, rpm
    3. Force, lbf
    4. Torque = force × moment arm
    5. Flowmeter reading × factor to give fuel flow rate in L/s
or
    Fuel volume at start, L
    Fuel volume at end, L
    Time for consumption of fuel volume, s
C. Primary uncertainty estimates at high and low speeds, taking into account observations
    of unsteady operation
    1. Speed: _____ rpm at speed of _____ rpm
    2. Force: _____ lbf at reading of _____ lbf
    3. Flowmeter _____ at reading of _____
    4. Flow volume _____ L at flow rate of _____ L/s
    5. Time increment _____ s
D. Calculated results
    1. Torque $T = F \times$ moment arm (assumed exact)
    2. Power $= P = 2\pi TN/33{,}000$, horsepower
    3. Fuel consumption $Vf =$ meter reading × factor or $\Delta V/\Delta \tau$
       $\Delta V = V_1 - V_2$; $V_1 =$ initial volume of fuel, $V_2 =$ final volume of fuel
       $\Delta \tau =$ time for consumption of ΔV
    4. Efficiency parameter $\eta =$ power ÷ fuel consumption $= P/Vf$. This parameter may be
       multiplied by an appropriate factor to take account of the heating value of the fuel.
E. Calculated uncertainties of results, performed for high and low speeds
    1. Torque: $w_T/T = w_F/F$ (moment arm assumed exact)
    2. Power: $w_P/P = [(w_T/T)^2 + (w_N/N)^2]^{1/2}$
    3. Fuel consumption Vf:

$$\frac{w_{Vf}}{Vf} = \left[\left(\frac{w_{\Delta V}}{\Delta V}\right)^2 + \left(\frac{w_{\Delta \tau}}{\Delta \tau}\right)^2\right]^{1/2}$$

$$\Delta V = V_1 - V_2$$

$$w_{\Delta V} = [(w_{V_1})^2 + (w_{V_2})^2]^{1/2}$$

    4. Efficiency parameter $\eta = P/Vf$

$$\frac{w_\eta}{\eta} = \left[\left(\frac{w_P}{P}\right)^2 + \left(\frac{w_{Vf}}{Vf}\right)^2\right]^{1/2}$$

F. Graphical presentations of results

    Performance parameters T, P, Vf, $\eta = \dfrac{P}{Vf}$ vs. N

    Presentation alternatives:
    1. Propeller type, one engine
    2. Propeller type, comparison curves for all engines
    3. One engine, different propeller types
    In addition, a useful presentation may be:
    4. N vs. throttle setting for each engine with different propeller types. This information
       may be useful in establishing settings for the radio control of throttle positions.

A summary sheet for collection of data and organization of calculations is shown in Table Example 16.2. The actual format should be arranged as a spreadsheet or other data entry form for the computer calculations and presentation of results.

Summary Comments

This example has also made use of the protocol for experiment design described in Tables 16.1 and 16.2. Since motivation for the tests originated with uncertain or conflicting specifications of engine performance, it is difficult to predict how the results will turn out and what value they may have in modifying the performance of the small drone aircraft. An informal reporting of the results will likely be sufficient for the aircraft manufacturer's use in dealing with the engine suppliers.

While anecdotal evidence has been used as a guide for propeller selection in the past, the actual power-torque performance curves may serve as a better basis for matching propeller to engine characteristics. One cannot predict results, but it is possible that an engine which previously exhibited poor performance relative to another engine may improve with better propeller selection and matching. If that is the case, the engine suppliers should appreciate sharing of the test results.

We should point out that the thrust characteristics of the propeller-engine combinations will also depend, to some extent, on the speed of the aircraft. To obtain experimental data on such thrust characteristics would require wind-tunnel data with a major increase in expense and complexity of the experiment design.

Example 16.3 | **DETERMINATION OF CAUSE OF PLASTIC CUP FAILURES.**   A small manufacturer of designer-styled styrofoam cups is experiencing customer dissatisfaction because bottoms of some of the cups fall out shortly after being filled with hot coffee. The problem is serious enough that sales may be lost unless the production problem is fixed. The cups enjoy a premium price because of novel design. A consulting engineer is called to examine the situation and suggest a course of action that may be used to determine the cause of the product failures.

The cups are produced by injecting styrofoam beads into a mold which is then heated with steam to fuse the beads together and form the cup. Other styrofoam products are manufactured in the same plant using the same steam boiler source for heating, but no failures have resulted in these other products. The client was asked about manufacture of other product items at the same time as the cup machine operation, with the possibility that the boiler may have been overloaded. No answer was forthcoming. The boiler manufacturer steadfastly assured the client that the boiler should be able to meet specifications of 120 boiler horsepower with perhaps 20 percent overload capability. A conference between the consulting engineer and company management suggests several possible causes for the cup fabrication problem:

1. The steam is not hot enough or does not flow in sufficient quantity to accomplish the fusing process in all of the cup machines. There are 12 cup machines in all.

2. Some or all of the cup molds are defective in design or construction.

3. Some or all of the machines may leak steam through faulty fittings. A check of all fittings is scheduled immediately.

4. The cause of item 1 may be failure of the boiler to produce steam at its rated capacity, or the design specifications on steam consumption for the cup machines may be incorrect, or both.

5. The syrofoam beads may be faulty. This possibility is rejected because no problems are encountered with other products when using the same stock material.

No tests have ever been conducted on the steam boiler or the cup machines to determine steam production and consumption. Therefore, the conclusion was reached that information on these quantities must be determined before any corrective action is taken. There are ancillary issues to be decided based on the results of such measurements and testing:

1. Was the boiler properly sized by the engineer designing the plant?

2. Was the boiler properly installed and does it meet specifications?

3. Do the cup machines consume steam at the rate specified in the design of the plant?

Answers to these questions may indicate legal liability for expenses to correct the problem, so there is a need to be reasonably confident in whatever experimental measurements are taken. In addition, the data may be used in the design of future production facilities, so confidence in the measurements is needed in this regard also.

Despite some cup failures, sales of the specialty products have been maintained and the production line is still operating at capacity, 24 h per day, and any disruption of the line may cause loss of revenue. Any experimental program must operate under the constraint of minimum downtime for the steam system. Based on the initial conversations, the engineer is asked to develop a plan to experimentally determine the performance of the boiler and cup machines. A simple schematic of the cup machine portion of the system is given in Fig. Example 16.3a.

After a tour of the plant, the engineer returns to management to obtain the following additional information:

1. Specifications for the boiler: 120 boiler hp at 100 psig saturated steam, 70°F water inlet. *Note:* 1 boiler hp = 33,479 Btu/h.

2. Design consumption rate of each cup machine: 10 boiler hp at 100 psig saturated steam.

3. Schedule and sequence of operation of the 12 cup machines.: 6 h operation, 2 h reload time, adjustment and cleaning, three times per 24-h day. The production schedule attempts

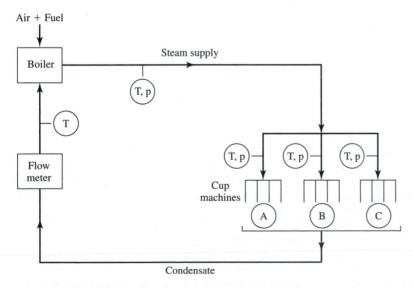

Figure Example 16.3a Schematic of steam flow for cup machines.

to maintain at least 8 of the 12 machines in operation all the time. The piping and cup machines are arranged so that the machines operate off a steam header in groups of four machines. It is not easy to change this piping, so the experiment must be designed to accommodate these groupings.

4. Except for a pressure gage and temperature indicator at the boiler outlet, and combustion temperature indicator, no instrumentation is presently installed on the boiler or available at the plant. Small (2-in-diameter) pressure gages are installed on the inlet lines to each cup machine.

5. Qualified technician support is available for installation of instruments and recording of data, except for the third shift, which operates from midnight to 8 A.M. Persons may be shifted to work on an overtime basis. To minimize costs, company personnel should be used for data collection as much as possible. The company will procure instrumentation unless something particularly expensive is required.

6. Management is asking for advice on the accuracy (uncertainty) to be achieved in the measurements, in case some legal liability action is taken, and for guidance in future design of expanded production facilities.

7. In view of the urgency of the matter, management requests that the tests be performed as soon as possible and that maximum advantage be taken of the 2-h "off" time for loading and maintenance of the cup machines.

8. The tests are to be conducted under the supervision of a registered professional engineer. (The consulting engineer is registered, so this is no problem.)

9. A proposal for accomplishing the needed tests, including time schedule, proposed report submission, and costs, is requested of the consulting engineer.

Design Protocol

The design protocol for this project is primarily influenced by items 6(O, P, Q) in Table 16.1. Not only is the client in a rush for the experiments to be conducted, but the tests must conform to the regular production time schedule without unnecessary disruption in that schedule. A detailed check with the production people reveals that the on-line and downtime of the groups of cup machines are only approximate for operation in 6- and 2-h time segments. At some times, all three groups of four machines are operating, while at other times only one group is operated. An average of two groups appears to be operating over each 24-h period.

Because of the grouping of banks of four machines operated off a common steam header, it will not be possible to measure individual steam consumption for each machine; only the average consumption for each bank can be determined. If the results indicate substantial differences between groups, then further study or examination of the setup may be proposed.

Since the boiler specification calls for a rated output of 120 boiler hp and each machine carries a tentative design specification of 10 boiler hp, it appears that operation of all 12 machines may produce an overload. Then, there could be a possible decrement in steam flow to an individual machine below that flow necessary for maintaining proper fusion of the feedstock material.

Acknowledging the desire of the client to minimize consultant expense and maximize use of his own personnel, the engineer proposes that the program be conducted by himself and two associates, working over 24-h periods, along with company personnel as needed. The client accepts this proposal. The decision is made that critical monitoring of pressure, temperature, and flow measurements will be performed by the consulting persons, while company personnel will be limited to strict adherence to a time schedule of operating sequence designed by the engineer.

Measurements to Be Performed

Since all steam flow leaving the boiler arrives at the cup machines (with no other production items), the flow to each group of four machines may be determined by a careful sequencing of operation and flow, temperature, and pressure measurement at the boiler itself. The consultant and his associates will take these measurements.

Referring to Fig. Example 16.3a, the specific measurements that will be taken are:

1. Condensate flow rate to boiler. A positive-displacement flowmeter calibrated in gallons is proposed, with readings to be taken at specified times. (A commercial-type water consumption meter is satisfactory for such measurement.)

2. Temperature and pressure of condensate entry to boiler.

3. Temperature and pressure at outlet of boiler. Enthalpy determination will be made if in the saturation region.

4. Combustion gas flow, temperature and pressure (for reference only).

5. Time in hours-minutes-seconds for above parameters; recorded by a digital clock.

The measurements to be performed by the shop personnel are the following:

1. Start time and stop time for each bank of four machines.

2. Pressure and temperature at inlet to each bank of machines recorded each hour. This measurement is not judged to be very accurate and is mainly for reference in case of substantial leakage or failure.

3. Observations of any quality control problems as stipulated by the production manager.

Timing and Sequencing of Operation and Measurements

The production manager suggests that the banks of machines not be operated longer than 6-h periods. If needed, the reloading maintenance operation can be performed in 1 h instead of 2. Average operation of 8 machines should be maintained over 24-h periods to ensure production goals.

During the tests, sequencing of operation of the banks of machines must be organized to provide performance information on the individual banks as well as joint operation of dual and triple banks of machines.

With the above requirements in mind, a schedule for operation of the banks of machines is proposed as shown in Table Example 16.3a. The groups are designated as A, B, and C. Appearance of a "1" in the table indicates that the machine group is operated at that hour. Appearance of a "0" indicates that the machine is shut down for reloading and maintenance. No bank of machines is operated for longer than 6 h. In a few cases only 1 h of downtime is available, e.g., from midnight to 1:00 A.M. on day 3 for group B and midnight to 1:00 A.M. on day 2 for group C. The timing sequence is arranged so that there are times when single groups are in operation, dual groups in service, and all groups producing product. Summaries of these sequences for each of the 4 days is given at the bottom of the table. Each of the single or dual groups occurs three times in each 24-h period. Operation of all groups (A + B + C) occurs for 8 h in each 24-h period. The average number of machines on line in any 24-h period is 7, slightly short of the 8-machine specification, but tolerable in that the experiment design allows for individual measurements of operations of banks A, B, and C.

Table Example 16.3a Proposed sequence of cup machine operation

	Day 1			Day 2			Day 3			Day 4			Day 5
Hour	**A**	**B**	**C**	**A**	**B**	**C**	**A**	**B**	**C**	**A**	**B**	**C**	
24 = midnight													
1	1	0	0	1	0	0	0	0	1	0	1	0	Retest of groups as indicated by preliminary data analysis
2	1	1	0	1	0	1	0	1	1	1	1	0	
3	1	1	1	1	1	1	1	1	1	1	1	1	
4	1	1	1	1	1	1	1	1	1	1	1	1	
5	1	1	1	1	1	1	1	1	1	1	1	1	
6	1	1	1	1	1	1	1	1	1	1	1	1	
7	0	1	1	0	1	1	1	1	0	1	0	1	
8	0	0	1	0	1	0	1	0	0	0	0	1	
9	1	0	0	1	0	0	0	0	1	0	1	0	
10	1	1	0	1	0	1	0	1	1	1	1	0	
11	1	1	1	1	1	1	1	1	1	1	1	1	
12	1	1	1	1	1	1	1	1	1	1	1	1	
13	1	1	1	1	1	1	1	1	1	1	1	1	
14	1	1	1	1	1	1	1	1	1	1	1	1	
15	0	1	1	0	1	1	1	1	0	1	0	1	
16	0	0	1	0	1	0	1	0	0	0	0	1	
17	1	0	0	1	0	0	0	0	1	0	1	0	
18	1	1	0	1	0	1	0	1	1	1	1	0	
19	1	1	1	1	1	1	1	1	1	1	1	1	
20	1	1	1	1	1	1	1	1	1	1	1	1	
21	1	1	1	1	1	1	1	1	1	1	1	1	
22	1	1	1	1	1	1	1	1	1	1	1	1	
23	0	1	1	0	1	1	1	1	0	1	0	1	
24	0	0	1	0	1	0	1	0	0	0	0	1	

Single and Dual Groupings

A	A	C	B
A+B	A+C	B+C	A+B
B+C	B+C	A+B	A+C
C	B	A	C

All days A+B+C

A, B, C = Groups of 4 cup molding machines each

The total number of 1-h samplings for the single and dual sequences over the 4-day period are:

Banks	Number of 1-h Samples
A	9
B	6
C	9
A+B	9
B+C	9
A+C	6

This number of samples should be enough to allow a reasonable determination of the average steam consumption of each bank of four cup machines.

A fifth day of testing is proposed as contingency in case of unforeseen breakdowns during the 4-day program.

Design Calculation for Flowmeter

The rated power of the boiler is

$$120 \text{ boiler hp} = (120)(33,479) = 4.02 \times 10^6 \text{ Btu/h}$$

The rated powers for the cup machines are

$$10 \text{ boiler hp} = 3.35 \times 10^5 \text{ Btu/h}$$

The enthalpies for rated output at 100 psig dry saturated steam and 70°F feedwater are

$$h_g = 1190.4 \text{ Btu/lbm} \qquad T = 338°F$$

(Preliminary observations indicate a boiler output temperature of about 350°F or 12°F superheat.)

$$h_f \text{ at } 70°F = 38 \text{ Btu/lbm}$$

The design flow rate for the boiler is therefore

$$m_s = \frac{4.02 \times 10^6}{1190.4 - 38} = 3488 \text{ lbm/h}$$

or 6.98 gal/min at 8.33 lbm/gal for the feedwater.

The corresponding flow for the cup machines is 0.582 gal/min, or 2.328 gal/min for each of the 4-machine groups.

Since the groups of cup machines will be operated for at least 1 h at a time, one may assume that at least 30 min will be available for measuring each quantity of water. Thus,

120 boiler hp is equivalent to 209.4 gal over a 30-min period

40 boiler hp (4 cup machines) is equivalent to 69.8 gal over a 30-min period

A positive-displacement meter selected has a stated accuracy of 1.0 gal (least count) or 2 percent of elapsed volume indication, whichever is greater. So, for these flows, the 2 percent figures are:

$$2\% \text{ of } 69.8 \text{ gal} = 1.4 \text{ gal}$$

$$2\% \text{ of } 209.4 \text{ gal} = 4.19 \text{ gal}$$

Therefore, the 2 percent figure will be taken as the uncertainty for the flow quantity measurement.

Uncertainty in Flow Rate Determination

The flow rate is determined from

$$Q = \frac{G_2 - G_1}{t_2 - t_1}$$

where the G's are the flow quantities in gallons at the beginning and end of the 30-min period. The t's are the recorded times at beginning and end of the measurement period. From the above flow calculations w_G/G is taken as 0.02, which enables a calculation of the uncertainty in $\Delta G = G_2 - G_1$. A reasonable estimate of the uncertainty for each time measurement may be about 10 s, taking into account reading time for the meters, etc. The uncertainty in Δt may then be calculated as about 14 s, and finally, an estimate made for the uncertainty in Q. Using the above figures and Eq. (3.2) gives an estimate of

$$w_Q \approx 2.7\%$$

This may be considered negligible for this kind of experiment.

Temperature and Pressure Measurements

A new bourdon tube pressure gage will be installed on the boiler outlet line. Thin adhesive-backed thermocouples will be installed on the same line and the condensate/feedwater inlet lines. Readings of the temperatures will be made with a hand-held digital thermocouple meter having a least count of 0.1°F. After installation, the thermocouples will be wrapped with insulating tape and covered with insulation. The same type of thermocouple will be installed on the inlet lines to the cup machines where they will be monitored from time to time by the production people. Considering the standard tolerances for thermocouple manufacture (see Table 8.3) the uncertainty for the temperature measurements is estimated at $\pm 2.5°F$.

The uncertainty in the pressure measurement is estimated at ± 2 percent of full scale for a 0 to 200 psig pressure gage, or ± 4 psig.

Determination of Heat Rate for Boiler

Using the nomenclature of Fig. Example 16.3a, the heat rate for the boiler is given by

$$q = m(h_{\text{out}} - h_{\text{in}})$$

Taking into account the uncertainties in the temperature and pressure measurements stated above, the uncertainty in the enthalpy difference is about ± 5 Btu/lbm or about 0.5 percent. Combined with the uncertainty of flow measurement of ± 2.7 percent, the uncertainty in the heat rate determination is less than 3 percent. This is a satisfactory figure.

Collection of Data and Calculations

Two proposed data sheets are shown in Figs. Example 16.3b and 16.3c. The boiler data sheet is organized in spreadsheet form so that the data may be entered into a computer for calculation of the desired results. The columns and formulas for these calculations are not shown because the display would be too large for printing here. Conversion of units will all be taken into account in the spreadsheet calculation formats. Depending on how busy he may be, the engineer recording the measurements may enter data into a computer during the test periods as well as listing them on the data sheet. *The hand entry data sheet must be retained in case of computer failure. Otherwise, the tests might need to be repeated, at increased cost to the client.*

	B	C	D	E	F	G	H	I	J	K	L	M
2							DATA SHEET FOR BOILER					
3												
4												
5		Date			Barometer			Plant Temp.		Engineer:		
6												
7		Combustion data										
8												
9	Time	Outlet	Outlet	Outlet	Feedwater	Feedwater	Flow	Time	Flow	Heat Rate	Machines	Comments
10	hr/min/sec	Temp	Press	enthalpy	Temp	enthalpy	meter	increment	rate	Btu/h	operating?	
11		F	psig	Btu/lb	F	Btu/lb	gal	min	gal/min	Boiler hp	Communicated?	
12												
13												
14												
15												

Figure Example 16.3b

				DATA SHEET FOR CUP MACHINE GROUPS				
	Date			Machine Group (A,B, or C)				
Time hr/min/sec	Inlet pressure psig	Machines Operating Yes/No	Operation Steady Yes/No	Communicated to Engineer? Yes/No	Product Quality Comments			Technician

Figure Example 16.3c

Note column L. The engineer must communicate with the production technicians to verify which groups of machines are in operation, that is, that they are following the design sequences. This communication will be accomplished with cell phones. Once the engineer is satisfied that a steady operating condition is in effect, data collection may begin. Thirty-minute time periods for the flow measurement are judged to be sufficient.

The data sheet proposed for the operating technicians is shown in Fig. Example 16.3c. No calculations will be performed from these sheets, so there is no need to display column and row headings. Assuming correct communication between the technicians and engineers, these data sheets will serve mainly as backup confirmation of the operating times for the three groups of machines. Observations regarding product quality may be coordinated with the measurements if something unusual occurs.

Data Analysis Procedure

As previously noted, individual measurements will be made of the steam consumption for each of the three groups of cup machines. In addition, the group consumption rates may be obtained from solutions of the simultaneous equations

$$(A + B)_{avg} = K_1$$
$$(B + C)_{avg} = K_2$$
$$(A + C)_{avg} = K_3$$

where the K values are calculated as an average value of the measurements taken for $A + B$, $B + C$, and $A + C$, e.g., at hours 2, 10, and 18 on the 1st day for $A + B$.

Finally, a check on the calculation may be made from measurements of

$$(A + B + C)_{avg} = K_4$$

where this K_4 value is taken over the large number of measurements of consumption with all cups operating. Standard deviations may be computed for each of the average or arithmetic mean values determined above using

$$\sigma_A = \left[\frac{\sum_1^n (A - A_{avg})^2}{n - 1} \right]^{1/2}$$

where n is the number of samples for A, B, and C involved in the determination of A_{avg}, etc.

Depending on the actual steam consumption rates observed, more reliance may be placed on some sets of measurements than on others.

Anticipated Results

Based on experience with similar equipment, the consultant expressed doubt that the boiler performance would fall short of specifications. No predictions about steam consumption for the cup machines were offered by anyone. Production persons acknowledged that the faulty cups could have been produced when the system was also delivering steam to another product line of styrofoam ice chests. All persons agreed that experimental determination of cup machine steam consumption rates was desirable, both from the standpoint of maintaining product quality and for purposes of design of future expansion of the production lines.

Report Requirements

The consultant proposed that the report for the project include:

1. The design of the experiment plan as described above
2. Calculation methods
3. Computer disk and summary tables of data and calculations of average steam consumption rates in accordance with the stated calculation methods.
4. Summary of measured results
5. Recommendations for:
    a. Solution of cup quality control problem
    b. Design of production line expansion
    c. Other matters that may arise

Proposed Budget

As mentioned previously, the client has asked for a budget and cost estimate for performing the planned experiments, analyzing the results, and reporting recommendations, in other words, a package or turnkey price. The cost of the project is mostly the result of hourly consulting fees for the engineers. A suggested format for the proposed budget is shown in Table Example 16.3*b*. If further studies are needed, they can be contracted for after the experimental program is completed.

Summary Comments

This experiment is rather simple in terms of the actual physical measurements that must be performed. Great precision is not required in the temperature and pressure measurements, and the flow determinations are made over a rather extended period of time to smooth out fluctuations in machine operation. There are no dynamic response considerations or need for elaborate analysis of data.

The experiment *design,* however, is rather involved because of schedule requirements and the need to avoid disruption of the production process. There is also a requirement for careful coordination between the engineers and the production staff at the client's facility. Maintaining a consistent product output forces a strict schedule upon the engineers and technicians for collection of data that will provide an accurate representation of the performance of the boiler and consumption rates for the cup machines. The sequencing of machine group operations presented in Table Example 16.3*a* is not the only schedule that would produce satisfactory data, but it should be a workable one.

Use of a data acquisition system would be ill advised for this set of experiments because of the low tolerance level for errors that could require repeating segments of the test program. As stated above, the plan calls for data collection by hand in a format for computer spreadsheet analysis. With the schedule as open as it is, there should be ample time for entry of the data into

Table Example 16.3*b*   Proposed budget for tests of cup machines

1. Preliminary expenses:
   Consultations
   (Dates) _____ h at $_____ /h = _____
   Analysis of problem and planning
   (Dates) _____ h at $_____ /h = _____
   Photocopy charges _____
2. Instrumentation (installed by client)
   a. Flowmeter _____
   b. Thermocouples _____
   c. Pressure gages _____
   d. Portable temperature indicators _____
   e. Miscellaneous supplies _____
3. Planned testing
   a. Senior engineer, 8 h/day, 4 days at $_____ /h = _____
   b. Two junior engineers, 8 h/day, 4 days at $_____ /h = _____
   c. Contingency, 5th day of testing, $_____
4. Preliminary analysis of data, 8 h at $_____ /h = _____
5. Consultation with client, 4 h at $_____ /h = _____
6. Final analysis of data, report, and recommendations, 12 h at $_____ /h = _____
7. Final consultation, 2 h at $_____ /h = _____
   Total project estimated cost _____

a portable computer also. If the spreadsheet program is checked and loaded in advance, it will then be an easy matter to produce a printout of preliminary results at the end of the first day.

This example illustrates how an experiment design may involve other considerations which override concerns of accuracy of measurements, instrument precision, and other technical factors.

SOLAR ASSISTED HEATING SYSTEM.   A manufacturer of solar collectors produces | **Example 16.4**

a unit to provide supplemental energy input for a heat pump system for residential housing. The manufacturer teams up with a producer of heat pumps to provide a combined system shown in the schematic of Fig. Example 16.4. The team subsequently obtains a small business government grant to conduct "demonstration" experiments with and without the solar assist with an objective of establishing the savings in electric power consumption.

Recognizing the commercial nature of the tests as well as the number of variables which may enter into evaluation of the final results, the solar collector–heat pump team contracts with an independent consulting engineer to design the experiment so that the results will be as unbiased as possible, and acceptable for commercial marketing purposes. The objective is to achieve an experiment design that will provide ample data for analysis without omission of important parameters. The actual data reduction and analysis will be accomplished later. The tests are expected to be conducted over a 13-month period so that all seasons of the year may be included in the time span. Minor modifications in the data collection may be expected from time to time as preliminary calculations are performed during the year.

The purpose of this example is show how an experiment can be designed to ensure all necessary data are collected without giving details of the eventual calculations that may be performed.

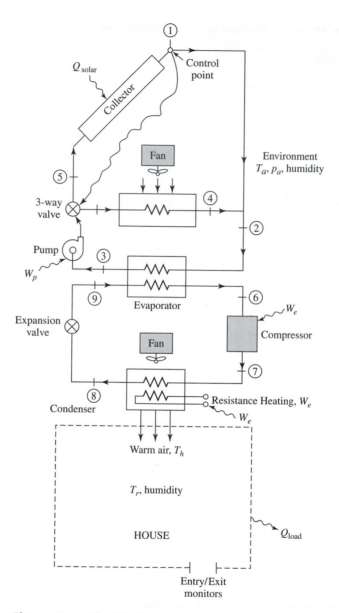

Figure Example 16.4

Two essentially identical houses of new construction, with new appliances, are available for the tests. One will be fitted with a heat pump all-electric system, and one with the heating cooling system that includes the solar assist indicated in the schematic diagram. While the objective of the tests could be the answer to a single question: "Does the solar assist pay for itself in reduced electric consumption?" the answer is not as simple as it may appear. The solar–heat pump team recognizes that much detailed information should be obtained to reach an understanding of the fine structure of the system behavior.

Consulting Table 16.2, it is obvious that the first step in the design protocol is a conference between the engineer and the solar–heat pump team to provide a list of questions to be answered by the experiments. This initial set of questions, comments, and tentative solutions is given as follows:

1. While the two houses are identical in construction and size, the actual energy consumption may vary because of (a) different temperature settings by the occupants and (b) different behavior patterns of the occupants, for example, frequency of entries and exits from the house, frequencies of showers and baths, laundry, lighting patterns. The test setup must be able to detect these differences, at least partially.

2. As with most heat pumps, some operation with electric resistance supplemental heating can be expected during the coldest time periods. The exact amount of such heating should be measured.

3. Total electric power consumption is influenced by many factors, such as cooling, laundry, frequency of hot water use, stereo and computer equipment, lighting, and more. The test should be able to indicate the power consumption directly attributable to the heating and cooling system and the savings associated with supplemental solar assist.

4. For reference, data should be provided on the amount of solar irradiation and the corresponding weather patterns (rainy, cloudy, or bright sunlight).

5. Detailed temperature and flow measurements should be obtained that will enable determination of performance of the heat pump and solar collector.

6. The measurements should be performed with a data acquisition system that can operate over 24-h periods for several days at a time without the physical presence of test personnel. Retention of the data on both printed strip charts and computer memory is desirable. The data stored in computer or data acquisition memory should be in a form that may be manipulated later with conventional data analysis methods, for example, spreadsheets.

7. While the tests are being performed over the 13-month period there should be minimal interference with or interruption of the daily activities of the residential occupants. Ideally, the residents should be unaware of the tests except when the personnel appear to collect data output.

8. Any maintenance or repair functions required for the heat pump, solar assist unit, or household appliances should be performed immediately, so that they will have minimal influence on the energy consumption results.

9. Climatological data (barometric pressure, temperatures, humidity) for the particular location should be collected, as well as that for the local weather station.

10. Some calculations and comparisons of energy consumptions should be performed at least once a month during the test program.

General Considerations

Consulting Table 16.1, we see that this type of experiment falls clearly in category 7 for "commercial" purposes, although category 8 "want to know" is certainly of ancillary importance. An argument can be made that the results of the experiments can be predicted from analysis, taking into account the insulation and construction of the house, known performance characteristics of the heat pump, and average climatological data for the particular locale. The advantage of a demonstration-type project is that the results include human variance factors that occur in real life, including responses to equipment failure and maintenance problems. We could pursue these arguments further, but that would be getting away from the objective here of designing the experiment to produce the desired results.

System Operation

Before going further, let us examine the schematic diagram of the solar assisted heat pump shown in Fig. Example 16.4. We assume the solar collector will be taken out of the system in the cooling mode so the operation is indicated only for the heating mode. The condenser, or heating unit, receives the hot refrigerant gases at point 7 and gives up heat to the house through a forced air finned tube heat exchanger. Supplemental electric resistance heating is provided downstream from the heat exchanger and is indicated at that point on the diagram. The liquid from the condenser is reduced in pressure in the expansion valve and enters the evaporator–heat exchanger at point 9, where it absorbs heat from a circulating ethylene glycol (antifreeze) solution. This solution is pumped through either the solar collector or fan-air heat exchanger, or both. The three-way valve regulates the flow of glycol so that all the flow is diverted to the air heat exchanger when the solar collector no longer provides heating, as at night. This regulation is controlled by a temperature-sensing element at point 1. In the cooling season the solar collector is taken out of the circuit. On the coldest days the presence of the solar collector may make it possible to operate without supplemental electric resistance heating.

Design Protocol

Clearly, there are several measurements necessary to satisfy the 10 test objectives described above. As our next step in the design process, we will specify and list the actions that can be taken to satisfy these requirements. We list them in the same order as above.

1. The different operating characteristics for the two houses and residents may be measured with the following parameters: (*a*) inside air temperature measured at the thermostat location; (*b*) approximate measurement of inside humidity; (*c*) determination of electric power consumption for nonheating and cooling appliances, including the hot water heater. This determination is most easily accomplished by measuring the power consumption for all elements of the heating and cooling system, and then subtracting the sum of these energies from the total power consumption, i.e.,

$$\text{Power}_{\text{non-heating cooling}} = \text{power}_{\text{total}} - \sum \text{power}_{\text{heating cooling units}}$$

The heating and cooling units are the compressor, glycol pump, resistance heater, and two air fans. The energy input to these units should be monitored continuously. (*d*) The entrance and exit traffic can be determined by using a sensor that records the opening and closing of each exterior door. It is possible to calculate (estimate) the heating or cooling load that results from such traffic.

2. The power input to the resistance heater should be monitored, as described above.

3. Covered in item 1.

4. A solar radiation-measuring device can be installed on the solar collector. Without personnel on the scene, it is difficult to record atmospheric conditions, but the solar radiation measurement itself should provide a good indication of cloudy skies, etc.

5. In addition to the power measurements indicated in item 1, and the indoor temperature and humidity, the temperatures at points 1 through 9 should be recorded continuously. The total flow rate delivered to the pump at point 3 and the flow through the collector at point 5 should also be measured. These measurements will permit an execution of the mass and

energy balances given by

$$m_4 = m_3 - m_5$$
$$m_2 = m_3$$
$$m_5 = m_1$$

Energy absorbed in solar collector $= m_5(h_1 - h_5)$

Energy absorbed in outside air fan unit $= m_4(h_4 - h_5)$

Total energy delivered to evaporator $= Q_{evap} = m_3(h_2 - h_3)$

Total heating delivered to house $= W_{comp} + W_{inside\ fan} + W_e + Q_{evap}$

In addition to the above measurements the outside air temperature and approximate humidity should be recorded.

6. A number of data-logger or data acquisition systems are available on the market that will accommodate the requirements of this experiment. Power interruptions are not expected, so the use of a noninterruptible power supply should be unnecessary.

7. This requirement simply means what it says. Personnel should be courteous to call before arriving to collect data. The data acquisition system should be installed in a separate external shed that may be accessed without entering the residence. The residents will probably be curious about the test results, but they should be advised to maintain their regular lifestyle and living patterns.

8. Since both houses will be new construction with all new appliances, breakdowns should not occur. That said, failures do happen and plans should be made to give priority treatment to any needed repairs. Obviously, the heating and cooling units should be repaired immediately if any problem should arise.

9. The need for collection of outside air temperature and humidity data has already been indicated. For reference, the high and low temperatures from the local weather station should also be recorded.

10. The personnel conducting the test will discover very soon that monitoring over 24-h periods collects a large amount of data. At the start, it may be prudent to make some of the energy consumption–balance calculations indicated above on a daily or weekly basis to be sure that everything is working properly. After that, monthly calculations should suffice.

Instrumentation

The proposed instruments are:

1. Temperatures: either thermocouples or resistance temperature detectors (RTDs) may be employed, depending on the selection of data logger.

2. Flowmeters: Turbine meters (Sec. 7.6) for the two glycol flow measurements are suggested. No flow measurements of the refrigerant internal to the heat pump are anticipated.

3. Electric power: Commercial recording meters are available to measure watt-hour electric energy as a function of time. These meters should be connected to measure
   a. Total electric energy consumption for the house.
   b. Energy consumption for the compressor–internal fan combination.
   c. Energy consumption for the external fan.
   d. Energy consumption for the supplemental resistance heater.
   e. Energy consumption for the glycol pump will be very nearly constant as long as it is in operation. Therefore, a single measurement of energy consumption may be made, along with recording of times of operation to determine energy consumption. This pump will not be involved for the house without solar assist.

The assumption is made that the pricing of electric energy for the residences is based on total energy (watt-hours) consumed and is not dependent on time of day usage or peak "demand." Demand charges are common for commercial installations, but rare in personal residences. If a demand charge is imposed by the electric utility, then the electric meter for the total house consumption must be able to record such information.

Evaluation of Results

This experiment is not one that should be expected to have clear-cut results. The measurements by themselves should provide indications of the savings in electric energy resulting from solar assist. Savings should indeed be observed, but the economic value of these savings may be difficult to appraise. Some questions that arise are:

1. Does the electric utility provide an incentive in pricing for installation of such energy conservation measures, as either a direct rebate applied to equipment purchase, or a rate adjustment?

2. Is there a possible tax break for installation of energy conservation measures?

3. Is a prospective user "environmentally sensitive" and willing to invest in measures that reduce power plant emissions?

4. How can one estimate the add-on cost of the solar assist in terms of future installations? Will the cost drop for an increase in the number of such installations? How much should the cost drop?

5. Are there ancillary uses for the solar collector, such as heating a swimming pool, even during times of the year when heating of the house is not required? Could the solar collector also be used to provide part or most of the heating for hot water all year? Should the ancillary uses of solar energy noted be planned from the start, including an increase in the size of solar collector? How should one apportion costs in these cases?

The point of the above remarks is to emphasize that this kind of experiment cannot be designed to answer the single question: "Does it pay to install a solar assisted heat pump?" The experiment can provide results that will be useful in obtaining at least a partial answer to the question. In terms of our design protocol and use of Table 16.2, this means that items 2, 3, and 5 will have incomplete answers at the end of a successful experiment.

Evaluation of Measurement Uncertainties

Except for gross mistakes in installation, the measurement readings for these experiments may be taken at face value. The uncertainties and variations of weather patterns and occupant behavior will likely override any errors or uncertainties that may be present in the measurements of temperature, flow rate, or electric energy consumption. This is not to say that one may be careless with the measurements. Gross mistakes do occur, for example, setting the data recorder for the wrong type of thermocouple or faulty installation of temperature sensors so that radiation effects significantly influence them. As a prudent matter, temperature sensors should be checked before installation and the combination of sensors and data logger should be examined to ensure that everything is operating properly. Other than these checks, execution of a formal uncertainty analysis is probably unnecessary.

Presentation of Results of Experiments

We have already indicated some of the energy balances that need to be performed, and noted possible difficulties in evaluating the economic results of the experiments. Using these energy balances, a reasonable set of technical results may be arranged as shown in Table Example 16.4. The results should be presented in both graphical and tabular form. Note the modifications called for in item 6. The engineer must evaluate the data for both residences

Table Example 16.4   Proposed presentation of results

---

1. Weekly kWh total electric consumption of residences with and without solar assist.
2. Weekly kWh heating and cooling electric consumption of residences with and without solar assist.
3. Weekly kWh non-heating and cooling electric consumption of residences with and without solar assist.
4. Actual weekly electric energy costs with and without solar assist.
5. Fraction of heating provided by solar assist on a weekly basis.
6. Modifications to heating and cooling loads of residences to account for differences in:
   a. Behavior patterns of residents (number of entrances and exits, and extra heating or cooling requirements resulting from differences in appliance or lighting use).
   b. Difference in thermostat settings by residents.

These adjustments should be made on the basis of accepted ASHRAE calculation methods.

7. Adjusted values of items 1 through 5 based on calculations of item 6.
8. Weekly average solar irradiation on collector.
9. Average weekly outside air temperature.
10. Average weekly inside air temperature for each residence.
11. Comments on unusual observations.

---

and make a judgment as to whether sufficient differences exist to justify this calculation. We cannot elaborate further except to say that adequate methods are available for effecting the calculations. A decision in this matter is best made by an engineer experienced in HVAC systems.

Summary Comments

This experiment is different from the other examples in that conclusive results are not expected even when the experiment goes well and is properly executed. It is different also because it extends over a long period of time and does not require close monitoring. It is still different from the others we have considered with the absence of a need for uncertainty analysis. Extension of the notions mentioned in item 6 of Table Example 16.4 would involve details of HVAC calculations that are beyond the scope of this presentation. Because the eventual evaluation of the results of the experiments involves factors that are not easily quantified, the experiment design protocol is left in an unfinished state. We shall rely on the ability of the reader to mentally extend the process in his or her own mind.

BURNING LEGAL TESTS.   A man incurs burns on the upper part of his body. About 6 months later he consults an attorney, who advises a lawsuit claiming that leakage of fuel from a butane cigar lighter was responsible for the burns. The suit is filed against the manufacturer of both the butane lighter and the canisters used for refueling. The suit alleges faulty design and asks for substantial damage awards. The attorney for the plaintiff is filing the case on a contingency fee basis.

   The attorney for the defense hires a consulting engineer to evaluate the faulty design claim and perform tests as needed to refute the claim. A brief summary of the facts regarding the case was established by the attorney as follows:

1. The plaintiff was indeed burned in the chest area and treated for the burns in the emergency room of a local teaching hospital. At that time, the plaintiff claimed that his T-shirt caught

Example 16.5

on fire while he was smoking a cigar, but medical records do not so state and the attending physician does not recall the matter after 6 months. The plaintiff was treated for the burns in two follow-up visits to the hospital.

2. The plaintiff claims that the butane lighter leaked fuel on his T-shirt during the filling process and that this fuel caught fire when he lighted a cigar. He was burned before he could douse the fire.

3. The plaintiff no longer has the T-shirt, and there is no record of its being presented at the emergency room.

4. The butane lighter and fuel canister are available for inspection by the defense attorney and his expert consultants. In addition, the defense attorney has purchased from a retail store like items for comparison in construction and materials.

5. The plaintiff has no medical insurance.

6. As mentioned above, the plaintiff waited 6 months before filing any kind of product complaint or contacting an attorney.

An initial conference between the defense attorney and the engineer is arranged to set the objectives of a test plan. As in most lawsuits the overriding objective is

Win the case and/or minimize damage awards and attorneys fees.

An insurance carrier is paying the defense fees based on the liability policies of the manufacturers. The insurance company will also pay expert witness (consulting engineering) fees.

The attorney-engineer conference covers items 1, 2, 3, and 6 of Table 16.2 and establishes some tentative objectives.

1. Examine the products for defects in construction or design and comment thereon. If appropriate, comment on measures that might be taken to correct any faults in design.

2. Attempt to duplicate the filling-leakage process that the plaintiff alleges as the cause of his burns.

3. Try to establish some careless behavior in the filling-leakage process by the plaintiff that might account for the burns, other than that explained in item 2.

4. Conduct an actual experiment filling the butane lighter with an actual fire promoted on cotton T-shirt material. Obvious safety measures should be taken such as having a fire extinguisher on hand.

5. The above evaluations are to be made with a minimum of written records and reported verbally to the defense attorney.

Design Protocol

The design of this experiment is directed at both fact finding and presentation of results which will be advantageous to winning the case. We will assume that the engineer will not be able to find anything defective in design or construction of either the butane lighter or the fuel canister. In planning for the filling-leakage-flame tests, some additional questions that may be asked are:

1. What fraction of the contents of the lighter, or what absolute quantity of fuel might be expected to leak, spill, or dribble out during the filling process?

2. What action must one take to cause the leakage to occur?

3. How long does it take for the fuel to evaporate from cotton T-shirt material so that it is no longer a flame hazard?

4. How easy is it to ignite the cotton T-shirt immediately after dripping with leaked butane?

5. Once the butane is ignited on the T-shirt, how easily is the flame extinguished? Will striking with bare hands do the job? Will the hands be burned as a result?

6. With deliberate effort, how easy is it to disperse a significant quantity of fuel onto T-shirt material for ignition? How long does it take—in seconds?

7. Assuming answers are obtained for all the foregoing questions, what does one do with the results?

These questions are not representative of a scientific inquiry at all, but suggest the following simple tests.

1. Perform a filling operation. Observe the amount of leakage of fuel. If none, or very small, repeat in a careless fashion to try to leak a significant amount of butane.

2. Observe where the leaked (forced or accidental) fuel goes. How can it get to a T-shirt being worn by the person filling the lighter? Would the person need to be lying down, rather than sitting or standing? Or, if the person has a vision problem, would it be necessary to hold objects close to the face and possibly drip fuel down on the T-shirt?

3. Dispense a significant amount of butane (one-fourth to one-half the content of the lighter) onto an old cotton T-shirt placed in a horizontal pan. Ignite immediately. Observe:
   a. How long does it burn?
   b. Can it be extinguished by striking with the hand? Take care not to burn the hand.
   c. Does the T-shirt continue burning after the fuel is consumed and if so, how intensely? Can it be extinguished by striking with the hand?

4. Repeat item 3, but delay ignition for a period comparable to the time between filling a lighter and lighting a cigar (10 to 30 s?).

Possible Results of the Proposed Experiments

We will assume three scenarios for results of the experiments outlined above:

1. Very little leakage during filling, and very difficult to cause leakage. When butane is intentionally placed on T-shirt material and ignited, the flame is not intense and is easily extinguished by striking with the hand.

2. Significant leakage during filling. Fuel burns intensely on cotton material and is not easily extinguished by hand action without incurring burns on the hand.

3. Little or no leakage during filling, but intense flame when fuel is ignited on cotton material.

There may be shadings of these choices, but they appear reasonable.

Analysis of Possible Results

If the result of scenario 1 is observed, or anything close to this result, the question must immediately be raised: How did the man burn himself? One plausible explanation is that he was smoking in bed and went to sleep, only to awaken with his chest on fire. In this event, the lighter or fuel canister would have nothing to do with the burns inflicted, unless they were lying in bed and caught fire also. But the physical evidence suggests no such burns on the lighter or canister.

If the result of scenario 2 is observed, the lighter–canister combination has the potential to set the man ablaze, and one would have to speculate how the fuel got on the T-shirt.

Depending on the degree which the test results may shade between scenarios 1 and 2, there are various speculations the engineer and attorney can make regarding the actual circumstances surrounding the burn incident.

Results of the Tests

For this example we state the results of the tests: Scenario 1 is observed with little or no leakage during filling of the lighter. Moreover, when fuel is placed on the cotton material and ignited, the flame is easily extinguished with the hands.

Presentation of Results

Item 3 of Table 16.2 is unusual in this example in that the engineer should present the results in the form of a physical demonstration with verbal report only, as requested by the defense attorney. Depending on how close the trial date is set, the attorney may wish to confront the plaintiff with the demonstration evidence, or delay a while. Either way, the engineer is alerted to the possibility of performance of the same tests in a court setting. If that should come to pass, the test must be carefully orchestrated, with obvious safety measures, and conducted in a legal format suggested by the attorney with trial judge approval.

The attorney for the plaintiff may want to question the engineer about the tests in a sworn deposition. Keeping in mind the overall object of winning the case and/or settling with *minimum* attorney and consultant fees, the engineer will probably be asked to put his efforts on hold until both sides negotiate a possible settlement. The nature of such a settlement depends on many factors beyond the scope of our discussion.

Summary Comments

The example considered here is a fairly simple legal case which can require experiments by a competent engineer. The protocol is straightforward, though: lay out the alternatives, examine each, conduct the experiments, and present the results. Note that the *need* for the experiments arises from two sources: (1) an examination of the physical evidence of the butane lighter and fuel canister that reveals no serious design flaws and (2) a need to explain the physical evidence of burns on the plaintiff's body which are central to the lawsuit. A plausible explanation has been offered here of "smoking in bed." Should the case go to trial, it would become subject to all the uncertainties and vagaries of the legal system. These uncertainties may not be evaluated with the methods of Chap. 3!

16.5 SUMMARY

After presenting the five diverse examples of experiment design and execution, we see that the use of such indefinite phrases as "sometimes," "depends on," "possibly," and "variable" in Tables 16.1 and 16.2 is not inappropriate. The need for uncertainty analyses in the experiment planning protocol and during analysis of data varies widely from the extensive use in Example 16.1 to essentially none in Example 16.5. Time schedule requirements vary from the relaxed annual project of Example 16.4 to the rigid hour-by-hour protocol imposed by the production schedule of Example 16.3. Precision measurement requirements vary from the rather strict needs of Example 16.1 to a simple demonstration described in Example 16.5. Budget restrictions are also variable, but have the common businesslike thread of required planning and avoidance of excessive expense for either personnel or equipment.

The reporting and presentation-of-results requirements also vary widely in the five cited examples. A rather detailed technical report is expected in Example 16.1, with perhaps a follow-on technical paper submission. Example 16.2 also requires a technical report, but without as much precision measurement. Example 16.3 requires an action-oriented report that management can use for process modification immediately. In contrast, the eventual report requirement for Example 16.4 is vague and indefinite and depends on evaluation of several factors outside the scope of the experiments. Finally, only a verbal report accompanied by a burning demonstration in required in Example 16.5.

The foregoing remarks contrast the widely different technical and reporting requirements for the stated examples, but do not mention the *one* factor or element of design protocol that projects a strong presence in *all* of the examples. There is always a strong need for detailed consultation to establish the objectives and goals of the experimental project, both at selected times before the experiments begin, and thereafter as the program develops.

Of course, the consultation needs are listed as items 1, 2, 3, and 6 in Table 16.2. In the consultations and discussions one should always be seeking a clear delineation of the *why* as well as the *how* for the experiment. In other words the need for the experiment should be clearly established. The *need* for each of the experimental programs described in the five examples varies markedly. Confusion regarding the objectives for the experiments can produce very poor results. It would not be enough to *demonstrate* the plastic cup failures through a reduction in steam flow to the molding machines, because the demonstration would not establish specific flow information necessary for correcting the problem. On the other hand, scientific data could be presented on the flame temperature for butane burning with air without offering any clear rebuttal of the T-shirt fire scenario alleged by the plaintiff in Example 16.5.

The reader will no doubt be able to propose other factors which should be involved in the design protocol, and suggest alternatives to the experiment outlines presented in the examples. Such alternatives are encouraged, and may in fact offer improvements over the procedures and plans developed here.

16.6   PROBLEMS

16.1. *Expansion of information in experiment design examples.* Examples 16.1 to 16.5 follow the design protocol suggested by Tables 16.1 and 16.2 in a general narrative way with more or less specific references to the table entries. Select one or more of the examples and make margin notations in the text which designate cell references for Tables 16.1 and 16.2. Then check off these items in a copy of each table. If there are serious omissions, suggest improvements that might be made in the protocol for that example.

16.2. In Example 16.3, a schedule for collecting the steam consumption data was specified which would not disrupt the manufacture of product. Can you suggest an alternate schedule that would do as well?

16.3. At the close of Example 16.4 some questions of economic significance are proposed. Can you offer other questions, or elaborate on the five items listed?

16.4. Example 16.5 presents a rather unusual experiment that an engineer might be asked to perform. If convenient, consult an attorney experienced in such matters (pro bono, of course) for comments.

16.7 REFERENCES

1. Doebelin, Ernest: *Engineering Experimentation: Planning, Execution, Reporting,* McGraw-Hill, New York, 1995.

2. Dhillon, B. S.: *Engineering Design: A Modern Approach,* McGraw-Hill, New York, 1996.

3. Burghardt, M. D.: *Introduction to Engineering Design and Problem Solving,* McGraw-Hill, New York, 1999.

4. Ullman, D. G.: *The Mechanical Design Process,* 2nd ed., McGraw-Hill, New York, 1997.

5. Eide, A., R. Jenison, L. Mashaw, and L. Northrup: *Introduction to Engineering Design,* McGraw-Hill, New York, 1998.

CONVERSION FACTORS AND MATERIAL PROPERTIES

Table A.1a Conversion factors[1]

Length
   12 in = 1 ft
   2.54 cm = 1 in
   1 μm = 10^{-6} m = 10^{-4} cm

Mass
   1 kg = 2.205 lbm
   1 slug = 32.17 lbm
   454 g = 1 lbm

Force
   1 dyn = 2.248 $\times$ 10^{-6} lbf
   1 lbf = 4.448 N
   10^5 dyn = 1 N

Energy
   1 ft $\cdot$ lbf = 1.356 J
   1 kW $\cdot$ h = 3413 Btu
   1 hp $\cdot$ h = 2545 Btu
   1 Btu = 252 cal
   1 Btu = 778 ft $\cdot$ lbf
   1 Boiler hp = 33497 Btu/h

Pressure
   1 atm = 14.696 lbf/in^2 = 2116 lbf/ft^2
   1 atm = 1.0132 $\times$ 10^5 N/m^2
   1 inHg = 70.73 lbf/ft^2

Viscosity
   1 cP = 2.42 lbm/h $\cdot$ ft
   1 lbf $\cdot$ s/ft^2 = 32.16 lbm/s $\cdot$ ft

Thermal conductivity
   1 cal/s $\cdot$ cm $\cdot$ °C = 242 Btu/h $\cdot$ ft $\cdot$ °F
   1 W/cm $\cdot$ °C = 57.79 Btu/h $\cdot$ ft $\cdot$ °F

| [1] See also inside cover.

Table A.1b   Conversion factors between common SI and English units

Physical Quality	Symbol	SI-to-English Conversion	English-to-SI Conversion
Length	L	1 m = 3.2808 ft	1 ft = 0.3048 m
Area	A	1 m^2 = 10.7639 ft^2	1 ft^2 = 0.092903 m^2
Volume	V	1 m^3 = 35.3134 ft^3	1 ft^3 = 0.028317 m^3
Velocity	v	1 m/s = 3.2808 ft/s	1 ft/s = 0.3048 m/s
Density	ρ	1 kg/m^3 = 0.06243 lbm/ft^3	1 lbm/ft^3 = 16.018 kg/m^3
Force	F	1 N = 0.2248 lbf	1 lbf = 4.4482 N
Mass	m	1 kg = 2.20462 lbm	1 lbm = 0.45359637 kg
Pressure	p	1 N/m^2 = 1.45038 × 10^{-4} lbf/in^2	1 lbf/in^2 = 6894.76 N/m^2
Energy, heat	q	1 kJ = 0.94783 Btu	1 Btu = 1.05504 kJ
Energy, heat flow	q	1 W = 3.4121 Btu/h	1 Btu/h = 0.29307 W
Energy, heat flux per unit area	q/A	1 W/m^2 = 0.317 Btu/h · ft^2	1 Btu/h · ft^2 = 3.154 W/m^2
Heat flux per unit length	q/L	1 W/m = 1.0403 Btu/h · ft	1 Btu/h · ft = 0.9613 W/m
Heat generation per unit volume	$\dot{q}$	1 W/m^3 = 0.096623 Btu/h · ft^3	1 Btu/h · ft^3 = 10.35 W/m^3
Energy per unit mass	q/m	1 kJ/kg = 0.4299 Btu/lbm	1 Btu/lbm = 2.326 kJ/kg
Specific heat	C	1 kJ/kg · °C = 0.23884 Btu/lbm · °F	1 Btu/lbm · °F = 4.1869 kJ/kg · °C
Thermal conductivity	k	1 W/m · °C = 0.5778 Btu/h · ft · °F	1 Btu/h · ft · °F = 1.7307 W/m · °C
Convection heat-transfer coefficient	h	1 W/m^2 · °C = 0.1761 Btu/h · ft^2 · °F	1 Btu/h · ft^2 · °F = 5.6782 W/m^2 · °C
Dynamic viscosity	μ	1 kg/m · s = 0.672 lbm/ft · s = 2419.2 lbm/ft · h	1 lbm/ft · s = 1.4881 kg/m · s
Kinematic viscosity, thermal diffusivity, and diffusion coefficient	ν, α	1 m^2/s = 10.7639 ft^2/s	1 ft^2/s = 0.092903 m^2/s

Table A.2 Properties of metals at 70°F (21.1°C)

Metal	Density		Specific Heat		Linear Coefficient of Expansion		Electrical Conductivity Relative to Copper	Thermal Conductivity	
	lbm/ft³	kg/m³	Btu/lbm·°F	kJ/kg·°C	°F⁻¹ × 10⁴	°C⁻¹ × 10⁴		Btu/h·ft·°F	W/m·°C
Aluminum, pure	167	2675	0.22	0.921	0.13	0.23	0.66	118	204
Red brass: 85% Cu, 9% Sn, 6% Zn	544	8714	0.092	0.385	0.10	0.18		35	61
Copper, pure	555	8890	0.095	0.398	0.09	0.16	1.00	223	386
Iron, pure	493	7897	0.108	0.452	0.065	0.12	0.18	42	73
Lead	710	11,370	0.031	0.130	0.16	0.29	0.08	20	35
Magnesium	109	1746	0.242	1.013	0.145	0.26	0.40	99	171
Nickel	556	8906	0.106	0.444	0.07	0.126	0.25	52	90
Platinum	1335	21,380	0.032	0.134	0.05	0.09	0.18	40	69
Silver	657	10,520	0.056	0.234	0.107	0.193	1.15	235	407
Stainless steel, 18% Cr, 8% Ni	488	7817	0.11	0.46	0.003	0.005	0.0234	9.4	16.3
Steel, structural	485	7769	0.11	0.46	0.07	0.126	0.12	33	57
Tin	456	7304	0.054	0.226	0.11	0.198	0.16	37	64
Tungsten	1208	19,350	0.032	0.134	0.02	0.036	0.33	94	163
Zinc	446	7144	0.092	0.385	0.18	0.32	0.31	37	64

Table A.3 Thermal properties of some nonmetals at 70°F (21.1°C)

Material	Density		Specific Heat		Thermal Conductivity	
	lbm/ft³	kg/m³	Btu/lbm·°F	kJ/kg·°C	Btu/h·ft·°F	W/m·°C
Asbestos sheet	60	961	0.2	0.84	0.083	0.144
Brick, common	110	1762	0.22	0.92	0.042	0.073
Concrete	140	2243	0.16	0.67	1.0	1.73
Fiberboard	15	240	0.5	2.09	0.027	0.047
Fiberglass	6	96	0.5	2.09	0.022	0.038
Glass, window	160	2563	0.16	0.67	0.50	0.87
85% magnesia pipe covering	15	240	0.2	0.84	0.042	0.073
Plaster	95	1522	0.25	1.05	0.27	0.47

Table A.4 Properties of some saturated liquids at 68°F (20°C)

Fluid	Chemical Formula	Density lbm/ft³	Density kg/m³	Specific Heat Btu/lbm·°F	Specific Heat kJ/kg·°C	Kinematic Viscosity ft²/s × 10⁵	Kinematic Viscosity m²/s × 10⁵	Thermal Conductivity Btu/h·ft·°F	Thermal Conductivity W/m·°C	Prandtl Number
Ammonia	NH_3	38.2	612	1.15	4.81	0.386	0.036	0.301	0.521	2.02
Carbon dioxide	CO_2	48.2	772	1.2	5.02	0.98	0.091	0.0504	0.087	4.10
Freon-12	CCl_2F_2	83.0	1329	0.23	0.96	0.213	0.020	0.042	0.073	2.17
Glycerin	$C_2H_3(OH)_3$	78.9	1264	0.57	2.39	1270	118	0.165	0.286	12.5
Mercury	Hg	847.7	13,580	0.033	0.138	0.123	0.011	5.02	8.69	0.025
Motor oil (typ.)		55.0	881	0.45	1.88	1000	92.9	0.08	0.138	10.0
Water	H_2O	62.4	1000	1.0	4.19	1.08	0.10	0.345	0.597	7.02

Table A.5 Properties of gases at atmospheric pressure and 68°F (20°C)

Gas	Formula	Molecular Weight	c_p Btu/lbm·°F	c_p kJ/kg·°C	$\gamma = c_p/c_v$	Thermal Conductivity Btu/h·ft·°F	Thermal Conductivity W/m·°C	Viscosity lbm/h·ft	Viscosity kg/m·s × 10⁵
Air		28.95	0.24	1.005	1.40	0.025	0.043	0.044	1.82
Acetylene	C_2H_2	26.02	0.38	1.591	1.25	0.0124	0.0215	0.024	0.99
Ammonia	NH_3	17.03	0.52	2.177	1.30	0.0142	0.0246	0.024	0.99
Carbon dioxide	CO_2	44.00	0.20	0.837	1.30	0.0095	0.0164	0.036	1.49
Freon-12	CCl_2F_2	120.9	0.15	0.628	1.14	0.0053	0.0091	0.031	1.28
Hydrogen	H_2	2.01	3.4	14.24	1.40	0.103	0.178	0.022	0.91
Methane	CH_4	16.03	0.54	2.26	1.30	0.0196	0.0339	0.027	1.12
Nitrogen	N_2	28.02	0.245	1.03	1.40	0.0150	0.026	0.044	1.82
Oxygen	O_2	32.00	0.215	0.90	1.40	0.0153	0.0265	0.048	1.98

Table A.6   Properties of air at atmospheric pressure (SI units)[1]
The values of μ, k, c_p, and Pr are not strongly pressure-dependent and may be used over a fairly wide range of pressures

T, K	ρ, kg/m^3	c_p, kJ/kg · °C	μ, kg/m · s × 10^5	ν, m^2/s × 10^6	k, W/m · °C	α, m^2/s × 10^4	Pr
100	3.6010	1.0266	0.6924	1.923	0.009246	0.02501	0.770
150	2.3675	1.0099	1.0283	4.343	0.013735	0.05745	0.753
200	1.7684	1.0061	1.3289	7.490	0.01809	0.10165	0.739
250	1.4128	1.0053	1.599	11.31	0.02227	0.15675	0.722
300	1.1774	1.0057	1.8462	15.69	0.02624	0.22160	0.708
350	0.9980	1.0090	2.075	20.76	0.03003	0.2983	0.697
400	0.8826	1.0140	2.286	25.90	0.03365	0.3760	0.689
450	0.7833	1.0207	2.484	31.71	0.03707	0.4222	0.683
500	0.7048	1.0295	2.671	37.90	0.04038	0.5564	0.680
550	0.6423	1.0392	2.848	44.34	0.04360	0.6532	0.680
600	0.5879	1.0551	3.018	51.34	0.04659	0.7512	0.680
650	0.5430	1.0635	3.177	58.51	0.04953	0.8578	0.682
700	0.5030	1.0752	3.332	66.25	0.05230	0.9672	0.684
750	0.4709	1.0856	3.481	73.91	0.05509	1.0774	0.686
800	0.4405	1.0978	3.625	82.29	0.05779	1.1951	0.689
850	0.4149	1.1095	3.765	90.75	0.06028	1.3097	0.692
900	0.3925	1.1212	3.899	99.3	0.06279	1.4271	0.696
950	0.3716	1.1321	4.023	108.2	0.06525	1.5510	0.699
1000	0.3524	1.1417	4.152	117.8	0.06752	1.6779	0.702
1100	0.3204	1.160	4.44	138.6	0.0732	1.969	0.704
1200	0.2947	1.179	4.69	159.1	0.0782	2.251	0.707
1300	0.2707	1.197	4.93	182.1	0.0837	2.583	0.705
1400	0.2515	1.214	5.17	205.5	0.0891	2.920	0.705
1500	0.2355	1.230	5.40	229.1	0.0946	3.262	0.705
1600	0.2211	1.248	5.63	254.5	0.100	3.609	0.705
1700	0.2082	1.267	5.85	280.5	0.105	3.977	0.705
1800	0.1970	1.287	6.07	308.1	0.111	4.379	0.704
1900	0.1858	1.309	6.29	338.5	0.117	4.811	0.704
2000	0.1762	1.338	6.50	369.0	0.124	5.260	0.702
2100	0.1682	1.372	6.72	399.6	0.131	5.715	0.700
2200	0.1602	1.419	6.93	432.6	0.139	6.120	0.707
2300	0.1538	1.482	7.14	464.0	0.149	6.540	0.710
2400	0.1458	1.574	7.35	504.0	0.161	7.020	0.718
2500	0.1394	1.688	7.57	543.5	0.175	7.441	0.730

| [1]From *Natl. Bur. Std. (U.S.)*, Circ. 564, 1955.

Table A.7   Properties of dry air at atmospheric pressure (English units)[1]

T, °F	μ, lbm/h · ft	k, Btu/h · ft · °F	c_p, Btu/lbm · °F	Pr
−100	0.0319	0.0104	0.239	0.739
−50	0.0358	0.0118	0.239	0.729
0	0.0394	0.0131	0.240	0.718
50	0.0427	0.0143	0.240	0.712
100	0.0459	0.0157	0.240	0.706
150	0.0484	0.0167	0.241	0.699
200	0.0519	0.0181	0.241	0.693
250	0.0547	0.0192	0.242	0.690
300	0.0574	0.0203	0.243	0.686
400	0.0626	0.0225	0.245	0.681
500	0.0675	0.0246	0.248	0.680
600	0.0721	0.0265	0.250	0.680
700	0.0765	0.0284	0.254	0.682
800	0.0806	0.0303	0.257	0.684
900	0.0846	0.0320	0.260	0.687
1000	0.0884	0.0337	0.263	0.690

| [1]From: *Natl. Bur. Std. (U.S.), Circ.* 564, 1955.

Table A.8 Properties of water [saturated liquid] (SI units)[1]

°F	°C	c_p, kJ/kg · °C	ρ, kg/m^3	μ, kg/m · s	k, W/m · °C	Pr	$g\beta\rho^2 c_p/\mu k$ 1/m^3 · °C
32	0	4.225	999.8	1.79×10^{-3}	0.566	13.25	
40	4.44	4.208	999.8	1.55	0.575	11.35	1.91×10^9
50	10	4.195	999.2	1.31	0.585	9.40	6.34×10^9
60	15.56	4.186	998.6	1.12	0.595	7.88	1.08×10^{10}
70	21.11	4.179	997.4	9.8×10^{-4}	0.604	6.78	1.46×10^{10}
80	26.67	4.179	995.8	8.6	0.614	5.85	1.91×10^{10}
90	32.22	4.174	994.9	7.65	0.623	5.12	2.48×10^{10}
100	37.78	4.174	993.0	6.82	0.630	4.53	3.3×10^{10}
110	43.33	4.174	990.6	6.16	0.637	4.04	4.19×10^{10}
120	48.89	4.174	988.8	5.62	0.644	3.64	4.89×10^{10}
130	54.44	4.179	985.7	5.13	0.649	3.30	5.66×10^{10}
140	60	4.179	983.3	4.71	0.654	3.01	6.48×10^{10}
150	65.55	4.183	980.3	4.3	0.659	2.73	7.62×10^{10}
160	71.11	4.186	977.3	4.01	0.665	2.53	8.84×10^{10}
170	76.67	4.191	973.7	3.72	0.668	2.33	9.85×10^{10}
180	82.22	4.195	970.2	3.47	0.673	2.16	1.09×10^{11}
190	87.78	4.199	966.7	3.27	0.675	2.03	
200	93.33	4.204	963.2	3.06	0.678	1.90	
220	104.4	4.216	955.1	2.67	0.684	1.66	
240	115.6	4.229	946.7	2.44	0.685	1.51	
260	126.7	4.250	937.2	2.19	0.685	1.36	
280	137.8	4.271	928.1	1.98	0.685	1.24	
300	148.9	4.296	918.0	1.86	0.684	1.17	
350	176.7	4.371	890.4	1.57	0.677	1.02	
400	204.4	4.467	859.4	1.36	0.665	1.00	
450	232.2	4.585	825.7	1.20	0.646	0.85	
500	260	4.731	785.2	1.07	0.616	0.83	
550	287.7	5.024	735.5	9.51×10^{-5}			
600	315.6	5.703	678.7	8.68			

[1] Converted from A. I. Brown and S. M. Marco, *Introduction to Heat Transfer*, 3d ed., McGraw-Hill, New York, 1958.

Table A.9   Properties of water [saturated liquid] (English units)[1]

°F	c_p, Btu/lbm · °F	ρ, lbm/ft³	μ, lbm/ft · h	k, Btu/h · ft · °F	Pr
32	1.009	62.42	4.33	0.327	13.35
40	1.005	62.42	3.75	0.332	11.35
50	1.002	62.38	3.17	0.338	9.40
60	1.000	62.34	2.71	0.344	7.88
70	0.998	62.27	2.37	0.349	6.78
80	0.998	62.17	2.08	0.355	5.85
90	0.997	62.11	1.85	0.360	5.12
100	0.997	61.99	1.65	0.364	4.53
110	0.997	61.84	1.49	0.368	4.04
120	0.997	61.73	1.36	0.372	3.64
130	0.998	61.54	1.24	0.375	3.30
140	0.998	61.39	1.14	0.378	3.01
150	0.999	61.20	1.04	0.381	2.73
160	1.000	61.01	0.97	0.384	2.53
170	1.001	60.79	0.90	0.386	2.33
180	1.002	60.57	0.84	0.389	2.16
190	1.003	60.35	0.79	0.390	2.03
200	1.004	60.13	0.74	0.392	1.90
220	1.007	59.63	0.65	0.395	1.66
240	1.010	59.10	0.59	0.396	1.51
260	1.015	58.51	0.53	0.396	1.36
280	1.020	57.94	0.48	0.396	1.24
300	1.026	57.31	0.45	0.395	1.17
350	1.044	55.59	0.38	0.391	1.02
400	1.067	53.65	0.33	0.384	1.00
450	1.095	51.55	0.29	0.373	0.85
500	1.130	49.02	0.26	0.356	0.83
550	1.200	45.92	0.23		
600	1.362	42.37	0.21		

[1] Converted from A. I. Brown and S. M. Marco, *Introduction to Heat Transfer*, 3d ed., McGraw-Hill, New York, 1958.

Table A.10 Diffusion coefficients of some gases and vapors in air at 25°C and 1 atm[1]

Substance	D, cm^2/s
Ammonia	0.28
Benzene	0.088
Carbon dioxide	0.164
Hydrogen	0.410
Methanol	0.159
Water	0.256

[1] From R. H. Perry (ed.): *Chemical Engineers' Handbook.* 5th ed., McGraw-Hill, New York, 1973.

Table A.11 Approximate total normal emissivities of various surfaces at 20°C

Surface	Emissivity
Aluminum, heavily oxidized	0.20–0.30
Aluminum, polished	0.09
Asbestos sheet	0.93–0.96
Brass, polished	0.06
Brick	0.93
Carbon, lampblack	0.96
Copper, heavily oxidized	0.75
Copper, polished	0.03
Enamel, white fused on iron	0.90
Iron, oxidized	0.74
Paint, aluminum	0.27–0.67
Paint, flat black lacquer	0.97
Steel, polished	0.07
Steel plate, rough	0.95
Tin, bright	0.06
Water	0.95

Table A.12   Area moments of inertia for some common geometric shapes

Rectangle

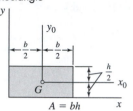

$$A = bh$$

$$I_{y0} = \frac{bh^3}{12}$$

$$I_{y0} = \frac{hb^3}{12}$$

Right triangle

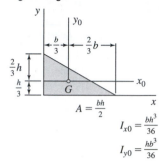

$$A = \frac{bh}{2}$$

$$I_{x0} = \frac{bh^3}{36}$$

$$I_{y0} = \frac{hb^3}{36}$$

Quadrant of circle

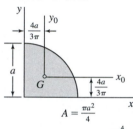

$$A = \frac{\pi a^2}{4}$$

$$I_x = I_y = \frac{\pi a^4}{16}$$

$$I_{x0} = I_{y0} = 0.054a^4$$

Circle

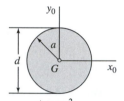

$$A = \pi a^2$$

$$I_{x0} = I_{y0} = \frac{\pi a^4}{4} = \frac{\pi d^4}{64}$$

(Polar) $J = \dfrac{\pi a^4}{4} = \dfrac{\pi d^4}{32}$

Quadrant of ellipse

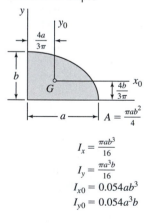

$$A = \frac{\pi ab}{4}$$

$$I_x = \frac{\pi ab^3}{16}$$

$$I_y = \frac{\pi a^3 b}{16}$$

$$I_{x0} = 0.054ab^3$$

$$I_{y0} = 0.054a^3 b$$

Ellipse

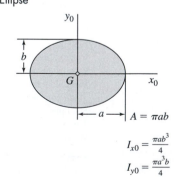

$$A = \pi ab$$

$$I_{x0} = \frac{\pi ab^3}{4}$$

$$I_{y0} = \frac{\pi a^3 b}{4}$$

B

DIGITAL IMAGING SYSTEMS

B.1 INTRODUCTION

Much of the technological advances of the past 100 years have followed a pattern of scientific discovery, high-cost use in military or space applications, and then development for low-cost consumer application. The development of high-performance aircraft for military use was followed by design of commercial jetliners; and improvements in jet-engine design were then transferred to the production of more efficient engines for powering commercial jets.

The development of computers followed about the same pattern. Giant vacuum-tube-powered machines were developed for government and military applications; and as solid-state and integrated-circuit electronics appeared, computer engineers were quick to apply the new technology to computers. Computing power increased, memory capabilities expanded tremendously, mass production for consumer use increased exponentially, and both purchase prices and operating costs declined in response. These cost reductions were fueled by consumer demand, or marketing efforts that appealed to a seemingly insatiable consumer desire for more and more computer power to drive computer games, flat-screen TVs, and an unending array of gadgetry.

Along came digital photography that developed with computing power and memory expansion—and their resultant reductions in cost because of mass production. In this case, the development did not follow the familiar pattern of military-government use transferred to commercial application. Instead, it was a piggyback development that occurred and was stimulated by the cost reductions in memory and the increase in computer power. The development of digital signal processing techniques also aided in this development.

Development of digital imaging techniques is still underway and difficult to keep up with. A discussion of currently available hardware would almost certainly be outdated the day it was published, so we will limit our discussion to an examination of terminology that is common to most parts of digital photography and does not change with the introduction of each new product. The reader will recognize several terms that apply to both emulsion (film) photography and digital systems, and we include them for completeness sake. While it would be entertaining to discuss portrait or landscape photography, we will limit our applications to those situations most likely to involve engineering experimentation.

B.2   DEFINITIONS AND UNITS

A schematic of a general digital imaging system is shown in Fig. B.1. The object to be photographed could be a piece of any sort of apparatus, smoke particles in an airflow stream, temperature-sensitive colored paints, a vibrating mechanism,

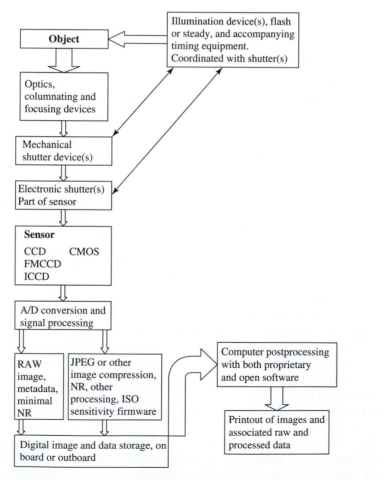

Figure B.1   Schematic of digital imaging system.

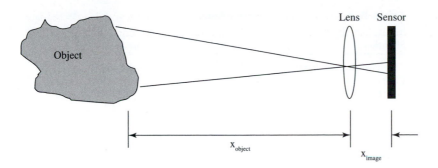

Figure B.2   Focal length $= x_{image}$ for $x_{object} \to \infty$

a crashing automobile, a shock wave in a supersonic flowstream, and so on. For simple applications, the focal length of the lens is related to object and image distances as indicated in Fig. B.2.

$$1/x_{object} + 1/x_{image} = 1/\text{focal length}. \qquad \text{[B.1]}$$

So, for an object a large distance away from the lens

$$\text{Focal length} = x_{image} \text{ for } x_{object} \to \infty$$

The f-stop, or f-number, of the lens is defined by

$$f\text{-stop} = \text{focal length}/\text{lens diameter} = FL/d_{lens}$$

For most commercial lenses the lens opening, or aperture, is mechanically adjustable, thereby permitting adjustment in the f-stop and the amount of light transmitted through the lens system to the sensor. The f-number specification for the lens corresponds to the maximum opening. The total amount of light energy that is transmitted to the sensor depends on the size of the lens opening and the time duration of the opening, or shutter speed. The amount of light energy transmitted to the sensor also depends on the brightness of illumination of the object being photographed.

The fundamental SI unit of luminous intensity is the candela (cd), defined as follows:

> One candela is the luminous intensity from a source that emits monochromatic radiation of 1/683 W/steradian at a frequency of 540 THz (wavelength of 555 nm, green)

From this are derived the units of luminous flux, lumens (lm)

$$1 \text{ lumen} = 1 \text{ cd-sr}$$

as well as illuminance, lux (lx)

$$1 \text{ lux} = 1 \text{ lm/m}^2$$

Clearly, the lux is a measure of the "brightness" of an illuminated area. Table B.1 gives some typical illuminance values.

Table B.1   Typical illuminance values

Situation	Illuminance, lux
Clear day, direct sunlight	30–120 k
Clear day, shaded area, indirect sun	10–25 k
Well-lighted office	300–500
Bright TV studio	1.0 k
Typical home living room	50
Clear night, full moon	0.3
Clear night, quarter moon	0.01

B.3   SENSORS

The heart of the digital system is the image sensor, which is usually a charge-coupled device (CCD) or complementary-metal-oxide semiconductor (CMOS). CCDs operate by creating charges in capacitive bins or wells that are proportional to the illumination imposed on that pixel. The charges are subsequently transferred to another region of the sensor chip where they are converted to voltage and read from the device into circuits where they may be digitized and stored in various types of memory devices, including direct storage in a computer. The larger the pixel size, the larger the amount of light that may be collected, and the potential for higher S/N ratio is higher for larger sensor sizes. Commercial CCD have pixel densities ranging from 1 to 100 MP/cm^2. The basic noise level of the sensor chip may be reduced by cooling the chip to low temperatures ($\approx -190°$C) using liquid nitrogen. Such cooling systems increase the cost considerably, but are warranted in applications involving very low light, such as astronomical photography.

It is possible to mask off part of the sensor (by increasing total sensor size) that may be used for readout at convenient speed necessary to minimize noise, while the cells exposed to light carry on their collection of data. Such a device is called a frame-transfer CCD. It obviously involves higher cost because of the doubling of the cell area.

It is possible to install an image intensifier in front of a CCD, thereby producing an intensified-charge-coupled device (ICCD). The control circuits employed for the intensifier enable an electronic shutter function on the order of 0.5 nanosecond.

CMOS devices are also designated as active-pixel sensors, because each pixel on the integrated circuit contains its own amplifier. This fact enables the circuit design to control each pixel individually, in regard to response to light, manipulation of noise reduction, and averaging of exposure parameters over the entire image.

B.4   SIGNAL PROCESSING

Once the signal has been captured, processing may begin. One has a choice of processing the raw signal direct from the sensor (after A/D conversion), which may involve quite large storage requirements or the order of 20 to 50 MB per image, or engaging in an image-compression process such as JPEG (Joint Photographer's Expert Group) which can reduce the image storage requirements by 50 to 80 percent. Other methods such as TIFF (Tagged Information File Format) may also be employed. Some noise reduction (NR) may also be executed before the digital information is stored. In general, the greater the amount of noise reduction employed, the smaller the resolution in the final image.

One advantage of digital imaging systems is that luminance data may be collected on an individual pixel basis and then examined for important variation by using subsequent computer processing. If only a simple image is required of an inanimate object, then digital imaging offers little advantage over its emulsion (film) counterpart. On the other hand, the ability to measure the variation in illuminance across a shock wave in a high-speed flow stream as visualized with a schlieren optical system is a notable advantage of the digital system. Immediate access to image data, without the need for photographic development, is another obvious advantage. Postprocessing of image data using both proprietary and locally designed software is also an advantage of the digital system.

B.5 EXPOSURE AND NOISE REDUCTION

The "noise" in an emulsion photographic system results largely from the grain size or "clumps" in the photosensitive film. The larger the grain, the larger the interference with the image and thus the smaller the final resolution. A finer grain results in less "noise" and better resolution. Unfortunately, a finer grain also results in a lower photosensitivity of the film, and the need for greater luminance to effect a properly exposed image. In a digital system the noise results from the basic noise level of the electronic-circuit components. The higher the illuminance on the photosensitive device, the higher the signal/noise ratio. When a JPEG image is produced in the digital imaging device, there is usually an NR process applied also, usually involving proprietary algorithms. For most RAW images, NR techniques are not applied and the matter is left to the user, employing outboard computer software.

It is clear that the S/N behavior of the system depends on a number of factors:

1. Illumination of the object being photographed
2. Time of exposure of the sensor (shutter speed)
3. Aperture opening to capture the light from the object
4. Sensitivity of the CCD or CMOS sensor and associated electronics to the illumination collected at the surface of the sensor

Item four might be thought of as "grain size." It is influenced by the basic sensitivity of the sensor device itself as well as the gain of the electronic amplifier circuits (see Sect. 4.7). The overall effect is described by the International Standards Organization (ISO) number, which is analogous to the film speed in emulsion photography.

The higher the number, the greater the sensitivity of the combined sensor-electronics arrangement. A sensor may be manipulated to obtain a higher ISO number. For example, an 8-MP sensor can be switched to operate as *two* 4-MP sensors thereby doubling the ISO performance number. Note, however, that the resolution has been reduced in half for a given size image present at the surface of the sensor.

When a rapid change in movement must be examined in an experimental setup, innovation is sometimes called for. Consider the problem of photographing the crash of an automobile traveling at a speed of 27 m/s (about 60 mi/h). A resolution in the photograph of 1 mm is desired. The approximate shutter speed needed is about $0.001/27 = 1/27,000$ s. This is a shutter speed that can be achieved. Suppose a sensor of 20 MP is available in 4×5 proportions, thereby resulting in dimensions of 4000×5000 pixels. Also, suppose the optical system will be focused on an area of the automobile of approximately 1 m^2. Either the 4000- or 5000-dimension pixel fits the 1-mm resolution requirement $(1/4000 < 0.001)$. The only remaining question concerns the illumination of the automobile. Because of the short shutter-speed requirement, either intense steady light or a flash source carefully synchronized with the shutter may be required. If the desired level of illumination is not easily attained, an alternate procedure may be examined of increasing the ISO number by either increasing the amplifier gain or grouping sets of pixels. The latter procedure is less desirable because it lowers the basic pixel resolution.

To see how f-stop (controlled by lens aperture), shutter speed, and ISO setting interact to govern the exposure, consider a camera set to produce a proper exposure of the sensor with

f-stop = 8

Shutter speed = 1/100 s

ISO number = 200

Since the f-stop varies inversely with the diameter of the lens aperture, it follows that the light admitted through the lens, which varies directly with area, will vary according to

$$\text{Light admitted at } f\text{-stop}_1/\text{light admitted at } f\text{-stop}_2 = (f\text{-stop}_2/f\text{-stop}_1)^2$$

[B.2]

Now suppose that it is determined that a shutter speed of 1/3000 s is required to stop the motion of the object being photographed. What adjustment(s) in the f-stop and/or ISO sensitivity must be made to ensure the same amount of light reaches the surface of the sensor? The amount of light varies inversely with the shutter speed, and the voltage output of the sensor varies directly with the ISO setting for a given amount of light striking the sensor. Thus,

$$\text{Light admitted at shutter}_1/\text{light admitted at shutter}_2$$
$$= \text{Shutter speed}_2/\text{shutter speed}_1 \qquad \textbf{[B.3]}$$

$$\text{Voltage output at ISO}_1/\text{voltage output at ISO}_2 = \text{ISO}_1/\text{ISO}_2 \qquad \textbf{[B.4]}$$

Combining all three effects,

Sensor output$_1$/sensor output$_2$

$$= (f\text{-stop}_2/f\text{-stop}_1)^2(\text{shutter speed}_2/\text{shutter apeed}_1)(\text{ISO}_1/\text{ISO}_2)$$

$$\text{[B.5]}$$

Therefore, to maintain the same sensor output for the two shutter speeds described in the foregoing example would require

$$1.0 = (f\text{-stop}_2/8)^2(1/100)/(1/3000)(200/\text{ISO}_2)$$

The operator of the device may choose to accomplish the objective by varying either or both of the parameters: $f\text{-stop}_2$ or ISO_2. Opening the lens aperture alone would require

$$f\text{-stop}_2 = (100/3000)^{1/2}(8) = 1.46$$

while adjusting ISO alone would require

$$\text{ISO}_2 = (200)(3000/100) = 6000$$

This is a rather high value of ISO and might also require activation of noise reduction algorithms in the electronic readout circuits. A combination of changes in both might be selected as $f\text{-stop}_2 = 2$ with ISO_2 calculated as

$$\text{ISO}_2 = (200)(3000/100)(2/8)^2 = 375$$

This is a more reasonable value and would likely not require activation of noise-reduction measures. Of course, all of the above calculations assume constant illumination of the object being photographed.

B.6 SUMMARY

With the foregoing sections in mind we may now summarize the applicability of digital imaging (photography) to experimental work. For that purpose we do not make a distinction between "still" photography or "video" or "motion picture" photography, since the latter is represented as a sequence of still images taken at a specified frame rate.

Regardless of the object to be photographed, some type of optics must be employed to capture the image and focus it on the sensor-electronics system. Such optics may be quite similar to that which would be used in an emulsion (film) photography system. Whatever the situation, adequate illumination must be provided on the object(s) to be photographed. Digital sensors usually require less illumination than photographic film. For photographing moving images, the proper balance must be achieved between illumination level and duration, shutter speed, and sensor sensitivity–noise level as specified by the ISO setting. For example, a particle-image velocimetry (PIV) setup might be used to measure the velocities of individual smoke particles in an airflow stream. The velocity is to be computed by measuring the particle displacement over a certain period of time. Illumination is provided by either a high-intensity spark discharge (≈ 10 μs) or by a pulse laser focused on a small-volume flow element. The object may be placed in a darkened enclosure so that a modest shutter speed (≈ 0.01 s) can be employed. Of course, this is the classic use of flash photography. An alternate is to employ a fast shutter speed (≈ 0.1 ms) and continuous illumination. High-speed rotating mirrors can also provide a means for obtaining high-speed sequential (motion picture) images. Once again, the obvious advantage of the digital imaging system is that it allows examination of multiple alternatives (trial and error) with almost instant comparison of results.

Digital imaging systems offer several advantages to the experimental engineer.

1. The ability to address and manipulate individual pixel outputs of the device
2. Sensitivities that exceed those available in emulsion photography
3. Immediate viewing of sensor outputs on computer monitors
4. Availability of both proprietary and stand-alone software for processing RAW images (direct output from sensor) to reduce noise and effect various image enhancements
5. Availability of numerous high-quality, hard-copy output devices (printers)
6. Availability of high-quality optics suitable for a variety of applications

INDEX

M